In-Flight Simulators and Fly-by-Wire/Light Demonstrators

Peter G. Hamel
Editor

In-Flight Simulators and Fly-by-Wire/Light Demonstrators

A Historical Account of International Aeronautical Research

Translated by
Ravindra V. Jategaonkar

Springer

Editor
Peter G. Hamel
Director (Retired)
DLR German Aerospace Center
Institute of Flight Systems
Braunschweig
Germany
e-mail: p.hamel@t-online.de
peter.hamel@alum.mit.edu

Translated by
Ravindra V. Jategaonkar
Senior Scientist (Retired)
DLR German Aerospace Center
Institute of Flight Systems
Braunschweig
Germany
e-mail: ravindra.jategaonkar@t-online.de

This book can be indexed, referenced, and cited as follows:
In-Flight Simulators and Fly-by-Wire/Light Demonstrators: A Historical Account of International Aeronautical Research, Edited by Peter G. Hamel, Translated by Ravindra V. Jategaonkar, and published by Springer, March 2017.

Revised, extended and translated from the original book in German:
"Fliegende Simulatoren und Technologieträger: Braunschweiger Luftfahrtforschung im internationalen Umfeld", edited by Peter G. Hamel and published by Appelhaus Verlag, Braunschweig, 2014, ISBN 978-3-944939-06-3.

ISBN 978-3-319-53996-6 ISBN 978-3-319-53997-3 (eBook)
DOI 10.1007/978-3-319-53997-3

Library of Congress Control Number: 2017933443

Printed on acid-free paper

This Springer imprint is published by Springer Nature
The registered company is Springer International Publishing AG
The registered company address is: Gewerbestrasse 11, 6330 Cham, Switzerland

Be more attentive to new ideas from the research world

George S. Schairer,
Former Vice President Research, Boeing (1989)

Foreword to the English Edition

For the first 40 years of aviation, most of the failures of the earliest attempts at powered, fixed-wing flight were associated with inadequate understanding of dynamic stability and control. Although Lanchester, Bryan, and Williams had already developed the theory of aircraft dynamics by the early 1900s, their work still had found negligible use for design purposes by as late as the mid 1940s. However, WWII brought demands for maneuverability so that aircraft dynamic stability and control has been a focus of attention ever since.

It was through the research and the technology demonstrators that are excellently reviewed, chronicled, and documented in this book that the aviation industry gained an understanding of aircraft stability and control. With that knowledge, aircraft designers have been freed forevermore from the constraints of the classical conception of stability and control associated with fixed stabilizing fins and manually movable surfaces for control. Fly-by-Wire artificial stability systems give the designer the flexibility to design an aircraft solely from the perspective of performance and ignore the classical stability requirements. The genesis of the development of artificial stability was in the pioneering work that was done primarily in the United States and in Germany (often in collaboration).

In the US, the early basic studies were conducted in the 40s and 50s by NASA and by the Flight Research Department of the Cornell Aeronautical Laboratory (CAL, now called Calspan) with totally different objectives. The scientists at NASA were seeking ways to improve the flying qualities of a particular existing aircraft with a control problem. Meanwhile, CAL's objective was to develop methods for measuring and describing the dynamic stability and control characteristics of any aircraft in flight. Both paths led to the invention of variable stability aircraft that could be used in the early stages of design of automatic control for stability augmentation. This book describes the multiple projects in US from 1947 until today that produced a progression of theoretical and experimental advances in aircraft dynamic stability and control.

In Germany, significant contributions to understanding dynamic stability and control began even earlier than did the work in the US, but there too, as with NASA, the initial work was largely concerned with correcting an existing control problem. Since WWII, important contributions have come mostly from the Institute of Flight Mechanics of DFVLR (since 1999 Institute of Flight Systems of the German Aerospace Center—DLR) in Braunschweig. That Institute, under the leadership of its Director, Dr. Peter G. Hamel, has established a worldwide reputation for its expertise in all the fields related to flying-qualities investigations and, in particular, the development and highly innovative use of their in-flight simulators of both fixed-wing and rotary-wing aircraft, and their novel applications of Fly-by-Wire and Fly-by-Light. Although autopilot functions with limited stabilization were already available on aircraft with mechanical flight control systems, stability augmentation and variable stability aircraft would not have been possible without Fly-by-Wire. Peter and his Institute are also to be credited with their pioneering work in the adaptation of the concepts of system identification to flight vehicles.

Aircraft system identification is a way to build the accurate mathematical model of an aircraft that is essential to designing augmented stability and automatic control for that aircraft.

An ever-increasing number of modern civilian and military aircraft is inherently aerodynamically unstable. However, the stability augmentation and automated control that are basic to all current aircraft designs ensures the comfort and safety of your flight.

Peter and his collaborators at the DLR Institute of Flight Systems achieved important advancements in the multiple technologies of stability augmentation and in-flight simulation decades before they were thought of anywhere else. These achievements have been particularly noted and praised by two well-known experts in the field, Mal Abzug and Gene Larrabee, in their 2002 book titled, "Airplane Stability and Control: A history of the technologies that made aviation possible".

Peter and the co-authors of this book have made a noteworthy contribution to the history of aviation. The book is the consequence of an enormous effort to cover the complete spectrum of international contributions in the evolution of artificial stability, variable stability aircraft, and in-flight simulation. It also addresses the development of the technologies of Fly-by-Wire and Fly-by-Light that made these developments possible. While this book presents an exhaustive account of variable stability and Fly-by-Wire research and demonstrations worldwide, it emphasizes the work in Germany and, in particular, at DLR's Institute of Flight Systems. The prominence of descriptions of related activities in Germany is understandable and appropriate considering that much of the pioneering work was done there and that this is the first time it has been so well documented. The extensive efforts in Germany that go back over 100 years and, in particular, the achievements at the Institute of Flight Mechanics contributed profoundly to the development of stability and control augmentation and the use of in-flight simulation. I can personally attest to the scope and value of contributions to the current state of understanding aircraft dynamic stability and control made by Peter and his Institute. I have known Peter G. Hamel for over 40 years and I have had the pleasure of collaborating with him in several projects related to the topics of this book. Through my personal knowledge, I can vouch, without hesitation that he is eminently well qualified to chronicle and evaluate the worldwide developments of these capabilities that have become essential to aviation.

This book will be of interest not only to novices, but also to practicing scientists and engineers and to those interested in aviation history. This comprehensive historical account is devoid of mathematical equations and deep theoretical discussions, but it is full of tales of innovative experiments and creative thinking, amusing anecdotes, and fascinating photos that I have no doubt the readers will enjoy. So, my advice to the hesitant reader is, if you are interested in gaining reliable knowledge about the origins, the innovators, and the evolution of stability augmentation, variable stability aircraft, and in-flight simulators as well as of Fly-by-Wire/Light this is the book for you.

Mountain View, CA, USA
September 2016

Irving C. Statler
Principal Engineer, Flight Research Department
Cornell Aeronautical Laboratory, 1946–1970
Director, U.S. Army Aviation Aeromechanics Laboratory, 1970–1985
Director, NATO Advisory Group for Aerospace R&D, 1985–1988
Research Scientist, Human Factors Research Division
NASA Ames Research Center, 1988–2008

Foreword to the German Edition

A beautiful definition of performance is: "A impersonates B, while C observes". Applied to the in-flight simulation, a highly sophisticated modified aircraft "impersonates" in real flight another vehicle in the sense of a "flying actor", while scientists and engineers "observe", that is, analyze the outcome of flight to make decisions.

One should be a bit careful while using the term "supreme discipline", no matter in which area, because thereby you implicitly degrade all other disciplines in the same field, or in other words assign them a lower importance. In the present case, however, it seems justified to speak of in-flight simulation possibly as a "supreme discipline" in aircraft construction, as it encompasses all other disciplines such as aerodynamics, flight mechanics, construction, structures, aircraft systems, and aeroelasticity. In order to get an in-flight simulator operational in the air, best experts in all these fields have to work together, as one encounters the boundaries of physical sciences and engineering techniques, which have to be fathomed and extended in this most challenging task. This was true in the past with fewer tools available to develop aircraft such as the technology demonstrator VFW 614 ATTAS, just like today when the demands on the prediction accuracy have increased with the availability of better numerical and experimental tools.

It goes to the credit of Peter G. Hamel as an initiator and editor of this book, supported by many who were involved then as well as even today, to have looked back and reappraised the technical history of in-flight simulation. For his contributions in this field, he was honored with the most prestigious award "Ludwig-Prandtl-Ring" of the German Society for Aeronautics and Astronautics. Prandtl was the one who has provided a scientific basis to flight science, which was still in the infant stages during the early twentieth century, through his work on the boundary layer theory and the construction of wind tunnels. Almost a century after Prandtl's fundamental work, the author looks back at the history of in-flight simulation and illustrates thereby the rapid development of flight physics.

This has a long tradition in Germany. In a broad sense, the work of the first flight scientist Otto Lilienthal may be interpreted in a figurative sense as contributions to in-flight simulation. He wanted to replicate the flight of birds with his gliders. Later the theoretical work by Ludwig Prandtl appeared, the aircraft profile and the flying wing by Hugo Junkers, the jet engine by Hans Joachim Pabst von Ohain, and the swept wing by Adolf Busemann, just to name a few. This book shows impressively how the efforts of scientists and engineers in this country have contributed not only with technical achievements to the success of aeronautics, but also established over a long period a highly successful unified community of research, education, development, production and operations. Only such a close cooperation in a fertile network guarantees a technically and socially valuable future in the days to come.

Aircraft have changed only a little in recent years. However, there is now an urgent need for fundamentally new aircraft to meet the growing demands for better cost effectiveness, environmental sustainability, and passenger comfort. These new devices need different properties than the aircraft flying today, and these properties must be simulated in advance, especially in flight. Hence, the need for simulations and in-flight simulators is higher than ever, and as such this book is not only a historical reappraisal, but also represents a mandate for the future.

The book you hold in your hands shows in an excellent manner the technical facets and the great efforts which are needed to successfully "act" as a problem solver in aeronautics. I would like to take this opportunity to sincerely thank the editor, Peter G. Hamel, not only for his technical expertise, but also for his tenacity on the long road from an idea to realizing the book in its impressive final form. To the readers, I wish much enjoyment in reading the book, many insights into the rich past, own perceptions of the current global state of the art, and inspiration derived thereby for the future.

Cologne, Germany
August 2016

Rolf Henke
Member of the Executive Board
German Aerospace Center (DLR)
President, German Society for Aeronautics and Astronautics (DGLR)

Foreword

In-flight simulation is the ultimate approach in applied flight sciences to assessment and evaluation of aircraft and other aeronautical systems, as it represents the most intense fusion of flight mechanics, flight control, flight systems technology, and flight testing. It is a versatile tool for flight research and aircraft industry alike. Starting with studies of future configurations up to the simulation of atmospheric phenomena including system influences, dependencies, and even failure cases can be addressed with in-flight simulation, reaching high technology readiness levels.

But is this approach still the best way to tackle the current and future questions in aircraft flight research and development? From the viewpoint of a research institution with strong links to aircraft industry and governmental partners, we can state today that in-flight simulation still has and most probably will ever have an important role to play in aeronautics. The reasons for this are many and I would like to mention just three of them.

Firstly, new configurations are about to enter the scenario of modern civil transport aircraft with possibly radical new features with no proven databases to rely on. Early in-flight simulations of potentially unstable configurations and, even more important, of the flight control systems will be an efficient, fast, and reliable way to establish the required confidence of engineers and management alike.

Secondly, the segment of unmanned flying vehicles is the fastest growing area in modern aeronautics. Autonomously operating vehicles will enter the world's airspace in the next decades. Manned and unmanned in-flight simulation will be one of the most powerful tools to prove the maturity of new designs and, not to forget, new legal approaches to certification.

Thirdly, international aeronautics industry starts to detach from national institutions and educational systems, leaving national authorities with the task of maintaining basic capabilities in aeronautical design and development. As the in-flight simulation places highest demands on the key competencies of aircraft modeling, control, and integration, it is best suited to ensure maintaining the expertise in these areas. This is all the more true in the future with new systems having ever-increasing complexity and safety standards. The implementation of in-flight simulation will help us to understand that technical capabilities will not be sufficient to make a flight vehicle design optimal but the knowledge on how to design and develop will be as important as the technical skills and includes a deep understanding of the humans, acting within the engineering, design, and decision processes.

There are many good reasons to pursue in the future this jewel of modern flight sciences. It is apparent from this book that a symbiosis of scientific excellence at universities, research institutions, industry capabilities, and political focus is mandatory in this pursuit. It is highly appreciated that the authors of this book present the highlights, achievements, and worldwide historical evolution of this aspect of flight engineering sciences. Thanks to the authors for this brilliant and one of the most vivid compilations of aeronautic achievements that has been given to international scientific communities!

Braunschweig, Germany
November 2016

Stefan Levedag
Director, Institute of Flight Systems
German Aerospace Center

Preface

More than sixty years of international research and development in the field of airborne simulation and electronic Fly-by-Wire flight control systems have left their marks on the advances in aeronautical system design. After all, two generations of aeronautical specialists and generalists have passed during this period. It is likely that this wealth of accumulated knowledge and experience in this complex field of aircraft systems may be difficult to trace, or may even be forgotten, or not efficiently used in the future.

A call from a former test pilot and chief of the flight department of the German Aerospace Research Center (DLR) in Braunschweig resulted in the friendly and collegial suggestion to summarize the knowledge and experience gained by DLR in the supreme discipline of flight research, namely in-flight simulation. This was the occasion to invite a small group of former and still active scientists and engineers from research, academia, and industry to participate in such a project. Embarking on the project on a small scale on February 6, 2013, it evolved into an undertaking of much wider scope, encompassing not only DLR activities in Germany but global view covering all international organizations.

Now, historical retrospect in technical fields is not particularly in demand since past experiences are quickly overtaken. Nevertheless, despite the rapid technological progress, historical technical accounts may become valuable resources and reference points for knowledge refreshing and long-life learning to avoid the pitfalls.

An objective of the book is, therefore, to look back on the development, testing, and utilization of in-flight simulators and Fly-by-Wire technology demonstrators. They have strongly contributed to the current international state of knowledge in designing and evaluating today's modern aircraft, free according to the slogan "without the knowledge of the past, one can neither understand the present nor shape the future".

For the first time, this book attempts to describe, in some depth, chronologically the global complementary research and development activities of in-flight simulation and associated electronic and electro-optical flight control systems (Fly-by-Wire/Light). This task is invariably associated with the risk that equivalent or similar research activities abroad are unintentionally overlooked or not adequately accounted for. Keeping this in mind, the book attempts to maintain a fair balance of presentation of global activities, to avoid any scientific autism.

The authors of this book try to give as objective a description as possible of the activities in this demanding field of research in experimental flight system technologies, with an increased degree of detail in the description of German research and development results. This is particularly evident in the sections of the chapters "In-Flight Simulator VFW 614 ATTAS" and "Helicopter In-Flight Simulator EC 135 FHS". This level of detailing is sometimes useful for the definition phase of a future project by providing the experience and the lessons learned from former project scientists at the beginning of a new project to minimize potential risks.

A further concern of the book is to pass on the knowledge and experience to aerospace students, young scientists, and engineers, thereby stimulating and accelerating the lifelong learning process without repeating mistakes that were made in the past.

The book is also intended as a landmark and reference book for aviation and technology enthusiasts who would like to get an overview of the historical evolution of in-flight simulators and Fly-by-Wire/Light technologies. Sufficient references in the individual chapters are given to the interested reader in order to allow a further deepening of individual scientific and technical aspects as required. For optimum visualization, the number and size of illustrations and graphics were not spared. The technical language was formulated in the most general way possible in order to achieve the desired readability.

Braunschweig, Germany
December 2016

Peter G. Hamel

Acknowledgements

If the contributors of this documentation have succeeded in portraying the history of research and development activities of supreme aerospace technologies over the past sixty years in an appropriate and illustrative form, it was possible only due to the cooperation and support of many aviation enthusiasts. To name each and every one, who have contributed directly or indirectly, in a small or bigger way, to the book is a difficult task, associated with the risk of overlooking one or the other. I take an easier way out, by thanking all of them for support of this endeavor. Nevertheless, few individuals and organizations need special mentioning.

In addition to the co-authors listed in the book, who participated as retirees or active persons from industry, research, and universities, without hesitation, fees or expenses, there were financial sponsors who had made the original book project in the German language realizable. In this respect I would like to express my special gratitude to Airbus Helicopters (*Wolfgang Schoder*), Diehl Aerospace (*Gerardo Walle*), Dornier Museum (*Bernd Sträter*), ExpertSystems (*Earon Beckmann*), German Aerospace Center (*Rolf Henke, Stefan Levedag*), and VDev Systems & Services (*Önder Bagci*).

Likewise, I extend my sincere thanks to the benevolent supporters and advisors *Ulrich Butter* (formerly Eurocopter), *Jürgen Klenner* (formerly Senior Vice-President Airbus), *Hans-Peter Reerink* (formerly CEO, Diehl Aerospace), *Hannes Ross* (formerly EADS), and *Jens Strahmann* (formerly Airbus).

I also owe special thanks to *Rolf Henke*, a member of the executive board of the German Aerospace Center, for valuable suggestions and moral support, and the two institute directors, *Stefan Levedag* and *Dirk Kügler*, who gratefully supported the book project and for the infrastructure granted at their institutes.

Acknowledgement is owed to the German Aerospace Center members *Stephan Bensberg, Hans-Jürgen Berns, Martin Gestwa, Uwe Göhmann, Hans-Ludwig Meyer, Michael Press, Detlef Rohlf, Ina Rüdinger, Berend van der Wall* and *Michael Zöllner*, who gave of their time and talents for the good of the book.

I would also like to express my special appreciation to *Roger Connor* (Smithonian Institution), *Jean-Francois Georges* (formerly CEO, Dassault Falcon Jet Corporation), *Lou Knotts* (CEO, Calspan), *Michael Stephan* (formerly Airbus Helicopters) and *Hans-Ulrich Willbold* (Airbus Group), who helped to close documentary gaps in the book by providing text and image material.

In preparing the English edition of this book, big merits go to my colleague and fellow campaigner *Ravindra Jategaonkar,* who was Senior Scientist at DLR. He did perform a remarkable *intelligent* and *critical* translation in order to balance the scope and phrasing within the individual chapters and to improve the overall book quality. Without his unstinted participation and substantial contributions, the English edition of our book would not have become possible.

From the United States, the country of origin to the evolution of advanced airborne simulation and Fly-by-Wire technologies, I had the pleasure to receiving very helpful advice from *Irving C. Statler,* a former leading scientist and member of the first hour of the famous Cornell Aeronautical Laboratory pioneering team who were designing and operating the world's first variable stability aircraft. *Irv* willingly submitted himself to the tedious work of reviewing and commenting on various chapters of the book in terms of contents, presentation, and clarity.

I am also sincerely indebted to *Gavin Jenney*, the former leading expert of the first hour in designing and testing a Fly-by-Wire system at the USAF Flight Dynamics Laboratory, and *Ed Aiken*, a proven NASA scientist of the first hour in variable stability rotorcraft research. Both helped me in the last minute to review a few chapters, confirm, and streamline our research results concerning specific historical aspects of US Fly-by-Wire and variable stability rotorcraft research.

I would like to gratefully acknowledge the US Air Force and Army, Calspan, NASA, DLR, JAXA and other international industry, institutions, and organizations. We retrieved several photographs and information from their open public Internet sources, which is credited to them.

Apart from the individual professionals mentioned in the foregoing, I particularly appreciate two other individuals. First, my wife *Hannerl*, who had to miss over an extended period of time the normal family life twice, once during the German edition during 2013–2014, and again during 2016. Second, *Padma*, wife of my collaborator and Translator of this book, for going through the same fate. Their support and understanding made the task bearable and have indirectly contributed to this volume.

Finally, I would like to acknowledge the interest of Springer International Publishing AG in publishing this book. We would like to extend our thanks to *Leontina Di Cecco* for the sovereign care and implementation of our wishes in the development of this book project. We also extend our appreciation to *Janet Sterritt-Brunner* of Springer Nature for the cover page design and *Gowtham Chakravarthy* and the Production Team of Scientific Publishing Services, Chennai, India for copy-editing the manuscript and efficiently incorporating the corrections.

Braunschweig, Germany
December 2016

Peter G. Hamel

Contents

Abbreviations and Acronyms

AAF	Autonomous Aerial Refueling
ACAH	Attitude Command Attitude Hold
ACVH	Attitude Command Velocity Hold
ACT	Active Control Technology
ACTIVE	Advanced Control Technology for Integrated Vehicles
ACTTA	Active Control Technology of Transport Aircraft
AD	Applied Dynamics
ADIRS	Air Data and Inertial Reference System
ADOCS	Advanced Digital Optical Control System
ADS	Air Data System
ADS-33	Aeronautical Design Standards 33
A/D	Analog to Digital
ADB	Autonomes Digitales Bediengerät (Autonomous Digital Control Panel)
AFB	Air Force Base
AFCE	Automatic Flight Control Equipment
AFCS	Automatic Flight Control System
AFDD	Aeroflightdynamics Directorate
AFFDL	US Air Force Flight Dynamics Laboratory
AFFTC	Wright-Patterson AFB and Edwards Flight Test Center
AFTI	Advanced Fighter Technology Integration
AFTPS	US Air Force Test Pilot School
AFWAL	Air Force Wright Aeronautical Laboratories
AGARD	Advisory Board of Aerospace Research and Development
AH	Attitude Hold
AHRS	Attitude and Heading Reference System
AIAA	American Institute of Aeronautics and Astronautics
AKH	Arbeitskreis Hubschrauber Technologien (German Working Group on Helicopter Technology)
ALCS	Augmented Longitudinal Control System
ALIAS	Aircrew Labor In-Cockpit Automation System
ALT	Altitude Navigation Mode
AP	Autopilot
APC	Aircraft Pilot Coupling
ARA	Avionics Research Aircraft
ASTRA	Advanced Stability Training and Research Aircraft
ATD	Advanced Technologies Demonstrator
ATLAS	Adaptable Target Lighting Array System
ATOL	Automatic Takeoff and Landing
ATRA	Advanced Technologies Research Aircraft
ATTAS	Advanced Technologies Testing Aircraft System
ATTHeS	Advanced Technologies Testing Helicopter System

AVA	Aerodynamische Versuchsanstalt (today: part of DLR)
AVEN	Axis-Symmetric Vectoring Exhaust Nozzle
AVRADCOM	Aviation Research and Development Command
AVT	Applied Vehicle Technology Panel of RTO
AWI	Alfred Wegener Institute for Polar and Marine Research
AW&ST	Aviation Week & Space Technology
BA/BAe	British Aerospace
BASE	Gust Alleviation and Elastic Mode Damping
BFS	German Federal Air Traffic Control Authority
BGT	Bodensee-Gerätewerk (today: Diehl Aerospace)
BMFT	German Federal Minister for Research and Technology
BMVg	German Federal Ministry of Defense
BWB	German Federal Office of Defense Technology and Procurement (today: Federal Office of Bundeswehr Equipment, Information Technology and In-Service Support—BAAINBw)
CAA	Civil Aviation Authority
CAL	Calspan; Cornell Aeronautical Laboratory
CAS	Control Augmentation System
CAS	Calibrated Air Speed
CASTOR	Combat Aircraft for Training, Operations, and Research
CCS	Configuration Control System
CCV	Control Configured Vehicle
CEO	Chief Executive Officer
CEV	Centre d'Essais en Vol
CFD	Computational Fluid Dynamics
C.I.T.	Cranfield Institute of Technology
CPDLC	Controller Pilot Datalink Communications
CPU	Central Processing Unit
CWS	Control Wheel Steering
DAL	Design Assurance Level
DARPA	Defense Advanced Research Projects Agency
DC	Drag Control
DEHS	Digital Electric Hydraulic Flight Control System
DFC	Direct Force Control
DFG	Deutsche Forschungsgemeinschaft (German Research Foundation)
DFL	Deutsche Forschungsanstalt für Luftfahrt (German Aeronautical Research Establishment, today: part of DLR)
DFVLR	Deutsche Forschungs- und Versuchsanstalt für Luft- und Raumfahrt (German Aerospace Research and Test Establishment, today: DLR)
DGPS	Differential Global Positioning System
DIGITAC	Digital Flight Control for Tactical Fighter Aircraft
DISCUS	Digital Self-healing Control for Upgraded Safety
DIVA	Dialog Orientierte Versuchsdaten Auswertung
DLC	Direct Lift Control
DLH	Deutsche Lufthansa
DLR	Deutsches Zentrum für Luft- und Raumfahrt (German Aerospace Center)
DLS	DME Based Landing System
DME	Distance Measurement Equipment
DSFC	Direct Side Force Control
DVL	Deutsche Versuchsanstalt für Luftfahrt (German Aeronautical Test Establishment, today: part of DLR)

D/A	Digital to Analog
EAP	Experimental Aircraft Program
EASA	European Aviation Safety Agency
EBF	Externally Blown Flaps
E-COCK	Experimental Cockpit (ATTAS)
EDP	Experimental Data Processing (ATTAS)
EECU	Electronic Engine Control Unit
EFA	European Fighter Aircraft
EFCS	Electronic Flight Control System
EFM	Enhanced Fighter Maneuverability
EFMS	Experimental Flight Management System
EGNOS	European Geostationary Navigation Overlay Service
EMA	Electro Mechanical Actuator
EPNER	École du Personnel Navigant d'Essais et de Réception (French Test Pilots' School)
EPR	Engine Pressure Ratio
EPV	Easily Piloted Aircraft
ERR	Experiment Regelrechner (Experimental Control Computer for ATTAS)
ESA	European Space Agency
ETPS	Empire Test Pilots' School
EU	European Union
EVP	Easily Piloted Aircraft
FAA	Federal Aviation Administration
FACT	Future Aircraft Control Testbed
FADEC	Full Authority Digital Engine Control
FADS	Flush Air Data System
FAST	Full Scale Advanced Systems Testbed
FBW	Fly-by-Wire
FBL	Fly-by-Light
FCLC	Flight Control Law Computer
FCS	Flight Control System
FCU	Fuel Control Unit
FDL	US Air Force Flight Dynamics Laboratory
FFS	Flugführungssystem (Flight Guidance System
FHS	Flying Helicopter Simulator
FIS	Flight Inspection Systems
FL	Flight Level
FLISI	Fliegender Simulator (Flying Simulator)
FMP	Flight Mechanics Panel of AGARD
FMoD	German Federal Ministry of Defense
FMS	Flight Management System
FRI	Russian Flight Research Institute (Gromov Institute)
FRC	Flight Research Center
FRL	Flight Research Laboratory, NRC
FTB	Flying Test Bed
FVP	Flight Vehicle Integration Panel of AGARD
GCP	Guidance and Control Panel of AGARD
GLA	Gust Load Alleviation
GND	Ground
GPAS	General Purpose Airborne Simulator
GPS	Global Positioning System
GRATE	Ground Attack Test Equipment,

HAIG	Hongdu Aviation Industry Group, China
HESTOR	Helicopter Simulator for Technology, Operations and Research
HDG	Heading Navigation Mode
HFB	Hamburger Flugzeugbau
HIL	Hardeware-in-the-Loop
HLH	Heavy Lift Helicopter
HMD	Helmet Mounted Display
HPP	Hydraulic Power Package
HSF	Hubschrauber-Schlechtwetter- Führung (Helicopter Adverse Weather Guidance)
HTA	Hermes Training Aircraft
HUD	Head-up Display
IBLS	Integrity Beacon Landing System
ICAO	International Civil Aviation Organization
ICP	Intelligent Control of Performance
ILA	International Luftfahrt-Ausstellung
IFCS	Intelligent Flight Control System
IFR	Instrumented Flight Rules
IFS	In-Flight Simulation
IFSTA	Integrated Flight Simulation Test Aircraft
IHTP	In-House test Platform
ILAF	Identically Location of Acceleration and Force Application
ILS	Instrument Landing System
IMC	Instrument Meteorological Conditions
INS	Inertial Navigation System
JDA	Japan Defense Agency
JSF	Joint Strike Fighter
KURS	Kurzzeit-Experimente unter reduzierter Schwerkraft (short-term experiments under reduced gravitational force)
LAMS	Load Alleviation and Mode Stabilization
LAT	Liebherr-Aero-Technik (today: Liebherr Aerospace)
LAMARS	Large Amplitude Multi-Mode Aerospace Research Simulator (USAF FDL)
LAPAZ	Luft-Arbeits-Plattform für die Allgemeine Zivilluftfahrt (Air utility platform for the General Civil Aviation)
LEM	Lunar Excursion Module
LFA	Luftfahrt Forschungsanstalt (German Aviation Research Institute)
LLRV	Lunar Landing Research Vehicle
LLTV	The Lunar Landing Training Vehicle
LuFo	Luftfahrtforschung Förderung (National Aerospace Research Program)
LVAC	Launch Vehicle Adaptive Control
LVDT	Linear Variable Displacement Transducers
MAC	Mean Aerodynamic Chord
MATV	Multi-Axis Thrust Vectoring
MBB	Messerschmitt Bölkow-Blohm
MCDU	Multi-Function Control and Display Unit
MDO	Multidisciplinary Design Optimization
MFCS	Model Following Control System
M.I.T.	Massachusetts Institute of Technology
MLS	Microwave Landing System
MoA	Memorandum of Agreement
MoD	Ministry of Defense
MoU	Memorandum of Understanding

NAVAIR	Naval Air Systems Command
NACA	National Advisory Committee for Aeronautics
NAL	Japanese National Aeronautical Laboratory
NASA	National Aeronautics and Space Administration
ND	Navigation Display
NPL	National Physical Laboratory
NLR	National Aerospace Laboratory, The Netherlands
NRC	National Research Council, Canada
NTPS	United States Naval Test Pilot School
NWB	Normal Windkanal Braunschweig (Low-Speed Wind Tunnel)
OLGA	Open Loop Gust Alleviation
ONERA	The French Aerospace Laboratory
PACT	Precision Aircraft Control Technology
PFCU	Primary Flight Control Unit
PIO	Pilot Induces Oscillation
PSA	Peek-Seeking Algorithm
PTU	Power Transfer Unit
RAE	Royal Aircraft Establishment
RASCAL	Rotorcraft Aircrew Systems Concepts Airborne Laboratory
RC	Rate Command
RCAH	Rate Command-Attitude Hold
RPC	Rotorcraft-Pilot-Coupling
RTC	Real-Time Clock
RTO	Research and Technology Organisation
RVG	Reglerversuchsgestell (Controller Test Stand)
SAC	Side-Arm Controller
SAE	Society of Automotive Engineers
SAFAR	Small Aircraft Future Avionics Architecture (EU Project)
SARA	Sikorsky Autonomous Research Aircraft
SCI	Systems Concepts and Integration Panel of RTO
SCAS	Stability Command Augmentation System
SEMA	Smart Electro Mechanical Actuator
SETAC	SEctorTACan with TACAN (Tactical Air Navigation) of the company SEL
SFCC	Secondary Flight Control Computer
SFCS	Survivable Flight Control System
SFTE	Chinese Flight Test Establishment
SGL/SIGRI	SGL Carbon Group
SHADOW	Sikorsky Helicopter Advanced Demonstrator of Operator Workload
SLS	Space Launch System
SRA	Systems Research Aircraft
SSE	Single or Separate Surface Excitation
STA	Shuttle Training Aircraft
STANAG	Standarization Agreement (NATO)
STO	Science and Technology Organization
STOL	Short Take-Off and Landing
STOVL	Short Take-Off and Vertical Landing
SV	Stabilité Variable (Variable Stability)
TAGS	Tactical Aircraft Guidance System
TALAR	Tactical Landing Approach RADAR (Early MLS flight path guidance system, USAF)
TAS	True Airspeed
TIFS	Total In-Flight Simulator

TMR	Thrust Measuring Rig
TNT	Tragflügel Neuer Technologie (Aerofoil New Technology)
TPS	Test Pilot School
TRL	Technology Readiness Level
TSRV	Transport Systems Research Vehicle
TST	Transonischer Tragflügel (Supercritical Transonic Wing)
TTA	Trajectory Training Aircraft
TU	Technical University
TU-BS	Technical University of Braunschweig
UAV	Unmanned Aerial Vehicle
UHCA	Ultra High Capacity Aircraft
USAF	United States Air Force
USNTPS	US Naval Test Pilot School
UTSI	Space Institute of the University of Tennessee
UTTAS	Utility Tactical Transport Aircraft System
VAAC	Vectored-Thrust Aircraft Advanced Flight Control
VCID	Voice-Controlled Interactive Device
VECTOR	Vectoring, Extremely Short Takeoff and belonged Landing, Control and Tailless Operation Research
VFW	Vereinigte Flugtechnische Werke
VISTA	Variable Stability In-Flight Simulator Test Aircraft
VMS	Vertical Motion Simulator
VRA	Variable Response Research Aircraft
VS	Variable Stability
VSA	Variable Stability Aircraft
VSC	Variable Stability and Control
VSRA	Variable Stability Aircraft Research
VSS	Variable Stability System
VTOL	Vertical Take-Off and Landing
VVS	Variable Stability Testbed
WADC	US Air Force Wright Air Development Centers
WTD-61	German Air Force Flight Test Center
ZKP	Ziviles Komponenten Programm
ZKR	Zentraler Kommunikationsrechner (ATTAS)
ZTL	Zukunfttechnik Luft (Future Technology in Aeronautics)

Introduction 1

Peter G. Hamel

The „Fly-by-Wire“ Problem of 1914:

„An aeroplane will last for two or three years in constant use, unless it is very often transported by the railway, when it suffers a certain amount of deterioration if not carefully covered and carried“.

<u>**First British Air Mail Pilot**</u>

Gustav Hamel et al. : „Flying - Some Practical Experiences“, Longmans, Green and Co (1914)

P.G. Hamel (✉)
Braunschweig, Germany
e-mail: p.hamel@t-online.de

P.G. Hamel (ed.), *In-Flight Simulators and Fly-by-Wire/Light Demonstrators*,
DOI 10.1007/978-3-319-53997-3_1

1.1 In-Flight Simulation as Ultimate Tool for Flight Systems Research

During the past five decades, sensor, actuator and image information systems (displays), in conjunction with control laws, provided important technologies to improve the flight performance and characteristics of aircraft and spacecraft. As a prerequisite for this, the revolution in the digital technology that took place in parallel led to an explosive increase in the computing power, which in turn enabled significant progress in the enhancements of features to improve flying qualities, automation, and monitoring for improved flight performance and safety. Figure 1.1 depicts this integration process with its associated developmental technological risks. It is obvious that the interdependency between the three basic elements the flight system techniques will mostly dictate the research focus. In order to achieve a proper balance between effectiveness and flight safety of the integrated systems, it is necessary to account for and to optimize the dynamic interaction between the aircraft, the pilots, and the systems [1, 2].

With the trend of increasing automation, it is the human-automation interaction that is not adequately understood and taken into account during the design process. The pilot-aircraft interactions entail well-trained skills, whereas the pilot-automation interactions pose cognitive workload that is not understood. As a consequence, it must be ensured that during pilot's control inputs through his control panels the presented information and the effect of automatic influence and decisions remain plausible for the pilot in the sense of flight physics. The description of the pilot-related performance potential/capabilities, with regard to the perception of the current flight and system situation, of the ability to work under changing flight and environmental conditions, and his decision-making process in critical flight conditions, represents one of the most complex research tasks for engineers, medical doctors and psychologists in the field of aviation science. In connection with the understanding of whether the pilot, flying an aircraft equipped with complex computer logics, will react correctly in an unfamiliar or unknown flight situation, the pilot represents a weak spot or in other words may symbolize the Achilles' heel of safe flight [3].

Particular importance is also placed on timely proof of the functionality and system safety of new technologies through flight tests. This objective is based on the demand for a timely and cost-effective review of the technical and economic risks associated with the development of operationalization of new methods or critical technologies. Thereby it is essential that today's development and life cycles of civilian or military flight systems cover a period of an engineer's life of about 35–40 years. Thus, there is also the risk of losing interdisciplinary know-how in the aeronautical engineering field. As a consequence, it calls for continuous research and industrial-political efforts to realize the anticipated developments or ongoing system improvements though demonstrator programs or in-flight simulations in reasonable time periods. In international terminology, this is termed as reaching of a technology maturity level (Technology Readiness Level—TRL), which is assigned a value of about 6, that means "*functional and test prototype in operational range*" (see also Sect. 6.1.2).

The interlinkages of flight system techniques depicted in Fig. 1.2 elucidates the individual steps to be followed in an ideal case during the new development or improvement of existing flight systems.

The limited usage of these technologies, due to, say, developmental, political or financial reasons, or of other research tools in related disciplines such as structures or propulsion technologies, culminates to the disastrous effects shown in Fig. 1.3. Such events and the resulting socio-political issues have become a world-acclaimed predicament. As such, *Norman Augustine* needs to be greeted [4].

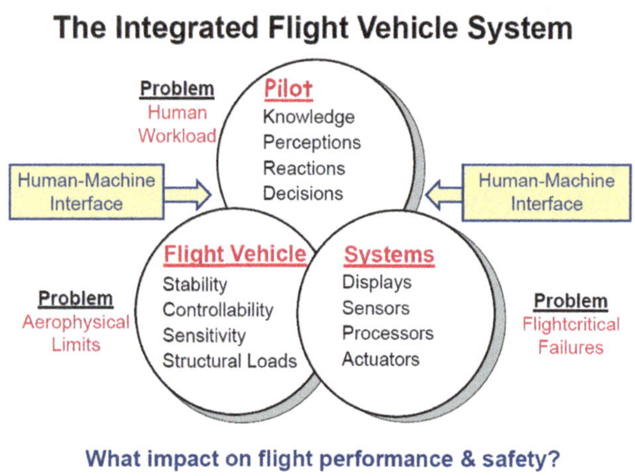

Fig. 1.1 Interactions in integrated flight systems

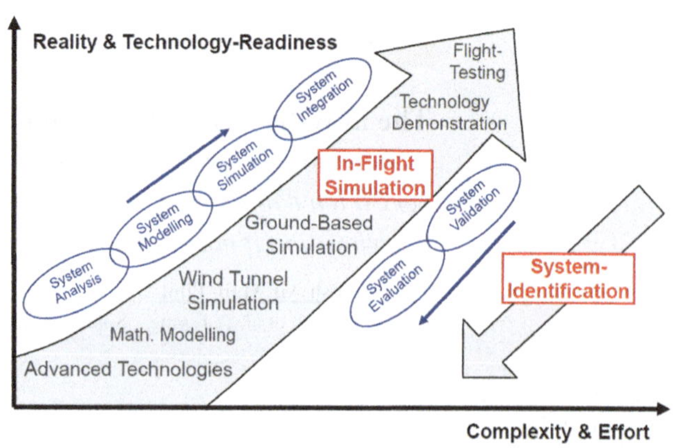

Fig. 1.2 The chain of research tools for flight vehicle system development

Project Delays & Cost Increases (%)

Selected examples

EF2000, NH90 and V-22	~ 10 years	~ 40-80%
A400M and F-35	~ 2++ years	~ 40-80%
A380 and B787	~ 2 years	~ 20-30%

What technical, managerial and political issues can be identified as the cause?

Fig. 1.3 Development risks and realities

The two elements shown in Fig. 1.2, namely the in-flight simulation, the supreme discipline flight testing, and the arts and science of system identification, a symbiosis of creativity and specialized knowledge, offer two versatile and experimentally oriented methods and are of particular value for the verification, optimization, and evaluation of flying qualities of manned or unmanned aerial systems with integrated Fly-by-Wire/Light flight control and information systems. But, it should also be pointed out that the human-in-the-loop ground-based simulation, plays, indeed, an indispensable role in a flight vehicle development program to minimize the more costly in-flight simulation.

A more detailed discussion and definition of the disciplines "in-flight simulation" and "system identification" will be given in Chap. 3. Both these disciplines represent also special, long-term focal points of research activities at the Institute of Flight Systems at the German Aerospace Center (DLR) in Braunschweig. An account of these efforts will be given in Chaps. 7–10.

1.2 Current State of Knowledge

There are a number of national and international, historical reports on the development of electronic flight controls for improving the handling and flying qualities of aircraft and helicopters [5–12].

The hitherto most detailed historical account related to airplanes with variable stability and in-flight simulation comes from one of the fathers of in-flight simulation, *Waldemar O. Breuhaus*, of the former Cornell Aeronautical Laboratory, the company which later became the Calspan Corporation in the USA [13]. Throughout this book, the name Calspan (CAL) will be used for all references to the company.

This historical account was later extended and supplemented by the Calspan expert *Norman Weingarten* with his years of experience [14]. A further well represented and detailed history of aerospace research at Cornell Aeronautical Laboratory and Calspan is given in [15]. In an extremely exciting book that goes beyond the scope of in-flight simulation, one comes across a highly readable autobiography of *William F. Milliken*, a former managing director of Cornell Aeronautical Laboratory (CAL). *William F. Milliken*, *Waldemar O. Breuhaus*, *Irving C. Statler*, *Robert P. Harper* and *Edmund V. Laitone* and others spearheaded at CAL the flight test research in aircraft dynamic response measurements, variable stability flight testing, the importance of test pilot judgements and closed-loop system analysis. Further, the book provides a historical overview of industrial flight testing and the use of aviation-related technologies in US automobile sport by way of Bugatti as an example [16]. Also, the special role of the NASA Ames and Dryden Research Centers (the latter since 2014: Armstrong Research Center) in this field can be easily traced through a few selected examples [17–19]. From the international book world two publications are known which describe experimental aircraft and to a limited extent also in-flight simulators, predominantly developed in the United States or Russia [20, 21].

The importance of in-flight simulation and their technological benefit was emphasized in the first international symposium during 1991 held in Braunschweig. Flight demonstrations with DLR′s in-flight simulators VFW 614 ATTAS and Bo 105 ATTHeS were presented then [22, 23]. A detailed discussion of this symposium took place in the international leading aviation magazine *Aviation Week & Space Technology* ("*Gathering of the In-Flight Simulation Fraternity*") [24].

1.3 The Book Layout

The current compendium is organized into three parts. The first, short part consisting of Chaps. 2 and 3 introduces succinctly the topics addressed in this collection, namely flying qualities background, basics, and benefits. The second part consists of Chaps. 4–6. It provides a brief account of predecessors in Germany in Chap. 4. This is followed by an exhaustive account of variable stability aircraft and in-flight simulators in Chap. 5, covering United States, Canada, England, France, Russia, Japan, China, and Italy. Chapter 6 provides, likewise, an elaborative account of Fly-by-Wire/Light Demonstrators, first from abroad, and in the latter half those from Germany. The third part of the book, consisting of Chaps. 7–12 focuses on the research and development activities in Germany in more detail. It aims at providing the readers with the inside information about these challenging projects to understand the intricacies, efforts required, and the outcome. Each of these chapters in all the three parts provides relevant technical literature to trace the historical developments, the evolution, and the current status in the fields of in-flight simulation and Fly-by-Wire/Light research.

References

1. Hamel, P. (Ed.): The Challenge of Flight Research—A Historical Account and Technical Guide, DLR-IB 111-99/02 (1999)
2. Levedag, S. (Ed.): Institute of Flight Systems—Status Report 2000–2008: Scientific and Technological Results (in German) (2008)
3. Advani, A.: Fixing the Aviators' Achilles´ Heel, Aviation Week & Space Technology, p. 58, March 24 (2014)
4. Augustine, N.: Augustine's Laws, AIAA, 6th edn (1997). ISBN-13: 978-1563472404
5. Howard, R.W.: Automatic flight controls in fixed wing aircraft—the first 100 years. Aeronaut. J. **77**, 533–562 (1973)
6. Oppelt, W.: A historical review of autopilot development, research, and theory in Germany. J Dyn Syst Measur Control. **98**, 215–223 (1976)
7. Hunt, G.H.: The evolution of fly-by-wire control techniques in the UK. Aeronaut. J. **83**, 165–174 (1979)
8. Abzug, M.J., Larrabee E.E.: Airplane Stability and Control, Cambridge Aerospace Series. Cambridge University Press, Cambridge (2002)
9. McRuer, D., Graham, D.: A flight control century: triumphs of the systems approach. AIAA J. Guidance Control Dyn. **27**(2), 161–173 (2003)
10. Carico, G.D.: Helicopter Controllability, Naval Postgraduate School, Monterey, CA, AD-A220 078 (1989)
11. Prouty, R.W., Curtiss Jr., H.C.: Helicopter control systems: a history. AIAA J. Guidance Control Dyn. **26**(1), 12–18 (2003)
12. Stiles, L.R., Mayo, J., Freisner, A.L., Landis, K.H., Kothmann, B. D.: Impossible to Resist—The Development of Rotorcraft Fly-by-Wire Technology, 60th Annual Forum & Technology Display of the American Helicopter Society, Baltimore, MD, USA, June 8–10 (2004)
13. Breuhaus, W.O.: The variable stability airplane. AAHS J. **36**(1), 30–55 (1991)
14. Weingarten, N.C.: History of in-flight simulation at general dynamics. AIAA J. Aircr. **42**(2), 290–304 (2005)
15. Burns, K.R., Milliken, W.F., Statler, I.C.: The History of Aerospace Research at Cornell Aeronautical Laboratory and Calspan, Vol. 4 in a series: The Flight Research Department, AIAA 2007–0350
16. Milliken, W.F.: Equations of Motion—Adventure, Risk, and Innovation. Bentley Publishers (2006)
17. Aiken, E.W., Hindson, W.S., Lebacqz, J.V., Denery, D.G., Eshow, M.M.: Rotorcraft In-Flight Simulation Research at NASA Ames Research Center: A Review of the 80's and Plans for the 90's, NASA TM 103873 (1991)
18. Shafer, M.: In-Flight Simulation Studies at the NASA Dryden Flight Research Facility, NASA TM-4396 (1992)
19. Borchers, P.F., et al.: Flight Research at Ames—Fifty Years of Development and Validation of Aeronautical Technology, NASA SP-1998-3300 (1998)
20. Markman, S.: One-of-a-Kind Research Aircraft: A History of In-Flight Simulators, Testbeds & Prototypes. Schiffer Publishing Ltd. (2004)
21. Gordon, Y., Komissarov, D.: Soviet and Russian Testbed Aircraft. Hikoki Publication Ltd. (2011)
22. Hamel, P. (Ed.): In-Flight Simulation for the 90's, International Symposium, July 1–3 (1991), Braunschweig, DGLR Mitteilungen 91-05 (Proceedings, Inquiries at: DLR, Institut für Flugsystemtechnik, P.O. Box 3267, 38022 Braunschweig) (1991)
23. Harper, R.P.: Overview: The Evolution of In-Flight Simulation at Calspan, in [22], Paper 1 (1991)
24. Mecham, M.: Airborne Simulation Expands, AW&ST, pp. 42–49, October 7 (1991)

Author Biography

Peter G. Hamel was the Director of the Institute of Flight Mechanics/Flight Systems of the German Aerospace Center (DLR/DFVLR) (1971–2001). He received his Dipl.-Ing. and Dr.-Ing. degrees in Aerospace Engineering from the Technical University of Braunschweig in 1963 and 1968, and his SM degree from M.I.T. in 1965. From 1970 to 1971 he was Section Head of Aeronautical Systems at Messerschmitt-Bölkow-Blohm in Hamburg. Since 1995 he is an Honorary Professor at the Technical University of Braunschweig and a founding member of three collaborative research centers at the University. He was Chairman of the National Working Group on Helicopter Technology (AKH) (1986–1994) and the appraiser for the National Aviation Research Program (LuFo) until today. He was also the Manager of DLR's Rotorcraft Technology Research Program and the German Coordinator for the former AGARD Flight Mechanics/Vehicle Integration (FMP/FVP) Panel. He is a member of the German Society for Aeronautics and Astronautics (DGLR) and of the American Helicopter Society (AHS), and a Fellow of AIAA. He is the recipient of the AGARD 1993 Scientific Achievement Award, of AGARD/RTO von Kármán Medal 1998, of AHS Dr. A. von Klemin Award 2001, and of the prestigious DGLR Ludwig-Prandtl-Ring 2007.

2 Flying Qualities—Some History

Bernd Krag and Bernd Gmelin

What looks good, flies also good.

B. Krag (✉) · B. Gmelin (✉)
Braunschweig, Germany
e-mail: bernd.krag@t-online.de

B. Gmelin
e-mail: bernd.gmelin@t-online.de

P.G. Hamel (ed.), *In-Flight Simulators and Fly-by-Wire/Light Demonstrators*,
DOI 10.1007/978-3-319-53997-3_2

2.1 Fixed-Wing Aircraft

Anyone who has ever watched a seagull, gliding effortlessly over the lake in the upwind along the bluff, is full of admiration of the ease and elegance with which he flies. The fine movements of wings and tail to correct the flight are not discernible. The bird is in an absolute balance with the wind, gravity, and lift. Thereby he conveys an impression of a perfect flier to the observer.

The first aircraft at the beginning of the last century were far away from such a perfection. Even the first aviation pioneers, who delved into flying 200 years back, had recognized what one must do to enable a flying vehicle cover a longer distance in an undisturbed gliding flight. To construct their models they oriented themselves mostly on the basic configuration of a bird with a wing in the front and an empennage arranged behind. With the center of gravity of the flight vehicle in correct position, these models flew stable and quite well. Technically, the expert then speaks of a "stable flight". In this case, stable does not mean "durable", but rather the ability of the aircraft to return automatically to its unperturbed initial condition in response to a disturbance. Also, the exotic tropical plant "Zanonia Macrocarpa", whose seeds show an extremely stable flight behavior, served as a prototype for a favorable aerodynamic design of a flying vehicle (see Fig. 2.1). The knowledge about what must be done to accomplish a stable gliding flight was essentially known to the aviation pioneers at the beginning of the last century. Design and numerical data required for construction of an aircraft, however, did not exist. One had to rather learn from the practical experience.

However, an airplane should not only fly straight and level in gliding flight, but must start from the ground, land again and above all fly in curves. With regard to the stability, one could rely on some existing knowledge. On the other hand, in the area of flight controls, the aviation pioneers had to tread unknown territories.

The Stork, whose flying skills *Otto Lilienthal* had studied intensively and had drawn his conclusions, served as a role model for his hang glider. Aerodynamic experiments by a specially constructed device provided him with the numerical data for the construction of his flying machine. His "normal apparatus" from the year 1894 even revealed the basic configuration of an aircraft. A horizontal and vertical surface were mounted some distance behind the wings. He needed both of these for the stability of his gadget about the vertical axis (directional stability) and about the lateral axis (longitudinal stability). The wings were mounted significantly upward (V-position), which provided sufficient inherent stability about the longitudinal axis (roll stability). However, *Lilienthal* steered his gadget by shifting of weight. As a consequence, the controllability was severely restricted and thus was the cause of his fatal crash in 1896.

The *Wilbur and Orville Wright brothers* followed another method. They had keenly followed the flight tests by *Lilienthal* and recognized that ensuring sufficient controllability about all the three axes is of pivotal importance. To achieve that, the flyer possessed aerodynamically effective elevator and rudder. For roll control about the longitudinal axis, which is necessarily required for coordinated curve flight, highly elastic wings were built and twisted. The horizontal tailplane was arranged in front of the center of gravity (see Fig. 2.2). Thereby the "Flyer" was no longer stable. Pilot had to intervene constantly in order to stabilize

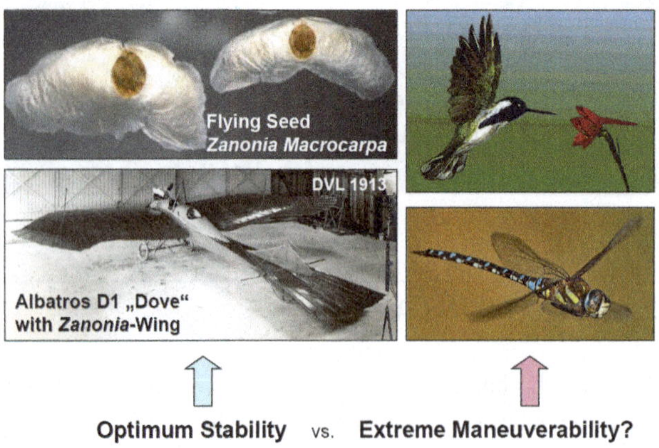

Fig. 2.1 Stability or controllability—where is the compromise? (*Credit* P. Hamel)

Fig. 2.2 The first fully controllable aircraft, the "Flyer III" in flight (1905), (*Credit* Deutsches Museum)

the flyer and keep it on the track. The *Wright* brothers believed that a skilled pilot must be in a position to continually balance the Flyer through an effective control. Too large an inherent stability is on the other hand more likely to be obstructive, when large disturbances must be compensated by control inputs. Probably the *Wright* brother had not adequately appreciated the importance of inherent stability for the flying [1].

With their desire for neutral stability, the *Wright* brothers were rather alone in the pioneer generation by aircraft manufacturers. Awareness was established that sufficient inherent stability is absolutely essential for safe flying. A pilot continuously struggling to stabilize the gadget can hardly perform any other task. Only much later the idea of reduced static stability (*Relaxed Static Stability*) was once again taken up during the development of highly maneuverable combat aircraft. These aircraft were, however, not stabilized by a pilot, but through a multi-redundant flight controller.

In the year 1909, *Louis Charles Blériot* had arrived at the classical basic configuration for the aircraft with his Type XI, which was mostly adopted for the aircraft construction thereafter, (see Fig. 2.3). This configuration is characterized by a front mounted motor with a tractor propeller and tailplane located at the rear for the longitudinal and lateral control. The roll control was not yet by ailerons, but through twisting the entire wing. The aircraft was apparently sufficiently stable about all the axes and allowed *Blériot* a smooth ride across the English Channel to England.

Aircraft development showed a rapid boom during the First World War. In Germany a variety of very different aircraft types were delivered to the Imperial German Army Air Service. Of course, there was no question of consistent and good flying qualities. It was a challenge for the pilots to fly many of these aircraft and they were hardly deployable. This diversity led the aircraft engineering department of the German Aeronautical Test Establishment (DVL) in 1917, to test several aircraft types for their flying qualities and to assess them by the pilots. Although it was not an objective assessment, a few characteristics emerged, which were then considered important and desirable [2]. These evaluations based on the pilot assertions, however, did not offer yet a reliable basis to improve the flying qualities through technical measures.

The aircraft which came closest in demonstrating the wishful good flying qualities was the biplane Fokker D VII constructed by Antony Fokker (see Fig. 2.4). Introduced in the year 1918, the Fokker D VII evolved as the most successful combat aircraft during the First World War. The aircraft was inherently stable, highly maneuverable and possessed good control surface effectiveness about all axes. One characteristic was particularly notable, namely, the aircraft recovered itself automatically from the dangerous spin as soon as the controls were released by the pilot. In the heat of an aerial combat, it was quite easy for the pilot to stall the aircraft and thereby enter into a spin.

Since the beginning of the aircraft development, the above problem encountered in flight was also addressed scientifically in parallel. It was the British mathematician *George H. Bryan*, who in the year 1911 formulated the problem of aircraft motion on a sound mathematical basis. *Bryan* formulated the equations of motion and introduced the concept of "stability derivatives". However, it was not possible yet to solve this system of equations.

Based on Bryan's equations, the flight scientists *Leonhard Bairstow* from the National Physical Laboratory (NPL) in the UK provided the first rudiments of stability analysis. He realized that the complex system of equations could be decoupled into a "longitudinal motion" and in largely decoupled "lateral motion". Solutions to these now simplified equations resulted in the flight mechanical eigenmodes of motion, which are commonly known to today's aeronautical engineer such as "Phugoid", "short period" and "Dutch roll"

Fig. 2.3 Safely over the English Channel: Monoplane Blériot, Type XI (1909). *Louis Blériot* as pilot, (*Credit* Deutsches Museum)

Fig. 2.4 The best fighter aircraft of First Word War, the legendary Fokker D VII biplane aircraft (1918), (*Credit* Deutsches Museum)

etc. To provide information about the order of magnitude of the stability derivatives, in the year 1913, *Bairstow* performed the first wind tunnel measurements on a model of Blériot-monoplane. A few years later, in the year 1916 commenced the wind tunnel measurements also with the same goal at the Massachusetts Institute of Technology (M.I.T.) in the United States under the leadership of *Jerome C. Hunsaker*. Although the measurements provided information about the static stability, predictions regarding the dynamic behavior and flyability were difficult to deduce. It was felt that a reasonable correlation between the wind tunnel measurements and flying qualities could be arrived at only through flight tests and in consultation with experienced pilots. These findings of the flight scientists were not utilized then by the aircraft manufacturers, as it was not possible yet to solve the mathematical problems without the aid of proper tools. The time was simply not yet ripe for that.

The lack of information about which parameters of an aircraft are most relevant to the flying qualities prompted the American aviation authority National Advisory Committee for Aeronautics (NACA) in 1919 to an extensive flight test program [3]. The objective was to establish a good correlation to the previous wind tunnel tests at M.I.T. The flight tests were performed mainly by the test pilot *E. T. "Eddie" Allen*, who later became one of the most distinguished test pilots of the USA. For these experiments, the aircraft were for the first time fitted with a simple measuring equipment to record the main parameters (control force, rudder position) during the flight. In this way, it was possible to determine reasons for the known poor flying qualities of Curtiss JN4H "Jenny". This laid the foundation stone for the flight test department of NACA, which excelled in the course of the following years with better and better instrumentation and excellent test pilots.

After the First World War, the flying qualities remained further on the topic of aeronautical research in Germany too. In the year 1926, the DVL Flight Department in Berlin-Adlershof dealt once again with flight performance tests. The aim was to gather reliable data for stability requirements. The tests pointed out the need for improving the stability characteristics of the then assembled aircraft and that the documents were not yet sufficient to predict, for example, duration of damping of oscillations. Nevertheless, a good balancing of elevator, rudder, and ailerons control forces was already demanded. The results of the DVL flying qualities experiments are reflected in the construction regulations for aircraft (BVF) published in 1928 [2].

Until the early nineteen-thirties, everything necessary was done to construct inherently stable, well controllable and safe aircraft. They were built on a wealth of experience and less on flight mechanical theories. The development engineers had then absolutely no options yet to predict the dynamic behavior of an aircraft. From the experience, one knew, however, what had to be done to suppress effectively such disturbing oscillations. For this reason, the development engineers hardly felt any need till mid-nineteen-forties for a detailed mathematical analysis of the flight performance.

Both civil and military aviation wanted aircraft to have flying qualities tuned to the respective requirements and with which an average pilot could cope well. For commercial aircraft, special emphasis was on instrument flight or the approach on a radio beam, whereas for military aircraft the maneuverability stood in the foreground. These are sufficient reasons for aircraft procurer to enquire about the flying qualities of a particular vehicle.

This topic led in 1940 to a large-scale research program at the NACA. The flight departments of NACA research centers at Langley Field (Virginia) and Moffett Field (California) conducted flight tests, with which stability and controllability of aircraft regarding the "flying qualities" were to be evaluated based on the pilot comments. After the outbreak of World War II, this study was greatly expanded with the participation of U.S. Army Air Corps and the U.S. Navy. Numerous civil and military aircraft were investigated under the leadership of NACA flight test engineer *Robert R. Gilruth*. Based on the flight physics, *Gilruth* developed a series of quantitative flying qualities criteria, which he correlated with the flight tests. His investigations resulted in a collection of flying qualities criteria, which were then stipulated as mandatory flying qualities guidelines for the aviation industry [4].

Whereas the flight test at Langley Field concentrated on flight test programs with subsequent pilot survey, on the other side of the continent the focus was already on determining the flying qualities of a future aircraft from wind tunnel measurements, before the pilots pointed out the deficiencies, which could be corrected only through highly time-consuming and costly efforts [5]. A very detailed wind tunnel measurements were carried out on different types of aircraft and these were correlated with those from flight tests.

Similarly, as in the case of NACA, the developments in Germany were directed towards quantitatively definable flying qualities guidelines. Essential experiments related to this aspect were carried out at the DVL in the nineteen-thirties by *August Kupper*. After *Kupper* passed away, *Karl-Heinrich Doetsch* was responsible for the continuation of the work. Though the use of instrumentation to measure important flight mechanical parameters, it became possible to obtain numerical data necessary for quantification. In the year 1943, the DVL, and almost at the same time the NACA, published a new draft of the flying qualities guidelines [6]. In these guidelines, special emphasis was placed on the verifiability of the stipulated flying qualities. Thus, the guidelines also contained a number of standardized testing instructions and directives for the assigned test pilots. As a result, the German aviation industry possessed a set of rules to design control surfaces and flight controls for a standardization the flying qualities to a large extent.

With the introduction of jet engines, the usable flight regime was extended up to the speed of sound. This resulted in new demands and requirements on the flight performance, flying qualities and on the flight safety. On approaching the speed of sound, the so-called compressibility effects were encountered, leading to hitherto unknown flight instability with reduced control capability. Although the new configurations such as sweptback and delta wing ensured acceptable flying qualities in the high subsonic range, they faced new problems at lower airspeeds, such as those during landing. The dilemma was how to manage the higher elevator and rudder forces resulting from the increased aircraft size and enhanced airspeeds

However, conventional aircraft configurations also showed occasionally stability problems during certain flight tasks. Already during the war, there were attempts to improve the deficient flying qualities with the help of a controller. Again, the fundamental contributions of *K.-H. Doetsch* on automatic control at DVL provided the practical basis for artificial stabilization of aircraft (see Chap. 4).

New aircraft configurations and the increasing air traffic forced an enhancement of flying qualities guidelines. The criteria should now be formulated task-specific (takeoff, landing, cruise flight, maneuvering, approach in a glide slope, etc.). Also, the different types of aircraft (transport aircraft, combat aircraft) should be evaluated differently. The military and the aeronautical establishments requested the NACA to deal with flying qualities of transport aircraft during instrumental approach and landing. The Navy needed specific criteria for a safe approach on an aircraft carrier. The US Air Force Air Materiel Command was interested in determining the stability derivatives of an aircraft from flight tests.

This also applied to the control forces. The control forces should be such that a safe "feeling" for speed and stability is conveyed to the pilot. Already in the nineteen-twenties, it was realized that the forces to be exerted by the pilots would no longer suffice to actuate the rudder of large aircraft. This problem was initially solved through the introduction of so-called auxiliary rudders. Auxiliary rudders, whether Flettner tab or spring-loaded auxiliary rudder (spring tab), were successfully installed worldwide in many aircraft types. The harmonization of control forces and adaptation on the prescribed guidelines led sometimes to serious problems. This resulted in numerous changes to the tailplane, until the desired characteristics was achieved. For some aircraft hundreds of flying hours were spent to adjust the rudder forces to the operational conditions [7].

The first aircraft with adjustable flying qualities (*Variable Stability Aircraft*) originated in 1947 out of this difficulty. Thereby the elaborate and expensive modifications to aircraft could be averted (see Chap. 5).

The development and the use of automatic course controls (autopilot) brought a new set of tools, attitude gyro, rate gyro, electrical and hydraulic servomotors, which could be used for other purposes as well. With such servomotors large rudder control forces could be exerted even without the help of auxiliary rudder. This was particularly important for the new generation of jet aircraft, which flew under high dynamic pressures and required large control forces. However, the pilot lost thereby the sensation for the control forces, which is an important criterion for handling qualities. This feeling had to be provided to him now artificially by a force feedback. Autopilot, stabilizing controller (*Stability Augmentation Systems, Response Feedback Control Systems*), servos, and artificial control forces contributed significantly to growing complexity of aircraft.

The correct designing of such complex systems needed a thorough flight mechanical analysis, before they could be integrated into a plane. Since the late nineteen-forties, new procedures for stability analysis and controller design were available in the USA. Using the root locus method developed by *Evans* and the *Bode* plots in the frequency domain one was able to model complex systems and carry out stability investigations. With the introduction analog computers, it became finally possible to investigate complex flight systems in a simulation. These methods were the result of years of research in the fields of electrical engineering, flight mechanics, and control engineering [8]. This way the errors and instabilities could be detected and eliminated at an early stage. The subsequent stability and system analysis on the digital computer was another important step towards flight safety and saved a lot of precious flight test time, and thereby avoided dangerous situations for test pilots [9].

The flying qualities analysis also benefited from the new procedures and methods of stability analysis. These advances and the outbreak of war in Korea forced the U.S. Air Force and Navy to a closer cooperation in the area of flying qualities. The result of these joint efforts was the newly revised flying qualities guidelines in the form of a "Military Specifications, Flying Qualities Piloted for Airplanes" during 1954 [10].

The initial flight controllers were still based on pure flight state feedback systems (*Response Feedback Control*). The controller is superimposed on the manual control through the pilot without affecting its function. As a result, the flying qualities could be improved and brought in accord with the guidelines.

New technologies and tasks led to significant expansion of flight domain. The flight at supersonic speed, intelligent weapons systems, challenging flight tasks such as flights close to ground-level (*Terrain-Following*) revealed the limits of mechanical manual flight control. Only with an elaborate flight controller, it was possible to meet the flying qualities guidelines. The complex flight controller could be readily realized, dispensing with the manual control through levers, push rods, and ropes. Instead of a mechanical connection to the inceptor, the hydraulically operated rudder actuation received now an electrical input. This new technology was termed as "Fly-by-Wire". New controller functions made possible the flight with reduced stability in order to achieve performance improvements. Security features, such as stick shaker and flight envelope limiter, ensured safety and avoided structural overloading. The pilot now flew a controller-aided airplane. An exhaustive account of national and international flight test demonstrators, which serve the purposes of testing, certification, and implementation of Fly-by-Wire technologies in the military and civil aviation, is provided in Chap. 6.

Forgoing the natural stability with the objective of improved flying qualities placed high demands on the integrity of the flight control system. A multiple redundant flight controller now ensures stability and makes sure that all functions required as per the flying qualities guidelines are restored. Aircraft which were designed for a spectrum of specific tasks and which could only be flown with the aid of a flight controller were the so-called *Control Configured Vehicle* (CCV) [11]. With the CCV Technology, the foundation stone was laid for all modern combat aircraft (see also Chap. 6).

Mathematical analysis, ground simulation, in-flight simulation, and flight test were the tools used in the nineteen-sixties and nineteen-seventies to investigate and to optimize the flying qualities of new aircraft types. The MIL-F-8785 provided a set of criteria, which allowed to verify objectively the flying qualities from flight test data. This alone was, however, not enough. Subjective evaluation by an experienced test pilot was additionally demanded (see Fig. 2.5). To account for this aspect a new policy "MIL-STD-1797" was published [12]. This guideline was based substantially on the MIL-F-8785, but contained additionally flight test methods and an evaluation table for a subjective evaluation by the test pilot. The *Cooper-Harper* Rating Scale [13] allowed a differentiated assessment of flying qualities and was a good supplement to the objective evaluation by the MIL.

To be more specific, the *Cooper-Harper* Rating scale is the subjective but structured measure to evaluate the handling qualities, in other words, the controllability of an aircraft in terms of pilot workload. For all practical purposes, it is the standard procedure employed over more than four decades. Basically, it aims at determining "those qualities or characteristics of an aircraft that govern the ease and precision with which a pilot is able to perform the tasks required in support of an aircraft role" [13]. In general, it is intended

Fig. 2.5 NACA test pilot *George E. Cooper* with *Hugh L. Dryden*, the Director of NACA Ames Research Center (1951)

for pilot-in-the-loop tasks. As such design and definition of appropriate tasks to be flown and evaluated is critical in the application of this rating scale. It is based on the pilot assessment immediately after performing a particular task, the rating varying from 1 to 10. The highest rating of "one" implies fulfillment of task with very little effort, whereas the lowest rating of "ten" implies cases which were not controllable in all the test phases, because major deficiencies were encountered and that improvements are absolutely necessary. The standard *Cooper-Harper* Rating scale is presented in Fig. 2.6.

New tasks, expansion of flight operations, monitoring systems were also reflected in the cockpit. As a result, the pilot behavior, his way of response, mental and physical ability to withstand workload gained importance. It was particularly a trait of digital flight control systems that resulted, as a consequence, in a new kind of flying qualities problem. Computational and signal transmission times led to time delays in the control system. After the input of a command by the pilot, the aircraft reaction did not result immediately, rather only after a certain delay. Thereupon the pilot intensified his input and retracted the command immediately once the aircraft reacted vehemently to the control commands. The now developing oscillation (*Pilot Induced Oscillation*—PIO) was very difficult to control and has led to accidents with a few of the latest generation of combat aircraft. Even in the case of commercial aircraft, the PIO problems were not unknown (see also Sect. 9.2.12).

Investigations of flying qualities involving the pilots were the order of the day. Also, however, well-equipped ground simulators were not adequate enough to reproduce the environment of a modern control-assisted aircraft in totality. Only the in-flight simulation offered the possibility of replicating the flight control of another aircraft and to investigate under real workload conditions. Concomitant with the flight tests, the PIO problem was analytically

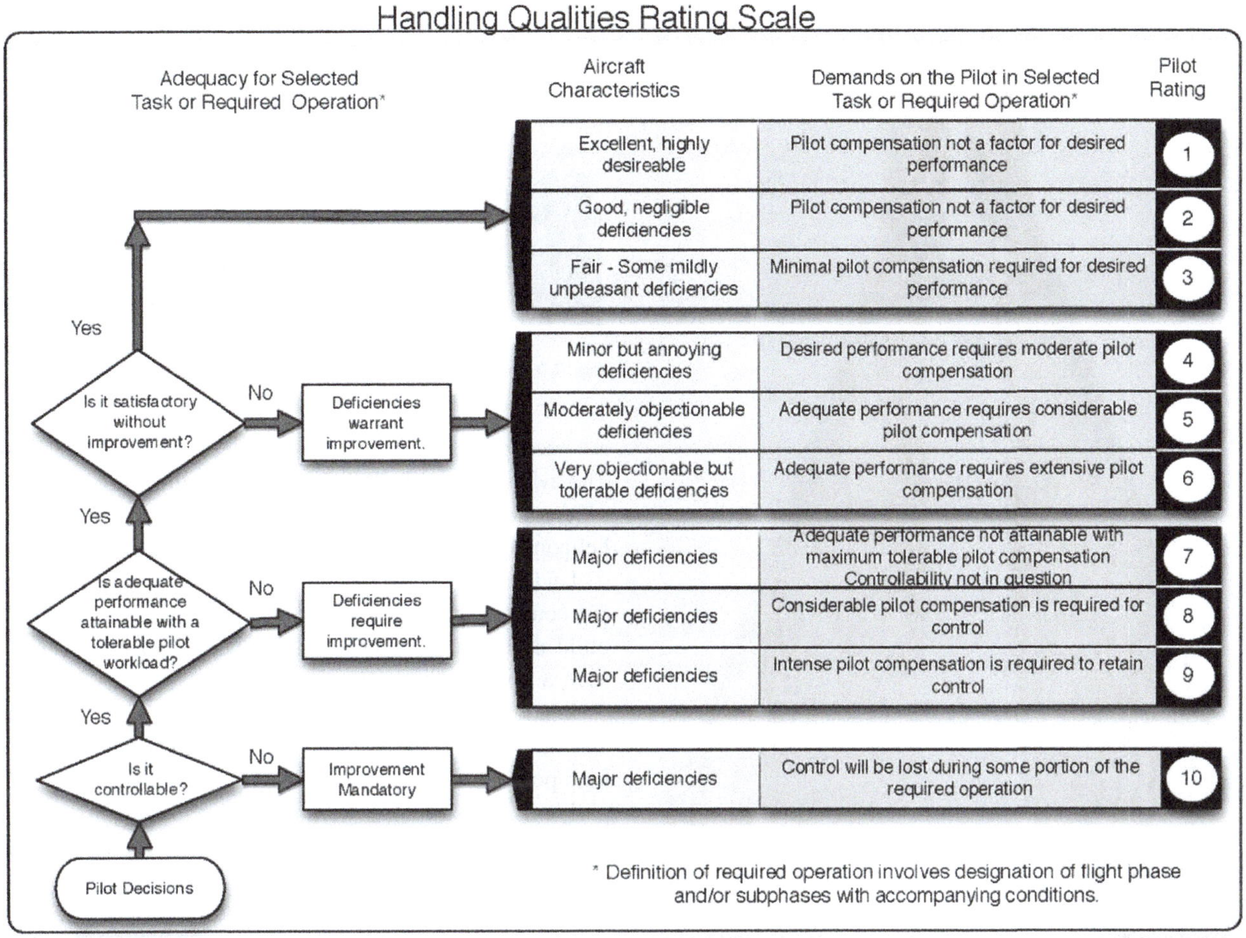

Fig. 2.6 Cooper-Harper Rating Scale

investigated. From the results of these investigations, strategies to adjusting the flight control could be derived so that PIO could be avoided.

2.2 Rotorcraft

The idea of rotary wing vehicles is almost as old as for the fixed-wing aircraft. In this case, too, a few paragons existed in nature, which provided orientation. The seeds of the sycamore (see Fig. 2.7), rotating in the presence of wind, can cover long stretches in a stable flight, whereas dragonflies and hummingbirds demonstrate the possibility of hovering flight in principle. The realization of a functioning rotorcraft, from the first ideas of *Leonardo da Vinci* (Fig. 2.8) to a viable helicopter, however, presented significantly more difficulties than those for the fixed-wing aircraft. After the maiden flight of the *Wright* brothers in the year 1903, another twenty years were needed until a so-called Autogyro or gyrocopter was built, flown and marketed by the Spaniard *Juan de la Cierva* (see Fig. 2.9). *De la Cierva* had solved two fundamental problems, namely (1) the control of high moments at the rotor blade root through an articulated connection of the rotor blades, the so-called flapping hinge, and (2) the steering of the Autogyro about the longitudinal and lateral axes by tilting the rotor head in the desired direction of flight (tilting hub control). However, in the case of the gyrocopter, the rotor is driven by the airflow, as such the flight vehicle requires an additional propulsion, for example via propeller, as in the case of fixed-wing aircraft. A hovering flight was therefore not possible and it took more than 10 years until the first halfway usable helicopter could demonstrate the hover and forward flight.

For a fixed-wing aircraft, a configuration (with front mounted engines, the lift-generating wings and the tailplane located at rear), which provided stability and good controllability was crystallized quite fast. In the case of helicopter, everything had to be principally accomplished via the rotating rotors. The rotor, the rotating wing, produces the lift as well as propulsion and control forces and moments which

Fig. 2.7 Sycamore seed (Acer pseudoplatanus)

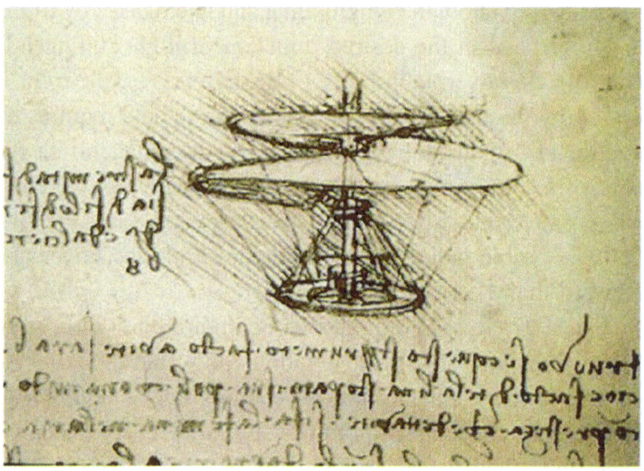

Fig. 2.8 Sketch of helicopter by *Leonardo da Vinci* (1490)

Fig. 2.9 Autogiro C 19 Mk IV of *Juan de la Cierva* (1926)

are needed for maneuvering. Not an easy task for a designer. The aviation pioneer *Wilbur Wright* saw this, however, quite differently, when he stated in 1909 that it is easy to build a helicopter, but otherwise it is a worthless device.

Unquestionably *Wilbur Wright* was not informed about the activities pursued at that time by many constructors to build a helicopter. Otherwise, he would not have come to such a misjudgment.

In addition to stiffness and vibration problems at the rotor and the lack of a light and high-performance motor for the high power, which is needed for hover, the flight mechanical problems, in particular, were in the foreground. It was necessary to cope with the turning moment on the fuselage, the reaction to the torque of the rotor drive. The problem of steering the helicopter for variations in lift and directions, with acceptable control forces, was not resolved for a powered rotor. The flight dynamics of the devices was highly unstable; the pilot had to intervene constantly to damp the arising oscillations and to keep the helicopter stable in the air. It is, therefore, not a surprise that the constructors arrived at quite different solutions.

For the development of an airworthy device, the evolution of scientific principles of rotor aerodynamics, rotor dynamics, vibration behavior and flight dynamics was essential and the developed theories had to be validated through experiments. In this respect, pioneering work has been carried out in Germany, which led to the construction of the first operational helicopter over here before the second World War [2, 14].

One of these pioneers was *Henrich Focke*. He drafted some important criteria that a practical helicopter design had to meet:

1. The ability of a safe landing after engine failure by means of autorotation.
2. Ensuring stability and control through the pilots, whereby the demands should not be higher than those for fixed-wing aircraft.

Fig. 2.10 The first ready for use helicopter Focke-Wulf Fw 61 (1936)

3. Ease of using the inceptors. Also, the control forces should be similar to those in the case of fixed-wing aircraft.
4. Acceptable flight performance in hover and forward flight.
5. Operational safety should be comparable to that of fixed-wing aircraft.

After numerous considerations and experiments, *Henrich Focke* opted for a largely symmetrical construction, with two counter-rotating rotors on cantilever arms, arranged right and left of the fuselage (see Fig. 2.10). With this the torque balance about the vertical axis was assured. To control the flight vehicle the incidence angle of the rotor blades was adjusted through a so-called swashplate on both rotors. Through raising and lowering the swashplate the incidence angle of the blades could be jointly adjusted (collective blade pitch control). In this way, climbing and descent, and through oppositely adjusting the swashplates the rolling motion of the helicopter could be controlled. By tilting the swashplates horizontal forces and moments could be generated for the longitudinal, lateral and yaw control of the helicopter. The requirements for a maneuverable helicopter were fulfilled for the first time with the Focke-Wulf Fw 61.

Here Wilbur Wright was wrong

Already as children, Wilbur and Orville Wright were interested in flying. Wilbur was 12 and Orville 8 years old when their father brought them a toy helicopter with rubber motor, a brand new invention then. Soon they built their own copies of the toy and these were actually their first motorized flying gadgets. When they were asked later what had triggered their fascination for flying, they always re-called this toy helicopter.

But soon they were looking for new challenges and Wilbur later was cited as follows:

> *Like all novices, we began with the helicopter (in childhood) but soon saw that it had no future and dropped it. The helicopter does with great labor only what the balloon does without labor, and is no more fitted than the balloon for rapid horizontal flight. If its engine stops it must fall with deathly violence, for it can neither float like the balloon nor glide like the aeroplane. The helicopter is much easier to design than the aeroplane but it is worthless when done.*
>
> *Wilbur Wright, Dayton, Ohio, January 15, 1909*

Fig. 2.11 Flettner Fl 265 helicopter in large wind tunnel at Chalais-Meudon (1941)

In April 1937, *Henrich Focke* together with *Gerd Achgelis* founded the company Focke-Achgelis. In 1938 the first major contract came from the German Lufthansa to develop a large transport helicopter with a payload capacity of 700 kg. Built on the Fw 61 concept, the Fa 223 was the first helicopter manufactured in serial production. During the war, it was deployed to transport operations in high mountains, for rescue operations, and for submarine surveillance.

In the year 1938, the inventor and aircraft designer *Anton Flettner* started with the development of helicopters. Like *Focke's* Fw 61, the successful *Flettner* helicopter had two counter-rotating rotors in a mostly symmetrical configuration. Unlike the Fw 61, these rotors were not mounted at cantilever arms, rather they were mounted directly on the fuselage at an angle of inclination of 12°. Because of the small distance, the two-bladed rotors were intermeshing. Collisions between the rotor blades were avoided through gearing. Similar to the Fw 61, the blade control was ensured by swashplates.

Prior to the flight tests, extensive wind tunnel tests were performed during August 1940 in the great French wind tunnel of Chalais-Meudon (see Fig. 2.11).

The Flettner helicopter showed high maneuverability and good controllability about all three axes and in all directions. With the Fl 265 a total of 126 flight hours were flown and over one thousand takeoffs and landings performed. No other helicopter worldwide at that time had achieved those hours in operation. Starting 1941 this helicopter was built as Fl 282 on a small scale and deployed by the armed forces for many different tasks (see Fig. 2.12). After the war, American pilots availed the opportunity to fly the Fl 282. Following this, they praised the stability and controllability of this helicopter, as being better than all other helicopters they had flown until then.

Independent of the development in Germany, the work was pursued on the construction of helicopters in other countries too. Already before the First World War, the French aircraft pioneer *Louis Charles Breguet* dealt with the construction of helicopters [15]. However, his preliminary designs, called Gyroplane, had difficulties even to takeoff from the ground. After that, he turned his attention to aircraft manufacturing. Not until 1930 his interest in the helicopter was rekindled.

Fig. 2.12 Flettner Fl 282 helicopter in operation (1943)

Together with a young engineer named *René Durand*, he developed a helicopter with two superposed rotors. The two rotors rotated in opposite directions, thereby the torque on the fuselage due to the rotors could be compensated. With the coaxial arrangement of the rotors, *Breguet* had arrived at another, almost symmetrical configuration, on whose basis a stable to fly and a controllable helicopter could be developed.

The rotors were designed two-bladed. Both featured a swashplate adjustable cyclic blade pitch control. The control around the vertical axis was effected by means of collective blade pitch control of the two rotors in opposite directions, thereby generating different amounts of torque moment, resulting in the desired yaw moment. For a straight level flight and for lateral motion, as in the case of other helicopter configurations, the rotor disks were inclined accordingly with the swashplates. To improve the stability, the helicopter was additionally equipped with a tailplane.

In the years 1934 and 1935, the helicopter was an object of numerous modifications. In June 1935 the helicopter took off for the successful maiden flight (see Fig. 2.13). In December 1935, with a flight time of 62 min the hitherto endurance record for helicopters was broken. The French aviation authority was impressed and placed in the year 1936 a contract with *Breguet* to develop a helicopter suitable for operation. The development progress was, however, slow and was interrupted repeatedly due to repairs and modifications. One difficulty was the ability for autorotation, an important requirement for the client. In the year 1939, the helicopter was damaged severely in an autorotation test.

The impending World War II put an end on the further development. The "Gyroplane Laboratoire" was destroyed by a bomb attack.

Born in Ukraine, American aircraft designer *Igor I. Sikorsky* had already commenced in 1930 to engage himself with the development of a usable helicopter. In the year 1931, he filed a patent for a helicopter, that already had all the essential features of the later model VS-300, namely a main rotor and a small vertical tail rotor for torque compensation and for control about the vertical axis. This asymmetrical configuration promised significant performance benefits; the technical capabilities available at that time, however, did not allow a successful construction [16].

After many attempts *Sikorsky* began in spring 1939 with the construction of an airworthy helicopter denoted VS-300 (Vought-Sikorsky) (see Fig. 2.14). The three-bladed main rotor was equipped with a cyclic rotor blade pitch control for vertical, longitudinal and lateral helicopter control. Besides the torque-compensation, the tail rotor provided also the yaw control. The first flight took place in September 1939 and was flown by *Sikorsky* personally.

Fig. 2.14 Experimental helicopter Sikorsky S-46/VS-300 (1941), (*Credit* Igor I. Sikorsky Archives)

Many alterations followed, mostly because of poor controllability. Among other things, the rotor control was replaced, in the meantime, by small rotors on side arms for pitch and roll control; whereas only the vertical motion was controlled by the main rotor. With this configuration the world record for longest flight duration, that was hold since 1937 by the Fw 61, could be broken in May 1941 with a flight time of one hour and 32 min. In December 1941 the helicopter attained its final configuration with one main rotor and only a single tail rotor, which is adopted in many helicopter designs till today.

The success of the VS-300 convinced the US Army. Early 1943, they placed an order for production of 100 helicopters of the type R-4 developed meanwhile (see Fig. 2.15). During the war, the R-4 helicopter was successfully deployed for rescue operations and for passenger transportation.

Henrich Focke, *Anton Flettner*, *Louis Breguet* as well as *Igor Sikorsky* have independent of each other arrived at

Fig. 2.13 Experimental helicopter Breguet-Durand Coaxial "Gyroplane Laboratoire" (1941), (*Credit* American Helicopter Society)

Fig. 2.15 Production version of Sikorsky S-47/R-4 in operation (1941), (*Credit* Igor I. Sikorsky Archives)

technical solutions for a workable and "easy" to fly helicopter. Their ideas were incorporated in numerous successful helicopter developments after the war.

While *Focke*, *Flettner* and *Breguet* concentrated on configurations with two main rotors, Sikorsky concentrated on the design based on one main rotor and tail rotor. The advantages of this configuration over the dual rotor helicopters became readily apparent: higher performance, greater flexibility in rotor design, cost-effective construction. But the drawbacks weighed heavily, namely asymmetrical flying vehicle with poor flying qualities, complicated aerodynamic and dynamic conditions on main rotor and tail rotor. Particularly because of the better performance, the configuration main-rotor tail-rotor was largely preferred for small and medium helicopters, whereas other configurations were adopted only for heavy helicopters and special flying vehicles (see Fig. 2.16). The request for better flying qualities was of secondary importance during the first few decades of the development. Top priority was to meet the demands, mostly higher flight performance, for the military and civilian operations of helicopters. As a result, recent activities are characterized by constant efforts toward acceptable flying qualities and reduction of undesired oscillations and vibrations. It is estimated that in many helicopters about 25–50% of the development time was needed to deal with these shortcomings in the flying qualities [2, 17].

Based on the research work at NACA, the first flying qualities guidelines for helicopters were published during 1951 in the American *Civil Air Regulations Part 6-Rotorcraft Airworthiness; Normal Category*. Under the heading *Flight Characteristics* it is stated there: "*It shall be possible to maintain a flight condition and to make a smooth transition from one flight condition to another without requiring an exceptional degree of skill, alertness, or strength on the part of the pilot,...*", in the section *Controllability*: "*The rotorcraft shall be safely controllable and maneuverable during steady flight and during the execution of any maneuver...*", and in the section *Stability*: "*It shall be possible to fly the rotorcraft in normal maneuvers...*" [18]. These general qualitative statements have been made more precise in the first quantitative specifications for flying qualities of military helicopters MIL-H-8501 from the year 1952 [19]. Without sufficient database and without detailed explanations, numerous quantitative requirements were formulated, amongst others for static and dynamic stability, for control forces and the helicopter behavior in autorotation. In the year 1961, MIL-H-8501 was revised and this specification was effective as MIL-H-8501A in US and other countries for over 30 years.

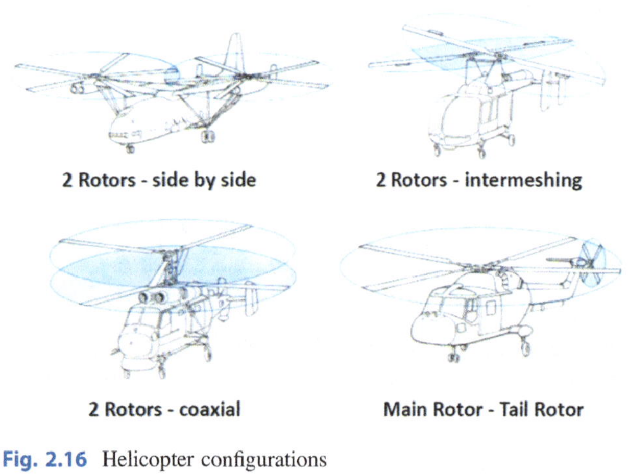

Fig. 2.16 Helicopter configurations

In the civil sector, essentially the qualitative criteria were retained and further developed (FAR 27 and FAR 29, EASA CS-27 and CS-29). However, the quantitative military criteria were often adopted as guidelines for design and certification.

For a long time, the specifications were considered as the targeted goal and not as indispensable requirement. This was partly due to the unavailability of design tools to meet soundly the stipulated criteria in the case of a new development, and on the other hand due to technical improvements, which made some criteria to appear as obsolete. As such, for example, the helicopter Bo 105 could not fulfill a few criteria related to dynamic stability or control coupling. The exceptional control-characteristics, namely fast control response and high effectiveness, of the newly developed hingeless rotor led, however, to flying qualities, which were rated by the pilots as very good. The helicopter was certified by the civil and later by the military authorities in many countries, despite its "shortcomings" compared to the valid criteria.

The expansion of the range of tasks, particularly for the military helicopters, the inadequacies of the flying qualities guidelines, and the availability of electronic systems for pilot assistance strongly necessitated new criteria. In order to compile a systematic database required for this, the US Army together with the NASA launched in 1975 a research program to which research organizations and institutions from Canada, England, and Germany (DLR) provided important contributions (see Sect. 8.4.1). Based on the data from ground-based simulations, especially, however, from flight tests, new flying qualities specifications for military helicopters, *Aeronautical Design Standard 33* (*ADS-33*), were finally published in the year 1988, and in the following years further refined [20].

The acceptance and applicability of the new, in many aspects revolutionary, flying qualities specification is based essentially on the systematic and reliable data base, whose evaluation led to new criteria and important insights about the relationships between subjective pilot handling qualities assessment ratings and quantified flying qualities parameters [21].

The in-flight simulator ATTHeS was a modified Bo 105 that, because of the Bo 105's control effectiveness, gave it the flexibility making it particularly well suited to make significant contributions to the currently available data base (see Chap. 8). As a result, important prerequisites were established for design and certification of helicopters, which are safe and "easy" to fly under all operating conditions and in all possible missions.

References

1. Perkins, C.D.: Development of airplane stability and control technology. J. Aircraft **7**(4), 291 (1970)
2. Herb, H.: The development history of flight mechanics, In German, Deutsche Luft- und Raumfahrt, DLR FB 67–44, July 1967
3. Perkins, C.D.: Development of airplane stability and control technology. J. Aircraft **7**(4), 293 (1970)
4. Gilruth, R.R.: Requirements for satisfactory flying qualities of airplanes, NACA Report 755, (1943)
5. Hartman, E.P.: Adventures in research—a history of Ames Research Center 1940–1965, p. 80. NASA History Series, Washington (1970)
6. Doetsch, K.-H., Friedrichs, G., Höhler, P., Lugner, H.: Redesign of flying qualities guidelines, in German, DVL-Bericht, March 1943
7. Ashkenas, I.L.: Twenty-five years of handling qualities research. J. Aircraft **21**(5), 290 (1984)
8. McRuer, D., Graham, D.: Eighty years of flight control: triumphs and pitfalls of the systems approach, J. Guid. Control. **4**(4) (1981)
9. McRuer, D., Graham, D.: A flight control century: triumphs of the systems approach. J. Guid. Control Dyn. **27**(2), 161–173 (2003)
10. Anon.: Military Specification, Flying Qualities of Piloted Airplanes, MIL-F-8785 (ASG), September 1954
11. Wünnenberg, H. (Ed.): Handling Qualities of Unstable Highly Augmented Aircraft, AGARD AR-279, May 1991
12. Anon.: MIL-STD-1797, Department of Defense, March 1987
13. Cooper, G., Harper, R.: The Use of Pilot Rating in the Evaluation of Airplane Handling Qualities, NASA TN D-5153, April 1969
14. Knobling, K.: First helicopters and rotor systems. In: Hirschel, E. H., Prem, H., Madelung, G. (eds.) Aeronautical Research in Germany, pp. 293–309. Springer Verlag, Berlin (2004)
15. Leishman, J.G.: A history of helicopter flight. University of Maryland, www.terconnect.umd.edu/Leishman/Aero/history.html
16. Sikorsky Historical Archives, www.sikorskyarchives.com
17. Padfield, G. D.: Rotorcraft handling qualities engineering, managing the tension between safety and performance, 32nd American helicopter society. Alexander A. Nikolsky Honorary Lecture, May 2012
18. Anon.: Civil aeronautics board: civil air regulations part 6-rotorcraft airworthiness. Normal Category, Washington DC, USA (1951)
19. Anon.: Military specification: helicopter flying and ground handling qualities; general requirements for, MIL-H-8501, November 1952
20. Anon.: ADS-33E-PRF, Aeronautical design standard, performance specification, handling qualities requirements for military rotorcraft. United States Army Aviation and Missile Command, Aviation Engineering Directorate, Redstone Arsenal, Alabama, March 2000
21. Pausder, H.-J., Ockier, C.: Flying Characteristics specifications for helicopters. In: Hirschel, E.H., Prem, H., Madelung, G. (eds.) Aeronautical Research in Germany, pp. 480–482. Springer-Verlag, Berlin (2004)

Author Biographies

Bernd Krag was a research scientist at the Institute of Flight Systems of DLR of Braunschweig (1972–2002). From 1993 to 2002 he was head of the Fixed Wing Aircraft department. Prior to joining DLR, he was a research assistant at the Flight Mechanics Branch at the Technical University in Braunschweig (1967–1972). He received his Dipl.-Ing. degree (1967) and Dr.-Ing. degree (1976) in Aerospace Engineering from the Technical University Braunschweig. His main research interest were control configured vehicles (CCV), active control, modeling and system identification, data bases for training simulators, and the wake vortex problem. Since retiring from DLR, his interest is in the history of aeronautics research in Braunschweig and aviation history.

Bernd Gmelin was a research scientist and head of the Rotorcraft Department at the Institute of Flight Mechanics of DLR at Braunschweig (1973–1999). From 1999 to 2006 he headed the DLR Helicopter Program under the German/French DLR/ONERA Common Rotorcraft Research Program. Prior to joining DLR, he was a researcher at the Institute for Rotorcraft in Stuttgart (1968–1973). He received his Dipl.-Ing. degree in Aerospace Engineering from the University of Stuttgart (1968). His primary interests are mathematical modeling, helicopter simulation in the wind tunnel, control systems, flying qualities, helicopter in-flight simulation. He is a member of the German Society for Aeronautics and Astronautics (DGLR) and of the American Helicopter Society (AHS). He is the recipient of the AHS Agusta International Fellowship Award (1996), DLR/ONERA Team Award (2006), and AHS Fellow Award (2008).

3 Modeling and Simulation—Basics and Benefits

Bernd Krag

Assisted by W. Mönnich

3.1 Introduction

When we speak of a "simulation", it implies replication of a process on a computer using mathematical models. Here, the term 'Process' implies everything that can be analyzed nowadays in simulations, for example production procedures in factories, worldwide financial transactions, transport by rail and road, and many more. All simulations are based on a mathematical model of the process being investigated. Simulation allows a detailed study of the object before it is realized. The parameters of the mathematical model can be varied to examine various aspects of the process. Model parameters can also be so adjusted that the simulation yields replication of the reality as accurately as possible. The simulation is also a mechanism by which future

B. Krag (✉)
Braunschweig, Germany
e-mail: bernd.krag@t-online.de

P.G. Hamel (ed.), *In-Flight Simulators and Fly-by-Wire/Light Demonstrators*,
DOI 10.1007/978-3-319-53997-3_3

developments can be predicted or evaluated. Simulation plays a key role in successfully preparing and carrying out a flight test program. The more precisely the mathematical model maps the reality, the more realistic is the simulation and the more meaningful are the results. For this reason it is clear that the development of highly accurate mathematical models is of great importance. "Modeling & Simulation" is and remains a challenge for scientists and engineers.

3.2 Simulation and System Identification

3.2.1 General

Simulation software can be bought nowadays for any and every PC. Who does not know the "Flight Simulator" or "Train Simulator"? These computer games show an amazing realism in the representation of the simulated vehicle and its surroundings. What on the PC is just a game, is a training tool for pilots, captains, train drivers, astronauts, and others in their professional life. But simulators are used not only for training or retention of the professionalism, but also for scientific purposes. Development simulators are employed in aviation research to investigate new flight control laws, displays, control elements, etc. for their usability. Experienced pilots review and evaluate the success of a new measure and decide whether a new device or a control law will be introduced or not.

In ground-based human-in-the-loop simulators mathematical models are used, which describe the behavior of the respective vehicle to control inputs and external disturbances. The simulation quality depends not only on how accurately the reality is represented by the mathematical model, but also on the entire simulation environment. That includes the aircraft cockpit or the locomotive driver cabin together with a complete set of instruments and control devices. Furthermore, the representation of the external view and possibly the reproduction of motion feeling and impressions are added. The mathematical model ensures the correct driving of the instruments, displays and motion systems. For the construction of a simulator it is important to define in advance the intended purpose. In many cases it is simply unnecessary to build a complex and expensive simulator, when they are not required by the training tasks at hand [1]. The greatest possible realism for a specific training task leads to, however, meaningful assessments and good training results.

Together with the development of an aircraft, the mathematical models for simulation are developed. In the case of an aircraft the models for the aerodynamic forces and moments are derived from wind tunnel measurements and/or CFD calculations (CFD = *Computational Fluid Dynamics*). In wind tunnel tests the model of the aircraft is mounted on a measuring balance and exposed to the airflow. For determination of aerodynamic forces and moments resulting from the rotational aircraft motion, either costly wind-tunnel models or corresponding results from CFD computations are needed, or otherwise flight tests are required. Flight tests are also necessary for older aircraft, for which the databases are either unavailable or not accurate enough. For the determination of model parameters from flight tests, methods of system identification are applied. Since aerodynamic forces acting on the aircraft during the flight cannot be measured directly, they need to be determined indirectly from the reaction of the aircraft to control inputs. A prerequisite for a successful system identification is a good measuring equipment and data recording.

3.2.2 System Identification

System identification is based on the comparison between flight test and its simulation with a mathematical model. Aircraft movement is excited through the pilot control inputs. The control inputs and the response of the aircraft are measured and recorded. In the subsequent analysis, the measured control inputs, such as a rudder position, are fed into the simulation model. The free parameters of the model, for example, the aerodynamic derivatives, are determined in such a way that the deviations between measured aircraft response and simulated model response to the same control inputs are minimized. A common method for solving this optimization problem is the so-called. maximum likelihood method (a term from probability theory, detailed explanation of which would go too far here; suffice it to state that, in most cases, the product of the error variances represents the cost function for the optimization [2]). The block diagram in Fig. 3.1 illustrates the entire identification process. The system identification is applied to flight test data for the development and validation of mathematical models having very high simulation fidelity. Also, such precise mathematical models are needed for the model inversion process of

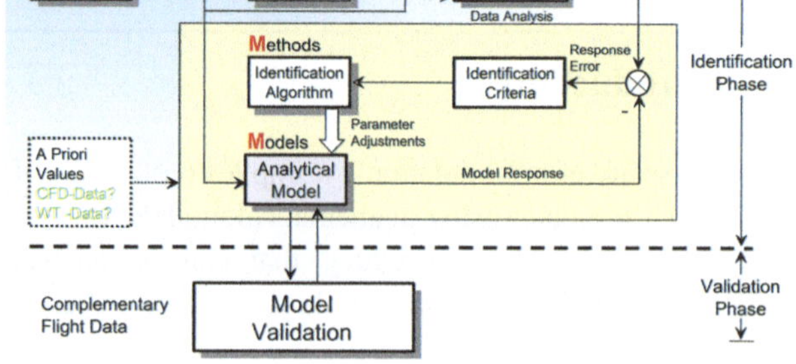

Fig. 3.1 The *quad-M*-principle of system identification (*Credit* Peter Hamel)

the so-called host aircraft of an in-flight simulator (see Sect. 3.3).

To successfully apply the so-called Quad-M-principle (Maneuvers, Measurements, Methods and Models) of system identification solid experience is required, such as that accumulated at the DLR Institute of Flight Systems over the past 50 years (see Figs. 3.2 and 3.3) [3, 4].

3.2.3 Ground-Based Simulation

It was a long way from the very first ground-based human-in-the-loop simulators to today's Level D training simulator (highest fidelity simulation), as utilized nowadays by the airlines. The first simulators ever were built for the training of pilots. Many aircraft from the pioneer generation

Flight Vehicle	Period	Type
Civil Aircraft	1970-1973	Dornier Do 27
	1975-1988	HFB 320-S1 FLISI
	1976	CASA C-212
	1978-1983	DHC-2 Beaver
	1981	A300-600
	1982-1985	Do 28 TNT OLGA
	1984	A310
	1986-1996	VFW 614 ATTAS
	1990	Dassault Falcon E
	1995	Grob G 850 Strato 2C
	1997-1998	Dornier 328-110
	1997	A300-600ST Beluga
	1997-1998	A330-200
	1999-2000	IPTN N250-PA1
	1999-2003	Dornier Do 128
	2000	VFW-614 ATD-EFCS
	2001	A340-600
	2002	Cessna Citation II
	2002, 2006	A318-121
	2004	Pitts S-2B
	2006	G180
	2007	A320
	2007	A380-800
	2010	Diamond DA42NG
	2010-2011	Falcon 20E
	2011-2012	Glider SB 10
	2013-2014	A350
	2013 ==>	A320 ATRA
	2014 ==>	Embraer Phenom 300
Military Aircraft and **Technology Demonstrators**	1976-1983	MRCA - Tornado
	1977	CCV F 104 - G
	1981	Do Alpha-Jet TST
	1984	Do Alpha-Jet DSFC
	1989-1997	C-160 Transall
	1993-1998	Dasa/Rockwell X-31A
	1997	EF 2000 Eurofighter
	2001	X-31A VECTOR
	2009-2010	A400M

Flight Vehicle	Period	Type
Aircraft Models	1978-1980	ATA free flight model
	1980	Do 28 TNT wind tunnel model
	1982-1983	Do 28 TNT free flight model
	2013 ==>	NumEx DLR F18
Helicopter and **Tilt Rotor Aircraft**	1975	Bo 105 S123
	1986-1989	Bell XV-15 tilt rotor
	1989-1990	AH 64 Apache
	1989-1990	SA 330 Puma
	1989-1995	Bo 105 S3-ATTHeS
	1998	SA 365 Dauphin
	1999 ==>	EC 135 FHS
	2007-2008	CH 53
Reentry Models	1989	OHB Falke
	1998	USERS
	2004	Astrium Phoenix RLV
Unmanned Aerial Vehicles	2008 ==>	ARTIS
	2010	UCAV (SACCON)
Rockets and **Projectiles**	1986	EPHAG
	1987-1988	EPHRAM
Aircraft Propulsion Systems	1992	Rolls Royce Tyne R.Ty.20 in C-160 Transall
	1997	Pratt & Whitney PW 119A in Dornier 328-110
	1997	Rolls Royce M45H MK501 in VFW 614 ATTAS
	1999	F404-GE-400 in X-31A
Aircraft Landing Gears	1996-1997	C-160 Transall
	1997	Do 328-100
	1997	VFW 614 ATTAS
	1998	Dasa/Rockwell X-31A
Miscellaneous	1985	Submarine, Class 206 U13
	1997	Aircraft pilot coupling
	1998	CombustionPlant RWE-VVA
	2000-2005	ALEX, Parafoil 252-7 Lite
	2002-2005	FASTWing
	2004-2007	Audi A8ABC
	2012-2013	Tragschrauber MTOsport

Fig. 3.2 System identification—experience at DLR (*Credit* Ravindra Jategaonkar)

Fig. 3.3 System identification—some recent contributions from DLR (*Credit* Ravindra Jategaonkar)

before the First World War were all but stable and good-natured (see Chap. 2). They demanded the constant attention of the pilot. There were numerous accidents caused by the lack of training. The opportunity to work with an experienced pilot to learn together in an airplane was rather the exception. Therefore, the importance of pilot training on ground, before the bold "Aviator" sat in an airplane, was recognized early. As such it is not surprising that just a few years after the historic flight of the *Wright* brothers, a few aircraft designers built training devices to protect both the pilots and the valuable aircraft.

The first viable ground-based simulator was offered in the year 1909 by the French aircraft company Société Antoinette. This apparatus (see Fig. 3.4), the *Antoinette Learning Barrel* ("Learning Drum") helped the pilots to fly the Antoinette VII monoplane. Student pilots at the flight school in Mourmelon-le-Grand found it necessary to use a training device with which the students could develop those reflexes which were needed to activate the control devices at the right moment in the right direction [5]. The apparatus consisted of two half barrels put over each other. A pilot's seat with the control wheels was mounted on the top. The entire assembly was unstable about all three axes and had to be constantly held in balance by the students. Thereby the simulation task was clearly defined for this trainee. Using the *Antoinette Learning Barrel* it was not yet possible for students to learn to fly, nevertheless they developed a feel for the aircraft reactions to control inputs.

Fig. 3.4 Antoinette "Learning Barrel" (*Credit* North American Museum of Flight Simulation)

Without going any further into the details of the history of development of ground-based human-in-the-loop flight simulators, it can be stated that with the development of new aircraft the need for corresponding training devices also grew. With the introduction of instrument flying during the late nineteen twenties, an appropriate training simulator was also at disposal. In the year 1929, *Ed Link* developed a simulator that provided the pilots a safe way to learn instrument flying. During the years from 1930 to 1950, the famous Link Trainer was built in large numbers and was used for pilot training in many countries around the world. The Link Trainer consisted of a cabin similar to that in an aircraft, but without outside view (see Fig. 3.5). The cabin could yaw up to 360°. It was supported on air-filled bellows which limited rolling and pitching motions. In any case the pilot sensed a reaction of the simulator on activating the control inceptors. More important were the instruments for the blind flight, following them, the pilot should "fly" on a predetermined course. On an evaluation table, the instructor could follow the course of the trainee pilot and over his microphone give instructions. The Link Trainer is considered to be a milestone on the road to the modern training simulator.

Modern training simulators for transport aircraft have achieved a simulation fidelity, which allows to retrain a pilot on a new aircraft type without further training flights (*Zero Flight Time Simulator*). In these ground-based simulators not only the cockpit environment is faithfully recreated, but the external view, the movement and the noise are generated too. The effort for the construction of such a simulator is indeed significant. The acquisition of high-fidelity training simulators is, however, worthwhile for airline operators, because the aircraft can be deployed to generate more revenue, while the pilot can be trained on the simulator day and night.

On the other hand, combat aircraft pilots are still predominantly trained in special training aircraft. According to the US Air Force, the training after the introduction of motion systems has even worsened. These motion systems were not able to deliver the acceleration impressions, which a fighter plane pilot is often exposed to. Also, due to a certain neural mismatch between system related time delays of motion and visual cues, training pilots sometimes became dizzy and felt sick (*Simulator Induced Sickness*—SIS). Furthermore, the US Air Force had lost many pilots due to becoming unconscious at high maneuvering loads, because these could not be trained in simulators (*g-Force Induced Loss of Consciousness*—G-LOC [6]). Only through the combination of simulator and centrifuge (*Authentic Tactical Fighting System*—ATFS) it was possible to improve the training success greatly.

3.3 In-Flight Simulation

No matter how good the ground-based human-in-the-loop simulator is, it cannot, however, reproduce the reality. There are always limitations with which one has to live with. To derive maximum benefits from a specific simulation task, the person entrusted with the execution of the task must also possess the necessary professional background (for example, test pilot, flight test engineer).

An aircraft which is converted into an in-flight simulator comes closest to the reality. The complete replica of a hypothetical new aircraft by converting an existing aircraft for the purpose of pilot training is almost impossible and is also not aspired in aeronautics. Depending on the simulation task, only sub-areas will be generally replicated. The special feature of the "in-flight simulation" compared to the ground-based simulation are the authentic vision and motion perceptions, which can be reproduced only by a real aircraft (host aircraft, Fig. 3.6). In addition to these physiological impressions the psychological effect is also important so that serious consequences of pilot actions can also be faced [7]. For the development of flying qualities criteria, aircraft with

Fig. 3.5 Link trainer (*Credit* Alberta Aviation Museum)

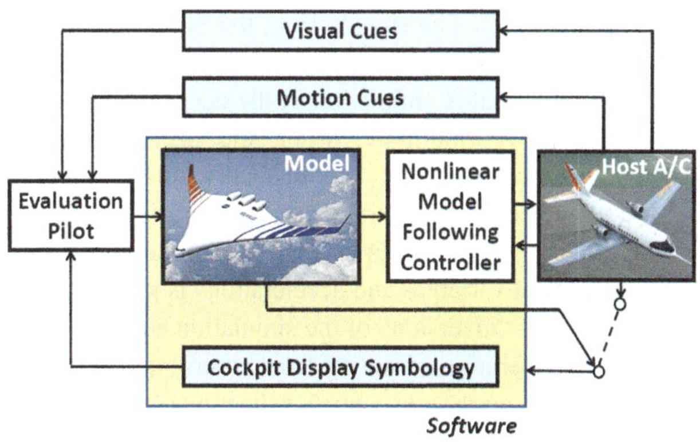

Fig. 3.6 Principles of in-flight simulation

variable stability (*Variable Stability Airplanes*) have proved to be successful (see Chaps. 5 and 7–10).

An airplane has, considered as a so-called rigid body, six degrees of freedom, namely three rotational (roll, pitch, yaw) and three translational (longitudinal, lateral and vertical). If the in-flight simulator should respond exactly about all the six degrees of freedom (6 DOF) like the aircraft being replicated, then accordingly as many, that is six, independent controls must be available. Normally, this is, however, not the case: an aircraft possesses usually three primary aerodynamic controls, namely elevator, aileron and rudder, and the thrust lever position or the thrust. Corresponding to their main effects, these four so-called control variables are suitable for motion simulation in the three rotational degrees of freedom (that is, the attitude angles and their temporal change) and in the longitudinal direction (that is, the airspeed). The replication of the vertical motion, and thereby also of the vertical load factor, as well as of the lateral motion will be only in limited agreement. If the motion is to be reproduced exactly in these translational degrees of freedom too, then appropriate additional control effectors are needed; for example, fast responding canard control surfaces or trailing edge flaps on the wing, so-called DLC flaps (*Direct Lift Control*), for the vertical motion or direct side force generators, for the lateral motion.

A so-called "model following controller" then ensures that the host aircraft replicates the behavior of the target aircraft in just as many degrees of freedom as the number of independent controls available [8]. In the other degrees of freedom, there are generally, give or take, marked differences in the motion. In the model following control, the procedure adopted is differentiated between the so-called implicit and explicit control.

In the case of implicit control, it is attempted, through static feedforward and feedback, to adapt directly the behavior of the host aircraft such that it behaves like the target aircraft.

In the case of explicit model following control, the model following controller includes in the dynamic feedforward an explicit simulation model (desired model) of the target aircraft (see Fig. 3.7). The signals from the pilot inceptors are connected only to the inputs of this simulation model. The main outcome of this simulation is the accelerations of the target aircraft. Furthermore, when the accelerations are simulated correctly, then automatically their integrals, namely the speeds and positions match too (as long as the initial values of these integrals match). As the relationship between control variables and accelerations is known for the host aircraft, by "inversion" of the simulation equation (from which the accelerations resulting from control deflections are calculated) the necessary control deflections can be calculated from the desired accelerations (which result from the explicit simulation of the target aircraft). Since the accelerations of the host aircraft depend not only on the control variables, but also on the particular flight condition characterized by the so-called state vector, additionally an estimation of this state vector is necessary for the calculation of the required control deflections. This is carried out in general using additional differential equations, which describe the dynamics of the so-called model following observer.

In the case of explicit model following control, the dynamics of the controlled overall system consist of the dynamics of the target aircraft, the dynamics of the model following observer and the so-called error dynamics. The error dynamics describe the temporal behavior of the model following error, no matter how it arises. Ideally, the error is reduced rapidly without overshooting. It can be shown that without additional measures the error dynamics are the same as those of the unregulated dynamics of the host aircraft. Since these dynamics usually exhibit some very slow and/or poorly damped elements such as the Phugoid or the Dutch roll mode, they must be changed through a feedback of the difference between the quantities estimated by the feedforward and the actual, that is measured, variables.

Even in a completely nonlinear case (that is, nonlinear model of the target aircraft, for example, a Level-D simulator model, and nonlinear equations of the host aircraft for the calculation of the control variables and for the model following observer), a very good model following quality can be achieved using this method. However, an iterative numeric inversion of the acceleration equations is necessary for this purpose.

An in-flight simulator is more than an airplane with a variable stability system. The US-literature, however, does not differentiate between the two. The in-flight simulator should convey an impression to the pilot that he/she is virtually flying another type of aircraft. This pertains not only to the visual and motion impressions, but also to the controllability. Despite all these efforts, the in-flight simulation is subject to limitations. One cannot simulate everything. The simulation of a flight at supersonic speed at low altitude using a subsonic aircraft remains problematic and the right cockpit environment cannot easily be realized. Likewise, to simulate a different type of aircraft, the whole database needs to be replaced, not to mention the necessary modifications in the cockpit for the new target aircraft. This is elaborate, time consuming, expensive and also safety critical. This can occasionally be realized more easily in a ground-based simulation. Excluded are then the acceleration impressions, which can hardly be realized realistically with a ground-based simulator. As such it is necessary to tradeoff between deployment of a ground-based simulation, a variable stability aircraft or an in-flight simulator with all the limitations outlined above [9].

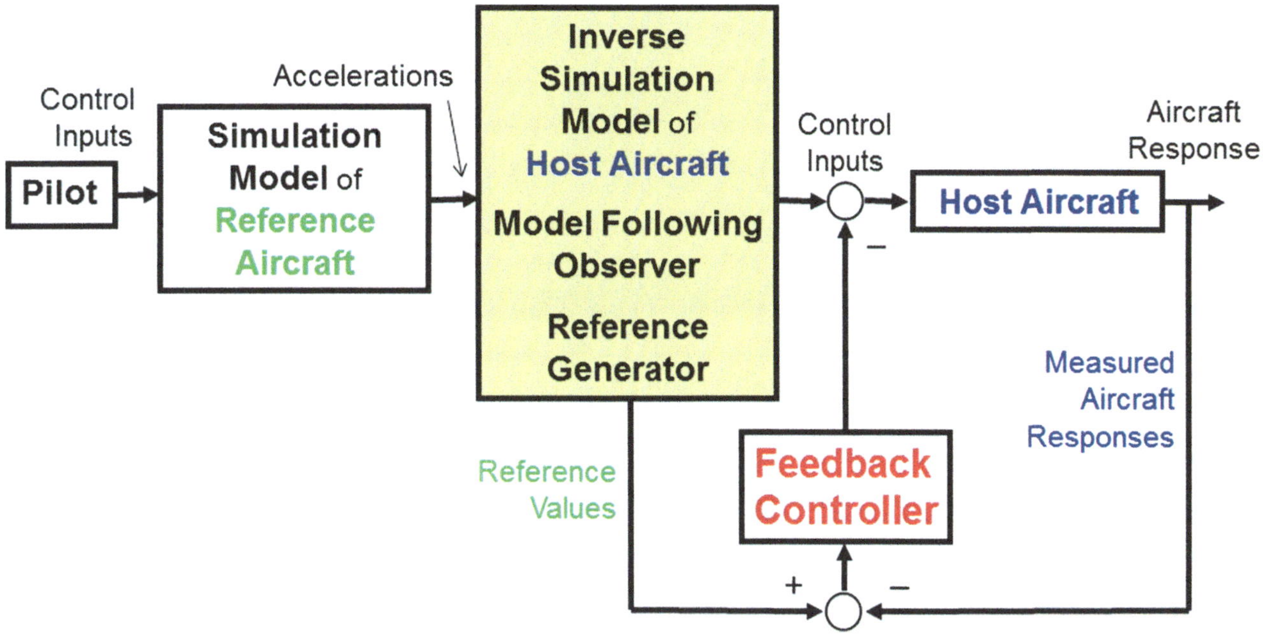

Fig. 3.7 Explicit model following control

Ideally, during the development of a new aircraft, the complete chain of simulation, that is, from of the ground-based simulation to in-flight simulation, is employed to keep the development risks as small as possible (see Fig. 1.2). With the increasing demand for unmanned flight vehicles, a renaissance of in-flight simulation is expected to test and to train flying such unmanned aircraft in civilian airspace with the so-called *Optionally Piloted Flight Vehicles* (aircraft controlled by pilot as needed) or with the so-called *surrogate aircraft* (substitute aircraft) (see Sects. 5.2.1.14 and 9.2.11)

References

1. Breuhaus, W.O., Harper, R.P.: The selection of tasks and subjects of flight simulation experiments. AGARD-CP-79, 1970, Paper 7
2. Hamel, P.G.: Advances in aerodynamic modeling for flight simulation and control design, pp. 7–50. GAMM-Mitteilungen, Band 23, Heft 1/2 (2000)
3. Hamel, P., Jategaonkar, R.: Evolution of flight vehicle system identification. AIAA J. Aircraft. **33**(1), 9–28 (1996)
4. Jategaonkar, R.V.: Flight vehicle system identification—a time domain methodology, 2nd edn. Progress in astronautics and aeronautics, vol. 245. AIAA (2015)
5. Greenyer, F.: A history of simulation—early days. Military simulation and training magazine (MS&T) **5** (2008)
6. Hornburg, H.M., Malone, M.D., Anderson, G.K.: A new approach to flight simulation. http://www.nastarcenter.com/a-new-approach-to-flight-simulation
7. Kidd, E.A., Bull, G., Harper, R.P.: In-flight simulation—theory and application. AGARD Report 368, April 1961
8. Mönnich, W.: Feedforward controllers for model following, in German. DLR-IB 111-1999/20, Status 2002
9. Abzug, M.C., Larrabee, E.E.: Airplane stability and control—a history of the technologies that made aviation possible. Cambridge aerospace series 14, pp. 35–40. Cambridge University Press, Cambridge (2002)

Author Biography

Bernd Krag was a research scientist at the Institute of Flight Systems of DLR of Braunschweig (1972–2002). From 1993 to 2002 he was head of the Fixed Wing Aircraft department. Prior to joining DLR, he was a research assistant at the Flight Mechanics Branch at the Technical University in Braunschweig (1967–1972). He received his Dipl.-Ing. degree (1967) and Dr.-Ing. degree (1976) in Aerospace Engineering from the Technical University Braunschweig. His main research interests were control configured vehicles (CCV), active control, modeling and system identification, databases for training simulators, and the wake vortex problem. Since retiring from DLR, his interest is in the history of aeronautics research in Braunschweig and aviation history.

Predecessors in Germany

4

Peter G. Hamel

The Control Engineering Approach to Artificial Stability

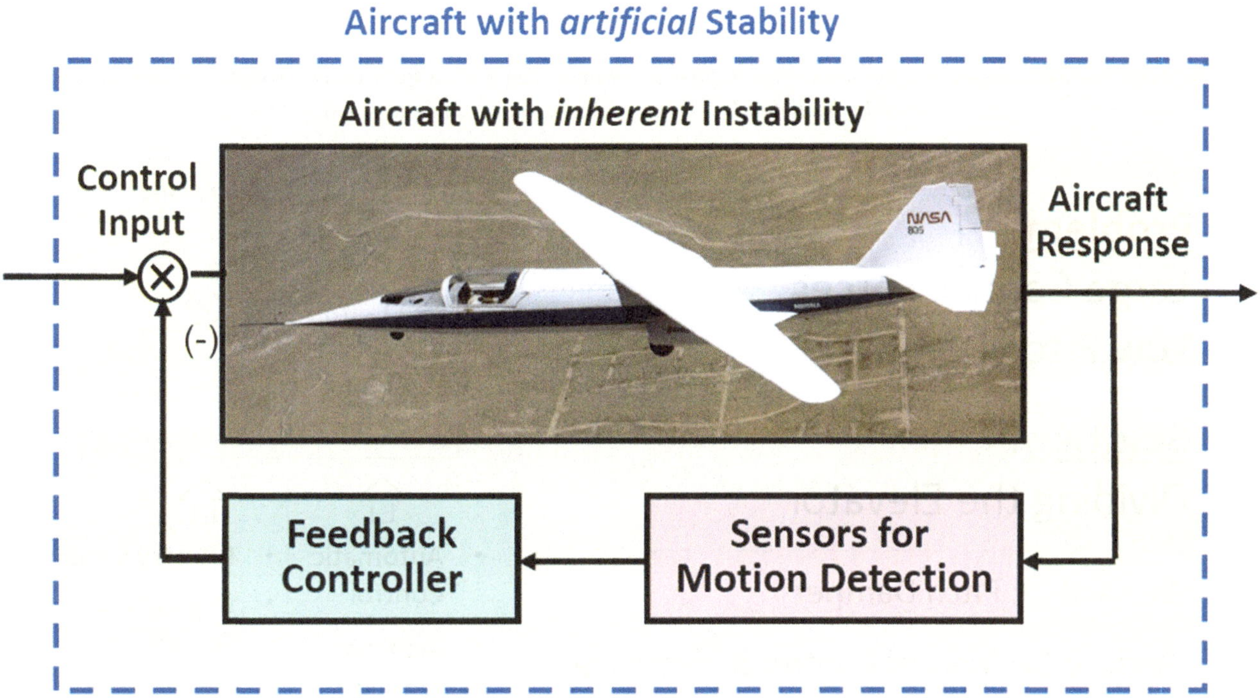

Keywords
Artificial stability • Yaw damper • Separate surface control

P.G. Hamel (✉)
Braunschweig, Germany
e-mail: p.hamel@t-online.de

P.G. Hamel (ed.), *In-Flight Simulators and Fly-by-Wire/Light Demonstrators*,
DOI 10.1007/978-3-319-53997-3_4

4.1 Artificial Stabilization

In the mid nineteen forties, gradual rethinking occurred in aircraft design. New requirements were made on the flight performance, flying qualities, and flight safety; for example, at higher airspeeds approaching the speed of sound compressibility effects were encountered, which led to previously unknown flight instabilities with reduced controllability. At the same time, a question of how to cope with the high forces acting on elevator and rudder with increasing size of aircraft and increased airspeeds had to be addressed. At that time the possible hydraulic or electric transmission for pilot assistance were discarded due to "a dangerous dependence of the operational reliability of such additional equipment" [1].

In the 7th scientific meeting of the regular members of the German National Aviation Academy held on September 20, 1940, the term artificial stability was employed for the first time by the aircraft designer *Ernst Heinkel* ("*The transition to artificial stability provided by an automatic control system would become inevitable necessary in the near future*"), by the Junkers engineer *Heinrich Helmbold* ("*The introduction of artificial stability appears inevitable at increasing airspeed*") and by *Eduard Fischel* ("*Evolving aircraft with high airspeed cannot be built inherently that stable anymore. Therefore, the installation of automatic control systems would become ever more imperative*") [2]. Thus, in contrast to the natural or inherent stability, they described the overall stability of an aircraft, which would be ensured through automatic control devices. Influences that deteriorate the natural longitudinal stability were initially attributed in the early 1940s to increased piston engine performance and the related amplified interference effect of the propeller slipstream as well as to the new airfoils for laminar boundary layer flow. Around the same time, compressibility effects with increasing airspeed were encountered.

In the debate during the aforementioned 7th scientific meeting, the famous aerodynamics engineer from Göttingen *Albert Betz* pointed out possible configuration changes to abate stability problems at increasing Mach numbers, for example, canard configurations and even variable sweep wings. The first systematic control theoretical approach for calculation of artificial stability "*through the intervening by*

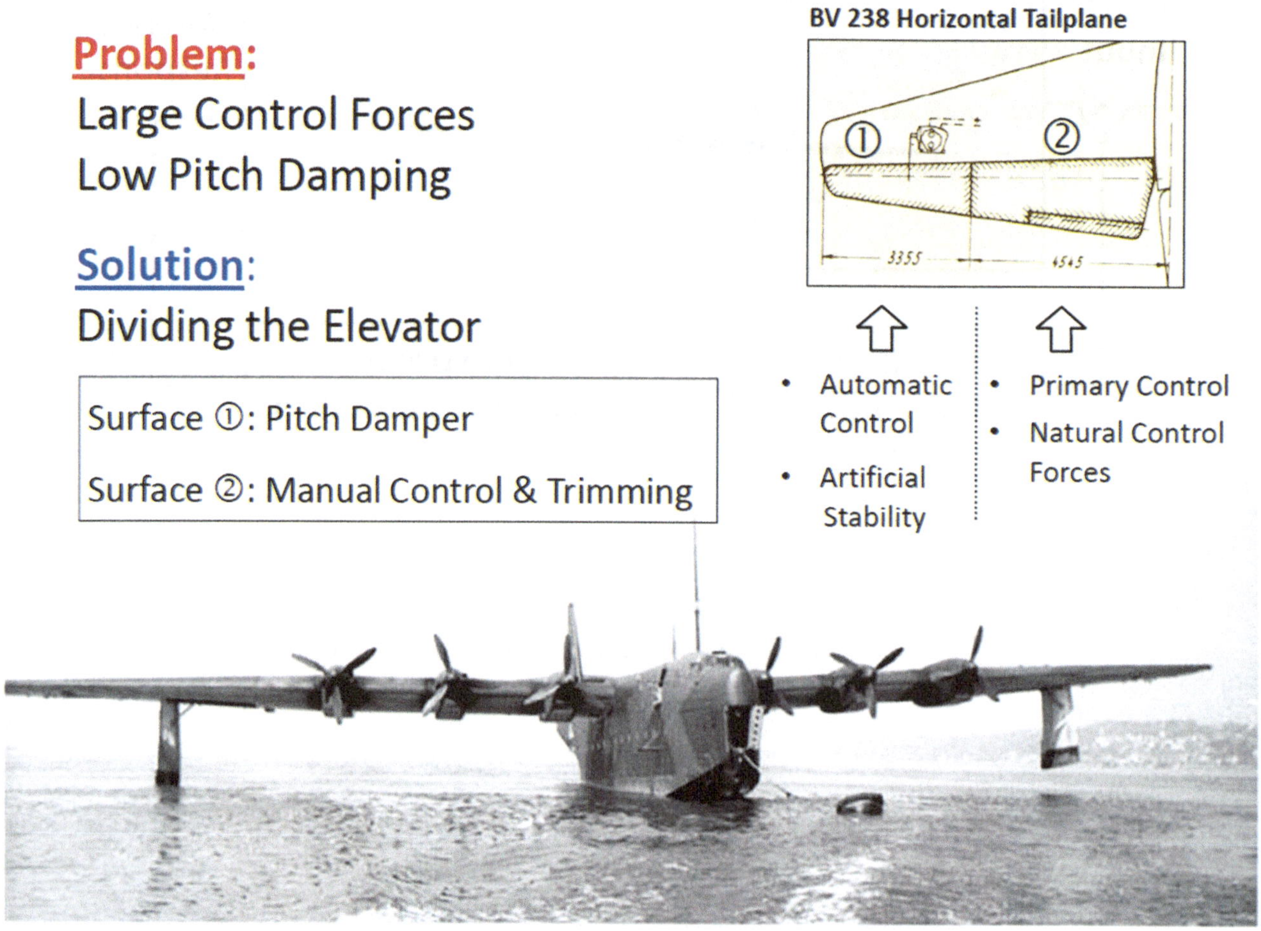

Fig. 4.1 Blohm & Voss BV 238—reduction of longitudinal control forces and improving flight stability

automatic control devices with the aim to increase inherent stability or generate lacking stability" is attributed to *E. Fischel* [3]. Thereby the cornerstone was laid for the calculation of flight control systems to produce variable flying qualities.

Yet another problem in the field of control and stabilization cropped up in the case of large aircraft due to the tremendous increase of control surface moments. While, for example, the airplane weight of a FW 190 from 3.2 tons increased to 93 tons of the large flying boat BV 238, that is, gone up about 30 times, already 117 times the rudder torque had to be applied in the case of BV 238 [4]. In the case of BV 222 (1940) and BV 238 (1944) *Richard Vogt* controlled the ever increasing growth of control forces by subdividing the elevator in an inner manual control surface ① with manageable natural control forces assisted by an auxiliary tab based on the Flettner principle and an outer control surface ② to artificially improve the pitch damping effectiveness (see Fig. 4.1). A somewhat different path was pursued in the roll control about the longitudinal axis. As seen in Fig. 4.2, the ailerons were divided in a small outer part ①, which was directly driven by manageable hand forces in the classical fashion. The inner and larger part of the aileron ② is trailed to the outer aileron, in turn, once again via a small Flettner auxiliary rudder requiring negligible additional control forces. With the method of control surface separation (*Separate Surface Control*), *R. Vogt* had achieved for the first time worldwide excellent flight test results for artificial enhancement of controllability and stability of large aircraft. In the year 1940, this technology was immediately incorporated into the series production of BV 222.

Finally, the yaw damper-flight experiments of *K.-H. Doetsch* and *EG. Friedrichs* in the year 1944 at the German Aeronautical Research Laboratory (Deutsche versuchsanstalt für Luftfahrt—DVL) in Berlin-Adlershof served to dampen the annoying snaking motion of the Henschel Hs 129 around the vertical axis. Also, *KH. Doetsch* used the principle of control surface separation for the rudder (see Fig. 4.3): little more than one-third of the upper rudder was separated and with the aid of a Flettner auxiliary rudder and yaw rate feedback the snaking motion was minimized ("*Overall, the tests showed the superiority of the automatism. It seems impossible to achieve such a favorable behavior by aerodynamic means*" [5]). Similar investigations were also carried out by the DVL towards the end of Second World War

Fig. 4.2 Blohm & Voss BV 238—reduction of lateral control forces

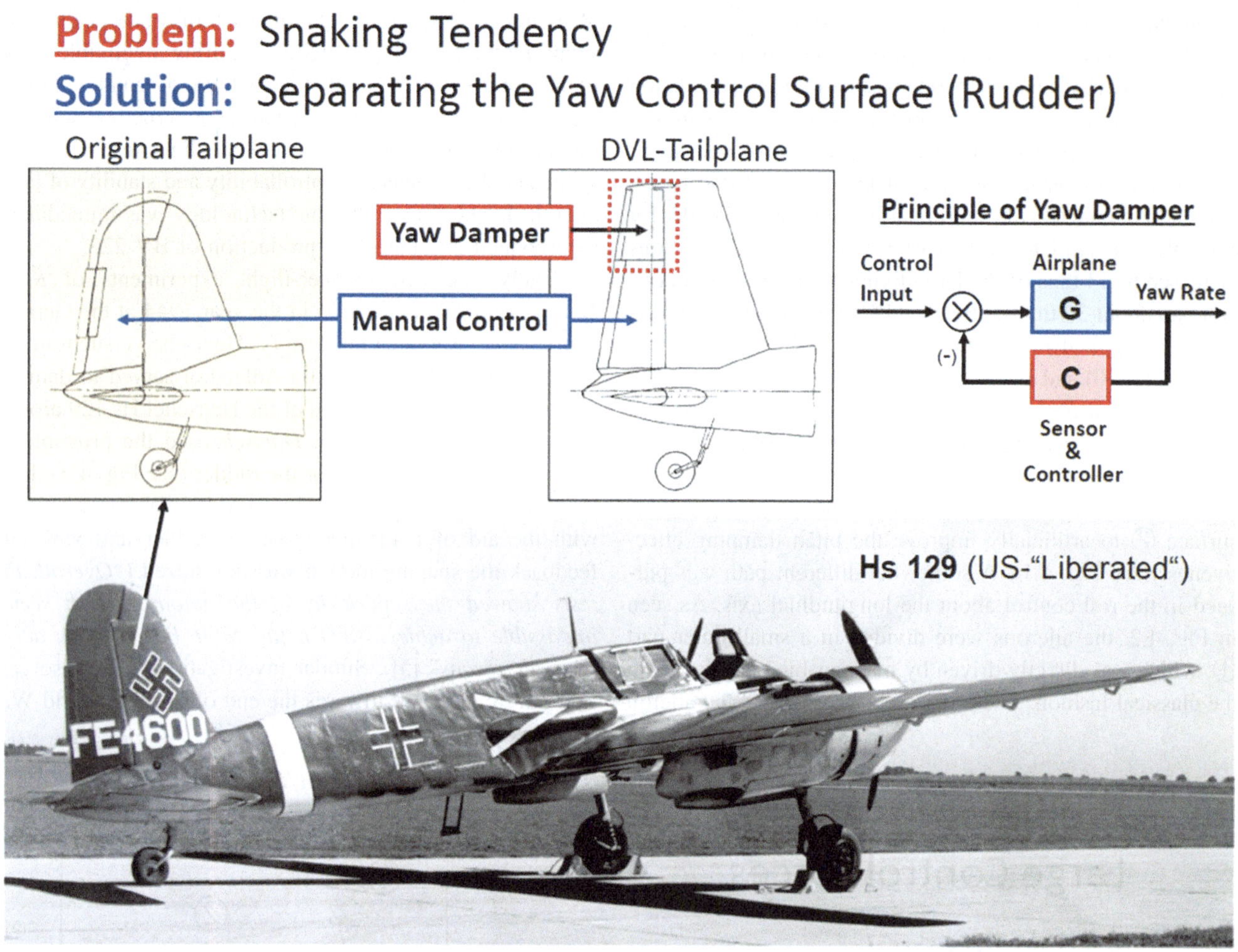

Fig. 4.3 Henschel Hs 129—artificial stability (DVL-yaw damper)

with a Me 262 test aircraft. They led to a strong reduction of yaw oscillations during target tracking (see Fig. 4.4, [6]).

4.2 From Aerodynamic Auxiliary Devices to Controller Assistance

From today's perspective, the aforementioned theoretical studies of *E. Fischel* and flight experiments by *R. Vogt* and *K.-H. Doetsch* to provide artificially controllability and stability were worldwide one of the first investigations to improve the aircraft flying qualities by automatic means. After the Second World War, both in foreign countries as well as after revival of the German aerospace research and industry, this method was deployed as an important tool for investigation and evaluation of the control and stability characteristics of the new types of aircraft, which were still in the development phase. Simultaneously, this also implied paving new ways to a more flexible aircraft design procedure with less aerodynamic crutches (auxiliary devices) such as wing fences (boundary layer fences) and vortex generators to influence separated flows and their effects on the flying qualities.

Extreme examples of such aids are highlighted in Figs. 4.5 and 4.6. Today, many effects, for example, aerodynamic instability for performance enhancement, can be compensated through specific flight control laws and strategies.

In the course of the time, research aircraft with variable flying qualities using analog computers evolved into the so-called in-flight simulators using digital electrohydraulic actuation systems and reliable digital computers. Therefrom the international research and development scenarios of revolutionary electronic or optical flight control technologies (*Fly-by-Wire/Light*) emerged without mechanical cables or control rods (see Chap. 6).

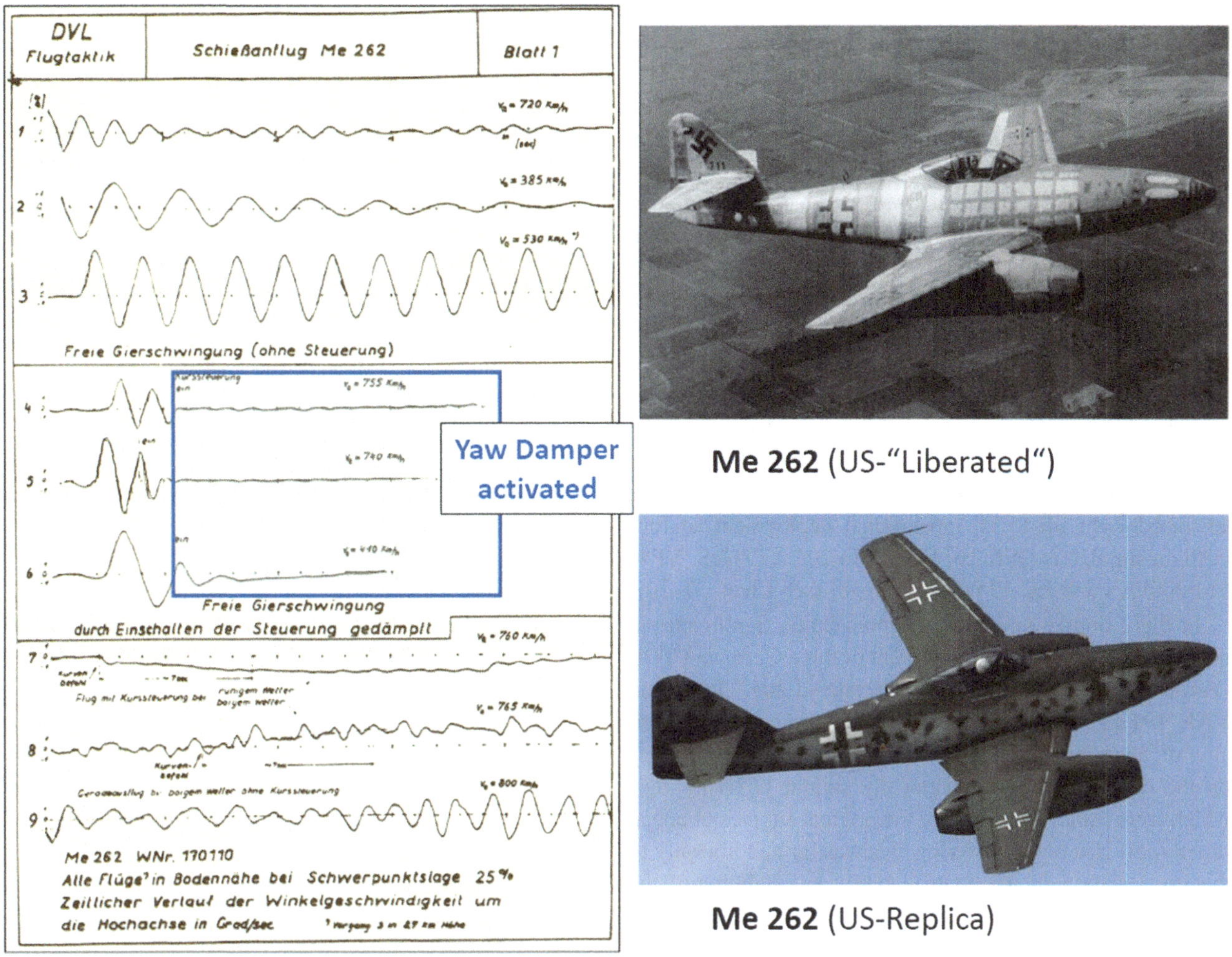

Fig. 4.4 Messerschmitt Me 262: artificial stabilization about the vertical axis through yaw damper

Fig. 4.5 Sukhoi Su-22UM3 K—oversized boundary layer fences

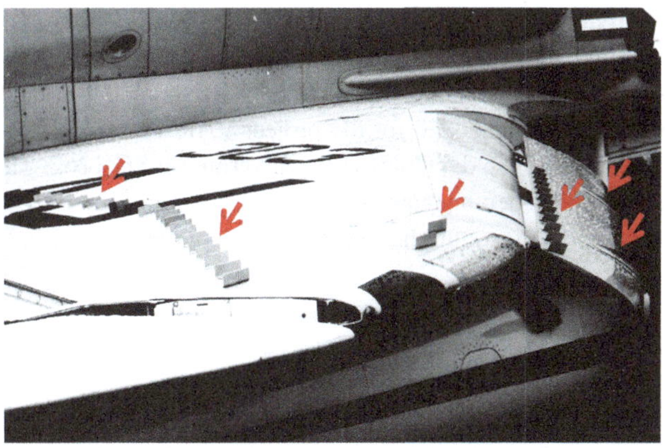

Fig. 4.6 Douglas A-4 Skyhawk—vortex generators on slats and wings

Flight vehicles with variable stability and controllability were investigated in Germany during the nineteen sixties and first flight tests with an analog computer on a Piaggio P149D at the DVL in Oberpfaffenhofen (see Sect. 7.1). The particulars of the beginnings of Fly-by-Wire research at DFL/DFVLR/DLR in Braunschweig is provided in Sect. 6.3.1. They were related to the national initiative "Variable Stability Test Beds" [7]. The objective was to convert three test vehicles based on a Fiat G 91T-3, a Bo 105 and a HFB 320 with variable stability characteristics. Since the early nineteen-seventies, this technology was developed by the DFVLR Institute of Flight Mechanics in Braunschweig (today DLR Institute of Flight Systems) with support from the DLR Institute of Flight Guidance and the national industry under project like conditions. Thereby the experimental technique of in-flight simulation could be advanced right up to the operational deployment of four in-flight simulators HFB 320 FLISI, Bo 105 ATTHeS, VFW 614 ATTAS and EC 135 ACT/FHS [8] (see Chaps. 7–10).

In-flight simulation acquired a more exotic significance in the course of industrial development of three German VTOL (*Vertical Take-Off and Landing)* aircraft VJ 101, Do 31 and VAK 191 [9]. As a consequence of this interest in VTOL aircraft, attention got focused during the early nineteen sixties on the development of dynamically similar hovering rigs, which were usually operated by the original engines of actual prototypes. Use of the hovering rigs was limited not only to the investigation of the controllability and stability of the VTOL aircraft in hover, but also to the pre-trials of 464 equipment components of the new aircraft projects [10] (see also Sects. 6.1.3.6 to 6.1.3.8). Many issues related to achieving good controllability during takeoff and landing, and to the transition to and from the horizontal, that is, aerodynamic flight. These investigations also delivered the first clues regarding the impact of the engine configuration on recirculation phenomena under different inflow conditions. Additionally, in 1969 a helicopter Bell 47 G with variable stability characteristics of the Canadian National Research Council was deployed for the project VAK 191 to describe and simulate the dynamics of the hovering rig SG 1262. The experiments demonstrated that the safety pilot of the Bell 47 G was able to monitor hazardous conditions (see Sect. 5.3.2).

References

1. Vogt, R.: Das Großflugboot BV 222 Wiking. Luftwissen **11**(3), 70–74 (1944)
2. Heinkel, E., Helmbold, H., Fischel, E.: Natürliche und künstliche Stabilität, Jahrbuch der Deutschen Akademie der Luftfahrtforschung, pp. 281–303 (1940/1941)
3. Fischel, E.: Die vollautomatische Flugsteuerung, Ringbuch der Luftfahrttechnik VE4 (1940)
4. Vogt, R.: Steuerung von Großflugzeugen im Hinblick auf die bei Blohm & Voss beschrittenen Wege, Schriften der deutschen Akademie der Luftfahrtforschung, Vol. 1080/44 g (1944)
5. Friedrichs, E. G.: Überlegungen zum Schießanflug, Deutsche Luftfahrtforschung, DVL, Untersuchungen und Mitteilungen No. 1336, 28 Aug 1944
6. Doetsch, K.H.: Deutsche Luftfahrtforscher nach 1945 in England, DGLR-Blätter zur Geschichte der Luft- und Raumfahrt, pp. 165–188 (1992)
7. Benecke, Th., Dornier, S., Bölkow, L.: Versuchsträger Variabler Stabilität (VVS)—Mittelfristiges Programm, 10 Dec 1969
8. Hamel, P. (Ed.): The Challenge of Flight Research – A Historical Account and Technical Guide, DLR-IB 111-99/2, Braunschweig, April 1999
9. Schumann, H.-G.: Flugsimulation als Hilfsmittel der Entwicklung und die Deutschen Arbeiten auf diesem Gebiet, Jahrbuch 1970 der DGLR, pp. 40–58 (1970)
10. Hamel, P.: In-Flight Simulation in Germany – Precursors of In-flight Simulation. In: Hirschel E., Prem H., Madelung H., G.: Aeronautical Research in Germany, pp. 608–612, Springer, Berlin (2004)

Author Biography

Peter G. Hamel was the Director of the Institute of Flight Mechanics/Flight Systems of the German Aerospace Center (DLR/DFVLR) (1971–2001). He received his Dipl.-Ing. and Dr.-Ing. degrees in Aerospace Engineering from the Technical University of Braunschweig in 1963 and 1968, and his SM degree from M.I.T. in 1965. From 1970 to 1971 he was Section Head of Aeronautical Systems at Messerschmitt-Bölkow-Blohm in Hamburg. Since 1995 he is an Honorary Professor at the Technical University of Braunschweig and a founding member of three collaborative research centers at the University. He was Chairman of the National Working Group on Helicopter Technology (AKH) (1986–1994) and the appraiser for the National Aviation Research Program (LuFo) until today. He was also the Manager of DLR's Rotorcraft Technology Research Program and the German Coordinator for the former AGARD Flight Mechanics/Vehicle Integration (FMP/FVP) Panel. He is a member of the German Society for Aeronautics and Astronautics (DGLR) and of the American Helicopter Society (AHS), and a Fellow of AIAA. He is the recipient of the AGARD 1993 Scientific Achievement Award, of AGARD/RTO von Kármán Medal 1998, of AHS Dr. A. von Klemin Award 2001, and of the prestigious DGLR Ludwig-Prandtl-Ring 2007.

5 Variable Stability Aircraft and In-Flight Simulators

Peter G. Hamel

P.G. Hamel (✉)
Braunschweig, Germany
e-mail: p.hamel@t-online.de

P.G. Hamel (ed.), *In-Flight Simulators and Fly-by-Wire/Light Demonstrators*,
DOI 10.1007/978-3-319-53997-3_5

5.1 Introduction

Especially in the light of innovative aircraft configurations with sweptback and delta wings, and without tail planes (tailless aircraft), that encountered compressibility effects in the transonic range, the question was raised, what kind of flying qualities can be expected of such an aircraft and how can they be improved, if the need arises. The pilot judgments were here especially called for, which demanded progressively new standards for the flying qualities guidelines. To generate the databases necessary for this, aircraft were needed whose stability properties could be varied through structural or flight control measures in a such way that a wide spectrum of flying qualities, as optimal as possible, could be evaluated for different flight tasks such as takeoff and landing or target tracking.

The aeronautical research in the United States of America started to deal increasingly with the problem of inadequate flying qualities of high performance aircraft in the mid nineteen forties. Although pioneering work in Germany during the Second World War preceded the work in the United States (see Chap. 4), it was the US National Advisory Committee for Aeronautics NACA at the Ames Research Center (today NASA) and the Cornell Aeronautical Laboratory (CAL, today: Calspan Corporation) in Buffalo [1, 2], who, independent of each other, converted the worldwide first two aircraft to experimental demonstrators with variable stability characteristics, and thereby took the leading role in this special field of aeronautical research.

Still, there was a fundamental difference in the objectives of the early work at NACA and at CAL in the beginnings. In order to gather some more background information about these early days, it was a unique opportunity to ask a contemporary witness, a leading scientist at Cornell Aeronautical Laboratory at that time, *Irving C. Statler*, a former Director of the US Army Aeroflightdynamics Directorate and of the Advisory Board of Aerospace Research and Development (AGARD) of NATO, to illuminate on these beginnings from today's perspective.

To quote *Statler*:

> When I joined the Cornell Aeronautical Laboratory (CAL) Flight Research Department in 1946, I was fortunate to become a part of a team that left a lasting legacy of understanding of aircraft stability and control. Most of the failures of the earliest attempts at powered, fixed-wing flight were associated with inadequate understanding of dynamic stability and control. The exploratory work at CAL during 1946–1947 turned out to be the genesis of the "variable-stability" aircraft and led to the development of inflight simulation. The histories of the accomplishments at CAL have focused on the activities directly associated with in-flight simulation for which CAL is best known and have not adequately addressed its origins.
>
> In 1946, the Flight Research Department at CAL was trying to find a reliable way to measure the flying qualities (that is, the dynamic stability and control characteristics) of an aircraft in flight. Bill Milliken and Ira Ross, who headed the Flight Research Department, believed that the key to understanding the dynamic stability of an aircraft and its controllability resided within the classical equations of motion. While Bill and Ira deserve full credit for having thought of this approach, the subsequent successful demonstration of the viability of the idea relied on the capabilities of the entire team, including, in particular, the analytical expertise of Walt Breuhaus, Dave Whitcomb, and Ed Laitone and the piloting skills of John Seal, Nello Infanti, Giff Bull, and Leif Larson.
>
> Ed Laitone and I developed the mathematics and the analytical techniques that were the foundations for the concept of measuring dynamic behavior in flight. We showed that the coefficients of the equation representing longitudinal motions of an aircraft could be identified with the stability derivatives that defined the longitudinal, fixed-control, short-period flying qualities of the airplane. The dynamics of aircraft motion could be expressed from the perspective of the frequency spectra of the responses to controlled inputs. Although the method was well known in analyses of other mechanical and electrical systems, this was the first time it was used to achieve understanding of aircraft dynamics.
>
> The U.S. Army Air Forces loaned CAL a B-25J bomber for the experiment that would demonstrate a way to measure dynamic stability characteristics in flight for the first time based on the mathematical analysis. The challenge was to develop a way to produce and measure precise control inputs and measure aircraft responses in flight. We tried having our test pilot put in sinusoidal elevator control, but that did not work. Honeywell donated to us an autopilot, which we modified to produce precise sinusoidal elevator motions over a range of frequencies and amplitudes.

CAL B-25J Flight Test Aircraft

> Air Force pilot Captain Glen Edwards (for whom the Edwards Air Force Base was later to be named) and Bill Milliken flew the first experimental flights with the B-25J using the modified autopilot to produce sinusoidal elevator motions. The aircraft responses in pitch angle, normal acceleration, and control forces to the elevator inputs were recorded on an oscillograph that were transcribed manually after each flight. The B-25J tests demonstrated that the mathematical representation agreed with the physical facts and maintained over a large range of control-input frequencies. For the first time, the frequency spectra of responses to control inputs were obtained in flight and used to measure an aircraft's dynamic stability characteristics and maneuvering behavior. This was not yet a variable stability simulation, but it is where it started.
>
> After we demonstrated that in-flight measurements of flying qualities could be done with reliability and repeatability, the question it raised was, 'What flying qualities does the pilot prefer?' In order to obtain information of pilots' opinions on what constituted good or bad flying qualities, we needed to be able to change the flying characteristics of the airplane in flight. Ed Laitone and I showed that, when control motions were made proportional to aircraft displacements, velocities, or

accelerations, it was mathematically equivalent to changing the stability derivatives that determined the flying qualities (I do not claim that this idea originated with us, but demonstrating its plausibility from the equations of motion was. In fact, Hamel has described the research of Heinkel and Fischel of using artificial stability in advanced aircraft already in 1940).

The success of the experiment with the B-25J was the beginning of the developments that evolved into the innovative concepts of enhanced stability augmentation and design specifications for handling qualities. This was the genesis of the "variable-stability" aircraft, which led to the development of in-flight simulation at CAL.

During about the same time, NACA also discovered the concept of the variable stability aircraft, but they arrived by way of solving particular stability and control problems of specific aircraft using augmented stability.

The methodologies for automatic manipulation of the flight controls in response to selected airplane motions were used by CAL to explore "handling qualities" while NACA used them to solve unacceptable flying qualities. These methodologies enabled development of variable stability simulation and in-flight simulation. However, more importantly, they laid the groundwork for the current use of automatic control to stabilize inherently unstable aircraft designs, to control stall and ride comfort, and to achieve prescribed operational capabilities. Artificial stability freed designers forevermore from the constraints of the classical approach that relied on aerodynamic means alone to achieve satisfactory dynamic stability and control using fixed stabilizing fins and manually movable surfaces for control.

As already pointed in Chap. 1, recollect that in this book the name Calspan (CAL) will be used for all references to the company though most of the pioneering was done when the organization was still the Cornell Aeronautical Laboratory (CAL). Correspondingly, the name NASA will be used for references to the organization though much of the work was performed when the agency still was called NACA.

NASA modified the aileron control function of a Grumman F6F-3 in such a way that different dihedral positions of wings could be simulated electronically (see Fig. 5.1). For the US Navy, Calspan fitted a Vought F4U-5 with an additional rudder, which was operated by servo control, independent of the manual actuation (see Fig. 5.2). The aim was to determine the optimum requirements for Dutch roll

Fig. 5.1 Grumman F6F-3 VS

Fig. 5.2 Vought F4U-5 (1948)

damping mode during landing approaches to an aircraft carrier. Both aircraft were single seated and in the case of a controller failure had to be switched to the basic (manual) control system by the pilot. This safety risk was later eliminated through the use of two-seater test aircraft with a so-called "Safety Pilot", who could take over the control of the aircraft in the case of an emergency, while the test or evaluation pilot focused fully on the experiment.

In the following decades, a variety of variable stability aircraft emerged at Calspan and NASA, whose flying qualities could be deliberately changed through autopilots and other flight controllers with limited control authority. With this, it was possible to change and to investigate the flying qualities in a systematic way, rather than relying on the results from flight tests with different types of aircraft. Quite soon variable stability measures were not sufficient enough to cover the spectrum of new aircraft configurations. The desire to predict the flying qualities of new types of aircraft was becoming increasingly obvious. Although analytical methods were meanwhile well developed, and also high-fidelity ground-based simulators became available, only flight testing provided the opportunity to carry out experiments and evaluations under realistic visual and motion cues. Consequently, the research aircraft were externally modified, for example, by additional control surfaces for generating lift, drag, and lateral forces, with which the new types of aircraft could be simulated in flight in all modes of motion with unlimited control authority. Special high-performance computers, initially analog and later digital in compact form, took over the task of emulating in real time the flight characteristics of the aircraft being simulated during actual flight. All these research aircraft, in conjunction with highly qualified test pilots, provided a solid base for the development of new flying qualities criteria and for the genesis of computer-controlled flight control systems known as Fly-by-Wire that would make it easier and safer to fly. With increasing control authority Fly-by-Wire systems would finally revolutionize the aircraft design process (see Chap. 6).

In the following, a relatively complete compendium is provided on the most important variable stability aircraft or in-flight simulators which emerged worldwide during the past six decades. Included are also some examples throughout this section that are not really representative of variable stability research aircraft and in-flight simulators, but these vehicles use augmented (or artificial) stability to solve particular dynamic problems. For example, review the following Sects. 5.2.1.6 EF-86E, 5.2.2.6 F9F-2, 5.2.3.1 XF-88A and 5.2.3.2 NF-104A.

From the successive sections, it becomes obvious which outstanding role Calspan (CAL) and NASA have played in the development and use of variable stability aircraft and in-flight simulators. In addition to extensive archives of the editor of this book and DLR as well as published technical literature, the attention is directed to the survey reports of Breuhaus [3] and the proceedings of the first International Symposium on in-flight simulation held in Braunschweig [4]. Also, three historic documentations about the important role of the NASA variable stability aircraft and in-flight simulators should be referred to as supplementary Refs. [5–7].

Subsequently, the unique European research expertise in the field of in-flight simulation of the German Aerospace Center (DLR) at the Braunschweig Research Airport will then be highlighted in separate Chaps. 7, 8, 9 and 10.

5.2 USA

5.2.1 CAL/Calspan

5.2.1.1 Vought F4U-5 VS (1948–1952)

The Sperry A-12 Autopilot of Vought F4U-5 aircraft was so modified that with the aid of a separated lower part of the rudder (damping rudder, Figs. 5.2 and 5.3) the lateral stability, that is, the directional stability and yaw damping, could deliberately be manipulated through feedback of sideslip angle and yaw rate. Through this superimposition, the yaw damping was artificially increased. In further studies it could be demonstrated that a nonlinear feedback controller leads to even better results for precision maneuvers. Such a nonlinear controller concept was also flown successfully by Calspan on a EF-86E (see Sect. 5.2.1.6).

In addition, the outer part of the landing flaps was replaced by separately controllable flaps, with which independent of the standard, hand-operated ailerons the roll and yaw dynamics could be artificially influenced through the superimposition of measured roll and yaw rates and angles of sideslip. Thereby for the first time, new lateral-directional flying qualities guidelines could be formulated for the US Navy through the evaluation of pilot assessments. The flight test program comprised of 160 flight hours.

Fig. 5.3 F4U-5 split rudder

5.2.1.2 Fairchild PT-26 (1948–1950)

As part of a research program on the stall and stability behavior of aircraft at higher angles of attack, a Fairchild PT-26 was equipped with the components of a production-version of a Sperry A-12 autopilot to stabilize the roll and yaw motion. An angle of attack vane was mounted on a long vertical boom just behind the cockpit. Other horizontal sensor booms for the measurement lateral motion data were mounted near the right and left wing tips (see Fig. 5.4). The autopilot was modified such that the roll attitude, as well as roll and yaw rates, could be fed back via pilot adjustable gains to the rudder and ailerons. In this way, steady and stable flight conditions were flown in the longitudinal mode with angles of attack up to $\alpha = 28°$, with fully separated flow well exceeding the condition of maximum lift at $\alpha = 15°$ (post stall flight regime). In this context, it is interesting to note that such investigations on the stability behavior of an aircraft with fully separated flow acquired special significance fifty years later (see Sect. 6.3.6).

5.2.1.3 Beechcraft C-45F (1951–1953)

The aim of this USAF research program was to alter the flight dynamics of the aircraft by artificial stabilization of the pitch, roll, and yaw axes. To this end, variably adjustable

Fig. 5.4 Fairchild PT-26 with vertical boom

Fig. 5.5 Beechcraft C-45F

Fig. 5.6 Douglas JTB-26B

signals of yaw rate, angle of sideslip and its rate of change, as well as yaw accelerations were fed to the hydraulic servo for rudder. Yaw rate and roll accelerations were applied to the servos for ailerons. The longitudinal acceleration signal was fed back to the elevator servo. Artificial control force feelings could be generated with continuously variable force gradients. With this Beechcraft C-45F, modified as described here, variable stability characteristics in the three primary control axes, that in pitching, rolling and yawing, were realized for the first time (see Fig. 5.5). Whereas for the in-flight simulation the control inceptors were mechanically disconnected from the control surfaces for the test pilots on the left-hand side, the control inceptors of the safety pilot in the right-hand seat remained permanently connected mechanically with the control surfaces. Other important contributions of this research program on in-flight simulation were the targeted development of electro-hydraulic actuation systems for activation of the aerodynamic control surfaces, and the introduction of the safety pilot concept with instantaneous access to the basic mechanical control system. Before the flight test program could be completed, the test demonstrator crashed during a routine landing.

5.2.1.4 Douglas JTB-26B (1951–1957)

In the early nineteen fifties, the analytical studies at Calspan pointed out that for flight speeds near the speed of sound (transonic range) and in the supersonic range, the flying qualities in the longitudinal mode deteriorate rapidly. The damping values of the short period mode converged to zero, and thereby the danger potential, called pilot-induced oscillations, increased accordingly. As such, flight test data were needed in this flight regime to adopt the guidelines for acceptable flying qualities in the longitudinal motion for both flight modes, namely short period and Phugoid. For this purpose, at the behest of the USAF, Calspan converted a Douglas JTB-26B (see Fig. 5.6), and almost simultaneously a two-seater F-94A (see Sect. 5.2.1.5), with a single-axis controller to generate variable longitudinal flying qualities. With both these testing platforms a variety of frequency-dependent flight test data was acquired, which led to definition of a new, worldwide utilized flying qualities criterion. The numerical range of this so-called *Control Anticipation Parameter*—CAP provides evidence of good or poor responsiveness of an aircraft in the longitudinal mode due to elevator step inputs.

From 1952, for the JTB-26B provided by the USAF a variable stability system was developed, which exclusively drove the elevator; in other words, the aircraft pitch response behavior could be artificially varied or modified. In the two-man cockpit, for the evaluation pilots on the right side the, pitch stick was disconnected from the mechanical basic system and replaced with an adjustable one with artificial control forces (*Artificial Feel*). The pitch stick control for the safety pilot on the left-hand side remained unchanged. Calspan engineers had devised a special feature specifically for the investigation of the Phugoid dynamics. Two small auxiliary control surfaces on the aft fuselage cone (see Fig. 5.7) were applied with freely adjustable signals of airspeed and longitudinal acceleration and thereby the normally weakly damped Phugoid dynamics could be artificially influenced, that is, damped. With this small variation, the pilot work load in landing approach on a predetermined trajectory, for example, ILS approaches, could be significantly reduced. Many reproducible pilot assessments were gathered up to 1957 for different combinations of frequencies and damping of the short period mode. This data base formed the basis for the first viable flying qualities guidelines USAF (MIL-Specs. 8785B).

By the end of 1958, the USAF research subsidies for flying qualities investigations drained. The USAF handed over the JTB-26 along with two further standard B-26B (see Sect. 5.2.1.9) to Calspan for further utilization.

5.2.1.5 Lockheed EF-94A (1952–1958)

Besides JTB-26 B, practically at the same time, at the behest of the USAF, Calspan converted a two-seater Lockheed F-94A to a single-axis elevator control system for generating variable flying qualities in the longitudinal mode (see Fig. 5.8). To prepare it as a Variable Stability Aircraft, the basic control system for the front seat evaluation pilot was

Fig. 5.7 JTB-26B auxiliary rudder surfaces

Fig. 5.8 Lockheed F-94A

mechanically disconnected and an electronic elevator control system introduced. On the other hand, for the safety pilot in the rear seat controls neither for rudder nor brake pedals were provided, with the result that he could not carry out takeoffs or landings. The objective of the program was to determine optimal flying qualities criteria for the longitudinal motion. The EF-94A, besides the B-367–80 (see Sect. 5.2.2.13), contributed to the emergence of the so-called. C*-flight control law, which was adopted in a modified form later in the flight control laws of the new generation of Fly-by-Wire commercial aircraft such as Airbus A-320/330/340/350/380-family and Boeing B-777/787 series. It was also used for simulation of specific aircraft projects before the first flight. These included the simulation and optimization of the longitudinal control law of the supersonic aircraft Convair XB-58 (Convair test pilot *B. A. Eriksen*: “*The EF-94A represented the XB-58 like meeting an old friend*”). The EF-94A was later replaced by the NT-33A with an F-94 radome to allow sufficient mounting space for analog computer systems and flight data recordings.

5.2.1.6 North American EF-86E (1953–1955)

It is worth noting that the system changes of the serial F-86E over the previous version of F-86A were significant steps towards progressive flight control augmentation. Thus, at the behest of the USAF, Calspan introduced a hydraulic aileron and elevator control system with artificial control force feeling, whereby the aerodynamic forces could no longer act directly on the control inceptors. The decisive change in the control system concerned the pitch control; the marginally effective elevator was now replaced by an electro-hydraulically adjustable horizontal tailplane. With this so-called *All-Flying Tail* the control problems due to local shock waves at higher subsonic speeds could be avoided. In contrast, the rudder was still operated by rods and cables, and as such hardly suitable for damping the typical swept-wing aircraft snaking oscillations around the yaw axis.

After the initial very promising investigations with a Vought F4U-5, Calspan extended the investigations to artificial stabilization of the yaw axis (yaw damping) on one of the EF-86E provided by the USAF (USAF Register Number (R/N) 50-588, see Fig. 5.9). To realize a nonlinear yaw damper, a servo-actuator was mounted on the fin, which steered an auxiliary tab of the rudder. The additional control movements were felt by the pilots on the pedals. Through these changes the aeroelastic stability properties of rudder changed to such an extent that, after initial flights, it led to fracture and loss of the rudder. Through skillful pilot

Fig. 5.9 North American EF-86E (AF50-588)

intervention the EF-86E could be brought to emergency landing and thereby a total loss could be prevented.

After the integration of a now irreversible electro-hydraulic rudder control system with artificial feel, the control signals of the pilots and of the yaw damper were directly fed to the rudder-surface. In doing so the sensitivity of the yaw damper (gain) was revoked with increasing sideslip angles. During the subsequent flight tests the high effectiveness of the nonlinear yaw damper was demonstrated [8].

"I lost my Tail" To improve the EF-86E lateral flying qualities CAL (Calspan) proposed to design a nonlinear automated control system that would make the aircraft feel very stiff to the pilot for small directional motions and make it responsive and agile for large yawing velocities. I was given the task of designing this nonlinear yaw-control system for the EF-86E.

Project leader Irving C. Statler (right), Jack Beilman (middle), and test pilot John Seal (left) in front of the EF-86E.

I learned from analog computer simulations that it would be necessary for the control system to operate the rudder at fairly high frequencies on occasion. However, the small servomotor that would fit inside the vertical tail to drive the rudder would not have sufficient power, so we had to devise something to balance the inertia of the rudder. We attached two arms to the front edge of the rudder extending ahead of the hinge line that held counterbalancing weights at the forward end of the arms. The weights had to be as small as possible and yet heavy enough to balance the weight of the rudder with the minimum length of the arms.

We pursued the initial flight tests cautiously by very gradually increasing the maneuvers. John would perform a scheduled test, then circle while we checked the data and gave him approval to proceed with the next test. One day, I was in the radio room as usual and gave John the go-ahead for the next test. Then we heard John say in a very calm voice "I think I just lost my tail." We held our breaths until we heard him say "I have it under control". We notified Buffalo Airport, they declared an emergency, and sent the fire trucks out, but Johnny made a nice normal landing and taxied the aircraft to the Flight Research Hangar. He had not lost the vertical tail, but he had lost the entire rudder including the counterbalancing arms and weights. Subsequently, pieces were reported found scattered over a large area of Williamsville. The public uproar plus the estimated cost of the repairs to the airplane were sufficient to bring my first (and last) flight-test project to an end [9].

Irving C. Statler

5.2.1.7 Lockheed NT-33A (1957–1997)

As the first experiences of simulating in flight the longitudinal motion before the first flight of the Convair B-58 Hustler with the NF-94A (see Sect. 5.2.1.5) were being gathered, soon thereafter a Lockheed T-33 was made available by the USAF, that was now modified in the three primary axes (pitch, roll, and yaw axes) to a system with variable stability and controllability (see Fig. 5.10). It was named NT-33A (the N implying non-removable equipment). Since the analog simulation equipment using computer with vacuum tubes could not be accommodated in the original T-33 fuselage, the T-33 fuselage nose was replaced through the bulky radome of the EF-94A (see Fig. 5.11). The first

Fig. 5.10 Lockheed NT-33A

Fig. 5.11 NT-33A with EF-94A nose

lucid description of the NT-33A including the underlying flight test philosophy followed in the year 1961 [10].

For the simulation of steep descent of the X-15 hypersonic aircraft, compared to X-15 the low drag aerodynamics of NT-33A had to be degraded sufficiently. Thereto, airbrakes mounted on the wingtip tanks were extended from time to time as variable drag generators (Drag Petals, see Fig. 5.12). That was, anyhow, a better solution than the triggering of a drogue in flight, as was initially tried with a North American F-100A. In addition, the front seat of the NT-33A was adapted to the cockpit of the X-15, whereby Calspan installed control sticks of the X-15 (see Figs. 5.13 and 5.14) on both sides, right for simulated reaction control (*Ballistic Control*) and left for the aerodynamic control in the atmosphere (*Aero Control*) during 60 s under zero-gravity. The *ballistic control* did not cause any aircraft reactions, but simulated only changes in the pitch attitude on the display.

Fig. 5.12 NT-33A during X-15 steep approach simulation

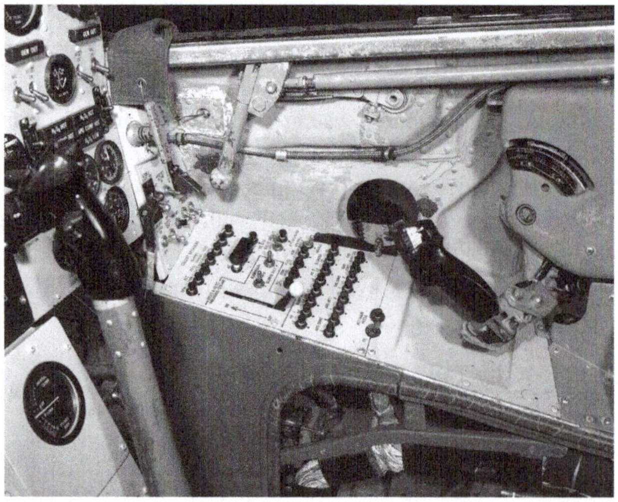

Fig. 5.13 X-15 (*right sidestick* Ballistic Control)

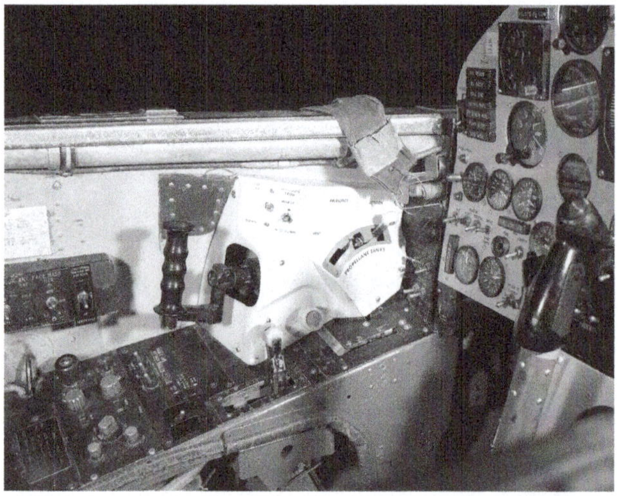

Fig. 5.14 X-15 (*left sidestick* Aero Control)

In the rear seat, the safety pilot had access to the basic mechanical flight control system of the NT-33A. The constantly changing flying qualities of the X-15 during the re-entry into the atmosphere were simulated through a programmable nonlinear function generator, which varied the gains of 32 measured state variables as well as flight control parameters of the flight controller. In May 1960, began the re-entry training and evaluation flights for the selected few X-15 test pilots, including *Neil Armstrong*. Soon he had prepared a remarkably foresighted documentation on the role and experience of in-flight simulation for the application domain of manned spacecraft [11].

"Here's your Stick" During one of the flights, with Neil Armstrong in the front seat, we were simulating failed dampers at something like Mach 3.2 and 100,000 feet altitude. Neil had great difficulty with this simulated undamped X-15 configuration and lost control of the airplane repeatedly. The safety pilot "*Nello*" *Infanti* had to recover from each one of these "lost-control" events using the controls in the back cockpit. *Infanti* later recalled that some of these recoveries were "pretty sporty". The ground crew was monitoring the test radio frequency as usual and followed these simulated flight control problems with great interest.

After landing, the NT-33A taxied to the ramp and *Howard Stevens* attached the ladder to the cockpits and climbed up to and talk to *Infanti* about the airplane status. I climbed up the ladderfront side to talk to *Neil*. He handed me his helmet knee-pad, got down from the cockpit and we talked about the flight and walked toward the operations building. As we arrived at the door *Armstrong* extended his right hand to grasp the door handle–but his hand still held the sidestick that he had broken during his last battle with the X-15 dampers-off simulation. I was unaware of any report of this incident during the flight and had not noticed the stick in Armstrong's hand when he exited the cockpit. Addressing the matter for the first time, Armstrong said–without additional comment—"*Here's your stick*!"

It developed that *Infanti* had been aware of the broken sidestick after it happened because *Neil* had held it up over his head in the front cockpit for *Nello* to see. After the debriefing, we took the broken sidestick to the NASA workshop where *Neil* found the necessary metal tubing and repaired the stick while I mostly watched him work. The sidestick was reinstalled and ready for the first flight the next morning. Really good test pilots fix what they break!

Jack Beilman, Calspan

Neil Armstrong - 5.8.1930 - 25.8.2012

The NT-33A became one of the most successful in-flight simulators worldwide. Over a period of 40 years, in addition to the X-15 project support and the simulation of reentry flight vehicles such as M2F2 and X-24A, primary data were flown with NT-33A for the USAF flying qualities requirements for highly control augmented aircraft (MIL-F-8785 and MILSTD-1797). Further, national aircraft projects such as A-9, A-10, F-15, F-16, F-17, F-18, F-117 and F-22, as well as international projects like TSR.2 (England), Lavi (Israel), JAS Gripen (Sweden) and LCA (India), were tested and evaluated before their respective first flights. Also, during the nineteen eighties, joint research programs were pursued with the DLR (see Sect. 12.3.2).

A total of 5200 flights accumulating 8000 flying hours were flown with the NT-33A at Calspan. A significant portion of flight hours were spent on the test pilot training at the Edwards Air Force Base (AFTPS). *Robert Harper*, the co-founder of the world famous Cooper-Harper flying qualities rating scale (*Pilot Rating Scale*, see Fig. 2.6) had, on behalf of Calspan, played a key role in structuring various flying qualities test programs for the AFTPS. In the year 1997, the NT-33A was handed over to the Air Force Museum at the Wright-Patterson AFB in Dayton, Ohio, as an exhibit with special merits.

5.2.1.8 Chance Vought F7U-3 (1958–1959)

Through the integration of two vertical "canard" control surfaces, above and below the fuselage nose of a Chance Vought F7U-3, carried out by Calspan at the behest of the US Navy, it became possible to stabilize the lateral-directional motion at high angles of attack, that is under separated flow conditions (*Post-Stall* flight envelope). With this unusual configuration sufficiently rapid yaw damping moments could be generated, which effectively prevented an uncontrolled buildup of critical lateral-directional flight conditions (*Post-Stall* Gyrations). The signal of the rate of change of angle of sideslip (beta-rate) was used in the controller feedback (see Fig. 5.15).

Fig. 5.15 F7U-3 with additional control surfaces

5.2.1.9 Douglas TB-26B (1959–1981)

After Calspan had demonstrated to the US Naval Test Pilot School (NTPS, Patuxent River, MD) the specific utilization potentials of Douglas JTB-26B (see Sect. 5.2.1.4) with variable flying qualities in the longitudinal motion, a special training program was worked out in 1960 for future flight test pilots. As a result, in addition to the theoretical knowledge of flying qualities evaluation, the test pilots could follow for the first time the practical demonstration in flight tests. As a consequence of the wide acceptance of these pilot trainings, the US Air Force Test Pilot School (AFTPS, Edwards AFB, CA) also introduced these training courses with the Douglas B-26B three years later. Due to the heavy demand, two of three B-26B aircraft loaned to Calspan by the US Air Force were converted in 1963 to TB-26B aircraft with variable stability characteristics in all three principal axes (3 degrees of freedom: pitch, roll, yaw). Finally, in the mid-1960s, the two engines could be driven with the aid of servo controls (*closed loop throttle servo*), and thus 4 degrees of freedom were at disposal for the first time for the in-flight simulation of a supersonic aircraft.

The two TB-26B (R/N N9146H and N9417H) were utilized for research and training programs till the late nineteen seventies (see Figs. 5.16 and 5.17). The third B-26 was used

Fig. 5.16 Douglas TB-26B N9146H

Fig. 5.17 Douglas TB-26B N9417H

in this period as "spare parts storehouse". One of them was lost in spring 1981 at the US Edwards Air Force Base due to a wing-fatigue fracture, the other is in the Air Museum of Edwards Air Force Base. They were successively replaced through timely procured and converted Calspan Gates Learjet Model 24 VS (see Sect. 5.2.1.13).

5.2.1.10 Convair NC-131H TIFS (1970–2011)

In addition to the aforementioned NT-33A, another in-flight simulator has played a particularly prominent role in the transport aircraft sector. On a former US Air Force C-131B transport aircraft (civil: Convair 340, built in 1955) extensive modifications were undertaken by Calspan during the late 60s (see Fig. 5.18). The original piston engines were replaced through propeller turbines with double the engine power. More striking conversions included the integration of vertical control surfaces for side force control on the wings and the attachment of an additional simulation cockpit (duplex cockpit) on the fuselage nose (see Fig. 5.19). Through installation of a comprehensive system to produce variable flying qualities, additional electronically controllable direct lift control (DLC) surfaces and a servo-controlled engine throttle control, an in-flight simulator was developed for the first time worldwide that permitted the manipulation of the aircraft dynamics in all the six degrees of freedom (3 rotational degrees of freedom: pitch, roll and yaw; 3 translational degrees of freedom: vertical, horizontal and lateral). Accordingly, the test aircraft was called TIFS (*Total In-Flight Simulator*) and due to the constructive changes carried out on a permanent basis, it was registered as NC-131H (civil: Convair 580) in the Air Force.

Fig. 5.18 NC-131H TIFS

Fig. 5.19 TIFS duplex cockpit (USAF)

The side-by-side seat duplex cockpit with an excellent exterior view, programmable displays, and replaceable or reconfigurable control inceptors with artificial force feel (see Fig. 5.19) was occupied by the test and evaluation pilots during flight experiments, whereby the flight safety was monitored by the safety pilot sitting left in the actual cockpit. He could turn off the simulation system and take over the command in the case of emergencies.

At Calspan (*Ed Rynaski*) the so-called Explicit Model Following Control principle was used on the TIFS for the first time, in contrast to the previously applied classical feedback control, also called Implicit Model Following Control (*Response Feedback*), see Sect. 3.3.

Not so well known is the second TIFS configuration (TIFS II), which was in a semi-finished stage, during the nineteen-seventies at the company Aerospace Lines (R/N N21466) with side force generators and a Boeing B707 nose, which was to be exclusively converted by Calspn for commercial applications and offered to NASA for utilization (see Fig. 5.20). This program was, however, cancelled due to cost reasons.

TIFS made its maiden flight in June 1970. The first full in-flight simulation took place on 10th June 1971 for the Rockwell B-1A project. Since then in the following 32 years continuous improvements were carried out to the simulation and cockpit systems, such as migration from "egg timer" to flat panel displays up to artificial view (*Synthetic Vision*), from analog computer components to compact and flexible digital computer components, and control columns (Wheels) to programmable sidesticks. Numerous utilization programs were flown, and aircraft projects such as X-29, X-40, B-2, YF 23, C-5 and C-17 as well as civil projects like Boeing 7J7 (later Boeing B-777 "*Triple Seven*"), MD-12X, SST (*Supersonic Transport*) were simulated in flight with the NC-131H before their respective actual first flight. Likewise, as in the British-French supersonic aircraft Concorde, flanking support programs were flown.

Based on a transatlantic cooperation program (MoU) between the USAF and the DFVLR (now: DLR, see Sect. 12. 3.2), DLR test pilot *Hans-Ludwig* ("*HaLu*") *Meyer* could participate in the flight test program to evaluate the flying qualities of a very large aircraft. An interesting outcome was the comparison of the control strategies of DFVLR pilot with an Air Force pilot. While the USAF pilot tried with larger, high-frequency control activity to compensate for any

Fig. 5.20 Convair 580 TIFS II with civilian registration number and Boeing 707 nose (*Credit* Peter de Groot)

smallest trajectory deviation, *HaLu* reached practically the same result with much quieter, low-frequency control inputs. The comparison was then interpreted informally as the mental difference between a Texan cowboy and a "cool" North German sailor.

In addition to the test pilot training for the Air Force Test Pilot School (AFTPS), TIFS served in a very special way the development and testing of flight control laws for the unpowered landing of the Space Shuttle. As in the case of the NT-33A, it was necessary here to generate additional aerodynamic drag for simulation of steep descents of up to 15° glide path angle. This was achieved on TIFS by deflecting the lateral force control surfaces in opposite directions.

On 17th November 2008, a piece of US aviation history was concluded. After more than 40 years and 2500 research flights, TIFS landed for the last time at the Wright-Patterson Air Force Base in Ohio (see Figs. 5.21 and 5.22). Thereafter, it received a special place of honor at the National Museum of the United States Air Force.

5.2.1.11 Bell X-22A VS (1971–1984)

The unusual Bell X-22A configuration is till today the only short and vertical take-off (V/STOL) aircraft, that was in the nineteen sixties already in development phase conceptualized as a flight test demonstrator with variable stability capabilities. The propulsion concept consisted of two pairs, one behind the other, encapsulated 2.1 m propeller devices (*dual tandem tilting ducted fans*), which were driven by four turboshaft engines in the rear stub wings. A total of 10 transmission units had to be integrated, to ensure safe flight even with an engine failure. The engine propeller encased in ducts could be tilted through 90° by hydraulic actuation systems for the transition from vertical takeoff to horizontal flight. All flight control and stabilization tasks, such as attitude and airspeed commands of the research aircraft were

Fig. 5.21 TIFS last flight on 7th November 2008 (*Credit* USAF)

Fig. 5.22 TIFS USAF team after the last flight (*Credit* USAF)

Fig. 5.23 X-22A R/N 1520

Fig. 5.24 X-22A R/N 1521

Fig. 5.25 X-22A during transition

Fig. 5.26 Decommissioned X-22A R/N 1521 (*Credit* Micha Lueck)

ensured through the collective adjustment of the blade pitch angle of the four turbofan propellers in conjunction with symmetrically or differentially operable control surfaces (*elevons*) in the wake of ducted propellers.

From the two X-22A vehicles which were built, the first (see Fig. 5.23, R/N 1520) crashed on 8th August 1966 after just 15 flights due to a hydraulic double failure. The second X-22A (see Fig. 5.24, R/N 1521) was equipped by Calspan with a computer-controlled simulation system for generating variable flying qualities. In the simulation mode the test pilot in the left seat could feel and evaluate the artificially modified control and stability properties in the hover and during the transition phase, whereas the safety pilot in the right seat could at any time switch over to the mechanical back-up control and take over the command.

In the period starting from the maiden flight on 19th May 1969 up to June 1970 a total of 400 vertical takeoff and landing flights, 200 short takeoff and landing flights, and finally 185 transitions between vertical and horizontal flight (Fig. 5.25) were executed for the US Navy Air Force and Army, NASA and FAA. In July 1970, the X-22A was handed over to Calspan for further research flights. These included test flights for the AV-8B Harrier II project and the development of a HUD (*Head-Up Display*) vision system. The X-22A was also employed at the US Naval Test Pilot School USNTPS for demonstration and training purposes and delivered important data for flying qualities specifications of future vertical and short take-off aircraft. These data ultimately contributed also to the success of the V-22 Osprey program. As a spin-off-product, a new air data sensor LORAS (*Linear Omnidirectional Resolving Airspeed System*) was developed for vertical and short takeoff aircraft (V/STOL) and helicopter, which is deployed today worldwide in series production helicopters. After a total of 500 flights with 405 flight hours, the X-22A was decommissioned in October 1984 and is now at the Niagara Aerospace Museum to admire (see Fig. 5.26).

Fig. 5.27 NCH-46A

Fig. 5.28 Calspan gates Learjet 24 N101VS

5.2.1.12 Boeing NCH-46 VSC (1972–1988)

A Boeing CH-46A was converted by the US Naval Test Pilot School (USNTPS) to a test helicopter NCH-46 VSC with variable flying qualities (*Variable Stability and Control*—VSC), which served the test pilot school until the mid-80s for educational and training purposes. By the end of 1972, the hover-autopilot from Sperry was modified in such a way that the series-servos in the pitch and roll axes could change the stability properties in these axes with limited authority. A variable stick force feel system was foregone. In 1981, Calspan modified the VSC computer system and the control laws. With these modifications the system reliability was increased, and thereby the flight test engineer who operated the VSC system onboard could be omitted. Besides the test pilot training, important contributions to the development of new flying qualities criteria and MIMO (*Multiple Input—Multiple Output*) flight control laws could be achieved (see Fig. 5.27). In 1988 the aircraft was returned to the US Navy.

5.2.1.13 Learjet Model 24D VS (Since 1981)

In the late nineteen-seventies, the life of the two B-26B (see Sect. 5.2.1.9) approached the termination and activities began at Calspan to convert a Gates Learjet (Model 24D, R/N N101VS) to a training aircraft with variable stability (VS) in the three primary control axes (pitch, roll, and yaw) (see Fig. 5.28). Only in this way continuity of training flights for the test pilot schools US Air Force (AFTPS, Edwards AFB, CA) and the US Navy (NTPS, Patuxent River, ML) could be maintained over longer periods. One of the main reasons for buying a Learjet was the structural robustness and also the high thrust to weight ratio.

Adequate space allowed the test and safety pilots, in addition to the time proven juxtaposition of cockpit seats, inclusion of an additional observer and a flight test engineer. The maneuverability was almost like that of a combat aircraft in subsonic flight range. New vanes were installed for the measurements of angles of attack and sideslip as well as electrohydraulic actuators were integrated for electronic control of the elevator, rudder, and ailerons. The right cockpit seat for the test or student pilots was provided with programmable operating elements, such as a centrally and laterally arranged stick (center stick and sidestick) and rudder pedals. The center stick and rudder pedals were freely programmable to produce a sufficient bandwidth of artificial control force (*Artificial Feel*).

At the beginning of VS system equipping, 64 parameters of an analog computer had to be programmed to define a particular configuration for an in-flight simulation. Alltogether, 128 different aircraft configurations were stored onboard, and each of these configurations could be selected by the safety pilot with a push of a button. Thereby the safety pilot in left cockpit seat remains in a position to intervene and take over control of the aircraft at any time on reaching the set simulation limits. The first flight of the Learjet with a system for the variable controllability and stability took place in February 1981. In later years, the analog computer of VS-system was replaced by a digital computer. Thereby the preparation and conducting of experiments could be further improved, especially in the areas of modeling and simulation in real time and flexible programming of the display imagery.

5.2.1.14 Learjet Model 25B/D VS (Since 1991)

After about ten years, because of the increasing demand for training flights with in-flight simulators at the international test pilot schools, another Gates Learjet (Model 25B, R/N N102VS) was procured by Calspan, that was initially equipped with a VS system similar to that in the Learjet Model 24B VS (see Fig. 5.29). The first flight took place in March 1991. Since then this Learjet is deployed as an in-flight simulator also in Europe, for example at the test pilot schools EPNER (France) and ETPS/NTPS (England). The VS system was in the meantime modernized and now includes an explicit Model Following system with MATLAB Simulink® software in conjunction with a so-called Real-Time *Autocode*.

In 2007, a third Learjet (Model 25D, R/N N203VS) was incorporated into the Calspan fleet. Finally, in 2014 a fourth Learjet (Model 25D R/N N304VS, formerly N515TE) was procured.

Fig. 5.29 Calspan gates Learjet model 25B N102VS (*Credit* Calspan)

Fig. 5.30 Calspan Gates Learjet model 25B N102VS with a refueling nose probe

Fig. 5.31 Calspan Gates model 25B N102VS during autonomous air refueling

With the Learjets of Calspan, many flight test programs were carried out for civil and military aircraft projects and their avionics components. In agreement with the FAA, since the beginning of 2000 training programs are offered for commercial pilots of scheduled and charter airlines, which should lead to a better perception and responsiveness under uncontrolled flight conditions (*Loss of control—LOC*). The Learjet training program (*Upset Recovery Training*—URT) is based on actually occurred aircraft accidents and it allows to illustrate and perform the same in the safer environment of in-flight simulation of chosen commercial aircraft with safety pilots under very realistic environmental conditions. Such training programs appear particularly useful with increasing automation of aircraft in connection with unexpected aerodynamic disturbances such as turbulence, wake vortices and icing, during faults or failures of aircraft engines, airborne and sensor systems as well as in operating errors and faulty pilot communications. Because of the importance of the global civil traffic aviation safety, the US Aviation Insurance Company *Global Aerospace* participates in this training program since mid-October 2013.

Further, noteworthy programs were flown with the Learjet N102VS as so-called replacement aircraft (surrogate aircraft). The program simulated an unmanned aircraft X-47B with the aim of an autonomous refueling task (*Autonomous Aerial Refueling*—AAR, see Figs. 5.30 and 5.31).

In conclusion, four variable stability Learjets continue the legacy of in-flight simulation (IFS) at Calspan. Each aircraft is equipped with a programmable Fly-by-Wire/Light control system that allows for modification of the dynamics of the base Learjet airframe. While the aircraft have been used for manned aircraft handling qualities evaluations for decades, the programmable nature of the VSS allows the aircraft to be used also as UAV surrogates to test the latest in unmanned aircraft technologies [12].

5.2.1.15 General Dynamics NF-16D VISTA (Since 1992)

The US Air Force commissioned in 1988 General Dynamics with the conversion of a General Dynamics F-16D series vehicle to an in-flight simulator with supersonic capabilities called NF-16D VISTA (*Variable Stability In-Flight Simulator Test Aircraft*). Calspan was subcontracted to install computer-controlled central and lateral side control grips for the test pilots and to develop a digital computer system for generating variable flying qualities (*Variable Stability System*—VSS). In the tandem cockpit, the test pilot sat in the front with programmable visual displays and behind him the safety pilot who also directed and monitored the test program. The integration of this VSS simulation system with the standard Fly-by-Wire computers of the F-16D base aircraft provided in the late 80s a special innovational boost. The development of the NF-16D (AF 86-048) was completed in April 1992 (see Fig. 5.32).

Meanwhile, from July 1993 to March 1994, the NF-16D VISTA was converted for the MATV technology program (*Multi-Axis Thrust Vectoring*) to test the controlled flight at high angles of attack using thrust vectoring and was tested at the Air Force Edwards Flight Test Center (see Sect. 6.2.2.18). Already in mid-1994 the variable stability system was reinstalled and was available again for training purposes and

Fig. 5.32 NF-16D VISTA (AF 86-048)

in-flight simulations at the Wright Patterson Air Force Base from January 1995.

The first comprehensive simulation program was flown in the year 1995 for the USAF project YF-22. Since then VISTA was equipped with other components such as a helmet mounted and head-up display (HMD, HUD) and employed in a variety of research and development programs. On October 1, 2000, the NF-16D was transferred to the test pilot school AFTPS at the Edwards Air Force Base. With the support of Calspan, there it continues to be operated further for test pilot training and selected research programs (see Fig. 5.33).

An adaptive flight controller that could enable pilots to save a damaged or out-of-control aircraft was tested on the F-16D VISTA in cooperation with the University of Illinois and the USAF Edwards TPS (1st flight August 26, 2016). The adaptive controller dubbed LI was designed as a backup safety flight control system (FCS) to augment a standard FCS in a conventional aircraft, or as the main control system for an unmanned aircraft.

The LI controller operates in real time to predict transient behavior by estimating an aggregate of uncertainties, rather than relying on the selection of preprogrammed gains, as do most other adaptive FCS. The controller includes a state predictor and a fast estimation law, which together approximate the dynamics of the aircraft in order to rate system uncertainties. These estimates are provided as input to a bandwidth-limited filter that generates a control signal to the FCS. The test and evaluation campaign covered 20 flight hours.

5.2.1.16 Sikorsky NSH-60B VSC (1992–2011)

Besides the hired in-flight simulators from Calspan (B-26B, NT-33A, X-22 VS, Learjet Model 24 VS), since the 70s the US Naval Test Pilot School (USNTPS) operated its own variable stability aircraft like the NCH-46 VSC, which was, however, returned to the US Navy in 1988. Since 1992, two Sikorsky H-60B with irreversible hydraulic flight control and autopilot equipment with 10% control authority were selected as successors. The simulation system (*Variable Stability and Control*—VSC) delivered by Calspan worked on the principle of response feedback, enabling the variation of flying qualities in the pitch, roll, and yaw axes. For this purpose, the hydraulic actuation systems and sensors of autopilot system were enhanced through installation of position transducers on control inceptors and rotor blades and a 3-axis pitch, roll, and yaw rate gyro package. The safety and instructor pilot sits on the left in the cockpit, and on the right-hand side the test or respectively the student pilot. From both places the VSC system could be operated and flown. Via a so-called configuration selection system (*Configuration Control System*—CCS), the digital computer provided for 146 different models (configurations) to choose from within a few seconds. To suppress the coupling of higher frequency rotor-controller-fuselage oscillations, extensive frequency response analysis and measurement signal filtering were performed during system testing. In safety-critical flight conditions, both pilots were reverted back to the basic control system. Because of increasing maintenance efforts on the helicopters, both the NSH-60B VSC were grounded in 2011 (see Fig. 5.34).

Fig. 5.33 NF-16D VISTA flight test preparations

Fig. 5.34 NSH-60B VSC

Fig. 5.35 NUH-60L of the US Navy test Pilot School

5.2.1.17 NUH-60L VSS (Since 2016)

As a part of the call for proposal, since July 7, 2013 a follow-up program was prepared at the U.S. Naval Test Pilot School (USNTPS) for its both NSH-60B VSC helicopters. It was originally planned to integrate the VSS system, at that time developed by Calspan for the NSH-60B (see Sect. 5.2.1.16) with an ability to modify the stability and control properties, with new hardware and software, in two UH-72 Lakota, to be designated as NUH-72 VSS. This helicopter type would have been ideally suited as host helicopter due to its high control effectiveness in the rotor system. Regrettably, however, the decision was taken by the customer that Calspan should carry out the conversion of two Sikorsky UH-60L VSS Black Hawk helicopters (NUH-60L, see Fig. 5.35), a helicopter type which flew for the first time already on October 17, 1974. NUH-60L is in operation since 2016.

The VSS system is designed for the three primary axes (pitch, roll, and yaw). These include intelligent electromechanical actuators (Smart Electro Mechanical Actuator—SEMA), again with a limited authority for expedious and cost-effective modification and certification. The in-flight simulators converted in such a fashion would receive a military certification (NAVAIR Flight Clearance), which is, however, based mostly on the civilian FAA type approval and thereby ensuring a more low-cost supply of spare parts.

5.2.2 NASA

5.2.2.1 Grumman F6F-3 VS (1952–1956)

As already explained in the introductory Sect. 5.1, NASA Ames Research Center played a pioneering role in implementing the first ideas of an in-flight simulation. With a converted Grumman F6F-3 VS (see Fig. 5.1, NACA 158) the worldwide first aircraft with a limited variable stability capability was created in the year 1948. The basic idea is traced back to the NASA engineer *William Kaufmann*, who critically followed up the heavy expenditure on the construction of three Ryan FR-1 Fireball aircraft configurations with different wing dihedral positions to determine the optimal lateral stability, presented by the rolling moment due to sideslip. This is an important flight mechanical parameter, which affects significantly the lateral flight behavior while initiating a turn. *W. Kaufmann* developed a concept for artificially influencing the rolling moment due to side slip in flight using tools based on control technology, which was patented in 1955. His system for varying the rolling moment due to side slip was based on actuating the ailerons as a function of the angle of sideslip. Later on, the electromechanical actuation of the rudder as a function of gyro-measured roll and yaw rates, as well as a function of the vane-measured angle of sideslip were added. The F6F-3 VS was deployed for a variety of flight tests to determine flying qualities criteria pertaining to the lateral-directional motion and for industrial assessment of the optimum rolling moment due to side slip in aircraft projects even before the first flight (see Fig. 5.36).

Fig. 5.36 F6F-3 and the NACA Ames team (1950)

5.2.2.2 Lockheed T-02/TV-1 VS (1952–1960)

In 1951, the NASA Langley Research Center received for research purposes a Lockheed TV-1 (BuNo 124933, redesignated since 1962: T-33B), which was fitted under the nose with a fixed vertical fin and a hinged control surface (see Fig. 5.37). The control surface was connected mechanically with a yaw rate gyro. The aim of this study was to investigate the influence of this variably adjustable damping device on the Dutch Roll (yaw oscillation) behavior at high airspeeds. As early as 1958 flight tests were undertaken for the first time with a sidestick controller in combination with an irreversible hydraulic flight control system.

5.2.2.3 North American F-86A VS (1950–1956)

To extend the recent variable stability investigations with the F6F-3 for determination of optimal flying qualities to the

Fig. 5.37 US Navy Lockheed TV-1 at NASA Langley

new generation of fast flying swept wing configurations, initially a North American F-86A (see Fig. 5.38, AF 47-609, NACA 135) and later an improved F-86E (see Sect. 5.2.2.7) as well as a YF-86D (see Sect. 5.2.2.5) were equipped at NASA Ames with variable stability (VS) systems. The F-86A VS system was limited to the yaw axis, and the rudder was operated via electrohydraulic means. During the period from 1952 to 1956, complementary in-flight simulations of an entire range of aircraft projects were completed with the F-86A and the F6F-3 to evaluate the flying qualities before the first flight. These included aircraft designs in different categories such as D-558-II, XF-10F, X-1 variants, B-58, XF-104, XF8U-1, F9F-9, XT-37, B-57D, T-38 and the large flying boat P6M with four turbojet engines on the wings.

5.2.2.4 Sikorsky HO3S-1 (1952–1958)

In the year 1952, a Sikorsky HO3S-1 (H-5F) helicopter was converted to the first helicopter with a variable stability and control potential, and since 1953 was operated by the NASA Langley Research Center (see Fig. 5.39, NACA 201). For this purpose, the pilots drove the electromechanical actuators through a modified autopilot with adjustable potentiometers, in parallel to the basic control in the pitch, roll, and yaw axes. In this way, the ratio of the stick displacement to attainable control moment (*control power*) and the damping in the pitch, yaw and roll axes could be varied. In an emergency, the safety pilot had direct access to the mechanical control system of the basic helicopter. With this flight vehicle, a first database used to determine the flying qualities requirements of helicopters was flown.

5.2.2.5 North American YF-86D (1952–1960)

NASA Ames upgraded a North American YF-86D with a simulation system for changing the flying qualities in the longitudinal mode (see Fig. 5.40, NACA 149). This

Fig. 5.38 F-86A (NACA 135)

Fig. 5.39 HO3S-1 (NACA 201)

Fig. 5.40 YF-86D (NACA 149)

included variable stick feelings such as force gradient and sensitivities using electro-hydraulic means. These control inputs were generated by the pilot using potentiometers and fed directly to the horizontal stabilizer. To change the stability behavior, the measurement signals of angle of attack and the yaw rate were fed back with variable adjustable gains to the horizontal tail. With this flight vehicle, unstable configurations were flown for the first time and handling qualities assessed. Such investigations became of special importance later in the development of Fly-by-Wire systems for unstable aircraft. Also at DFVLR corresponding investigations were carried out in the late 70s with the HFB 320 FLISI (see Sect. 7.3.12).

5.2.2.6 Grumman F9F-2 (1954–1955)

The flying qualities of a Grumman F9F-2 Panther (BuNo 122560) were modified by NASA Langley using an autopilot with limited authority (see Fig. 5.41). The roll and pitch attitude signals could be introduced with the aid of an analog computer. Also, testing of a sidestick device in conjunction with various artificial stick forces (adjustable spring or damping characteristics) was a part the NASA Langley research program. Originally built as a F9F-3, this test aircraft had a Pratt and Whitney J42 turbojet power plant, hence the designation change. This research aircraft served long enough at Langley to witness the change from the NACA to NASA on October 1, 1958.

5.2.2.7 North American F-86E (1957–1959)

While the variable stability system of the F-86A (NACA 135) was limited to the yaw axis, the North American F-86E could be driven electro-hydraulically both in the yaw and the roll axis via computer and thus the stability and control parameters of the lateral-directional motion could be varied (see Fig. 5.42, NACA 157). Because the F-86E compared to the F-86A was also equipped with a servo-controlled electro-hydraulic horizontal stabilizer as default, the aerodynamic parameters of the longitudinal motion could also be varied based on control engineering measures. Through separate servo actuators on the left and right ailerons, they could also be activated symmetrically. This meant that by combining symmetrical aileron and horizontal tailplane deflections the flight in the turbulent air could also be simulated. The main objective of these investigations focused on the identification of critical stability parameters in the longitudinal and lateral-directional motion, for which the flying qualities were still judged to be acceptable by the test pilots during both high-speed flights at high altitude and landing approach.

5.2.2.8 Grumman F-11F-1F (1960–1961)

The Grumman F-11F Tiger became worldwide famous, because on September 21, 1956, it shot down itself after overtaking a projectile fired from its own onboard cannons. In the research at NASA Langley Research Center, the F-11F-1F with a powerful J-79 jet propulsion had proved more useful (see Fig. 5.43). The longitudinal control system (pitch control) of the supersonic aircraft was provided with a variable steering assistance system (*Variable Control and Response Feel*), which could be influenced by five flight measured variables, namely control surface deflections and their rate of change, vertical acceleration, pitch rate and pitch acceleration. The flight test results showed that only a

Fig. 5.41 F9F-2 at NASA Langley

Fig. 5.42 F-86E (NACA 157)

Fig. 5.43 F11F-1F at NASA Langley

Fig. 5.45 JF-100C with NASA Ames team

well-matched ratio of vertical and pitch accelerations ensured optimal steering assistance with adequate flight stability.

5.2.2.9 North American JF-100C (1960–1972)

After the development of the F6F-3 at the NASA Ames Research Center, the concept of in-flight simulation was evolutionarily continued with the aforementioned NASA F-86 Series (F-86A, YF-86D, F-86E). Finally, in the year 1960, a North American F-100C was converted to an in-flight simulator JF-100C (see Fig. 5.44, AF 31709, NASA 703), initially for the X-15 reentry and landing simulation. The research aircraft was subsequently transferred to the NASA Flight Research Center, which was later named after the NASA researcher *Hugh Dryden* and from 2014 has been renamed after the NASA test and astronaut pilot *Neil Armstrong*. The JF-100C had, like for the aforementioned NASA variable stability aircraft, the flight critical disadvantage of single seat capability, which posed special challenges to the test pilots during system failures.

The X-15 flying qualities were realized using two mobile analog computers, which, on the ground, when connected with the JF-100C also enabled ground-based simulations as a system check before the actual in-flight simulation. With the JF-100C also safety-critical flight conditions with roll and yaw damper failures were investigated. After the aircraft returned to Ames Research Center in the year 1964, there were further investigations on the effectiveness of a direct lift control (DLC) to improve precision approach procedures for air refueling (see Fig. 5.45).

5.2.2.10 Bell X-14A/B (1960–1981)

The VTOL experimental aircraft Bell X-14A, looking somewhat unusual (see Fig. 5.46, NASA 234), was constructed in an extremely short period of time partially using construction components from two Beechcraft aircraft, namely 35 Bonanza and T-34A Mentor. The first flight in hover took place on February 17, 1957. The first complete VTOL cycle with transition to the wings supported flight was performed on May 24, 1958.

The X-14A had an open cockpit, two turbojet engines with thrust vectoring arranged in parallel and a programmable analog computer to generate variable flying qualities. During the period from 1961 to 1971, extensive in-flight simulations for extracting flying qualities criteria for flight in hover and during transition phase were performed at the NASA Ames Research Center. These flight tests played later an important role in the development of British vertical takeoff aircraft Hawker P.1127, from which the V/STOL fighter aircraft Harrier emerged. Also, it should not be forgotten that in 1965 *Neil Armstrong* tried out and trained on the X-14A the flying qualities of the Apollo Lunar Module (*Lunar Excursion Module*—LEM) (see Fig. 5.47 and Sect. 6.1.3.3).

Fig. 5.44 JF-100C (NASA 703)

Fig. 5.46 X-14A (NASA 234)

Fig. 5.47 X-14A with *Neil Armstrong* (center) 1965

Fig. 5.48 X-14B (NASA 704)

In the year 1971, the X-14A was fitted with new engines, a new digital system for in-flight simulation and was designated X-14B (see Fig. 5.48, NASA 704). With this version, a new model following system (*Variable Stability Control Augmentation System*—VSCAS) was tested for attitude changes during precision hover with V/STOL aircraft. About 25 test pilots flew the X-14B. During the last flight in 1981 with the project pilot *Ron Gerdes*, a programming error led to saturation of the roll-actuators with unstable oscillations, which resulted in a crash. He came out without injuries from that and flew later in Braunschweig the helicopter in-flight simulator Bo 105 ATTHeS of the German Aerospace research Center DLR under the MoU Helicopter Flight Control (see Sect. 12.3.3).

5.2.2.11 Boeing-Vertol YCH-46C (1962–1975)

After the initial experiences and successes with the HO3S-1 as an experimental helicopter with variable stability and control capabilities, in the year 1962, the US Army made available to the NASA Langley Research Center a Boeing-Vertol YHC-1A BV (Army 58–5514) with tandem rotor assembly. It was soon renamed as YCH-46C and equipped with an efficient system for generating variable stability and control characteristics (see Fig. 5.49, NASA 533). The 4 degrees of freedom, namely pitch, roll, yaw, and vertical dynamics, were driven by electro-hydraulic actuators in parallel to the mechanical control system. For the actual in-flight simulation, the then new method of explicit model following control was applied.

Fig. 5.49 YCH-46C (NASA 533)

With the YCH-46C an extensive flight test database was flown for the development of flying qualities guidelines for helicopters. In the year 1968, at the NASA Langley Research Center the project VALT (*VTOL Approach and Landing Technology*) was launched, for testing special landing approach procedures (*Decelerating Approaches*) under poor visibility and low cloud levels (*Zero Visibility, Zero Ceiling*). In the year 1968 the worldwide first fully automated landing on a pre-assigned target point was performed. The utilization of YCH-46C ended in the year 1974 after 12 years of service with 685 flight hours and impressive flight test results [13].

5.2.2.12 Lockheed C-140 JetStar GPAS (1965–1977)

In June 1964, at the behest of NASA Dryden Research Center, Calspan equipped a Lockheed C-140 JetStar, labeled GPAS (*General Purpose Airborne Simulator*) with an electro-hydraulic flight control system to modify the flying qualities (see Fig. 5.50, NASA 14, later: 814). The aircraft was fitted with a 4-axes simulation system (pitch, roll, yaw, and thrust modulation) and was delivered in November 1965 to the NASA Dryden Flight Research Center. In summer 1971, the landing flap system was supplemented with direct lift control (DLC) devices. With the help of an explicit model following system, different configurations to be simulated could be programmed on an analog computer. The test pilot sat on the left seat and had at disposal a set of programmable, transport aircraft specific control input and vision systems (displays). The safety pilot on the right had access to the basic system of the JetStar. A flight test engineer in the cabin operated the simulation system. After a demonstration flight for *Wernher von Braun*, he called the

Fig. 5.50 C-140 JetStar (NASA 814)

Fig. 5.51 Boeing 367-80 at NASA langley

procedure "*Dial-a-Model*", a term which was later cited often by many to illustrate the flexibility of the model following principle.

The range of applications of GPAS included initially in-flight simulations and training flights for the XB-70 Valkyrie Mach3 large aircraft project, investigations on the rolling behavior of transport aircraft, and fundamental research on the effect of visual and motion cues in the general simulation techniques on the typical example of XB-70. Further investigations from the early 70s addressed the influence of controller strategies to damp the aircraft reactions in turbulent air. In 1972, studies were initiated at NASA to develop a special in-flight simulator for the *Space Shuttle* project. To obtain a database, in-flight simulations of the approach and landing behavior of the Space Shuttle were carried out with GPAS in 1977. These support activities yielded a substantial contribution to the development of the *Shuttle Training Aircraft*—STA (see Sect. 5.2.2.14).

5.2.2.13 Boeing 367–80 SST (1965–1966)

To investigate the approach and landing properties of a large transport and supersonic aircraft (*Supersonic Transport*—SST), at the behest of NASA Langley, the prototype of the Boeing 707 with the designation B-367–80 (*Dash 80*) was converted to an in-flight simulator (see Fig. 5.51). From May to October 1965, the test pilots from Boeing, NASA, the US regulatory authority FAA (*Federal Aviation Administration*), and airlines flew a database on the stability and flight control parameters for the flight at low speed, which should serve certification requirements of future supersonic transport aircraft with variable sweep or double delta wings. Thereby, the Boeing flight mechanics engineer *Phil Condit* probably must have made special important contributions, because in 1992 he was nominated as Chairman of the Board of Directors and in 1996 the CEO (Chief Executive Officer) of Boeing.

5.2.2.14 Grumman C-11A STA (1976–2011)

A major challenge, and at the same time a special honor, for selected NASA astronaut pilots was to command a *Space Shuttle* "Orbiter" space vehicle. This manned space transportation system is practically a bluff body with an unfavorable glide ratio of 4.5. It is not an easy-to-handle flight vehicle: it is the world's largest, heaviest and fastest space transport glider, in other words a sailplane. The Space Shuttle, classified as an extremely sluggish vehicle, flies unpowered after the re-entry into the atmosphere, and has to be landed safely on the very first approach by the pilot, who had not yet adapted himself to the influence of earth's gravity after several days of weightlessness in space. To reduce the risks during landing, a comprehensive training program was developed with the aid of an in-flight simulator, which possessed roughly the same flying qualities as the space shuttle. Thus, the C-11A *Shuttle Training Aircraft*—STA evolved, based on a Grumman Gulfstream G-2 business aircraft equipped with variable flying qualities. For the Space Shuttle Program of NASA spread over more than 30 years, a total of 4 STA training aircraft were manufactured, two in 1976, another in 1985 and finally the fourth one in 1990 (see Fig. 5.52).

Fig. 5.52 Shuttle Training Aircraft (STA)

The Gulfstream G-2 was selected due to its maximum achievable altitude of about 12,000 meters and a large fuel reserve, which enabled a total of 10 full landing approaches. The required stabilized steep descent with an over 20° glide path angle was possible only through massive reduction of the so-called *L-over-D-ratio* (Lift over Drag) by increasing the aerodynamic drag of the Gulfstream II. This was accomplished through an additional (third) pair of wing flaps and through in-flight activated thrust reversal of the two engines.

In the cockpit, on the left-hand side, the controls and instruments of the Space Shuttle (see Fig. 5.53) were duplicated for training pilots, while for the safety pilot on the right-hand side the original control and instrumentation of Gulfstream II STA were retained (see Fig. 5.54).

Astronaut pilots, who were to fly the *Space Shuttle* for the first time, must have successfully achieved at least 500 steep descents with subsequent landing (see Fig. 5.55). On an average, the STA operation accumulated about 500–600 flight hours annually. On December 2, 2003, during a flight in the reverse thrust simulation mode, the right engine thrust reverser, tailpipe and cowling had separated from an STA. The safety pilot could regain control with the mechanical backup system of the training aircraft and made a normal landing. With better engine fittings in place, the STA fleet could resume their practice shuttle landings on January 12, 2004. With the last landing of Space Shuttle on July 21, 2011, the flight operation of STA training aircraft at NASA was also discontinued.

Fig. 5.53 Space Shuttle cockpit (Endeavour)

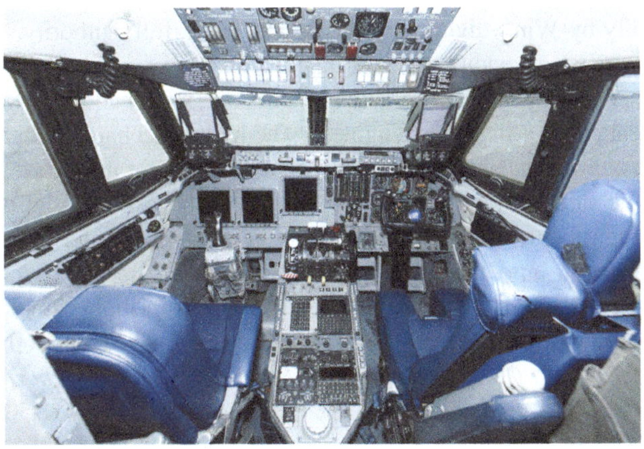

Fig. 5.54 STA Shuttle training cockpit (left)

Fig. 5.55 STA Shuttle approach simulation

5.2.2.15 Boeing CH-47B (1979–1989)

As a successor of YCH-46C, the NASA Langley Research Center, supported by the US Army, received a Boeing CH-47B Chinook. The helicopter had received under the so-called TAGS program (*Tactical Aircraft Guidance System*) a triple-redundant digital Fly-by-Wire system to demonstrate advanced flight control concepts. The CH-47B was developed to an in-flight simulator to further support the VALT Program (*VTOL Approach and Landing Technology*) (see Fig. 5.56, NASA 544). It was ultimately transferred to the NASA Ames Research Center in the year 1979.

Fig. 5.56 CH-47B VALT (NASA 544)

Fig. 5.57 CH-47B (NASA 737)

Fig. 5.58 YAV-8B VSRA (NASA 704)

At Ames Research Center, the TR-48 analog computer was replaced by two digital computers and was equipped with a programmable stick force simulation (*Force-Feel System*) and additionally with a programmable color display (see Fig. 5.57, NASA 737). With this equipment, the CH-47B, somewhat sluggish in its flight dynamics, was deployed for many flight experiments, and extensive flight test databases were generated to support new flying qualities guidelines. In developing new flight controller concepts for a helicopter on the basis of so-called MIMO *(Multi Input—Multi Output*) systems, the importance of unmodeled rotor dynamics was realized. As such for identification and validation of mathematical models incorporating higher-order rotor dynamics, extensive flight tests were additionally undertaken. Based on these findings, jointly with the German Aerospace Research Center (DLR) in Braunschweig, represented through *Gerd Bouwer*, a revised model following controller was developed, which resulted in significantly improved in-flight simulation accuracies [14] (see also Sect. 12.3.3). In 1989, the CH-47B was returned to the US Army, where it was converted to a CH-47D version.

5.2.2.16 British Aerospace YAV-8B VSRA (1984–1997)

To date, last V/STOL research aircraft at NASA Ames Research Center is based on a V/STOL prototype AV-8A Harrier of the US Marines. It was handed over by US Marines to NASA in 1984 with the aim to develop concepts of flight control and flight status display (*Displays*) for the next generation of V/STOL Aircraft. The now under the name YAV-8B VSRA (*V/STOL Systems Research Aircraft*) operated experimental aircraft received a digital Fly-by-Wire flight control system for the primary axes pitch, roll, yaw, and thrust vector control (thrust magnitude and direction) as well as a programmable *Head-Up Display* (see Fig. 5.58, NASA 704).

Extensive flight experiments to develop optimal flight control laws for a variety of missions were performed. This included the design of a three-axis airspeed command control system (*translational rate command system*), which turned out to be favorable during precision hover and vertical landing. In conjunction with the NASA VMS ground-based simulator (*Vertical Motion Simulator*), flying qualities criteria for future STOVL (*Short Take-Off and Vertical Landing*) aircraft were developed and new flight control laws and display symbologies were designed for precision approach and landing. These and many other flight test results were incorporated till 1997 in the F-35 development program (*Joint Strike Fighter*—JSF).

5.2.2.17 Sikorsky JUH-60A RASCAL (Since 1989)

As a successor to the CH-47B, NASA Ames received in 1989 the Sikorsky UH-60A Black Hawk, which was originally upgraded by Boeing to Fly-by-Light test vehicle ADOCS (*Advanced Digital Optical Control System*) (see Sect. 6.2.2.13). In the subsequent years, extensive modifications were carried out. These included a programmable Fly-by-Wire flight control system with full authority, a variety of additional instrumentation, active and passive sensors and the installation of another workstation in the cabin for a flight test engineer. The hydro-mechanical flight control system of the UH-60A was at disposal as before to the safety pilot as *back-up*. The first flight tests focused on flight control laws to increase flight agility and safer flying through automatic limitations of flight envelope (*Carefree Maneuvering*). The actual preparation for in-flight simulation with this research helicopter, now designated JUH-60A RASCAL (*Rotorcraft Aircrew Systems Concepts Airborne Laboratory*), began in 2006 (see Fig. 5.59, NASA 750).

Fig. 5.59 JUH-60 RASCAL (NASA 750)

Gradually the proven principle of explicit model following control using linear and nonlinear models and frequency-dependent feedback as controller architecture was introduced in the in-flight simulation.

After more than 20 years of research activities with the JUH-60, a computer problem was encountered, which was hardly noticed in the initial phase. As a part of system modifications, the circuit board of 32-bit onboard computer developed by Boeing for the Fly-by-Wire flight control system was aging and became prone to errors. Documentaries needed to repair were missing. Consequently, at not so insignificant expenses and considerable time, the so-called *Reengineering* was performed to keep the systems operational.

Toward the end of 2012, fully autonomous flight tests close to the ground were carried out with the JUH-60A to test new technologies that facilitate autonomous helicopter flight without a crew. The JUH-60A flew in an altitude range of 60–120 m above the ground. The airspeed during these flights was limited to 40 knots (75 km/h).

5.2.2.18 McDonnell Douglas F/A-18A FAST (2012–2014)

FAST

The highly instrumented and with a powerful digital FBW flight control system equipped research aircraft F/A-18A FAST (*Full-scale Advanced Systems Testbed*, NASA 853, see Fig. 5.60) was deployed at NASA *Armstrong* Research Center in the recent years for a variety of research programs. With the onboard computer system (*Airborne Research Test System Computer*) available to users, different and extensive user programs could be very efficiently implemented. About two particularly interesting flight test programs will be reported hereinafter. Incidentally, because of the multiple structural changes to the structure such as a hump on the fuselage, it was nicknamed "Frankenstein" by the NASA researchers.

Fig. 5.60 FA-18A FAST (NASA 853)

LVAC

Can a rocket maneuver like an airplane? Can an aircraft serve as a substitute for a maneuvering rocket? Precisely, that is what the NASA investigated under the flight test program LVAC (*Launch Vehicle Adaptive Control*). For the medium-term development of the manned US space transport system SLS (*Space Launch System*) being developed, the Fly-by-Wire onboard computer system of the F/A-18A was programmed in such a way that it could simulate the flight conditions of the SLS at takeoff and ascent before the first flight of the SLS rocket and could check the control algorithms. In the event that something went wrong during the experiment, the F/A-18A safety pilot could passivate the SLS Simulation system and take over control of the test vehicle.

The flight test program LVAC enables the simulation of the autonomous flight control system of the SLS and its required reactions to compensate unexpected disturbances during the ascent of the rocket such as wind effects, structural vibrations or fuel sloshing. The NASA carried out for the first time the testing of a flight control system which could autonomously identify and compensate such interferences encountered during the actual flight. The first flight tests for real-time adaptation of the SLS autopilots took place on November 14–15, 2013 and included 40 test scenarios of SLS relevant trajectories. This allowed extensive checking of the autonomous flight control system until the planned first flight of the SLS in 2017 for uncertainties and risks under realistic flight conditions. As a project illustration, Fig. 5.61 shows a SLS wind tunnel model in the transonic wind tunnel of NASA Ames Research Center.

Fig. 5.61 SLS wind tunnel model

ICP

In the year 2012, a flight test program in its first phase was completed with the F/A-18A FAST and it was demonstrated that by modifying the flight control laws the fuel consumption in cruise flight could be reduced. The research program, known under the acronym ICP (*Intelligent Control of Performance*), employed an optimization algorithm PSA (*Peek-Seeking Algorithm*), with which the aerodynamic control surfaces were tuned such that the aerodynamic drag could be reduced. The exploitation potential for future civil aviation purposes is quite interesting.

5.2.3 US Industry

5.2.3.1 Mc Donnell XF-88A (1954)

During the USAF flight test of the McDonnell XF-88A Voodoo prototypes (see Fig. 5.62) it turned out that the lateral-directional flying qualities were highly unsatisfactory.

Fig. 5.62 XF-88A (AF 46-525)

Thus, to mitigate the Dutch Roll behavior and the roll coupling potential, initially, a yaw damper using yaw rate gyro signals was switched on the rudder. Later through additional inclusions of sideslip angle and roll rate signals to the ailerons, variable flying qualities in the lateral-directional mode of motion could be produced. It was the aim to define, from the perspective of test pilots, tolerable flight mode limits in the case of system failures.

5.2.3.2 Mc Donnell NF-101A (1963–1964)

A little known, short flight test program of the US Air Force Flight Dynamics Laboratory (FDL) embodied a certain limited in-flight simulation of the analog electro-hydraulic flight control system in conjunction with a sidestick for the X-20 Dyna Soar orbiter project on the NF-101A (AF 53-2422) (see Figs. 5.63 and 5.64). In the X-20 project, based on the ideas of *Eugen Sänger*, at that time a highly innovative Fly-by-Wire flight control system (FCS) was conceptualized by Boeing and Honeywell. Citing Ref. [15]:

> The X-20 Flight Control System was designed to satisfy attitude control and stability requirements throughout the hypersonic-subsonic flight regime for the glider and glider/transition (abort) configurations.
>
> Among the unique features employed in the FCS were the completely Fly-by-Wire techniques wherein all signals to or from the FCS computer were electrical. In other words, all pilot

Fig. 5.63 X-20 Dyna Soar mockup

Fig. 5.64 NF-101A (AF 53-2422)

> control signals and all signals to the aerodynamic, thrust vector, and reaction controls were electrical. Another unique feature pertains to the manner that stability was provided to an aerodynamically unstable configuration consisting of the glider and abort rocket. This was accomplished by means of simultaneous aerodynamic and thrust vector control.
>
> The FCS employed dual and triple redundancy to attain a high order of reliability. Redundancy was also carried out, as far as practical, in the electrical connectors, isolation of electrical wiring bundles, electrical circuit mechanical isolation, and environmental isolation by careful orientation of mechanically sensitive components.

Within the framework of NF-101A in-flight simulation, the main focus was on optimization and evaluation of flying qualities of the aerodynamically unstable X-20 in low-speed flight. The promising X-20 project was discontinued in the year 1964.

5.2.3.3 Convair NF-106B VST (1968–1971)

At the Martin Marietta company two Convair F-106B delta wing aircraft (AF 57-2519 and -2529), commissioned by the USAF, were converted to variable stability aircraft (see Fig. 5.65). They were deployed by the USAF Test Pilot School at Edwards Air Force Base with the label NF-106 VST for advanced pilot training. The main modifications included an analog computer, which performed the model following computations for selected flight conditions, and an electro-hydraulic stick force simulation system, which reproduced correctly the control forces for the configuration being simulated in the rear cockpit of the test pilots. The safety pilot in the front cockpit remained during the in-flight simulation connected with the basic flight control system. Through the so-called *Stability Parameter Programmer*, he selected for the test pilots the various configurations to be simulated with different flying qualities depending on the altitude and Mach number. Thus, over the period from 1968 to 1971 various flight missions were performed for investigation and evaluation of the dynamic behavior of reentry vehicles (*Lifting Bodies*) such as the X-20 Dyna Soar, the X-24A, the Space Shuttle, and also a few X-15 configurations. The model following controller was designed applying the classical root locus method of *W. R. Evans*. Problems were encountered whenever the structural vibrations of the NF-106A were excited by the injection of acceleration signals. Due to the significant maintenance efforts, relatively few flight hours were gathered over a period of almost three years.

Fig. 5.65 NF-106B VST

Fig. 5.66 S-76A SHADOW

5.2.3.4 Sikorsky S-76A SHADOW (1983–1995)

In preparation for the RAH-66 Comanche project of the US Army, during the mid-nineteen eighties a S-76 was equipped by Sikorsky as an in-flight simulator with an additional cockpit in the fuselage nose (see Fig. 5.66). The research helicopter was called SHADOW (*Sikorsky Helicopter Advanced Demonstrator of operator workload*). The test equipment included a digital Fly-by-Wire system with a programmable simulation computer, corresponding sensors for determination of the flight condition, and a programmable display. Two safety pilots in the original cockpit of S-76 could at any time take over the helicopter for safety reasons. Additionally, programmable three- and four-axes sidesticks were developed, with which by deflections pitch and roll commands were initiated, and by rotating and vertical pulling or pushing the yaw and heave motion (bob-up and down) of the helicopter were controlled respectively. The main objective was to test new technologies, for example, vision and sensor systems as well as control laws, which enabled safe flight under poor visibility conditions, also close to the ground without overstraining the pilot.

5.2.4 US-Universities

5.2.4.1 Ryan Navion (1952–1954)

Already in the years from 1952 to 1954, commissioned by the US Air Force Wright Air Development Centers (WADC), the Princeton University had undertaken flight tests on a Ryan Navion to assess the influence of springs and

Fig. 5.67 L-17A Navion—1948

masses (*bob weights*) in order to obtain a constant stick force in the flight control system on the longitudinal motion. Both spring and bob weight, when employed in a right manner, are used to supply the pilot with what he wants It was demonstrated that the classical longitudinal dynamics, such as the Phugoid and Short Period modes, could be changed into aperiodic motion forms (see Fig. 5.67).

5.2.4.2 Piasecki HUP-1 (1958–1960)

A so-called Variable Stability System (VSS) for the Piasecki HUP-1 (S/N 15, R/N N4015A) tandem rotor helicopter was realized by the Princeton University through modification of the autopilot and was limited to the roll and yaw axes of the helicopter (see Fig. 5.68). Little is known about research results. FAA registration was cancelled on June 20, 1963.

5.2.4.3 Ryan Navion VRA (1968–1988)

In the nineteen sixties, a Ryan Navion (R/N N91566) was converted by the Princeton University to a test aircraft with variable flying qualities. This so-called *Variable Response Research Aircraft* (VRA) was used for research and education purposes untill the end of nineteen eighties (see Fig. 5.69). The test or evaluation pilot sits in the cockpit on the left hand side. The Navion VRA used a digital Fly-by-Wire control system to activate the electro-hydraulic actuator systems using full authority in all control axes. Additionally, a flap system was integrated for fast regulation of lift (*Direct Lift Control*—DLC). The position of the side force generators (*Direct Side Force Control*—DSFC) on the middle wing was determined by systematic wind tunnel tests at NASA, so that lowest possible coupling effects arose. After more than 20 years of flight research with the Navion VRA, and except for the absence of the lateral force generators almost similar sister aircraft Navion ARA (*Avionics Research Aircraft*), both the aircraft went in the inventory of the Space Institute of the University of Tennessee.

Fig. 5.68 Princeton HUP-1

Fig. 5.69 Princeton Navion VRA

5.2.4.4 Ryan Navion VSRA (Since 1989)

In the year 1988, the Space Institute of the University of Tennessee (UTSI) took over from the Princeton University both the Ryan Navion research aircraft with variable stability equipment and since then operates them under the name of *Variable Stability Research Aircraft* (VSRA, R/N N55UT, R/N N66UT, see Figs. 5.70 and 5.71). While with the

Fig. 5.70 UTSI Navion (R/N N55UT) with side force generators

Fig. 5.71 UTSI Navion (R/N N66UT)

Fig. 5.72 NRC Bell H-13G airborne simulator

R/N N66UT five degrees of freedom could be influenced, the R/N N55UT could generate side forces through two lateral force control surfaces, centrally placed on the wings, and thereby all six degrees of freedom could be exploited for the in-flight simulation purposes.

5.3 Canada

5.3.1 Pioneering Contributions of NRC Aerospace

In the early nineteen-sixties, after termination of the national supersonic fighter program *Arrow*, the National Aeronautical Establishment (NAE) of the Canadian National Research Council (NRC) in Ottawa shifted its aeronautical related research priorities to the area of rotorcraft and to vertical and short take-off (V/STOL) technologies. The Canadian aerospace industry was already very successful in the development and marketing of small transport and commuter aircraft such as the *Beaver* and *Twin Otter*, and as such, they wanted to technologically further expand the international utility-potential of such V/STOL aircraft.

To improve the flying qualities of future V/STOL aircraft, for the first time, helicopters with variable flying qualities were developed since 1961 at the NRC Flight Research Laboratory (FRL) [16]. At first, two Bell H-13G helicopters as long-term loans from the US Army were available, which were converted to flight vehicles with variable stability flying qualities. The principle of explicit model following control for generating reproducible, variable flying qualities was applied for the first time (see also Chap. 3).

Today, the NRC Aerospace in Ottawa possesses, besides the DLR in Braunschweig and the US Army/NASA Ames Research Center, the core competencies in the field of rotorcraft in-flight simulation. The Canadian research helicopters are presented below.

5.3.2 Bell H-13G (1962–1968)

The first helicopter with limited variable stability characteristics in the three primary axes of pitch, roll, and yaw was realized in the early 60s with a Bell H-13G alias Bell-47G loaned from the US Army (see Fig. 5.72). For this purpose, various stability and control parameters could be selectively modified through the potentiometers on an analog computer, which was located between pilot controls and autopilot system. The test pilot on the right side in the cockpit translates his commands by means of control sticks and pedals, whose control forces were variably adjustable. The main feedback variables in the model following control were the measured rotational rates about the primary axes. Up to 14 measurement signals could be recorded on a data recording device (*Recorder*). The electro-pneumatic servo motors, which operated in parallel with the mechanical control system of the host helicopter, drove the rotor blades and the tail rotor with almost full authority. And yet the servo actuators could be overridden in emergencies by the safety pilots seated left in the cockpit. The simulation system could also be switched off via a button on the control grips. Due to the lack of space, a part of the test equipment was accommodated in containers mounted outside.

With this research helicopter systematic investigations to assess flying qualities of V/STOL aircraft were carried out for the first time. Also, in-flight simulations were undertaken with Bell H13G of NRC to evaluate the flying qualities behavior of the German VTOL hovering rig SG-1262 (see Sect. 6.1.3.8) in the course of preparation for the first flight of the VTOL prototype VFW VAK191.

5.3.3 Bell 47G3 (1966–1970)

Unlike the Bell H-13G, besides controlling the three rotational degrees of freedom (pitch, roll, yaw) the Bell 47G3B-1

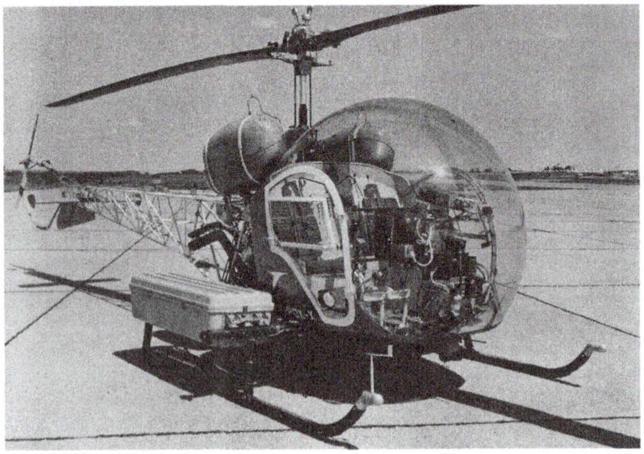

Fig. 5.73 NRC Bell 47G3B-1 airborne simulator

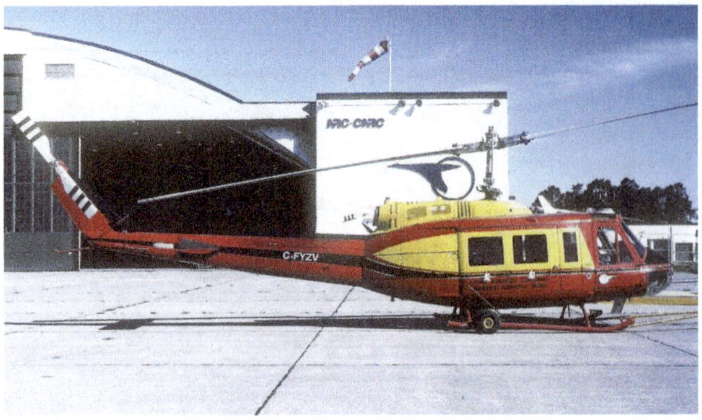

Fig. 5.74 NRC Bell 205A-1

was provided with an additional vertical control option through collective blade pitch deflections (see Fig. 5.73). Again, due to the lack of space, a part of the experimental equipment was accommodated in external containers. The extensions of the analog computer capacity were carried out at the NRC, and to reduce system lags and time delays the electro-pneumatic regulators were replaced through electro-hydraulic actuators. The allocation of pilot seats and roles remained unchanged compared to the Bell H-13G.

The test vehicle was deployed for basic investigations in the field of V/STOL flying qualities requirements as well as for simulating selected V/STOL aircraft projects. So the influence of directional stability (*weathercock stability*) on the flying qualities during landing approach was examined and comparatively assessed under visual and instrumental conditions. Project support was provided, for example, for the V/STOL tilt wing research aircraft Canadair CL-84. Thereby the impact of delays and errors in the flight control system on the handling qualities was simulated even before the first flight.

5.3.4 Bell 205A-1 (1970–2012)

A substantially improved in-flight simulator, especially with regard to the control capability, was realized with the procurement of a Bell 205A-1 (see Fig. 5.74). The helicopter was provided with a single-channel full authority Fly-by-Wire system, an improved hybrid computer (digital-analog), a programmable control unit and a sidestick. As in the case of the predecessor Bell 47G3, the three rotational degrees of freedom (pitch, roll, yaw) and the vertical motion could be influenced. The test pilot on the right was provided with displays for real and artificial visual representations of the environment.

Extensive research programs in the area of development of flying qualities criteria for V/STOL and STOL flight vehicle were flown with the Bell 205A-1, which have also been incorporated in the V/STOL flying qualities requirements issued by the Advisory Group for Aerospace Research and Development (AGARD). Likewise, important contributions were made to the US Army flying qualities requirements for control augmented helicopters (*Aeronautical Design Standard for Helicopter Handling Qualities—ADS-33C*). Project support was provided, for example, for the pre-flight testing of the critical flight envelope of the British VTOL research aircraft Short SC.1, for the test pilot training with the propeller-driven V/STOL tilt wing aircraft Canadair CL-84, for the assessment of a pitch attitude controller of the Hawker Siddely Harrier, and for the optimization of the tail rotor of the Sikorsky-76 helicopter. Not to go unmentioned the extensive in-flight simulations to integrate programmable multi-axis controller units and sidesticks in conjunction with different control concepts such as Rate Command/Attitude Hold (RC/AH).

5.3.5 Bell 412 ASRA (Since 2001)

With The Bell 412 ASRA (*Advanced Systems Research Aircraft*) the NRC has at disposal, for the first time, an in-flight V/STOL simulator with four rotor blades (see Fig. 5.75). ASRA was in turn equipped with a four-axes digital Fly-by-Wire system and extensive Avionics. The test pilot operates a programmable multiaxial sidestick. The Fly-by-Wire system was first commissioned on February 9, 2001.

The testing of a control strategy, that should simplify in a special way the different maneuvering requirements on a helicopter under conditions of poor visibility, proved to be an interesting utilization program. While so far the pilot had to select different control strategies depending on the flight

Fig. 5.75 NRC Bell 412 ASTRA

task, attempts were made for the first time with a so-called "Super-TRC" command system (*Super-Translational rate Command*) to automatically select in each case the correct control strategy through continuous recording of the control activity of pilots. For flight close to the ground in poor visibility a cautious control activity of the pilot is required (small amplitude, low frequency), hence, the Fly-by-Wire system commands horizontal speed inputs in conjunction with an increased attitude stabilization. In contrast, if the pilot calls for fast and large changes in the course (large amplitudes, higher frequencies), the system provides attitude rate commands in conjunction with reduced damping. In conclusion, it is to be noted that the automatic monitoring of the pilot intentions related to stability or agility is an appealing approach for future man-machine requirements not only in the field of aviation.

5.4 England

5.4.1 Short SC.1 (1957–1971)

The Short SC.1 (Registration number XG900) was a single-seat low-wing aircraft, that was designed as a tailless delta aircraft (without horizontal tailplane). The propulsion for vertical flight consisted of four vertically mounted Rolls-Royce RB108 engines in the fuselage and a RB.108 in the tail for forward flight. The SC.1 was designed for hovering and low-speed flight and for the intermediate transitions phase. The control and stabilization were carried out in low-speed flight via reaction jets in the fuselage nose, tail and at the wing tips. Swiveling of the fuselage-mounted engines about the lateral axis of the machine allowed a change in direction of the thrust vector. The SC.1 had the first Fly-by-Wire system, which was introduced in a vertical takeoff aircraft (*Vertical Take-Off and Landing*—VTOL), and thereby provided the potential of in-flight simulation.

Fig. 5.76 Short SC.1

The first vertical takeoff took place on October 25, 1958, the first transition flight on April 6, 1960 (see Fig. 5.76).

The single-channel Fly-by-Wire system allowed systematic flight tests with variable stability in the low-speed regime. Due to the lack of system redundancy, the flight experiments close to the ground were associated with a higher risk and even led to a fatal crash following a system failure. The flight tests were continued and served to identify desirable flying qualities in different phases of flight as a function of variable control effectiveness and damping parameters, and also under the effect of the engine dynamics and the ground effect. Thus, a first solid database for future flying qualities criteria for VTOL aircraft could be created.

5.4.2 Beagle Bassett VSS (1973–2014)

The UK Empire Test Pilot's School (ETPS) in Boscombe Down received in June 1973 a Beagle Basset CC.1 (R/N XS743) as a training aircraft that was equipped by the Cranfield Institute of Technology (C.I.T.) with a system for changing the stability and control characteristics (*Variable Stability System*—VSS) (see Fig. 5.77). This aircraft

Fig. 5.77 Beagle Basset CC.1 XS743

conceptualized for commuter flights was also utilized by *Prince Charles* to acquire his pilot license for aircraft with two engines. The analog computer of the VSS receives the electrical commands from the right control column of student pilots, disconnected from the mechanical control, and passes them on in a suitable form to the aerodynamic control surfaces. Through the potentiometers of the board analog computer, the training pilot in the left seat can now vary the flying qualities to a limited extent. Thus, the Bassett aircraft could be felt as having flying qualities, for example, like that of a large sluggish transport aircraft or of an airplane with system failures. Should the training pilots make serious operating mistakes, the VSS system switches off and the instructor can take over the command with the mechanical basic controls. The aircraft, finally with equipment of antiquarian value, has completed 40-years of successful deployment.

5.4.3 BAE VAAC Harrier T.2 (1985–2008)

To provide a solution to the original three-hand problem, namely the simultaneous operation of control stick, throttles and nozzle angle while maneuvering the British vertical takeoff Hawker Harrier aircraft, in the year 1985 the Cranfield Institute of Technology (C.I.T.) equipped a two-seater Harrier T.2 training aircraft (XW175) with a digital, two-channel duplex Fly-by-Wire system and a new rear experimental cockpit (see Fig. 5.78). The front cockpit retained its original equipment for the safety pilot. In the subsequent years, with this equipment various cockpit systems, variable control strategies and software versions could be checked for acceptance by the test pilot. After the retrofitting, dubbed as VAAC Harrier, the experimental aircraft (*Vectored-Thrust Aircraft Advanced Flight Control—VAAC*) was operated by QinetiQ, the successor of RAE/DERA.

The VAAC Harrier provided support in the development of the Lockheed Martin F-35B V/STOL version of the US-American JSF (*Joint Strike Fighter*) development program. For this purpose, the rear cockpit of the test pilot was equipped with two control units (*Inceptors*) to replicate the so-called *Unified Control Mode* of the F-35B. Future F-35B pilots operated with the left hand a linear thrust adjustment device for controlling the forward and reverse speeds and with the right a sidestick for vertical and lateral control commands.

5.4.4 BAE ASTRA Hawk (Since 1986)

As a supplement to the classic Beagle Basset VSS, in the early 80s, the conversion of a two-seater BAE Systems Hawk T.1 to an in-flight simulator began at the Empire Test Pilot's School (ETPS) in cooperation with the Cranfield Institute of Technology (C.I.T.) (see Fig. 5.79). The flying qualities of this research and training aircraft Hawk ASTRA (*Advanced Stability Training and Research Aircraft*) can be changed during the flight by the instructor and safety pilot in the rear cockpit. At the same time, he could monitor the simulation system or turn off the same in an emergency. In the front cockpit for training pilots, the mechanical control was disconnected from the host system and replaced through electronically connected control sticks in the center and on the right side with variable force or deflection characteristics. The same applies to the rudder pedals with variable force and deflection characteristics. Likewise, the imagery in the Head-up Display could be changed with the aid of computers. The actual Fly-by-Wire system employs electro-hydraulic rotary shaft drives, which establishes the

Fig. 5.78 BAE VAAC Harrier (*Credit* Andrew Dickie)

Fig. 5.79 ASTRA Hawk

connection to the aerodynamic control surfaces and are driven by the simulation computer. With the simulation system switched on, virtually the entire flight envelope of ASTRA Hawk with Mach numbers up to 0.7, load factors of −1.75 to +6.5 g and roll rates of 200°/s could be flown. Since the first flight in 1986, it was available at the ETPS 1990 onwards for training purposes.

5.5 France

5.5.1 Dassault Mirage IIIB SV (1967–1990)

Even before the first flight of the Anglo-French supersonic airliner Concorde 001 (first flight: March 2, 1969), a two-seater Dassault Mirage IIIB (S/N 225) was equipped at the CEV (Centre d'Essais en Vol) with special Fly-by-Wire electronics in the flight control system to change its flying qualities in such a way that the resulting flight mechanical properties and the control behavior corresponded to that of the Concorde (see Fig. 5.80). With this Mirage IIIB (labeled "*Stabilité Variable*"—SV) extensive in-flight simulations were performed for the Concorde Project. With its special simulation capabilities, it was employed also for the development of the analog Fly-by-Wire flight control system with multiple redundancies for the first European fighter plane Mirage 2000 with a fully electronic flight control. Furthermore, it served the French test pilot school EPNER as a training aircraft. Today, it is in the Saint Victoret Museum in Marseille.

5.5.2 Dassault Mystere/Falcon 20 CV (1978–1982)

Since the mid-70s, at the behest of DGA (Direction Générale de l'Armement), the first production aircraft the Dassault Mystère/Falcon (S/N 401, F-WMSH) was converted to an in-flight simulator in all six axes (labeled "*Characteristics Variable*"—CV). Besides the initial analog Fly-by-Wire

Fig. 5.80 Mirage IIB (S/N 225)

Fig. 5.81 Mystere-Falcon 20 CV without and with side force control surfaces (*Credit* Jean Francois Georges)

system, later to be equipped with two digital computers, and an additional direct lift control system to the flaps (*Flaperons*), aerodynamic control surfaces were mounted vertically in the center wing section for direct side force control (*Travelons*) (see Fig. 5.81).

An onboard computer was to cater for the in-flight simulation capabilities in all 6 degrees of freedom. The simulation pilot sat on the left in the cockpit with a control station decoupled from the basic mechanical controls. The scope of simulation included the entire flight envelope and was limited in the first phase to the 3 rotational degrees of freedom and 2 translational degrees of freedom, that is, to the longitudinal and vertical acceleration. In the second phase, it was planned to generate lateral accelerations by means of the *Travelons*.

The main scope of application was the testing of digital flight control systems and the creation of civilian certification criteria for future commercial aircraft with reduced stability that promised increased performance. For the first application of an in-flight simulation, the Dassault Mercure project was selected, with a mathematical model structure being quite familiar to the Dassault company. These flight demonstrations occurred as yet without the active deployment of *Travelons*. That was to take place later with considerable technical efforts and financial expenditures. Despite the good quality of these first results, the project in 1982 was terminated due to financial reasons.

Apparently, the ability to responsibly manage the in-flight simulators is beneficial for the personal career, because it is worth to point out that the Mystere/Falcon 20 CV project manager *Jean-Francois Georges* was appointed in 1995 as the Chief Executive Officer (CEO) of the Dassault Falcon Jet Corporation, similar to what happened in the case of the Boeing 367–80 SST in-flight simulator (see Sect. 5.2.2.13).

5.6 Russia

5.6.1 Tupolev Tu-154 M FACT (Since 1987)

For the sole purpose of the simulation of the Russian manned space vehicle Buran, four Tupolev Tu-154B passenger aircraft were simultaneously converted to in-flight simulators at the Research Center Zhukovsky of the Flight Research Institute (FRI), also called *Gromov* Institute named after a famous Russian test pilot. These aircraft served during the beginnings of the 1980s the development and optimization of the Buran flight control laws with subsequent test pilot assessment of the flying qualities and training of the cosmonaut pilots [17]. Similar to the NASA C-11A STA (*Shuttle Training Aircraft*), for realization of the simulation of the steep descent of the manned Buran space shuttle, the thrust reversal of both laterally mounted jet engines of T U-154B had to be used in flight. The resulting aerodynamic interferences between the deflected thrust stream and the tail plane had to be adapted in the modeling of the Tu-154B host aircraft. As is generally known, the Buran Project was discontinued due to the technological and economic difficulties. A copy of the Buran Space Shuttle, being equipped with four auxiliary engines on the rear fuselage conducted flight tests for the validation of the landing capability and cosmonaut training (see Fig. 5.82), is now part of the Aviation Museum in Speyer.

Since 1987, the FRI fitted the prototype of the Tu-154 M (temporarily also being designated as Tu-164) with a Fly-by-Wire system with the aim of building a general purpose in-flight simulator to support the development of new aircraft in addition to testing new flight control, operating and display systems, the verification and extension the flying qualities criteria, and to establish certification procedures for future Fly-by-Wire aircraft. The Tu-154 M (R/N RA-85317) was finally designated FACT (*Future Aircraft Control Testbed*, see Fig. 5.83). From the left cockpit seat of the Tu-154 FACT, the test pilot operates the Fly-by-Wire system, whereby the operating elements, in terms of stick force characteristics (*Feel Characteristics*) as well as the display imageries (see Fig. 5.84), could be adapted in flight with computer assistance. The safety pilot on the right-hand monitors the conventional mechanical flight control system and the classical flight instruments of the host aircraft. In the case of system errors or at flight envelope boundaries, he can switch off electrically the simulation system or take over the manual control by pressing the control column and transfer the aircraft into the basic flight mode. The TU-154 FACT, in the meantime, also supported transport and commercial aircraft projects such as An-70, Il-96 and Tu-204 and for the development and testing of sidestick devices, displays and filters for the suppression of so-called PIO (*Pilot Induced Oscillations*).

Fig. 5.82 The Buran brigade

Fig. 5.83 Tu-154 M FACT

Fig. 5.84 Tu-154 M FACT simulation cockpit (left)

5.6.2 Sukhoi Su-27 ACE (Since 1990)

The Sukhoi Su-27 ACE (*Advanced Control Experiment*), modified to an in-flight simulator by the Flight Research Institute FRI, became economically very attractive as the number of aircraft prototypes needed for a new aircraft design and to reach a production-readiness state could be reduced. A variety of equipment components such as a digital Fly-by-Wire system, sensors, and cockpit

Fig. 5.85 Su-27 ACE

components are available on the Su-27 ACE (S/N 24-05 or code LMK-2405, identification code "05", see Fig. 5.85). As such, imageries for the displays and alternative configurations of sidesticks, could be optimized relatively easily for special applications. Also, pre-testing of flight control algorithms (*control laws*) could be carried out and their influence on the flying qualities be assessed by the test pilot. The single-seat cockpit probably appears to be disadvantageous, as it does not facilitate separating testing functions and safety measures and thereby leading to the higher workload for the test pilots.

5.6.3 Yakovlev Yak-130 (Since 1999)

As a part of a joint Russian-Italian Trainer Development Program between Yakovlev and Aermacchi, the prototype Yak/AEM-130D (R/N RA-43130) made its first flight on April 25, 1996. In the year 2000, the partnership was dissolved and the project was further developed in two separate national programs, e.g. the Yakovlev Yak 130 and Aermacchi M-346 (see Sect. 5.9) training aircraft with variable flying qualities respectively. Through a Fly-by-Wire flight control system in conjunction with special flight control laws uncontrollable flight conditions can be prevented (*Flight Envelope Protection*). An outstanding feature of this training aircraft is the option to adopt during flight the flying qualities of selected types of aircraft such as MiG-29, Su-27, Su-30, F-15, F-16, F-18, Mirage 2000 Rafale, Typhoon. For this purpose the safety and instructor pilot selects on the onboard computer the corresponding computer model for the aircraft to be simulated (*Dial-a-Model*) in each case. The first production prototype made its maiden flight on May 19, 2009 (see Fig. 5.86).

Thus, a project idea of DLR from the year 1984, namely to routinely introduce the potential of in-flight simulation for training of young pilots (CASTOR, see Sect. 11.2), became a reality after a quarter of a century.

Fig. 5.86 Yak-130

5.7 Japan

5.7.1 Kawasaki P-2H VSA (1977–1982)

At the behest of the Japan Defense Agency (JDA), the Kawasaki Heavy Industries (KHI) converted a Lockheed P2 V-7 (License: P-2H) Neptune to an in-flight simulator P-2H VSA (*Variable Stability Aircraft*). The first flight took place on December 23, 1977. The essential modification features included the installation of a Fly-by-Wire flight control system with a mechanical backup for the safety pilots, replacement of the mechanical control system through an electrically driven control wheel with variable force gradient for the test pilot seat on the right hand side, and a powerful onboard computer to carry out all simulation operations. The outer landing flaps were converted for direct lift control, while the aerodynamic side force generator was integrated vertically on the top and underneath the center wing (see Fig. 5.87). In addition, two perforated airbrakes below the fuselage (*Ventral airbrakes*) were installed. Thus, in its role as an in-flight simulator, it could be driven with 5 degrees of freedom independent of each other (pitch, roll,

Fig. 5.87 P2H VSA

yaw, as well as vertical and lateral translations). In an emergency at any time, the safety pilot sitting left in the cockpit had an access to the mechanical control system of the host aircraft. While initially the flight tests were conducted to determine the effectiveness of the direct lift and side force generators in the operational flight regime, after a period of two years of testing by the test pilot school the P-2H VSA was transferred to JMSDF (*Japanese Maritime Self Defense Force*) for the training of future pilots. During this time, steep approach procedures making use of the air brakes were tested up to 7° glide path angle. The aircraft was decommissioned in 1980.

5.7.2 Beech B-65 VSRA (1980–2011)

A Beech B-65 was converted at the National Aeronautical Laboratory NAL (now Japan Aerospace Exploration Agency JAXA) to a test vehicle with variable stability (*Variable Stability Aircraft Research*—VSRA) in five degrees of freedom (no side force generator, see Figs. 5.88 and 5.89).

Fig. 5.88 B-65 VSRA of NAL

Fig. 5.89 B-65 VSRA simulation cockpit, *right*

Fig. 5.90 STOL research aircraft ASUKA

For the utilization of the test demonstrator, the model following control principle was adopted. It played a significant role in the pre-testing and evaluation of flying qualities of the four-engine STOL research aircraft ASUKA (see Fig. 5.90). Later it was deployed mainly for the exploration of the flight in the atmosphere. The last flight took place on October 26, 2011.

5.7.3 Dornier Do-228-202 MuPAL-α (Since 1987)

A Dornier Do-228-200 was equipped by JAXA (formerly NAL) and Kawashai and designated Do-228-202 MuPAL-α (***Mu**ltipurpose **A**viation **L**aboratory;* αεροσκάφος = aircraft) with a digital two-channel (duplex) Fly-by-Wire system (see Fig. 5.91). Elevator, rudder, and ailerons, as well as both the engines, were driven electrically. The landing flaps on the trailing edge were modified with fast-moving three-part, individually controllable auxiliary flaps to enable direct lift control (DLC) and thereby more precise in-flight simulations. The test pilot on the right-hand side in the cockpit actuates the control column with artificial feel. Via electrical taps the input signals were fed to the freely programmable FBW computer and forwarded to the electro-mechanical actuation system in accordance with the desired model following control laws. Thereby the response

Fig. 5.91 MuPAL Alpha

of the aircraft to both pilot as well as gusts inputs can be manipulated through control laws in the forward and feedback loops.

An additional experimental cockpit is located in the fuselage. It is particularly suitable for landing approach simulations at high altitudes. In critical flight situations, the safety pilot on the left hand side in the cockpit can switch back to the mechanical control system of the host aircraft. The Fly-by-Wire system was first activated in flight on November 11, 1999. Since April 2000 the aircraft is in operational service.

5.7.4 Mitsubishi MH-2000 MuPAL-ε (Since 1999)

Since March 2000, JAXA (formerly NAL) operates the research helicopter and in-flight simulator MH-2000A MuPAL-ε (***Mu**lti**p**urpose **A**viation **L**aboratory;* ελικόπτερο = helicopter). It is based on the indigenously developed Japanese helicopter MH-2000A of Mitsubishi Heavy Industries (see Fig. 5.92). The experimental equipment consists of a data acquisition system for providing extensive flight test data, a cockpit display with a high-performance graphics computer to generate variable imagery (see Fig. 5.93), a video recording system for recording images of the environment for navigation purposes and a Fly-by-Wire system for in-flight simulation purposes.

The sensor equipment includes a hybrid DGPS/INS system (*Differential Global Positioning System/Inertial Navigation System*), which merges satellite position data with that from an inertial platform and an ultrasonic velocity sensor for the low speed and hover range, which determines the flight speed components in all directions.

Fig. 5.92 MuPAL Epselon

Fig. 5.93 MuPAL Epselon simulation cockpit, *left*

Impressive with this research helicopter is the concentrated and consequent utilization in a timely manner for various programs like in-flight simulations to evaluate flying qualities and electronic display systems as well as to improving the fidelity of ground-based simulators, GPS-supported in-flight simulations, autonomous flying, noise abatement approach procedures, and the exploration of the helicopter response behavior due to atmospheric or wake vortex turbulence.

5.8 China

5.8.1 Shenyang JJ-6 BW-1 (Since 1989)

The earliest development of an aircraft with variable stability characteristics began in China in the late 70s at the AVIC-Chinese Flight Test Establishment (SFTE) in Xi'an. A two-seater training aircraft Shenyang JJ-6 (Export Version: FT-6) was equipped with a single-axis analog Fly-by-Wire system. In a tandem-seat assembly, the test pilot occupied the front and the safety pilot the rear seat. In an emergency, the safety pilot could access the mechanical back-up control.

The so upgraded test demonstrator, which was based on the Russian Mig-19 Farmer fighter aircraft, crashed in December 1984. The subsequent development of a single-axis digital flight control system for the longitudinal motion for another JJ-6 led to a successful first flight on April 22, 1989 (see Fig. 5.94). The Fly-by-Wire flight control system and the flying qualities of a two-seater aircraft Xian JH-7A could be tested with this research aircraft designated as BW-1.

Fig. 5.94 Shenyang JJ-6 BW-1

Fig. 5.95 HAIG K8V IFSTA

5.8.2 HAIG K8V IFSTA (Since 1997)

After another 8 years, the first flight of the training aircraft K8/JL-8 (S/N 320–203) took place at SFTE on June 25, 1997. It was designed in China (Hongdu Aviation Industry Group—HAIG) and named K8V IFSTA (*Integrated Flight Simulation Test Aircraft*) and was equipped with a three-axis digital duplex flight control system. Main parts of the test equipment were installed in an integrated container mounted underneath the fuselage (see Fig. 5.95). The flight control system still with a mechanical back-up can simulate the flying qualities of up to eight aircraft types with different parameter sets.

Future planning envisages an expansion of K8V to a five-axis digital flight control system, that means the provision of additional lift and drag/thrust control devices in order to provide the test pilots a particularly realistic simulation capability. In this context, it should once more be cross referenced, that the training aircraft Yak-130 that resulted from the Russian-Italian joint project and the Alenia Aermacchi M-346 could be equipped with programmable variable flying qualities (see Sects. 5.6.3, 5.9 and 11.2).

Fig. 5.96 Alenia Aermacchi M-346

5.9 Italy

Alenia Aermacchi M-346

After several years of development of a common training aircraft Yak-130, the partnership between Aermacchi and Yakovlev ceased in the year 2000 and both the companies followed their separate ways in the further development.

The Italian version was now called Alenia Aermacchi M-346 Master and performed its maiden flight in 2004. In the tandem arrangement, the student pilot (front) and the instructor sat one behind the other. The aerodynamics were designed for high maneuverability with extended wing leading edges and variable camber wing profile (see Fig. 5.96).

One of the main features of the M-346 is the four-channel (quadruplex) Fly-by-Wire flight control system from BAE Systems, which principally allows the in-flight simulation of other types of aircraft such as Eurofighter, Rafale, F/A-18 Super Hornet, MiG 29 for training purposes.

This reconfiguration is, however, controversial because some of the man-machine experts are of the opinion that the focus should not be on the type-specific training, rather on the handling of generic flying qualities, situational awareness and system or mission management.

References

1. Weingarten, N.C.: History of in-flight simulation at general dynamics. AIAA J. Aircr. **42**(2), 290–304 (2005)
2. Burns, K.R., et al.: The history of aerospace research at cornell aeronautical laboratory. AIAA 2007-0359. Jan 2007
3. Breuhaus, W.O.: The variable stability airplane. AAHS **36**(1), 29–55 (1991)
4. Hamel, P. (ed.): In-flight simulation for the 90's, international symposium, 1–3 July 1991. Braunschweig, DGLR Mitteilungen

91-05 (Proceedings, Inquiries at: DLR, Institute of Flight Systems, P.O. Box 3267, 38022 Braunschweig); (1991)
5. Pearcy, A.: Flying the Frontiers—NACA and NASA Experimental Aircraft. Airlife Publishing Ltd., England (1993)
6. Borchers, P.F., et al.: Flight Research at Ames—1940–1997. NASA/SP-1998-3300, (1998)
7. Aiken, E.W., Hindson, W.S., Lebacqz, J.V., Denery, D.G., Eshow, M.M.: Rotorcraft in-flight simulation research at NASA Ames research center: a review of the 80's and plans for the 90's. NASA TM 103873. Aug 1991
8. Bull, G., Kidd, E.A.: Air-to-air tracking with linear and non-linear yaw damping. WADC Technical Report 55-223. June 1955
9. Statler, I.C., Beilman, J.L.: Dynamic lateral stability through non-linear automatic control. CAL-Report TB-698-F-1 (AFTR-6712), (1952)
10. Kidd, E.A., Bull, G., Harper, R.P.: In-flight simulation—theory and application. AGARD report 369. April 1961
11. Armstrong, N.A., Holleman, E.C.: A review of in-flight simulation pertinent to piloted space vehicles, AGARD Report 403. July 1962
12. Schifferle, P.: Variable Stability Learjet In-Flight Simulator Capabilities. Calspan Corporation (2014)
13. DiCarlo, D.J., Kelly, J.R., Sommer R.W.: Flight investigation to determine the effect of longitudinal characteristics on low-speed instrument operation. NASA TN-4364. March 1968
14. Hilbert, K.B., Bouwer, G.: The design of a model-following control system for helicopters. AIAA Guidance and Control Conference. Seattle, WA (1984)
15. McDonald, E.H., Farris, J.A.: The X-20 flight control development, R&T Division AF systems command. WPAFB, SEG TDR 64/8. June 1964
16. Gould, D.G., Daw, D.F.: A Historical Account of the Research Activities of the Flight Research Laboratory of the National Research Council (1943–1984). NAE MISC 59. Feb 1985
17. Gordon, Y., Komissarov, D.: Soviet and Russian Testbed Aircraft, Hikoki Publications Ltd. (2011)

Author Biography

Peter G. Hamel was the Director of the Institute of Flight Mechanics/Flight Systems of the German Aerospace Center (DLR/DFVLR) (1971–2001). He received his Dipl.-Ing. and Dr.-Ing. degrees in Aerospace Engineering from the Technical University of Braunschweig in 1963 and 1968, and his SM degree from M.I.T. in 1965. From 1970 to 1971 he was Section Head of Aeronautical Systems at Messerschmitt-Bölkow-Blohm in Hamburg. Since 1995 he is an Honorary Professor at the Technical University of Braunschweig and a founding member of three collaborative research centers at the University. He was Chairman of the National Working Group on Helicopter Technology (AKH) (1986–1994) and the appraiser for the National Aviation Research Program (LuFo) until today. He was also the Manager of DLR's Rotorcraft Technology Research Program and the German Coordinator for the former AGARD Flight Mechanics/Vehicle Integration (FMP/FVP) Panel. He is a member of the German Society for Aeronautics and Astronautics (DGLR) and of the American Helicopter Society (AHS), and a Fellow of AIAA. He is the recipient of the AGARD 1993 Scientific Achievement Award, of AGARD/RTO von Kármán Medal 1998, of AHS Dr. A. von Klemin Award 2001, and of the prestigious DGLR Ludwig-Prandtl-Ring 2007.

6 Fly-by-Wire/Light Demonstrators

Peter G. Hamel

With contributions to Sect. 6.3 from Hermann Hofer, Bernd Krag, Robert Luckner, Lars Peter, Reinhard Reichel, Gunther Schänzer, Johannes Tersteegen, and Horst Wünnenberg

P.G. Hamel (✉)
Braunschweig, Germany
e-mail: p.hamel@t-online.de

P.G. Hamel (ed.), *In-Flight Simulators and Fly-by-Wire/Light Demonstrators*,
DOI 10.1007/978-3-319-53997-3_6

6.1 Introduction

6.1.1 Background

In the case of conventional aircraft, the pilot control commands from the control devices in the cockpit are transmitted directly to the aerodynamic control surfaces such as elevator, aileron, and rudder via mechanical connections consisting of cables and rods, or by means intermediary boosters. For very large aircraft with high control forces and large structural deformations, rotary shafts (torsion bars) are also employed. To this end, *Richard Vogt* had done pioneering work in the early nineteen forties during the development the large flying boat Blohm & Voss BV 222 and BV 238 (see Sect. 4.1, Figs. 4.1 and 4.2).

In the case of Fly-by-Wire (FBW) or Fly-by-Light (FBL) technology, the mechanical connections are replaced by electrical or electro-optical signal cables (copper or fiber optic cables). The pilot inputs from the control devices are tapped by position transducers (for example, potentiometers) and over the signal lines routed further to an onboard computer. In the onboard computer, these signals are processed together with measured flight condition data in such a way, that over electrohydraulic or electromechanical actuation systems (actuators) the aircraft control surfaces are so operated that the desired flight behavior is achieved. Fly-by-Wire/Light systems are therefore electronic or electro-optical control and stabilization systems, which facilitate automatic flight by a computer controlled augmentation system which also relieve the pilots in their tedious routine work.

Consequently, a Fly-by-Wire/Light flight control system decouples the pilot from the forces acting on the control surfaces and flaps. In the simplest case, an artificial feel can be created on the controls devices through the use of spring and additional masses, which conveys an impression to the pilot of a direct link with flight control surfaces. The artificial forces on the pilot control devices (for example, control column—*Stick*), however, can also be simulated using sensors and servomotors that provide pilots the so-called tactile information about the aircraft behavior (*Active Sticks*). Such "cues" can be of importance to pilots, especially in flight critical regimes such as stalled flight conditions, flight close to ground or control surface hardovers. Research at DLR Braunschweig related to active stick for rotorcraft is presented in Sect. 10.4.3.

The mechanical decoupling of controls devices (inceptors) and control surfaces necessitates, however, an emergency system in the case of energy supply or hydraulic failures. This is usually safeguarded by battery backup systems (for the next 10 min after a failure) and a so-called *Ram Air Turbine* (pull-out wind turbine) or APU (Auxiliary Power Unit). Fly-by-Wire signals are sensitive to electromagnetic interference, and therefore all data transmission cables are elaborately shielded. In the case of electro-magnetically insensitive optical fiber technology (Fly-by-Light), this problem is rarely faced. The two In-Flight Simulators VFW 614 ATTAS (Chap. 9) and EC 135 FHS (Chap. 10) represent successful examples of using Fly-by-Light technologies, whereby in the ATTAS system an optical fiber data bus was used for onboard computer communications as early as 1982 (see Sect. 9.1.5).

Susceptibility of air data sensors to errors is considered particularly as flight critical for computer aided flight control systems. Icing and moisture have thereby played a critical role and resulted in uncontrolled flight conditions with crashes (see Fig. 6.1, [1, 2]). Also, civilian aircraft have faced this difficulty several times (XL Airways/Air New Zealand A320 on November 27, 2008, Air France A330-200 on June 1, 2009, United Airlines B757-200 on October 20, 2013, Turkish Airlines B777-300 on January 16, 2014).

6.1.2 The Beginnings

The beginnings of the flight control system development and its operational utility go back to even before the First World War. Already in 1914, the Sperry Gyroscopic Stabilizer for the roll and pitch attitude was impressively demonstrated in the Glenn Curtis flying boat at an aviation event in Paris on the banks of the Seine [3]. In 1928 a Junkers W 33 was equipped with an "Automatic Pilot" of the German Boykow company to increase damping in the all three axes. The first automatic blind landings (Siemens Aircraft Factory) with He 111 and Ju 52 followed in the year 1940.

The then world's largest 8-engined Russian large "propaganda" aircraft Tupolev ANT-20 Maxim Gorky (First flight June 17, 1934) took an unusual course of development (see Fig. 6.2). Through an illuminated advertisement and public address system, this aircraft, equipped with telephone and radio station, working and passenger spaces, as well as photo laboratory and movie theater, was to inform the population on the merits of the communist party and to demonstrate the world the superiority of Russian research advancements and technologies. It was one of the first aircraft worldwide equipped with an autopilot using electromechanical actuators to operate the elevator and rudder for pitch and yaw control. Also, altitude hold and flight direction control was made possible. For the first time, both 120 V DC and AC current were applied for electrical power supply.

Just a day before the fatal crash on May 18, 1935, the only foreign pilot who participated in a ANT-20 flight was the famous French pilot *Antoine de Sainte-Exupery*. A six

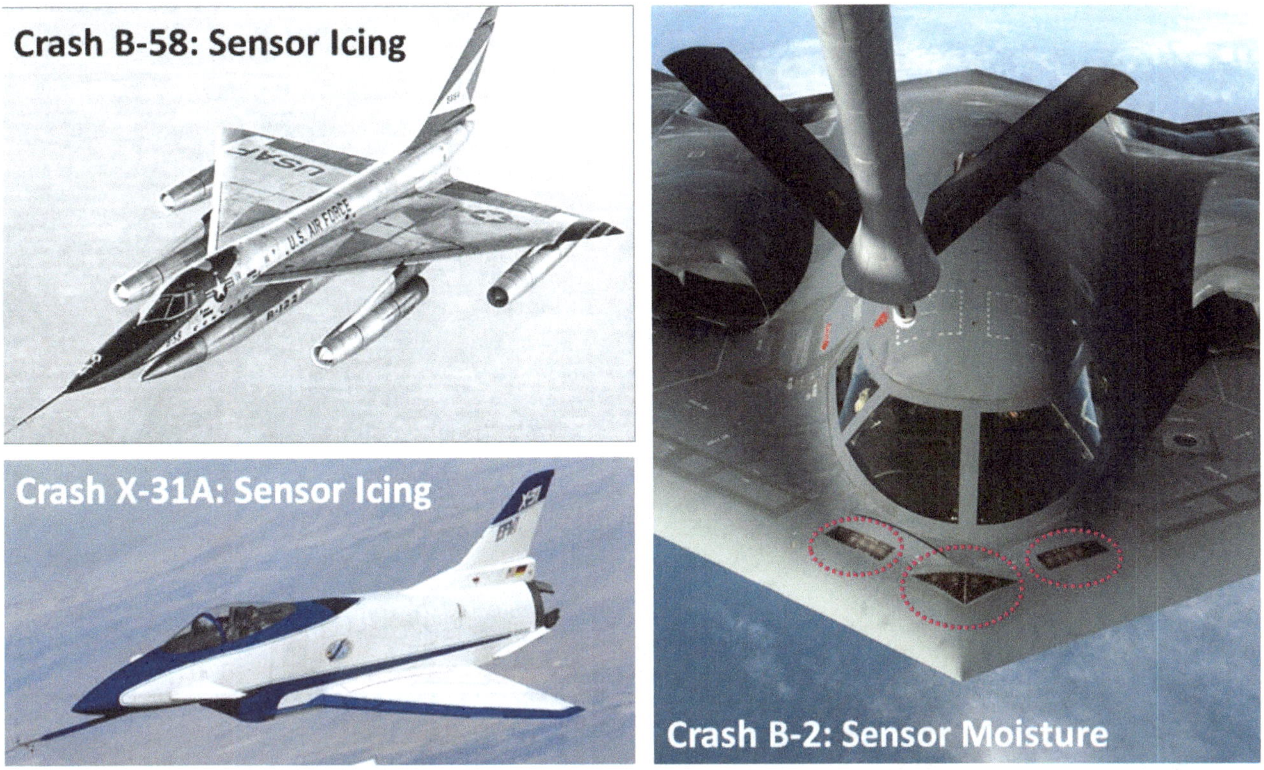

Fig. 6.1 Moisture and icing on air data sensors

engine replacement aircraft designated ANT-20bis was developed in Kazanwas and deployed by Aeroflot from 1940 onwards for passenger flights within Russia. Also, this aircraft crashed on December 14, 1942 when the pilot had left the cockpit and most probably a passenger had accidentally switched off the Autopilot. This is probably also one of the first cases where people had relied on the Fly-by-Wire automation and a human error had brought the aircraft in an uncontrolled flight condition. This issue of the human-machine interaction with increasing automation still represents a special area of common interest (*Loss of Control*, see Sect. 5.2.1.14).

Already in 1943, the Honeywell C-1 autopilot was used on a regular basis in the B-17E bomber series (see Fig. 6.3). For this purpose, the C-1 autopilot used analog electrical signals, which, based on sensor information, were transmitted to the control surfaces. In order to improve the precision bombing, the so-called *Automatic Flight Control Equipment* (AFCE) was introduced. It was a system in which the Norden bombsight controlled the aircraft during the final bomb run via a link with the C-1 autopilot. During a bomb run, the bombardier guided the plane very precisely with the aid of the AFCE without evasive maneuvering until the bombs were released. The autopilot provided a stable attitude and steady heading for the aircraft. Since the autopilot had no altitude sensor, adjustments had to be watchfully made to the aircraft's altitude by the pilot referring to the altimeter on his instrument panel. Only after the bomb run the pilot again would assume control of the plane.

The first integrated hydraulic servo actuator packages, in which an electric drive motor for the hydraulic pump with hydraulic accumulator, servo valve, and hydraulic actuator were integrated in a single unit, dates back to the development of autopilot controls of various German aircraft during Second World War such as Me-110, Do 17, He 111 and Ju 88. For directional (course) control the rudder was driven by the autopilot using compact parallel servomotors. Until the end of war, such integrated servo packages in all axes were tested successfully for course and attitude control. A corresponding Siemens unit was five years later re-tested at the USAF Wright Air Development Center [4].

Around the same time so-called Mistel (mistletoe) piggyback aircraft were developed in Germany to hinder the onslaught of Soviet troops shortly before the end of second World War, for example by destroying bridges on the river Oder. Control of the trailer combination was effected by the pilots of the Fw-190 or Me-109 fighter aircraft attached on the upper, who operated and monitored via cable and electric servomotors the control rods of the leaderless bomber aircraft Ju-88G, which operated the aerodynamic control surfaces (see Fig. 6.4). Shortly before the target the connecting supports and electrical control connections were clipped off.

Fig. 6.2 ANT-20 Maxim Gorkii (*Credit* Russian Museum, St. Petersburg)

Fig. 6.3 Boeing B-17E

Fig. 6.4 Captured piggyback Ju-88H and Fw-190A8 "Mistel"

From this moment on the bomber aircraft was steered autonomously towards the target.

A first computer-controlled flight control system, with attitude and position information from a so-called inertial platform, consisting of accelerometers and angular rate and position gyroscopes, was already available for the thrust vectoring graphite rudders of the A4 (V2) rocket during the Second World War. It employed one of the first analog computers, that modeled differential equations of the flight control laws and forwarded the control commands in the form of electrical signals via a cable to the integrated servo packets on the graphite rudders [5]. With this system, the rocket was controlled and artificially stabilized during the ascent into the atmosphere.

With the availability of analog computers and in combination with new control theoretic tools for stability analysis such as the Frequency Response method (*Hendrik W. Bode*) and the Root Locus method (*Walter R. Evans*) the technical premises for the systematic development of flight control systems were established, initially with limited controller authority [6, 7].

Limited control authority allowed the pilot to intervene directly in the flight control loop in the case of failures or emergencies. Since the end of the nineteen sixties, with increasing reliability of flight control components, such as onboard power supplies, sensors, analog and later digital computer blocks and finally electromechanical or electro-hydraulic actuation systems, the development of digital Fly-by-Wire flight control systems with full authority was initiated [8, 9]. The same is valid for the automatic code generation for the controller and monitoring software, which uses partly different software development platforms, and serves to increase system reliability or eliminate systematic errors. For this purpose, different software structures and algorithms are introduced (Dissimilar Software).

In this section, only those Fly-by-Wire aircraft will be presented, which served exclusively the development, demonstration, and deployment of FBW technologies with a technology maturity level TRL not exceeding 6 (*Technology Readiness Level* 6: "Functional and test prototype in operational range"). Nevertheless, historical first flights of FBW production aircraft will also be highlighted, those which achieved TRL 9 (TRL 9: "Qualified system with proof of successful deployment").

In the United States first yaw dampers were designed in the late 1940s for the Boeing XB-47 Stratojet swept-wing aircraft and the Northrop YB-49 Flying Wing with considerable success [10]. Since then the notions of stability augmentation or augmented aircraft are in colloquial usage. With the installation of the yaw damper in the Boeing B-47 family (first flight December 17, 1947), the Boeing Aircraft Company succeeded in the world's first integration of a production-line FBW component. From historical point of view, it may be remarked here that the decision to use a swept wing in the B-47 for the first time can be traced back to the visit of the Boeing chief aerodynamics engineer *George S. Schairer* to the German Aviation Research Institute (Luftfahrt-Forschungsanstalt—LFA) in Braunschweig-Völkenrode after the occupation by US troops in May 1945. He discovered extensive wind tunnel data which revealed the superiority of the swept wing over a straight wing [11, 12]. However, flight stability problems at low speeds and at high altitudes associated with the swept wing were then still unknown to Boeing. In this respect, it took another two years to get a grip on problems such as the so-called Dutch roll oscillations by implementing a production-line yaw damper. A related, technically excellent, publication is available that should be read as a classical reference by every aspiring aerospace systems engineer [13]. As is commonly known, the basic idea and principle of a yaw damper also originated in Germany (see Sect. 4.1). Another pioneering Fly-by-Wire research program with a Boeing B-47E will be discussed in Sect. 6.2.2.4.

In this context, the fundamental work of the Avro Canada company should not be overlooked. Already on March 28, 1958 Avro Canada conducted the first flight of the advanced supersonic combat aircraft CF-105 Arrow (see Fig. 6.5) equipped with a rudimentary FBW system with artificial force feedback, in which the pilot's input was detected by a series of pressure-sensitive transducers in the stick, and their signals were sent to an electronic control servo that operated the valves in the hydraulic system to move the various flight controls. The FBW system employed a multi-axis stability augmentation system. At that time there were no technical problems in the FBW flight control system, which could not be mastered [14]. The Canadian government directed the destruction of all prototypes and engineering information after the U.S. dropped their purchase agreement with Canada for the aircraft. The first flight and the construction of a total of 5 prototypes fell already in oblivion, when finally after more than another decade the British-Franco Concorde, the first civilian production-line aircraft with an analog FBW system, made its maiden flight on March 2, 1969. In addition to the Concorde, the US General Dynamics (now Lockheed Martin) F-16 (first flight January 20, 1974) and the European Airbus A320 (first flight February 22, 1987) went into series production for the first time worldwide. A mechanical backup control system, consisting of rudder control and the horizontal stabilizer, was still available with the A320.

In the military sector, the whole application potential of digital Fly-by-Wire flight control technology was later introduced in the highly unstable "stealth aircraft" Lockheed F-117 Nighthawk (first flight on June 18, 1981) and Northrop B-2 Spirit (first flight on July 17, 1989). These

Fig. 6.5 Avro Canada CF-105 Arrow

aircraft have been optimized with regard to the lowest possible identification by microwave detection systems (RADAR) and for this, in turn, had to accept inadequate aerodynamics and flying qualities. These shortcomings could be compensated through the flexible use of Fly-by-Wire technologies. Such control theoretic "Remedial Measures" on high-performance aircraft can be summarized under the term *Control Configured Vehicles* (CCV) [15].

This CCV approach, namely the integration of different disciplines such as aerodynamics, aeroelasticity, structural dynamics, flight mechanics, and control technology as well as propulsion technology is also referred to as Multidisciplinary Design Optimization [16]. The further development, maintenance, and utilization of such highly computerized aircraft design tools is one of the greatest challenges in the aeronautical research and design community. FBW/L technology enabled the practical implementation of CCV.

Applications of full authority Fly-by-Wire flight control systems to military and civil rotorcraft must be carried out with special caution, as unexpected couplings can occur between the high bandwidth dynamics of the electronic flight controller and hidden structural dynamics of the flexible rotor-body system. The importance of accurate modeling of the high-frequency characteristics of the rotor and fuselage dynamics and their coupling effects as well as of sensor and actuator dynamics cannot be overstated.

Although military Fly-by-Wire helicopter are in series production and in operational usage for quite some time, special challenges are faced in their civil certification. This can be readily seen in the context of the formerly planned certification of the first civil Fly-by-Wire helicopter which was built by the Russian JSC Kazan Helicopter Plant as a variant of the light helicopter "Ansat". The civil certification of the four-channel digital FBW control system (KSU-A) encountered "an unexpected obstacle", as no commercial FBW helicopter was hitherto certified anywhere in the world. No standard and established requirements existed for certification of such a full-authority Fly-by-Wire helicopter. In order not to depend on the terms of "Ansat's" certification with Fly-by-Wire controls, the decision was made to retrofit the helicopter for the world market with a traditional hydro-mechanical flight control system. This Ansat configuration was certified in 2013 by the Aviation Register of the Interstate Aviation Committee (IAC-AR), whereas the Fly-by-Wire version "Ansat-U" has been produced only as a primary military training and research helicopter for Russian flight academies.

Also for these reasons the US Federal Aviation Administration (FAA) had proposed special conditions for the certification of the first civil Fly-by-Wire series-produced helicopter Bell (BHTI) Model 525 Relentless in the United States:

> We propose special conditions for the BHTI Model 525 helicopter. This helicopter will have a novel or unusual design feature associated with Fly-by-Wire flight control system (FBW FCS) functions that affect the structural integrity of the rotorcraft. The applicable airworthiness regulations do not contain adequate or appropriate safety standards for this design feature. These proposed special conditions contain the additional safety standards that the Administrator considers necessary to establish a level of safety equivalent to that established by the existing airworthiness standards.

The FAA issued in 2016 a set of criteria that must be applied to demonstrate compliance with these special conditions for rotorcraft equipped with full-authority Fly-by-Wire systems, autopilots, stability augmentation systems, load alleviation systems, flutter control systems, fuel management systems, and other systems that either directly or as a result of failure or malfunction affect structural performance [17].

In the following the world's most important Fly-by-Wire/Light technology demonstrators are reviewed, those which have reached a Technology Readiness Level (TRL) of about 6 ("*functional and test prototype in operational range*" (see also Sect. 6.1.2). From Sect. 6.2.2 it becomes obvious that due to the abundance and variety of research and test programs in the United States, the period from 1960 to 1990 can be considered as the golden age of Fly-by-Wire research [18].

The most important flying VTOL (*Vertical Take-Off and Landing*) test rigs with electronic flight control systems developed worldwide, will be discussed in the following Sect. 6.1.3. Section 6.2 covers the international Fly-by-Wire/Light technology demonstrator programs. Finally, Sect. 6.3 deals with German Fly-by-Wire research in more detail, including the world's first demonstration flight with a digital FBW flight control system on June 12, 1967 (see Sect. 6.3.1).

6.1.3 Flying Bedsteads

6.1.3.1 Introduction

Beginning of the 1950s and 1960s, an unusual family of hovering rigs was developed for preliminary studies pertaining to control and stability of this special class of VTOL vehicles. Since the aerodynamic control forces are hardly exerted in hover condition, the VTOL vehicles have partially pivoting nozzles and reaction nozzles on the exterior of the vehicle in order to apply control moments about all axes (*Reaction Control*). The reaction control was usually generated from hydrogen peroxide or by tapping compressed air from a compressor of a jet engine. Due to their exotic appearance, such flying racks were referred to as "flying bedsteads" by the British public. In the 1950s and 1960s, the reaction control nozzles were driven by electrical signals, whereby the flight control laws were reproduced using analog electrical circuits. Thus, it was a precursor of Fly-by-Wire flight control systems, which in a broad sense also facilitate a limited in-flight simulation potential.

Also in the Federal Republic of Germany, technological uncharted territories in all related disciplines were explored with the development the three VTOL aircraft Do 31, VJ 101C, and VAK 191B. They deliberately served the goal of bringing up the German aviation industry to the world standards two decades after the World War II [19]. For example, under the leadership of *Waldemar Möller*, the Bodensee-Gerätewerk (BGT, today Diehl Aerospace) developed electronic flight control systems and flight control laws for all the three German VTOL programs (see Sects. 6.1.3.6 through 6.1.3.8). For attitude control in the hover mode, which was manually no longer controllable for the pilots, high demands on the reliability of the systems were postulated. They led to a doubling, sometimes even tripling of the most important components such as sensors, signal processors and actuation systems. These redundancy concepts constituted an important prerequisite for the later development of the Tornado Fly-by-Wire system components [19]. A historical review and evaluation of the German VTOL development programs were presented at a DGLR Symposium on March 31, 2000 in the aircraft base (Flugwerft) Schleißheim of the Deutsches Museum. Important contemporary witnesses such as *Rolf Riccius*, *Helmut Schubert*, *Gero Madelung*, *Gerhard Kissel*, and *Rolf Staufenbiel* as well as test pilots *Niels Meister* and *Dieter Thomas* provided an insight into these pioneering days [20].

Later, autonomous flying robots were introduced internationally, which were investigated and tested since 2000 in many variants in the preparation of lunar missions. Two representative examples are the autonomous Lunar Landing Vehicles ("landing robots") *Mighty Eagle* and *Morpheus* of

Fig. 6.6 Mighty Eagle in flight

Fig. 6.7 Morpheus test flight

NASA. Figure 6.6 shows *Mighty Eagle* during a test flight on November 26, 2013, while *Morpheus* (see Fig. 6.7) with a payload of 500 kg absolved a free flight test on August 9, 2012, after a few failures. This official first free flight occurred with Morpheus lifting off under full power, but after just a few seconds the vehicle tumbled over and crashed to the ground. The vehicle was virtually destroyed. The Inertial Measurement Unit (IMU), which supplies data to the flight computer, had failed.

It took just eight months after the crash till the debut of the second Morpheus "Bravo" lander. Having learned many lessons from the previous vehicle's testing, this lander looked the same but 70 upgrades had been made to both the vehicle and the ground systems to improve reliability and operability.

The Bravo vehicle successfully attempted its first free flight on December 10, 2013: "*It rose spectacularly like a ball balancing on the end of a straw. It flew almost flawlessly, rising to about 50 feet in altitude, hovering in place before moving to a landing spot approximately 23 feet from the takeoff spot, landing after about 50 sec of powered flight. Less than a week later, Morpheus flew again, going higher, faster, and further in an 82 sec free flight*" [21].

NASA's Morpheus lander proved that it was quite capable of successfully navigating a hazardous field after finding a safe landing spot during a night-time. This night flight on May 28, 2014, was the 14th and last free flight for Morpheus free-flight test conducted at Kennedy Space Center.

6.1.3.2 Rolls Royce TMR "Flying Bedstead" (1953–1957)

The Rolls-Royce Thrust Measuring Rig (TMR) was an experimental VTOL aircraft from the 1950s to enable the development of vertical takeoff vehicles in England, with a thrust-to-weight ratio of about 1.25:1. The engines themselves were at that time objects of research. The first flight took place on July 3, 1953. The experiments resulted in the

Fig. 6.8 Rolls-Royce thrust measuring rig TMR

development of the Rolls-Royce RB108 engines for the VTOL demonstrator aircraft Short SC.1 (see Sect. 5.4.1).

The TMR consisted of a tubular steel rig structure with two Rolls-Royce Nene 101 jet engines mounted vertically opposite to each other, whereby about 9% of the thrust was directed as bleed air through pipes to the reaction nozzles for platform stabilization (see Fig. 6.8). The first test rig (XA314) flew for the first time, still tethered, on July 6, 1953. The first free flight followed on August 3, 1954. This was followed by another 240 tethered and free flights until December 1954. After the transfer to the Royal Aircraft Establishment (RAE), this test vehicle crashed on September 16, 1957. The second test rig (XA426) first flew on November 12, 1956, but was also lost on November 28, 1957 during a tethered flight after a collision with a fatal outcome for the test pilots. A museum specimen, composed of residues from the two rigs, is all that is left over and located in the Science Museum in South Kensington in London (see Fig. 6.9).

6.1.3.3 Bell Lunar Landing Research and Training Vehicles (1964–1972)

Lunar Landing Research Vehicle (LLRV)

The British idea of hovering rigs was adopted by NASA in the nineteen sixties within the Apollo program for the development and testing of the Lunar Landing Module (*Apollo Lunar Module*). With this *Lunar Landing Research Vehicle* (LLRV), the required stability and controllability was determined that assured a precise approach and safe landing of the lunar module in the weak gravitational field of the moon. The first flight of LLRV took place on October 30, 1964, and the last one on November 30, 1966. In this time period, the two LLRV successfully performed a total of 204 flights without any mishap (see Fig. 6.10) [22].

Lunar Landing Training Vehicle (LLTV)

The Lunar Landing Training Vehicle (LLTV) was derived from the LLRV and was to provide the Apollo astronauts with enough training opportunities before landing on the Moon with the so-called Lunar Excursion Module (LEM). Especially the gravitational force of the Moon, which is only a sixth of the Earth's gravitational force, could be well simulated with the training device. The electronic control system for the LLTV was developed by Bell Aerosystem for NASA. It had two identical duplex channels, which were constructed from analog computer blocks using transistors. In 1967 the digital technology was not yet mature enough. A total of three training units were built (see Fig. 6.11).

The LLTV was very unstable, particularly at large sideslip angles. On May 6, 1968, the astronaut *Neil Armstrong* could at the last moment eject himself during an uncontrollable training flight. In spite of that, he had no concerns to sit again in a replacement vehicle shortly afterward. *Armstrong* was convinced that there was no better way to prepare for the

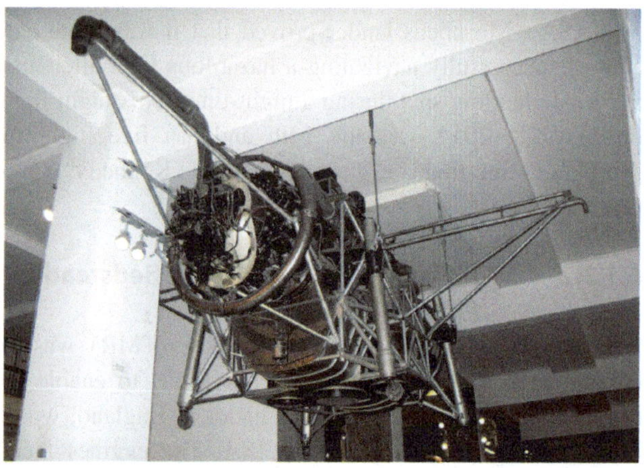

Fig. 6.9 Rolls-Royce thrust measuring rig TMR in Science Museum, London

Fig. 6.10 NASA Lunar Landing Research Vehicle (LLRV) with cockpit casing

Fig. 6.11 NASA Lunar Landing Training Vehicle (LLTV) without cockpit casing

forthcoming Lunar landing challenge. Because of his unwavering calm, he was also called "Ice Commander". The indispensable role which was played by the Lunar training devices is highlighted at best by the following statement of the astronaut *Bill Anders*: "*In my view, the LLTV was a much undersung hero of the Apollo programs.*"

6.1.3.4 FRI LII Turbolet (1956–1959)

The hovering rack *Turbolet* (also called as Toorbolyot—*Turbo-flyer*) was developed by the Russian Flight Research Institute LII (FRI—*Gromov* Institute) and based on the model of Rolls Royce Flying Bedstead TMR. As in the case of the TMR, the Russian hovering rig had a four-legged structure with a jet engine installed in the middle, pointing vertically downwards, and four reaction nozzles on side arms. The pilot's seat was cased. The first flight took place in 1956 with *Yuri Garnayev* (see Fig. 6.12).

The utility of this hovering rig was closely connected with the development of the Kolesov RD-36 lift engines for short and vertical takeoff and landing purposes. Special Mig-21, MiG-23, and Su-15 variants were developed with lift engines for demonstrating a STOL (*Short Take-Off and Landing*) capability. The first Russian vertical takeoff aircraft Yak-36 emerged ten years later in 1967. The Turbolet

Fig. 6.12 Turbolet in Flight

hovering rig is now on display in the Russian Monino Aviation Museum (see Fig. 6.13).

6.1.3.5 NAL VTOL Flying Test Bed (FTB)

The first free flight of the VTOL technology demonstrator *Flying Test Bed* (FTB) of the Japanese National Aeronautical Laboratory (NAL) took place on December 15, 1970 (see Fig. 6.14). It was built by Fuji Heavy Industries. Two vertically mounted JR100F lift engines manufactured by the Ishikawajima-Harima company were used for the propulsion system. The experimental data were conceived for a vertical takeoff project on the basis of the first jet aircraft Fuji T-1 built in Japan after the Second World War.

The FTB was inherently unstable. Stability could only be ensured manually or automatically about the pitch and roll axes through reaction nozzles on the side arms. Yaw control and stabilization about the vertical axis was ensured with the aid of hydraulically actuated tilt nozzles in the lift engines (see Fig. 6.15).

Fig. 6.13 Turbolet in Monino-Museum

Fig. 6.14 NAL VTOL Flying Test Bed (side view)

Fig. 6.15 NAL VTOL flying test bed (front view)

6.1.3.6 Hovering Rig for EWR VJ 101C (1962–1963)

It was pointed out in Sect. 4.2 that already during the early 1960s the analog Fly-by-Wire technologies and in-flight simulation played an important role in the development of hovering rigs in the preparations for the German vertical takeoff programs VJ 101, Do 31 and VAK 191. The attention was focused on the use of hovering rigs, which were usually driven with the original engines of the actual prototypes, for the optimization of the controllability and stability of the projected VTOL vehicles during hovering.

The hovering rig, shown in the upper part of Fig. 6.16, emerged as a simulation device for the VJ 101 C during its development. The uncased steel tube body of the hovering

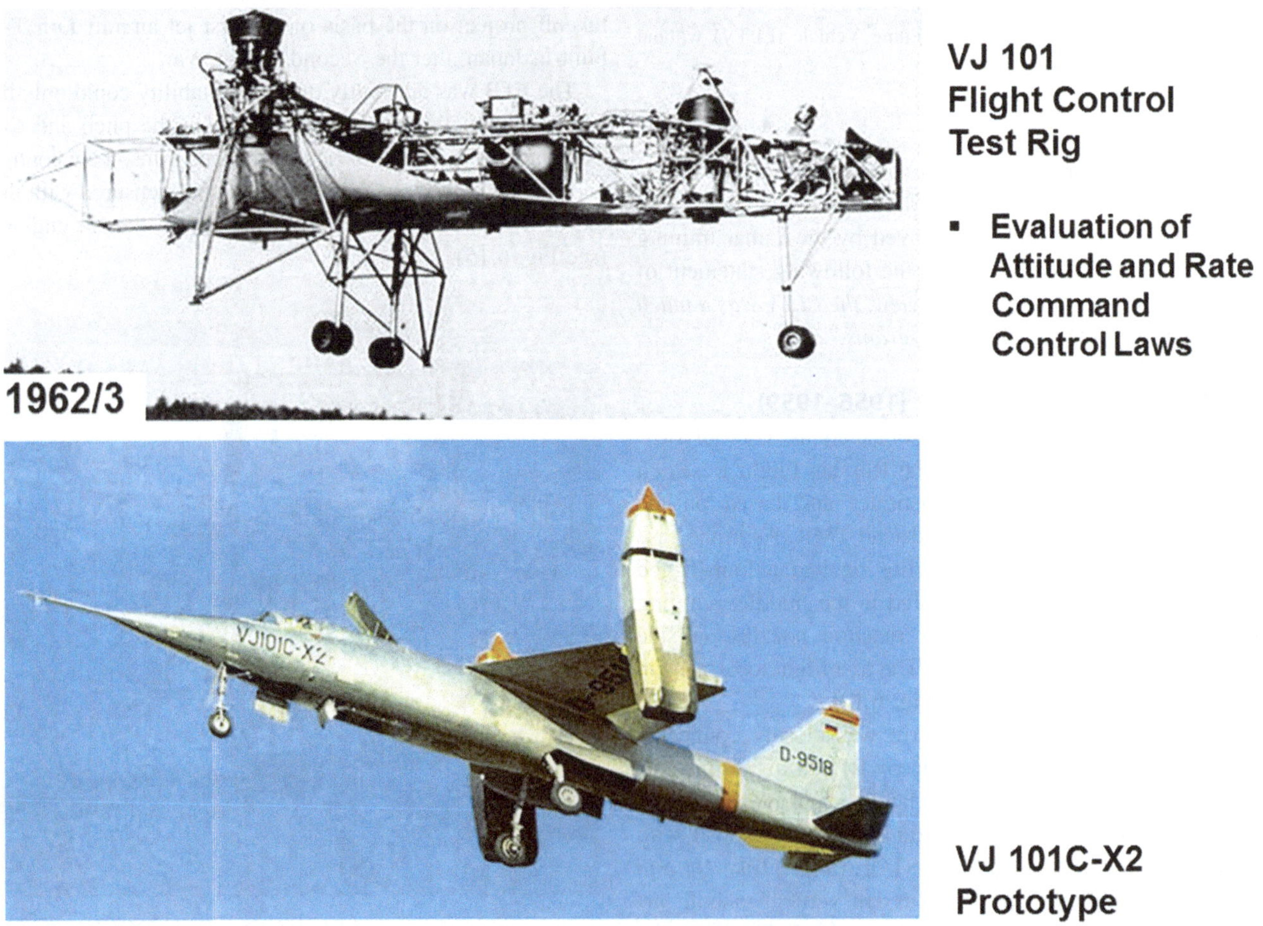

Fig. 6.16 VJ 101 hovering rig

rig with its lateral arms supports three Rolls-Royce lift engines RB 108, at same distances from the center of gravity as in the case of the planned aircraft. Two of them are mounted pivotable on the boom ends and one in the hull in front of the pilot's seat.

Since 1961, vertical takeoff and landings, as well as hovering and control in all three axes including an altitude hold mode, were tested with the hovering rigs. The control commands were carried out by thrust modulation and yawing motions around the vertical axis by swiveling the engines at the wing tips. The maneuvers could be executed both manually and with automatic control inputs (autopilot). The hovering rig also provided valuable services in the introductory training and retraining of test pilots for the VJ 101 C. In March 1962 the hovering rig flew freely for the first time without restraints [20, 23].

6.1.3.7 Test Stand for Dornier Do 31 (1964–1965)

The starting point in the development of the first German vertical takeoff transport aircraft Do 31 was a controller test stand (RVG—Reglerversuchsgestell) for drawing up the specification for the flight controllers and the control kinematics. This tubular lattice rig with four lift engines was initially tested on a column on which it was mounted on gimbals with three degrees of freedom. With a three-axis autopilot system, developed by Bodensee-Gerätewerk (BGT, today Diehl Aerospace), the control properties were optimized taking into account the interaction between pilot and control system during VTOL operations (see Fig. 6.17). With that, the test pilot *Karl Kössler* could perform the first free flight on April 21, 1964. The last flight of the RVG for system testing took place on June 4, 1965 [24].

6.1.3.8 VAK-191 Hovering Rig SG 1262 (1966–1969)

The hovering rig SG 1262 was developed to enable the field testing of the hovering control system of the VTOL fighter aircraft VFW-Fokker VAK 191B. A very good overview of the VAK 191B program is found in Ref. [23].

The initial flights in hover were carried out as tethered flights. The pre-testing of the analog-electrical flight control system, fine-tuning of the flight control laws, and the system architecture related to reliability requirements were carried out initially on the hovering rig. The risks during the development and associated costs could be reduced. The first

Fig. 6.17 Do 31 controller test stand RVG

Fig. 6.18 VAK 191 hovering rig

free flight of the SG 1262 took place in the year 1966 and in 1968 an impressive demonstration at the International Aviation Exhibition ILA in Hannover went off smoothly (see Fig. 6.18).

The utility of the hovering rig was limited not only to investigations of controllability and stability of VTOL aircraft in hover. It included later on also the pre-testing of Fly-by-Wire components for the MRCA Tornado project.

6.2 International Demonstrators

6.2.1 England

6.2.1.1 Vickers V.663 Tay-Viscount (1950–1959)

The twin-engine Vickers-Armstrong Tay-Viscount (VX217) was the worldwide first civil aircraft that operated a flight control system with an electrical signal transmission (see Fig. 6.19). First flight of the Tay-Viscount was on March 15, 1950, and in the same year demonstration flights at the International Airshow in Farnborough (SBAC Show) became a world sensation. For the test aircraft, Boulton Paul (a company that was later absorbed into the Dowty Group) developed in 1952 a three-axis analog-electrical flight control system with duplex redundancy, that is, the electrical signals from pilot station right up to the aerodynamic control surfaces were duplicated in each of the three control axes (elevator, aileron, and rudder) for flight safety reasons. The second control station in the cockpit remained for safety reasons mechanically connected to the control surfaces via cables and control rods. The first flight with this system took place in 1956 and constituted the world's first flight demonstration of a three-channel Fly-by-Wire system. In the course of two years, 20 test flights were performed with 21 h of flight time, half of them in Fly-by-Wire mode. The outcome of this program turned out later to be significantly

Fig. 6.19 Vickers Tay Viscount (*Credit* Mike Dowsing)

beneficial for the analog flight control system of the first production version of the supersonic airliner Concorde.

6.2.1.2 Avro 707C (1956–1966)

Following the Tay-Viscount experiments, with the development of the two-seater Avro 707C (WZ744) the British Royal Aircraft Establishment (RAE) began in September 1956 the first systematic flight investigations for electrical signal transmission at higher airspeeds (see Fig. 6.20). It involved an analog simplex flight control system from Fairey with artificial stick force feel (*Artificial Feel*), wherein the safety pilot could revert to the mechanical flight control in the case of system failures (*mechanical backup*). The Fly-by-Wire flight tests lasted until September 1966 and comprised of 200 flight hours. This test demonstrator made important contributions to the development of the Avro Vulcan bomber and also to the Anglo-French Concorde program [25, 26].

6.2.1.3 Short S.C.1 (1957–1971)

See Sect. 5.4.1.

6.2.1.4 Hawker Hunter T. 12 (1972–1982)

The two-seater Hawker Hunter T.12 (XE531) is considered as a logical continuation of the research in the field of Fly-by-Wire flight control systems that began at RAE in the 1960s with an Avro 707C (see Fig. 6.21). Actually, this aircraft was converted to test an innovative HUD (Head-Up Display) system for the British supersonic aircraft project TSR.2. After project discontinuation, it served the RAE as an experimental vehicle for an analog Fly-by-Wire system developed, once again, by Boulton Paul. In this case, it involved a so-called three-axes-quadruplex architecture, that is, the signal transmission in each of the three control axes was increased fourfold to meet the required system reliability. The cockpit of the test pilots was equipped on the right side with a sidearm inceptor (*Sidestick*). On the

Fig. 6.21 Hawker Hunter T.12 (*Credit* www.flickr.com-Irish 521)

left-hand side in the cockpit, the safety pilot had access to mechanical backup flight control system at any time [25, 27]. XE531 crashed on takeoff from Farnborough on March 17, 1982, after the engine's 11th stage compressor disc disintegrated.

6.2.1.5 BA Jaguar ACT (1981–1984)

To demonstrate the proof of concept of the so-called active flight control systems (*Active Control Technology*—ACT), the British Aerospace (BA) Jaguar ACT (XX765) was equipped with an innovative digital Fly-by-Wire system (FBW) with a quadruplex system architecture, that is, it was steered for the first time worldwide through four computer systems, operating independently of each other, and without mechanical backup (first flight on October 20, 1981). Through enlargement of the front wing section and a different weight distribution, the aircraft was later destabilized in order to achieve an increased maneuvering agility [28]. Good flying qualities could be ensured through artificial stabilization (see Fig. 6.22). The FBW Jaguar program was concluded in September 1984 after 96 flights. The

Fig. 6.20 Avro 707C

Fig. 6.22 British-Aerospace Jaguar ACT

Fig. 6.23 British-Aerospace EAP (*Credit* Fergal Goodman)

experience gained during this program proved later advantageous to the Eurofighter program.

6.2.1.6 BA EAP (1983–1989)

The EAP (*Experimental Aircraft Program*) of British Aerospace (BA) was a many faceted technology demonstrator. After the project initiation in May 1983, the first flight of this test aircraft (ZF534) was carried out on August 8, 1986 with RB.199 Mk 104D engines salvaged from the multinational Tornado program. During the very first flight, the aircraft climbed to an altitude of 9150 m and accelerated to Mach 1.1. Until the end of 1989, the EAP technology demonstrator completed 209 flights with a total 155 flight hours and in the course reached speeds up to Mach 1.6 (see Fig. 6.23).

The research program of the EAP was oriented towards the development of the future *European Fighter Aircraft*—EFA (Eurofighter) and eventually served the purpose of timely providing the proven technologies with a focus on aerodynamics, construction procedures, structures and flight control for the future Eurofighter Typhoon. This included the development of a digital-electronic, four-channel (quadruplex) Fly-by-Wire flight control system, and the optimization of the aerodynamic configuration with a high degree of instability.

6.2.1.7 BAC One-Eleven FBL (1984–1994)

Upon approval of government funding in 1980, a prototype One-Eleven (R/N G-ASYD) was allocated as an Active Control Technology (ACT) demonstrator aircraft and BAe (today: BAE Systems) thus began its own fundamental industrial research into investigating relaxed stability, maneuver load control, and sophisticated gust alleviation systems. Hitherto it had largely relied on the Royal Aircraft Establishment (RAE) at Farnborough [29].

In the mid-nineteen-eighties, an artificial stability system was tested on the One-Eleven, which consisted essentially of a series actuator tied into the existing elevator control circuit, the digital control augmentation system (CAS) built by BAe, and a water ballast system for moving the center-of-gravity. After nine flights the program was concluded and had demonstrated that reductions of 25% in the tail area were possible.

Maneuver load alleviation was achieved by deflecting the One-Eleven's outboard ailerons during turns and pullouts so as to reduce wingtip lift, shifting the aerodynamic loads inboard. Perhaps the most significant piece of work being carried out on G-ASYD is in the alleviation of gust loads.

On February 13, 1993, the first flight of the BAC One-Eleven Aircraft (R/N G-ASYD) took place equipped with a Lucas Fly-by-Light (FBL) spoiler actuation system (Fig. 6.24). The system, which gives lift dump, airbrake and roll assist control to the aircraft, was tested over the aircraft's normal operating envelope, together with failure and emergency conditions. The flight of over 2 h duration covered 27 flight test points, whereby no discrepancies were noted.

The system controlled the inboard spoiler panels of the aircraft which were fitted with A320 modified spoiler jacks. It was developed as a direct replacement for the mechanically signaled system that is normally installed. The system included a pilot operating panel, a central computer, two fiber optic harnesses, two digital smart actuator controllers and two hydromechanical servo actuators. Also included were two rotary position sensors to pick up the pilot control inputs.

On March 1, 1993, following CAA approval, the BAC One-Eleven began in-service operations as a liaison aircraft for BAe with an active FBL system. By the end of May, the aircraft had flown approximately 200 h (175 flights), in daily service between Filton and Liverpool, UK, and Toulouse, France. The flight trials ended when the aircraft was

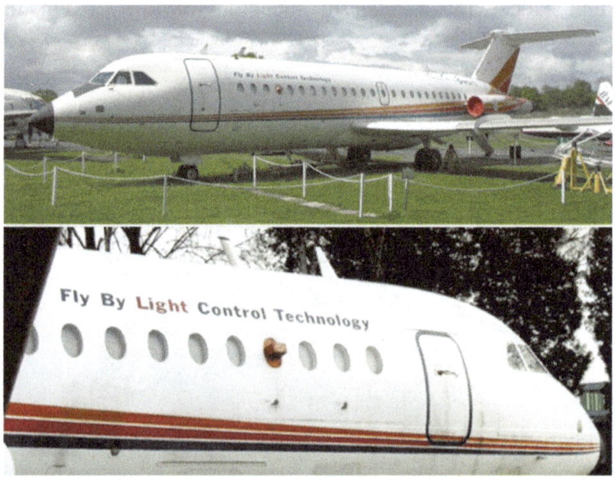

Fig. 6.24 BAC One-Eleven (*Credit* Ian Haskell and Andrew Simpson)

withdrawn from service in October 1993 with 470 flight hours logged on the so-called Smart FBL system [30].

Both, the Lucas BAC One-Eleven FBL spoiler control system and the fully integrated Liebherr Eurocopter EC 135 FHS FBL flight control system (see Chap. 10) are the most significant Fly-by-Light development contributions in the secondary and primary flight control systems respectively.

6.2.2 USA

6.2.2.1 Northrop YB-49 (1947–1949)

One of the most unconventional postwar aircraft was the flying wing aircraft Northrop YB-49 (first flight October 21, 1947, see Fig. 6.25). Although it seemed attractive to fully accommodate the engines, flight control components and payload in the wings, thereby promising significantly enhanced performance in terms of range and payload, the automatic flight control system had to be developed first to get a grip on the unstable yaw oscillations about the vertical axis.

Since, in contrast to a conventional aircraft with a tail plane, the rudder was not available in the present case, double flaps (*Clamshell Flaps*) on the wings were used for yaw damping. They were synchronously deflected up and down respectively on one side to generate additional drag at a wingtip and thereby enabling a yaw moment. With alternative combinations of these double flaps, rolling moments could be generated or even increases in drag to achieve speed control. Even then the general flying qualities were evaluated to be not adequate and the program was terminated. It took another 40 years until the above concept was once again adopted in the Northrop B-2 flying wing program, and from this time on with advanced digital Fly-by-Wire technology the program was steered to a successful production.

Fig. 6.25 Northrop YB-49

Fig. 6.26 North American F-107A (AF 55-5120)

6.2.2.2 North American F-107A (1958–1959)

A complex spoiler-slot system above and below the wing surfaces served as an aileron replacement for roll control of a North American F-107A (AF 55-5120). The aircraft already had one of the first Fly-by-Wire systems in the longitudinal control which was termed *Augmented Longitudinal Control System* (ALCS). With the aid of an air data system, the pilot could generate pitch rate commands by means of the ALCS. About 40 flight experiments were performed with the F-107A at the NASA and it was deployed to refine amongst others the control stick mounted on the side (*Sidestick*) for the planned X-15 hypersonic program (see Fig. 6.26).

6.2.2.3 Boeing NB-52E LAMS (1966–1969)

During a low-altitude test flight on January 10, 1964, a Boeing B-52H (AF 61-023) lost its fin and rudder due to extreme severe turbulence over a mountainous area (see Fig. 6.27). Six hours later after the fin-rudder-loss, the B-52H could be landed. These and similar incidents clearly showed that sensitive structures of large aircraft react under difficult atmospheric conditions not only with fatigue and increased susceptibility to vibrations, but may also lead to a complete disintegration of structural components.

Against this background, first control-theory based technical measures were undertaken towards the end of the nineteen-sixties by the US Air Force to actively suppress dynamic gust load on the structure and using Fly-by-Wire technologies. Thus, within the framework of the LAMS Research Program (*Load Alleviation and Mode Stabilization*) a Boeing B-52E (AF 56-0632) was renamed as NB-52E and equipped with electrohydraulic actuation systems for the existing and additional wing flaps, which were driven via analog computers (see Fig. 6.28). For this purpose accelerometers and angular rate gyros were mounted at several locations along the fuselage length, whose measurement signals during flight responded to turbulent air. These signals were processed in the analog computer and

Fig. 6.27 Boeing B-52H (AF 61-023) on January 10, 1964

Fig. 6.28 Boeing NB-52E LAMS (AF 56-0632)

provided the control commands to the actuation system with correct phase angles. In this way, structural loads and vibrations were attenuated. It could be demonstrated that the lifespan of aircraft structures can be significantly increased through such active measures.

6.2.2.4 Boeing JB-47E-FBW (1966–1969)

In the present case, the goal was the demonstration of an analog Fly-by-Wire system in the pitch axis of a Boeing B-47E (AF 0-32280) and its advantages over the conventional mechanical control system (Ref. [8], see Fig. 6.29). The first flight took place on December 14, 1967. After 45 flying successful hours in direct FBW mode with a

Fig. 6.29 Boeing JB-47E FBW demonstrator (*Credit* Gavin Jenney)

conventional control column, in a second trial phase, the conventional steering column was replaced through a side-mounted control inceptor (*Sidestick*) with adjustable output gradients. Besides the pilot input signals in the pitch and roll axes, the pitch rate and acceleration signals were blended additionally, which are also known as C* (*C-Star*)—control laws in the international aviation community (see also Sect. 6.2.2.7). The C* criterion characterizes a pilot input command, which is essentially composed of a demanded pitch rate (attitude change) and a vertical load factor (vertical acceleration at the pilot's seat). In simple terms, at low speeds the pitch rate prevails, whereas at higher airspeeds load factors are increasingly demanded. In many variations this control command principle has become accepted worldwide in civil aviation.

In further 34 flight hours, it was demonstrated that the flying qualities of the B-47E compared to the original version could be significantly improved with such a FBW configuration. Finally, in a third experimental phase, the single-channel elevator actuator (Simplex) was replaced by an electrohydraulic actuator with quadruple redundancy (quadruplex) and self-monitoring. In the concluding 12 test flights with a total of 18 flight hours, three different types of control channel errors were introduced. They were correctly detected and compensated by the actuator. After the last flight on November 21, 1969, the retired JB-47E FBW was placed in the US Air Force Museum at Wright Patterson Air Force Base until 2003. It was later cut up for scrap.

This proven and redundant analog electrohydraulic Fly-by-Wire flight control system in the roll and pitch axes and with a mechanical backup control for the yaw axis as well as the implementation of a sidestick corresponds basically to the digital electrohydraulic Fly-by-Wire flight control system of the Airbus A320. This flight control system was to revolutionize civil transport aircraft operations about 20 years later with the first flight on February 22, 1987. In contrast, Boeing furnished the B737, B757, and B767 further on with conventional flight control system and ventured 7 years later the jump into the civilian Fly-by-Wire world with the B777 (first flight June 12, 1994). This could raise an interesting aspect, namely did Boeing want the Europeans to face the expected struggle with the international certification bureaucracy pertaining to a new, flight safety critical technology?

6.2.2.5 Boeing-Vertol BV 347 (1969–1974)

As part of a cooperation program between Boeing-Vertol and the US Army, a CH-47A Chinook (US Army 65-07992) was converted to a technology demonstrator under the designation BV 347. The key elements of the conversion included a fuselage extension by 110 inches, 4-bladed tandem rotors, retractable landing gear, electrohydraulically adjustable auxiliary surfaces (tilt wings) with trailing edge flaps, and an analog-electrohydraulic Fly-by-Wire system (see Fig. 6.30). The slope of the tilt wings changed automatically with increasing load factors in order to relieve the rotors. They were aligned vertically while hovering and horizontally in forward flight. The first flight of the BV 347 took place on May 27, 1970 at the Vertol Flight Test Facility in Pennsylvania.

Subsequently, an extensive flight test program was carried out to determine the influence of the tilt-wings on the flying qualities, flight performance, and the vibration behavior. In preparation for the Boeing US Army XCH-62 Heavy Lift Helicopter (HLH) program, the BV 347 now without tilt-wings served as HLH demonstrator during 1971–1973 (see Fig. 6.31). The so-called *Dual-Fail Operational Triplex* Fly-by-Wire system with full control authority, developed for HLH, was successfully tested on the BV 347 with different, reconfigurable and mission-oriented flight control laws. On the basis of this upgrade, the BV 347 was declared as the first Fly-by-Wire helicopter of the United States in the beginning of 1972.

Easily flyable manual and automatic landings as well as highly precise attitude hold maneuvering could be carried out with the FBW system. It was also possible to suppress external loads oscillations with the FBW system [31]. Despite the positive experience with the HLH Fly-by-Wire system gathered with the BV 347, the further development of the Boeing XCH-62 was discontinued in October 1974 due to gearbox problems, even before the first flight planned in August 1975.

Fig. 6.30 Boeing-Vertol BV 347 technology demonstrator

Fig. 6.31 Boeing-Vertol BV 347 as HLH-Fly-by-Wire demonstrator

6.2.2.6 Boeing NB-52E CCV (1971–1974)

Another research program with the NB52E (AF 56-0632) was based on the LAMS program (see Sect. 6.2.2.3), that focused on the integration of CCV technologies to improve the flight performance, flying qualities, and particularly the ride comfort. The modification on the test vehicle, jointly implemented by Boeing and the US Air Force Flight Dynamics Laboratory (AFFDL), served to reduce structural loads and to suppress control surface flutter during flight through strong turbulence (see Fig. 6.32). Three additional active-controlled small surfaces called canards, two horizontal, one vertical, located in front of the main wing and close to the nose, were installed on the fuselage. This canard control surface triple was connected through the onboard computer with sensors distributed throughout the aircraft (see Fig. 6.33). In the case of strong turbulence the sensors (angular rate gyros and accelerometers) would register sudden changes in altitude and attitude as well as accelerations. The sensor signals were passed on to the onboard computer, which acted via calculated control laws on the canard control surfaces and conventional control surfaces in such a way that the response of the elastic aircraft due to turbulence was significantly attenuated.

6.2.2.7 Lockheed C-141A FBW (1971–74)

With the conversion of a C-141A (AF 61-2779) from 4950th Flight Test Wing at the USAF Wright Patterson Air Force Base to a dual-redundant Fly-by-Wire flight control system in the pitch and roll axes, an explicit research program was initiated to determine optimal control strategies to improve flying qualities of a large transport aircraft (see Fig. 6.34). The FBW system from Honeywell was operated in parallel to the manual flight controls using a sidestick from the copilot seat on the right-hand side. The FBW system architecture and operation, as well as flight control laws, were similar to that of JB-47E-FBW system (see Sect. 6.2.2.4). Special variants of the C* control law criterion were investigated in flight for

Fig. 6.32 Boeing NB-52E CCV (AF 56-0632) with canard control surface triple

Fig. 6.33 Boeing NB-52E CCV with canard control surfaces on fuselage

Fig. 6.34 Lockheed C-141 FBW (AF 61-2779)

the longitudinal motion in combination with different side-stick characteristics [8].

6.2.2.8 Vought F-8C DFBW (1971–1985)

In the early 1970s, the NASA had already acquired extensive knowledge of the principles of analog Fly-by-Wire technologies through the development and operation of In-Flight Simulators (see Sects. 5.2.1 and 5.2.2). So it was only logical to step into the digital world of electronic flight controls in the course of the experimental program F-8C *Digital Fly-by-Wire* (DFBW). It was a lucky chance that *Neil Armstrong* presented a proposal to utilize the digital computer, the inertial platform (determines attitude and position) and other components, which were originally developed for the Lunar Excursion Module (LEM) within the Apollo program ("*I just went to the Moon with one*") [32].

The conversion of the F-8C (NASA 802) began in 1971 and was completed in 1972 (see Fig. 6.35). This included initially, besides the single-channel digital Apollo FBW system, additionally a three-channel Triplex analog system (see Fig. 6.36) from Sperry as a backup. Also the mechanical flight control of the host aircraft was retained as an emergency backup system. The first flight took place on May 25, 1972 at the NASA Dryden Flight Research Center in Edwards, California. In 1973, after 42 flights, the mechanical backup system was in the meantime eliminated

Fig. 6.35 Vought F-8C DFBW

to demonstrate the reliability of the FBW system. From August to October 1973, the prototype of the sidestick for the F-16 development program was successfully tested.

In the next phase, the by-now obsolete Apollo computer was replaced through a self-monitoring, three-channel digital computer system from IBM and in the period from 1976 to 1978 tested without any fault occurrences in about 30 flights. In the subsequent years from 1979 to 1981, investigations for improving the reliability of sensors through so-called analytical redundancy were of special importance. Thereby sensor errors could be detected and compensated through a special digital computer software. As a direct result, the number of sensors could be reduced.

During approach and landing testing of the Space Shuttle prototype Orbiter, a problem was detected in the man-machine interface. The flight control system did not respond as quickly to control inputs as expected by the astronaut pilot. When the control commands with increasing amplitude and frequency were inserted, the system tended to become unstable. This behavior and similar incidents are called *Pilot Induced Oscillations* (PIO) or also *Aircraft Pilot Coupling* (APC), see also Sects. 2.1 and 9.2.12. With the development of an adaptive control filter that monitors the pilot activity and suppresses high-frequency control inputs, it became possible to avoid such coupling effects of the Orbiter. This NASA filter was subsequently integrated into the Space Shuttle.

From 1982 to 1984, the possible influence of systematic programming errors on the FBW system redundancy was investigated. The effect of a programming error in one of the control channels was substantially reduced through programming the same software functions for each of the digital computers by different experts. Likewise, self-monitoring using the redundant *Backup* software in each, by now "intelligent" computer systems, proved to be effective.

Through the 13-year lasting F-8C DFBW program, NASA could demonstrate for the first time the principal advantages of an integrated digital flight control system without a mechanical backup through practical flight test demonstrations. The FBW system was the forerunner of the digital flight control system of NASA's Space Shuttle Orbiter, which made its first flight in the atmosphere on August 12, 1977. Until the last flight on December 16, 1985, a total of 211 test flights were performed in this F-8C DFBW demonstrator program [32].

Fig. 6.36 F-8C Apollo Hardware Integration

6.2.2.9 McDonnell Douglas YF-4E SFCS/CCV (1972–1979)

Within the US Air Force project Survivable Flight Control System (SFCS), the prototype McDonnell-Douglas YF-4E Phantom (AF 62-12200) was modified in mid-1969 into a Fly-by-Wire research aircraft to evaluate the potential benefits in a high-performance, fighter-type aircraft (Fig. 6.37). The SFCS YF-4E was intended to validate the concept that dispersed, redundant Fly-by-Wire flight control elements would be less vulnerable to battle damage, as well as to improve the performance of the flight control system and increase overall mission effectiveness.

A quadruple-redundant analog computer-based three-axis Fly-by-Wire flight control system with integrated hydraulic

Fig. 6.37 McDonnell Douglas YF-4E CCV (AF 62-12200)

servo-actuator packages was incorporated and sidestick inceptors were added to both the front and back cockpits. Roll control was just Fly-by-Wire with no mechanical backup. For initial testing, the Phantom's mechanical flight control system was retained in the pitch and yaw axes as a safety backup. The most visual notable changes included automatically activated leading edge slats and horizontal canard surfaces above the engine intake for shifting forward the lift center and thus destabilizing the basic aircraft flight mechanics (see Fig. 6.37).

On April 29, 1972, the YF-4E SFCS flew for the first time. The mechanical flight control system was used for takeoff with the pilot switching to the Fly-by-Wire system during climb-out. The aircraft was then flown to Edwards AFB for a variety of additional tests, including low-altitude supersonic flights. After the first 27 flights, which included 23 h in the full three-axis flyby-wire configuration, the mechanical flight control system was disabled. First flight in the pure Fly-by-Wire configuration occurred January 22, 1973. The overall flight-test program included more than 100 Fly-by-Wire flights.

Also, the so-called PACT program (Precision Aircraft Control Technology) was executed with this test vehicle, with the aim to demonstrate the advantages and adequate reliability of Fly-by-Wire systems and of CCV technologies and to envisage these as an integral part of future aircraft designs. The first flight took place on April 29, 1974. After 34 flights the program was completed in the year 1979 and the YF-4E was handed over to the US Air Force Museum.

With the SFCS and PACT programs, the course was also set for the successful development of the YF-16, the world's first serial production combat aircraft with Fly-by-Wire flight control [8].

6.2.2.10 Vought YA-7D DIGITAC (1973–1991)

The Flight Dynamics Laboratory (FDL) of the US Wright Patterson Air Force Base, besides NASA's Dryden Flight Research Center (now: Armstrong Flight Test Centre), was an early supporter of digital flight control systems, as analog computers available then were inflexible, too heavy, and with high power consumption. The aim of the DIGITAC program (*Digital Flight Control for Tactical Fighter Aircraft*) was to demonstrate that faster system improvements (*Upgrades*) in hardware and software were possible through digital technology. After several studies, the concept of digital flight control on a current aircraft was demonstrated. The program was performed in three phases.

In the first phase (DIGITAC I), a Vought A-7D Corsair II (AF 67-14583) from Honeywell Avionics System was fitted with a 2-channel digital flight control system with a so-called BIT function (*Built-In Test Function*) for onboard computer self-monitoring and was designated YA-7D DIGITAC (see Fig. 6.38). Electronic and data recording equipment were housed in an outer container under the right wing. In February 1975 the first flight from a total 92 flights took place at the Edwards Air Force Base. Different flight control laws for reduction of pilot workload during different flight tasks such as formation flying, target or landing approach could be tested for the first time. Since 1976 the experimental aircraft was available also to the US Air Force Test Pilot School AFTPS enabling the test pilots candidates to gain experience in dealing with the new digital technology.

In 1979, under DIGITAC II another upgrade followed, especially the integration of a digital multiplex system, with which now all the subsystems were connected by means of copper or fiber-optic cable (data bus). Each signal was assigned a timeframe, in which it was available over the data bus and could be read simultaneously by all subsystems. This data bus made a vast number of connecting cables superfluous and led to considerable reductions in weight and maintenance efforts. The flight tests continued until 1981. Thereafter, the test vehicle was again available to the AFTPS.

The first flight with digital flight control system equipped with a Fly-by-Light data link took place on March 24, 1982. The system proved to be extremely reliable. The DIGITAC III Program began in 1988 with the installation of a more powerful onboard computer and the standard programming language Ada (Ada is a structured programming language of Honeywell Bull, which has proven itself especially in the use of safety-critical technical systems). The aircraft was then returned to AFTPS for educational and training purposes and remained there until July 1991.

6.2.2.11 General Dynamics YF-16 CCV (1975–1977)

The analog Fly-by-Wire flight control system of the General Dynamics prototype YF-16 (AF 72-1567) was converted to a digital system by the US Air Force Flight Dynamics Laboratory in December 1975. The aim was to test the performance benefits of digital CCV technologies under operational conditions (see Sect. 6.1.2). By integrating two

Fig. 6.38 Vought YA-7D DIGITAC (*Credit* John Bennett)

vertical foreplanes (*Canard*: auxiliary surfaces located in front of the main wing) as additional control surfaces underneath the engine intake, "decoupled" maneuvers could be flown (see Fig. 6.39). This meant, for example, trajectories in the vertical or horizontal plane without pitch or respectively roll attitude changes (flat turn without rolling).

The first flight took place on March 16, 1976. After a total of 87 flights with 125 flight hours, the flight tests were terminated on July 31, 1977 following a hard landing.

Fig. 6.39 General Dynamics F-16 CCV (AF 72-1567)

6.2.2.12 Boeing YC-14 (1976–1979)

The Air Force Boeing YC-14 Short Take-Off and Landing (STOL) jet transport technology demonstrator flew for the first time on August 9, 1976 (Fig. 6.40). Two prototypes were built with the second aircraft flying in October 1976. The YC-14 is noteworthy in that it was the first aircraft to fly with a fault-tolerant multichannel redundant digital Fly-by-Wire flight control system.

A mechanical backup flight control capability was retained. The full authority triple redundant digital Fly-by-Wire flight control system, designed by the British Marconi Company, computed the pitch, roll, and yaw commands that were used to control the elevator, aileron, and rudder actuation systems. The reconfigurable computer architecture divided the basic control path into three sub-functional elements with these elements replicated to provide fault tolerance. The internal element redundancy management function was intended to detect and isolate faulty elements and perform the necessary reconfiguration. The input signal selection methodology was intended to guarantee that all three computers used the same numbers and thus produced identical output values. During normal operation, the overall system output value was selected as the mid value of the three individual values. The system would continue to operate in the event of a failure of one computer by taking the average of the output of the two remaining computers. If they disagreed, both were disabled and the aircraft reverted to the backup manual control system.

The YC-14 was also noteworthy in that it used optical data links to exchange data between the triply redundant computers. The optical communications medium was chosen to eliminate electromagnetic interference effects, electrical grounding loop problems, and the potential propagation of electrical malfunctions between channels. Optical coupling was used to maintain inter-channel integrity. Each sensor's output was coupled to the other channels so that each computer had data from each of the other sensors. Identical algorithms in each computer were used. They consolidated the data, enabling equalization and fault detection and isolation of the inputs. The computers were synchronized to avoid sampling time differences and to assure that all computers were receiving identical data inputs. A similar computer architecture with optical data-links was used at about the same time within the German VFW 614 ATTAS In-Flight Simulator program (see Chap. 9).

Fig. 6.40 Boeing YC-14 (AF 72-1873)

The Fly-by-Wire system was designed to ensure that all computers used the same sensor input values and should, therefore, produce identical outputs. However, a significant fault in the digital flight control software was encountered during flight testing that had not been detected during ground laboratory tests. The software fault resulted in the incorrect tracking of control law computations in each of the three flight control channels, with each channel performing signal selections on a different set of values. This resulted in different input data for the three channels. Although the discrepancies between each channel's inputs were small, the cumulative effect led to large tracking errors between flight control channels when airborne.

Fig. 6.41 General Dynamics F-16 AFTI (AF-75-0570)

Fig. 6.42 General Dynamics F-16 AFTI without canards

6.2.2.13 General Dynamics F-16 AFTI (1982–1983)

To further expand the experience gained with the F-16 CCV and to use the same in an integrated system with additional avionics, a development version of the F-16 (AF 75-0750) was made available by the USAF. Designated as F-16 AFTI (*Advanced Fighter Technology Integration*) the test vehicle was fitted with vertical canards at 30 degrees, which were already used in YF-16 CCV program, and a thick dorsal spine in which the avionics components and triple-redundant digital flight control system were housed (see Fig. 6.41). The first flight took place on July 10, 1982.

Besides the already conducted test in the YF-16 CCV program for direct force control, that is maneuvering in a particular direction without attitude change, avionics functions for reduction of pilot workload stood in the forefront. Thus, pilot functions to activate avionics components could be effected via a voice input system VCID (*Voice-Controlled Interactive Device*). Here 256 word-commands could be identified. Likewise, helmet integrated vision systems (*Helmet-Mounted Sighting System*) were tested, in which alignment of radar or infrared sensors pointing on a target was carried out by head movements.

After 108 flights until July 1983 and a concluding testing of electromechanical actuation systems on the F-16 AFTI without canard control surfaces (see Fig. 6.42), the aircraft was handed over to the USAF Museum at Wright Patterson AFB.

6.2.2.14 Boeing JUH-60A Light Hawk ADOCS (1982–1989)

At the behest of the US Army, during the early 1980s, Boeing developed a JUH-60A Light Hawk technology demonstrator ADOCS (*Advanced Digital Control Optical System*), with the aim of demonstrating the feasibility and performance capability of an optical digital flight control with full authority (see Fig. 6.43). It was hoped that such a system would result in weight reductions, insensitivity to electromagnetic interference, and significantly improved flying qualities coupled with reduced pilot workload [33]. The ADOCS flight control architecture consisted of two components, the primary flight control system PFCS and the automatic flight control system AFCS. The PFCS established a highly reliable direct digital connection between the pilot and the rotor-actuation systems (*Direct Law*) in analogy and as a substitute for classical mechanical flight control. In contrast, based on the principles of the model following control, the AFCS delivered optimal flight behavior for selectable flight tasks such as hover, high speed flight or flights under low visibility conditions. The processing of flight condition dependent sensor data, such as position, attitudes, rotation rates, and accelerations, required for the controller support, was carried out via an optical data bus.

Besides the usual pedals for yaw control and the stick for collective rotor blade pitch control, the ADOCS equipment contains a freely-programmable 4-axes sidestick for the right pilot seat. For security reasons, the production-version mechanical flight control of the host helicopter UH-60A is retained as a backup for the safety pilot on the left-hand side.

Fig. 6.43 Boeing JUH-60A Light Hawk ADOCS

In the testing programs, different sidestick variants were checked out, whereby the so-called 3 + 1 *Controller Configuration* turned out to be particularly promising, namely pitch, roll and yaw control inputs through the right-handed sidestick and collective rotor blade pitch control (climb or descent) over the left-handed collective control stick.

With the reconfigurable digital optical flight control system, the ADOCS demonstrator program yielded good results for the selectable flight task and environmental-conditions dependent control lows. In a concluding appraisal of the ADOCS system based on a demonstration campaign from April to September 1987 with 75 guest pilots and 126 flight hours, it was ascertained that in all flight tasks both the controllability and the stability behavior of the UH-60A basic version was significantly improved, especially about the three primary axes, that is pitch, roll, and yaw [34].

The ADOCS flight control architecture was later on the basis for the technologically-innovative Boeing-Sikorsky RAH-66 Comanche project. The first two prototypes underwent an extensive flight testing from 1996 to 2004. However, before the series production started off, and after about 7 billion US $ were spent, the project was abandoned. The reasons for this were cost escalations, weight and transport problems. Furthermore, there was also a strategic military realignment of US Army after the Cold War, whereby due to the apparent global changes an increased future demand for unmanned aerial systems emerged. All the same, a second career as an In-Flight Simulator JAH-60A RASCAL lay ahead for the JUH-60A Light Hawk at the US Army/NASA at Ames Research Center from 1989 (see Sect. 5.2.2.17).

6.2.2.15 Grumman X-29 (1984–1991)

Forward swept wings have at high speeds the same drag-reduction advantages as the swept back wings. Furthermore, as opposed to swept back wings, adequate natural flight mechanical stability in the rolling and yawing motion is guaranteed for low-speed flights. Already during the Second World War, after extensive wind tunnel tests, Junkers had arrived at such a configuration, which went down in the aviation history as Ju-287 making its first flight on August 16, 1944 (see Fig. 6.44). In the redevelopment phase of the German aviation industry after the Second World War, there was still the HFB-320 Hansa Jet (first flight April 21, 1964) having slightly forward swept wings and very good flying qualities, which was built in small numbers. One specimen, converted as an In-Flight Simulator FLISI, was eventually utilized over a long period of time at DFVLR in Braunschweig, (see Chap. 7).

For decades, the forward swept wings, however, did not achieve any noticeable success, since the aeroelastic deformation and torsion during maneuvering proved to be structurally hardly manageable (*Aeroelastic Divergence*). It took

Fig. 6.44 Junkers Ju-287 (1944)

another 40 years to overcome the problems of wing deformation through stiffer and lighter components and new construction procedures utilizing new materials such as carbon fiber reinforced materials.

With the development of two Grumman X-29, in cooperation with the DARPA (*Defense Advanced Research Projects Agency*) and the NASA, a technology program was launched in 1984, and carried out until 1991, with the aim to generate a database for integration of advanced composite structures and for computer-aided Fly-by-Wire flight control systems for future generations of aircraft with complex wing and control surface combinations. The X-29 had an extremely stiff, forward swept wing with inboard and outboard flaps. Closely coupled canards were mounted in front of the main wings and near the engine inlet area, and flaps were fitted to the rear (strake flaps). This wing-control surface combination imparted the test vehicle extremely high agility with an instability of up to 35% MAC, (see Fig. 6.45a, NASA 003). In this respect, a complex, highly reliable triple redundant digital flight control system for artificial stabilization of the entire system was indispensable. Each of the three digital control computers had an analog backup. Upon breakdown of one of the computers, the other two functional computers took over. In the case of failure of two computer systems, the complete system operations were switched over to the analog system. Upon failure of one analog component, the two remaining analog computer took over.

The first flight of the first X-29 (NASA 003) took place on December 14, 1984 at the Flight Test Center of the USAF Edwards Air Force Base. The mission of the second X-29 (see Fig. 6.45b, NASA 049, first flight May 23, 1989) was focused on the controlled opening of flight envelope at high angles of attack up to 60 degrees with separated flow. Drag reduction up to 20%, Mach numbers of 1.7 and, thanks to the forward swept wing, excellent flying qualities even in

First prototype.

Second prototype with anti-spin devices.

Fig. 6.45 Grumman X-29

flight regimes with separated flow (without thrust vector control) were ascertained from the flight test program. After a total of 436 flights, the program was concluded on September 30, 1991. Both the aircraft are now displayed in the USAF museums of Wright-Patterson AFB and Edwards Flight Test Center (AFFTC) to venerate.

6.2.2.16 Lockheed NC-141A EMAS (1985–1986)

The test vehicle Lockheed NC 141A (AF 61-2775) was equipped by Lockheed with an electromechanical Servomotor EMAS (*Electro Mechanical Actuator System*) from Sundstrand for the left aileron (see Fig. 6.46). The aim of the investigations was to demonstrate in flight test for the first time the feasibility of such an electrical drive as a substitute for the classical electrohydraulic actuator in a primary aircraft control. The EMA aileron drive consisted of a dual electrical motor with separate electric power supply and two-channel electronic monitoring and control electronics. The comparison of performance with the right, unmodified conventional aileron showed only minimal deviations [35].

Fig. 6.46 Lockheed NC-141A EMAS (AF 61-2775)

The experimental aircraft was maintained and operated by the 4950th Test Wing of the USAF at Wright-Patterson AFB, and accumulated in February 1986 nearly 13 h of flight test time, whereby peak currents of up to 12.5 amperes were measured. The program was aborted after an EMAS-error with consequential damages.

Even today, electromechanical actuators are subject of international research under terms like "*Power-by-Wire*" and "*All-Electric Aircraft*" (see also Sect. 6.2.2.20).

6.2.2.17 Rockwell/DASA X-31A EFM (1990–1995)

See Sect. 6.3.6.

6.2.2.18 Boeing B-757 ARIES (1992–2006)

As a replacement for the avionics and flight guidance testbed Boeing B-737-100 TSRV (*Transport Systems Research Vehicle*), NASA Langley Research Center procured in 1994 a Boeing B-757-200 (see Fig. 6.47, NASA 557) acronymed ARIES (*Airborne Research Integrated Experiments System*). In contrast to the B-737-100 (NASA 515), the B-757 received Fly-by-Wire equipment in summer 1992 for the preparatory development of a digital Fly-by-Wire system of the Boeing B-777. It could be flown from the right co-pilot's seat in parallel to the normal mechanical control. Thus, virtually all control inputs, control laws, and the corresponding response behavior of the B-777 were tested and optimized before its first flight.

6.2.2.19 General Dynamics F-16D MATV (1993–1994)

During July 1993 to March 1994, the In-Flight supersonic Simulator F-16D VISTA (AF 86-0048, see Sect. 5.2.1.15) was converted for the MATV technology program (*Multi-*

Fig. 6.47 Boeing B 757 ARIES during taxi test on snow

Fig. 6.48 General NF-16D MATV with anti-spin devices on tail

Axis Thrust Vectoring) to enable testing of controlled flight at high angles of attack by means of thrust vectoring and tested at the Edwards Air Force Flight Test Center. For this purpose, the variable stability system, consisting of digital Simulation computer and programmable control sticks, was temporarily dismantled and replaced by test components for the MATV program. The main component of the MATV program was the thrust vectoring AVEN (*Axis-Symmetric Vectoring Exhaust Nozzle*) from General Electric, almost ready for series production, with which the thrust could be deflected in any arbitrary direction using a ring control of the nozzle lips (see Fig. 6.48).

The F-16D MATV used the standard flight control computer from series-production, which is a part of the four-channel digital flight control system of the F-16. It ensured the artificial stabilization of unstable host aircraft. To perform flights with separated flow at high angle of attack, it also provided the modified control laws through a swift selection of pre-programmed controller gains (*Dial-a-Gain*). The MATV program was initiated with high priority by the USAF, and the tactical utility of thrust vector control could be demonstrated under quasi-operational conditions [36].

Fig. 6.49 McDonnell Douglas NF-15B ACTIVE

6.2.2.20 McDonnell Douglas NF-15B ACTIVE (1994–1999)

Through the ACTIVE program (*Advanced Control Technology for Integrated Vehicles*) spread over several years, the NASA Dryden Flight Test Center pursued the goal of improving the performance of an integrated digital flight control systems using a combination of aerodynamic controls and thrust control by deflection of the exhaust stream. Likewise, the tools for the integrated aircraft design were to be fine-tuned and verified through flight test data. For this project, a McDonnell Douglas F-15B USAF was equipped with a digital FBW flight control system and a powerful engine with axially symmetrical thrust vector control, whereby the thrust nozzles could be adjusted in any direction. The test vehicle was designated NF-15B ACTIVE (see Fig. 6.49, NASA 837).

From 1996 to 1998, supersonic flights up to Mach 2 were demonstrated with an integrated thrust vector control as well as flights at high angles of attack up to 30° with yaw stabilization by thrust vector control. With the aid of FBW onboard computer, combined optimal settings of aerodynamic control surfaces and thrust nozzles could be determined, that resulted in a decrease of the total drag, more specifically to increase in airspeed by of Mach 0.1 at Mach 1.3 at an altitude of about 10,000 m.

6.2.2.21 NASA F-18 SRA (1996–1997)

During 1997, NASA *Dryden* (since 2014: *Armstrong*) Flight Research Center had evaluated a single electrohydrostatic actuator installation on the NASA F-18 Systems Research Aircraft (SRA, NASA 845, Fig. 6.50) loaned from the U.S. Navy.

The electrohydrostatic actuator, provided by the U.S. Air Force, replaced the F-18's standard left aileron actuator and was evaluated throughout the aircraft's flight envelope up to speeds of Mach 1.6. Numerous mission profiles were

Fig. 6.50 F-18 SRA

Fig. 6.51 Lockheed C-141A Electric Starlifter (AF 61-2776)

accomplished that included a full series of aerobatic maneuvers. The electrohydrostatic actuator accumulated 23.5 h of flight time on the F-18 SRA between January and July 1997.

Key technologies investigated with the F/A-18 SRA included advanced so-called Power-by-Wire concepts and Fly-by-Light (fiber optic cable) systems, as well as electric-powered actuators and advanced flight-control computer software. Power-by-wire and electric-powered actuators aim to eliminate cumbersome hydraulic lines in favor of more versatile wires and fiber-optic control cables [37].

6.2.2.22 Lockheed C-141A Electric Starlifter (1996–1998)

After the NC-141A EMAS program was aborted in the year 1986, it took another 10 years until the advances in the digital electronics and in the field of magnetic materials research motivated the US Air Force to replace the conventional electrohydraulic aileron control on a Lockheed C-141A by electromechanical actuators. Whereas during the EMAS feasibility study (see Sect. 6.2.2.16) only the left aileron drive was modified by Sundstrand, this time Lockheed-Martin carried out the electrification and integration of both ailerons in a specially designed digital FBW flight control system. The integrated drive packages IAP (*Integrated Actuation Packages*), developed by Lucas Aerospace, replaced not only the hydraulic actuators but also the elaborate hydraulic connecting lines to the central hydraulic power supply. With 7 kilowatts of electrical power, each of the two dual-channel Duo-Duplex actuation packages could separately operate an aileron.

The aim of this so-called *Electric Starlifter* program was to gather experience with this Power-by-Wire technology with regard to the potential of possible savings and system reliability under operational flight conditions. Thus, more than 1000 flight hours in long haul operation were flown with the C-141A by the 418th Flight Test Squadron of Edwards Air Force Base during the period from 1996 and July 1998 (see Fig. 6.51, AF 61-2776).

Also in Germany, within the framework of the VFW 614 ATD program (see Sect. 6.3.7), integrated actuator packages for future Power-by-Wire applications were developed and tested, among others by Liebherr Aerospace.

6.2.2.23 Boeing/EADS X-31 VECTOR (1998–2003)

See Sect. 6.3.8.

6.2.2.24 McDonnell Douglas NF-15B IFCS (1999–2008)

In the case of test demonstrator McDonnell Douglas NF-15B IFCS (*Intelligent Flight Control System*), it dealt with the same test vehicle with which the NASA-ACTIVE-program was carried out (see Fig. 6.52, NASA 837, see Sect. 6.2.2.20). The main objective of this project at NASA Armstrong Flight Research Center was an improvement of the FBW flight control systems under normal and malfunction conditions. The adaptive and fault-tolerant controllers designed for this purpose were based on so-called neural networks, which could increase the air safety and probability of survival of civil and military aircraft. The software of such adaptive and self-learning neural networks detected possible changes in the control and stability behavior in real time and reconfigured the remaining FBW flight control system for the recovery to a stable and controllable flight condition.

6.2.2.25 Sikorsky X2 TD (2005–2011)

The Sikorsky X2 Technology Demonstrator (TD) was a coaxial rotor design (counter-rotating rotors) that aimed to retain good hover performance as well as cruise speeds up to 460 km/h (250 kts). It was built at Schweizer Aircraft, a Sikorsky subsidiary. The key performance factors included speed but also low vibration, low pilot workload, and low acoustic signature (see Fig. 6.53).

Technologies used in the X2 TD design included coaxial rigid rotor blade designs with high lift-to-drag ratio, active vibration control, and a FBW system with advanced flight control laws. FBW technology concentrated the cyclic, collective and pedals on a single side-arm controller (SAC).

Fig. 6.52 NF-15B IFCS

Fig. 6.53 Sikorsky X-2 Speedster

The X2 TD featured a conventional collective stick, a SAC cyclic, and pedals for yaw control. Yaw is induced in a coaxial-rotor helicopter by increasing blade pitch in one rotor and decreasing it in the other, resulting in a differential torque which could be augmented by a rudder at higher speeds. The flight control laws for improved handling qualities have been particularly conducive to a low pilot-workload and single-pilot operation.

In order to demonstrate and evaluate these basic Fly-by-Wire capabilities sufficiently in advance, a surrogate Schweizer 333 helicopter was equipped with the X2 Fly-by-Wire system. The system was programmed with advanced flight-control laws and linked to the main rotor and engines to enable basic maneuvering capabilities. First flight took place on November 4, 2005 and subsequent flight testing was focused on reliability tests of the triply-redundant system.

First flight of the X2 TD was accomplished on August 27, 2008. On September 15, 2010, Sikorsky's design goal for the X2 was achieved with a horizontal flight speed of 250 kts (290 mph; 460 km/h). On July 14, 2011, the X2 TD completed its final flight and was retired after accumulating 22 h over 23 test flights. With the end of development, the X2 will be followed by its first application, the S-97 Raider high-speed scout and attack helicopter.

Fig. 6.54 Gulfstream CV550 AFC Demonstrator (*Credit* Preston A. Henne, Gulfstream, 2008)

6.2.2.26 Gulfstream G550 AFC Demonstrator (2006–2008)

The goals of an Advanced Flight Control (AFC) program at Gulfstream Aerospace was to demonstrate emerging flight control technologies such as Fly-by-Wire, Fly-by-Light, and Fly-by-Wireless as well as innovative actuation systems in flight and to determine the benefits in capability, performance, reliability, and cost for next-generation business-type aircraft.

A Gulfstream G550 (N532SP) was retrofitted as a dedicated demonstrator aircraft (see Fig. 6.54). First flight with Fly-By-Wire control demonstration using rotary electromechanical actuators (EMA) for the outboard wing spoilers took place on September 26, 2006, resulting in improved high-speed stability, roll performance, and ride comfort. The rotary EMA were provided by Parker Aerospace. Further, the inboard and mid-wing spoilers were modified with EMA supplied by Smiths Aerospace. The electrohydraulic FBW elevator control from Parker Aerospace was flight tested for the first time on May 16, 2007, and an electrical backup hydrostatic Actuator (EBHA) for the elevator on October 8, 2007. Thales supplied the flight control computer. The FBW system was flown from the right-hand seat whereas the left-hand seat retained the mechanical controls of the basic aircraft.

A Fly-by-Light flight control demonstration was completed on February 27, 2008. A fiberoptic harness was successfully used to transfer pilot control inputs from a flight control computer to the mid-wing spoilers. Electrical-optical connector technology, electromagnetic interference shielding, and manufacturing and installation concepts were assessed.

Gulfstream Aerospace also demonstrated on September 18, 2008, for the first time, wireless signaling (Fly-by-Wireless) for a primary flight-control surface in a civilian or military aircraft. The wireless control architecture included an internal wireless bus transmitter and external receiver at the EMA interface of the mid-wing spoilers. The units communicated using "direct sequence spread spectrum modulation and coding technology" (Gulfstream). It offered an additional channel of communication for redundancy, which increases system safety.

6.2.2.27 Sikorsky S-76 SARA (Since 2013)

The Sikorsky Innovation Group has developed a set of hardware and software capabilities to support autonomous flight of unmanned or optionally piloted rotorcraft. Within this so-called MATRIX™ program, a higher level, second generation of autonomous flight capabilities were developed. The program conducted its first test flight on July 26, 2013 on a S-76 fitted with Fly-by-Wire flight controls and multiple sensors for situational awareness. The S-76 demonstrator helicopter is dubbed SARA (Sikorsky Autonomous Research Aircraft, see Fig. 6.55).

The program's objective was to develop both a set of software applications and a "pallet" of hardware and software systems that could be "ported" or integrated on an existing aircraft or introduced with a new aircraft. The MATRIX™ technology can be easily adapted for other aircraft systems as well.

Within the DARPA-sponsored ALIAS program (*Aircrew Labor In-Cockpit Automation System*), a new level of automation into existing military and commercial aircraft has been developed and integrated to enable to operate rotorcraft with reduced onboard crews. ALIAS seeks to leverage advances in autonomy that reduce pilot workload, augment mission performance, and improve aircraft safety and reliability. Sikorsky utilized its MATRIX™ Technology to develop, test, and field hardware and software systems that significantly improved optionally piloted and piloted vertical take-off and landing (VTOL) aircraft. Sikorsky had later installed the MATRIX™ Technology also on a Fly-by-Wire and optionally piloted UH-60MU helicopter.

The first phase of the ALIAS program was completed on May 24, 2016, with the demonstration of a 30-mile autonomous flight using the Sikorsky S-76 SARA technology demonstrator. This flight highlighted the ability for an operator to plan and execute every phase of an autonomous mission with a tablet device. During the demonstration, a ground station crew monitored the progress of the ALIAS-enabled Sikorsky S-76 SARA.

Fig. 6.55 Sikorsky S-76 SARA (*Credit* Sikorsky)

A second phase of the ALIAS program will focus on continued maturation of the initial ALIAS system with additional flight tests, enhancements to the human interface and transition to additional rotorcraft to demonstrate ALIAS portability.

6.2.3 Russia

6.2.3.1 Introduction

To make more accurate inquiries and reliable assertions about Russian or Soviet developments in the field of Fly-by-Wire technologies, the readily accessible sources in the form of bilateral cooperation such as that between the DLR and the Russian Flight Research Institute (FRI, *Gromov* Institute), books and websites could be uncovered only during the last two decades [38]. Three technical books in English providing good historical overviews are now available for referencing [39–41]. Details about the DLR-FRI cooperation are given in Sect. 12.2.2.

6.2.3.2 MiG YE-63/T (1961–62)

The first steps towards the integration of additional, more adjustable aerodynamic surfaces to improve the aircraft performance were undertaken by the design office Mikoyan towards the end/beginning of the fifties and sixties with the MiG YE-6/3T (see Fig. 6.56). For this purpose, a pivoted delta canard surface (*Destabilizer*) was installed in the front fuselage area of a MiG-21F, which was aligned for the respective flow conditions. However, it was only possible to achieve a marginal destabilization in the longitudinal motion (pitch axis). A total of 56 flights were performed [41].

6.2.3.3 MiG YE-8 (1960–1963)

In the next step, significantly improved flight performance in supersonic range was achieved during 1962 with a special MiG YE-8, likewise based on a MiG-21. The canards were subsonic freely floating, while they were set fixed in the supersonic range. Both of the aforementioned MiG test aircraft, however, did not incorporate yet an actively controlled Fly-by-Wire flight control system.

The first of the two technology demonstrators YE-8-1 made its first flight on April 17, 1962 (see Fig. 6.57) and the second specimen YE-8-2 on June 29, 1962, gathering a total of 13 flights (see Fig. 6.58). The project was aborted in 1963 after the YE-8-1 disintegrated in the air after an engine explosion at Mach 1.7.

Fig. 6.56 Mig YE-6/3T

Fig. 6.57 Mig YE-8-1

Fig. 6.58 Mig YE-8-2

Fig. 6.59 Sukhoi 100LDU

6.2.3.4 Sukhoi 100LDU (1968–74)

As part of the Fly-by-Wire system development program for the large supersonic Sukhoi T4, the first Russian aircraft with an analog quadruplex Fly-by-Wire flight control system including a mechanical backup (first flight in August 1972), a two-seater supersonic training aircraft Su-7U was converted to a Fly-by-Wire technology demonstrator Su 100LDU (see Fig. 6.59). During the period from 1968 to 1971 special emphasis was placed on the impact of actively driven, destabilizing canards.

It should be noted that the Sukhoi 100LDU Fly-by-Wire demonstrator later (1973–1974) contributed also to the development the prototype Sukhoi T-10, the first Russian Fly-by-Wire aircraft without mechanical backup (first flight: May 27, 1977). After significant modifications, the successful T-10 flight test program provided the basis for the Sukhoi Su-27 employing the first operational Fly-by-Wire system in Russia. Today, a Su-27 version serves the Flight Research Institute (FRI, Gromov Flight Research Institute) as an In-Flight Simulator (see Sect. 5.6.2).

6.2.3.5 Sukhoi L02-10 (1968–84)

To generate direct control forces for special lateral precision maneuvers, additional actively driven control surfaces were needed besides the rudder. Various options for such control surfaces were implemented on a Su-9 (see Figs. 6.60 and 6.61). The research aircraft modified in this way and designated Su L02-10 was operated over several years (1972–1979) by the Flight Research Institute FRI.

Fig. 6.60 Sukhoi L02-10 with two additional vertical control surfaces

Fig. 6.61 Sukhoi L02-10 with one additional vertical control surface

6.2.3.6 Sukhoi Su-15 CCV (1980–82)

A Su-15 (S/N 1115328) was specially equipped with CCV capabilities by Sukhoi. The Gromov Flight Research Institute investigated variable stability and control characteristics sidestick inceptors (see Fig. 6.62). This Fly-by-Wire research aircraft crashed on November 11, 1982 [41].

6.2.3.7 Sukhoi T10-24 CCV (1982–87)

One of the first Sukhoi Su-27 production aircraft, the T10-24, was converted in the early 1980s to another CCV demonstrator. It was equipped with canards, mounted ahead of the main wing, to improve the maneuverability at high angles of attack (see Fig. 6.63). The Canards were automatically adjusted with increasing angles of attack, increasing the maximum lift in maneuvering flight.

The landing speed could also be reduced through the canards, which was particularly beneficial during operations on aircraft carriers (see Fig. 6.64). The first flight with canards was performed in May 1985 by the famous test pilot *Victor G. Pugachov*, the first demonstrator of the so-called “Cobra” maneuver. Some versions of the Su-27, such as the Su-27 M from which later the Su-35 emerged, were later equipped with standard canards. Another version, the Su-37, received, in addition, a thrust vectoring system. The Sukhoi T10-24 was lost on January 20, 1987 [40].

Fig. 6.62 Sukhoi Su-15 LL CCV (*Credit* Rob Schleiffert)

Fig. 6.63 Sukhoi T10-24 CCV conversion (*Credit* Jefim Gordon)

Fig. 6.64 Sukhoi T10-24 (*Credit* Jefim Gordon)

6.2.3.8 Mil Mi-8T FBW (1985–1990)

From the 12,000 pieces of Mil Mi-8 helicopters which were manufactured, just a single Mi-8T (Hip C) model was retrofitted with an electronic flight control system for testing the application potential of future Fly-by-Wire technologies for rotorcraft (see Fig. 6.65). Although this was just an experimental setup with a VUAP-1 standard autopilot having limited control authority, it was possible to investigate and evaluate the optimal flying qualities with this device using a data link to a ground station providing for variable controller settings. It could be demonstrated in 1989 that through the use of two sidesticks (left and right on the test pilot’s seat), the pilot workload could be significantly reduced during precision maneuvers [41].

Fig. 6.65 Mil Mi-8T (*Credit* military-today.com)

Fig. 6.66 Su-47 Berkut

6.2.3.9 Sukhoi Su-47 Berkut (1990–2008)

This technology demonstrator initially designated as S-37 was the Russian counterpart to the US X-29 project. The first flight took place on September 25, 1997, thus thirteen years later compared to the X-29 (see Fig. 6.66). Even the Russians had finally succeeded in changing the destabilizing torsion of the 45° forward swept wing principally into a less critical wing bending through a specially arranged *composite* fiber structure (US: "aeroelastic tailoring"). The test vehicle was renamed in the year 2002 as Su-47. It provided important insights into the controllability of a three-surface-configuration (canards, main wing and tail plane) by means of a complex digital Fly-by-Wire flight control system. This included the optimization of flying qualities in extreme flight regions while ensuring sufficient artificial stability. The ailerons of the forward swept wing provided sufficient controllability under separated flow conditions at high angles of attack.

6.2.3.10 Mig 1.44 (1994–2000)

The official unveiling of the Mig 1.44 technology demonstrator on January 12, 1999 ended years of mystery about this project (see Fig. 6.67). It featured thrust vectoring (pitch and yaw) for high-angle-of-attack (AOA) super maneuverability and supersonic cruise without afterburning of its Lyulka Saturn AL-41F turbofans with round 3D vectoring nozzles. The T/W ratio of the clean aircraft is not less than about 1.33. The aerodynamic delta wing configuration with all-movable close-coupled canards provided favorable vortex interactions during high AOA maneuvering, delaying both boundary layer separation (stall) and vortex core breakdown, commonly found at high angles of attack.

Almost full-span hinged flaps were mounted on the wing leading edges, while flaps, while on the trailing edge large inboard and outboard flaperons were attached. Whereas the test aircraft did not have a conventional horizontal tail, structural beams behind each wing carried outward sloping upper fins with inset rudders and below the beams vertical

Fig. 6.67 Technology Demonstrator Mig 1.44, Blue 1 (*Credit* Tyler Rogoway)

underpins with additional mini-rudders. Between the beams and the adjacent engines secondary elevators were placed. A single rectangular variable engine inlet was placed below the forward fuselage with the upper lip fully variable for supersonic flight. The lower inlet lip could be hinged down in high-alpha flight.

In all, 16 flight control surfaces were linked to the advanced Avionika KSU-142 digital Fly-by-Wire flight control system which provided artificial stability to the aircraft at subsonic speeds as well as high maneuvering agility. The aircraft was equipped with a glass cockpit.

After lengthy ground tests, the test aircraft manufactured in the early 1990s made its first high-speed run in the late 1994 with Mikoyan's Chief Test Pilot *Roman Taskayev* at the controls. Just as the test program began to pick up, it was again put on hold as the Mikoyan design bureau did not have enough funds to purchase the remaining components still missing on the demonstrator. This became the main factor in the indefinite postponement of the program for the next few years.

On February 29 (sic!), 2000, the Mig 1.44 performed its initial flight at the hands of Vladimir Gorboonov. During the 18-minute flight, the aircraft reached a maximum height of 1000 m (3300 ft) and reached speeds of 600 km/h (370 mph).

After a second 22-min test flight on April 27, 2000, engineers probably uncovered some problems, since there

Fig. 6.68 The defunct Mig 1.44, now: Blue 144 (*Credit* Max FOXBAT Bryansky Russian APT)

Fig. 6.69 Dassault Mirage IIING 01

were no reported flights thereafter and the program was canceled. The technology demonstrator aircraft MiG 1.44 has been put in long-term storage in the hangar of the Gromov Flight Research Institute. 15 Years later, on August 25, 2015, the defunct Mig 1.44 returned from retirement for a rare public viewing at the Moscow Air Show (MAKS 2015, see Fig. 6.68).

The MiG 1.44 design lived on in China's J-20 large stealth fighter, which some analysts think may even have benefited directly from the Mig 1.44's design and development. The two aircraft do have some remarkably similar design cues and proportions such as similarly styled delta wing canard configurations and a V-shaped tail section with closely mounted engines.

6.2.4 France

6.2.4.1 Dassault Mirag E IIING (1982–1984)

A Dassault Mirage IIIR designated as Mirage IIING (*Nouvelle Generation*) with a modified delta wing and additional, permanently attached canards above the engine intake served as a Fly-by-Wire and avionics technology demonstrator. The first flight took place on December 21, 1982. With the aerodynamic modifications and the Fly-by-Wire system, this unique aircraft was the best within the Mirage III family in terms of performance. All the more bleak is the sight and the state of the technology demonstrator as seen in Fig. 6.69. The canards are still mounted on the fuselage above the engine intake.

6.2.4.2 A300B FBW Demonstrator (1983–1985)

Already in 1978, Aerospatiale undertook flight tests with a company's own Concorde to explore technologies for electrohydraulic flight control. While the spoilers, landing flaps (*Flaps*), and slats of the A310 could already be electrically driven through a digital computer, the concept and the benefits of a Fly-by-Wire control system for the elevator and aileron surfaces were validated through flight tests with Concorde. Thereby, the rudder and the trimmable horizontal stabilizer functioned as a mechanical backup. At the same time, in combination with electrohydraulic elevator and rudder control system, a sidestick mounted in the left pilot seat was tested as a substitute for the centrally mounted control column being operated with both hands. With a digital computer the C*-control laws (see Sect. 6.2.2.4) were generated for precise flight path control, which also proved to guarantee optimal flight performance in the turbulent air.

At this time, there was no specific project to utilize the results of this demonstrator program. Also, the supersonic Concorde configuration as such was of little relevance for the future of commercial aircraft. Nevertheless, it provided an initial promising database on the effectiveness of future Fly-by-Wire flight control systems in conjunction with sidestick inceptors.

With the launching of the A320 project, the concept of digital Fly-by-Wire flight control with sidestick operation was to become a reality. For this purpose, Airbus modified the A300 (S/N 3) as a Fly-by-Wire demonstrator with a left and right hand sidestick on the two pilot seats respectively. Within two flight test campaigns (1983 and 1985), the special benefits of Fly-by-Wire flight controls, such as flight envelope protection including limiting the angle of attack during low speed approaches, could be impressively demonstrated to the general public [42] (see Fig. 6.70).

The first flight of the commercially successful A320 took place two years later on February 22, 1987. The introduction of C* like standard flight control laws proved to be highly successful in the operational service. Virtually all configurations within the by now very extensive Airbus family of A319/320/321/330/340/350/380 have almost identical flying qualities, despite large differences in their gross weights. Thus, the training procedure of airline companies for the pilots could be significantly simplified.

Fig. 6.70 Airbus A300 Fly-by-Wire technology demonstrator

Fig. 6.72 Mitsubishi T-2 CCV

6.2.4.3 Aérospatiale SA.365N1 Dauphin 2 FBW (1989–2001)

In 1989, the conventional mechanical control system of the prototype Aérospatiale SA.365N1 Dauphin 2 C/N 6001, F-WZJJ) was expanded by a Fly-by-Wire flight control system. A lateral sidestick was provided for the right-hand seat of the test pilot, while the mechanical backup was retained for the safety pilot in the left cockpit seat. The first flight in the Fly-by-Wire mode took place on April 6, 1989. The FBW flight tests focused on the electronic decoupling of the control axes and various rate and attitude control laws resulting in good and "forgiving" flying qualities (*Carefree Handling*). The experience was integrated directly into the development of the European NH-90 (NATO Helicopter 90), worldwide the first helicopter to be equipped with a production-version Fly-by-Wire system.

After Aérospatiale merged together in 1992 with the rotorcraft division of Messerschmitt Bölkow-Blohm forming the new company Eurocopter S.A., the Dauphin received a new registration F-WQAP. Later, the Fly-by-Wire system of the Dauphin received two additional actively controllable aerodynamic control surfaces, namely a horizontal stabilizer and a rudder. With these changes, an improved directional stability with reduced tail rotor performance and increased payloads could be achieved. The last flight took place in March 2001. Today, the experimental helicopter is displayed in the British "The Helicopter Museum" (see Fig. 6.71).

Fig. 6.71 Eurocopter FBW Dauphin (*Credit* Peter Clarke)

Under the framework of bilateral consultations (Memorandum of Agreement—MoA) between Eurocopter and DLR, it was agreed upon that thereafter the flight experiments with electronic/electro-optical flight controls (FBW/L) were to be primarily carried out on the newly developed Fly-by-Light Helicopter DLR EC 135 FHS (*Flying Helicopter Simulator*) at DLR in Braunschweig. In the meantime, a first successful flight test program for Airbus Helicopters (formerly Eurocopter) was carried out (Project ACT IME, see Sect. 10.4.1).

6.2.5 Japan

6.2.5.1 Mitsubishi T-2 CCV (1983–1986)

At the behest of the Japanese Defense Agency JDA and its subordinate Institute TRDI (*Technical Research and Development Institute*), a supersonic training aircraft Mitsubishi T-2 was equipped with an experimental, triple redundant (triplex) digital FBW flight control system. With three additional control surfaces (vertical fin under the fuselage, two horizontal canard surfaces above the sides of the engine inlets, see Fig. 6.72) it was designated T-2 CCV enabling the testing of the future CCV technologies (see Fig. 6.73). The first flight took place on August 9, 1983. Up to 1986, a total of 183 flights were performed. However, there is little known about the flight test results.

6.2.5.2 Kawasaki BK-117 FBW (Since 1994)

A Bolkow-Kawasaki BK-117 was converted by the ATIC Institute (Advanced Technology Institute of Commuter Helicopter) with the aim of making easier and safer the flying a single rotor, unstable helicopter under blind conditions by incorporating Fly-by-Wire technologies (see Fig. 6.74). This included, besides the triple-redundant digital Fly-by-Wire system, an active sidestick and programmable cockpit displays for optimal information extraction. In the case of a complete failure of the three digital onboard

Fig. 6.73 Mitsubishi T-2 CCV with additional control surfaces

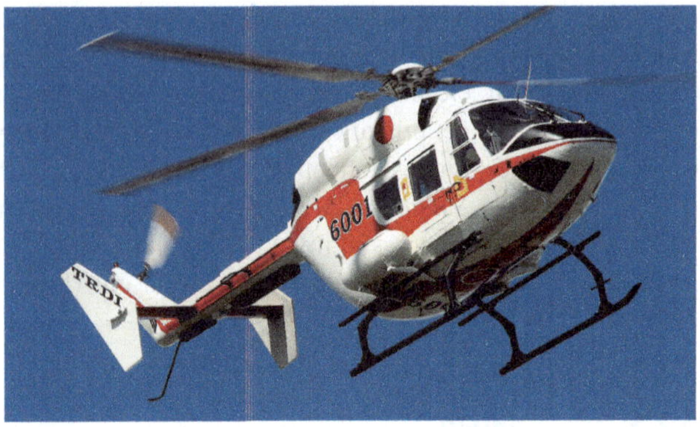

Fig. 6.74 Kawasaki BK-117

Fig. 6.75 Mitsubishi ATD-X Shinshin (*Credit* Getty Images)

Fig. 6.76 ATD-X twin engine thrust vector paddles (*Credit* AFP screencap)

computers, running asynchronous, an analog backup flight control was available. Three default control laws could be selected, namely Rate Command-Attitude Hold (RCAH), Attitude Command Attitude Hold (ACAH), and Attitude Command Velocity Hold (ACVH). Thereby the flight tasks under visual as well as under bad weather conditions could be performed. The first FBW flight took place in September 1999 [43].

6.2.5.3 Mitsubishi ATD-X Shinshin (Since 2008)

The Mitsubishi ATD-X Shinshin was developed by the Ministry of Defense Technical Research and Development Institute (TRDI) for research purposes. ATD-X is an acronym for "Advanced Technology Demonstrator—X" (see Fig. 6.75). The ATD-X is used as a technology demonstrator and research prototype to evaluate domestic advanced technologies for a fifth generation stealth fighter aircraft. The aircraft features a 3D thrust vectoring capability, promising greater maneuverability than other stealth planes. The experimental thrust vector control system was realized by three paddles on each engine nozzle similar to the system used on the Rockwell/DASA X-31 (see Fig. 6.76, see also Sects. 6.3.6 and 6.3.8).

The ATD-X made its maiden flight on April 22, 2016. Among the innovative features of the ATD-X, which also carries the official military designation X-2 (JAF 51-0001), is a Fly-by-Light flight control system. Substitution of wires by optical fibers allows faster data transfer and immunity to electromagnetic disturbances. A further feature is a so-called "Self Repairing Flight Control Capability" (自己修復飛 行—Fix it myself flight), which allows the aircraft to automatically detect failures or damage in flight control system and aerodynamic control surfaces. With this capability, the control commands are redistributed by the computer to the remaining control surfaces and regain controlled flight.

6.2.6 China

6.2.6.1 Shenyang J-8 ACT (1977–1990)

After preparatory studies, which began in 1977, a Shenyang J-8I was converted in 1988 to a Fly-by-Wire testbed and designated as J-8 ACT. First flight was conducted on June

Fig. 6.77 Shenyang J-8II ACT (*Credit* Weimeng)

24, 1990. There aren't any reliable information or pictures available concerning this project.

6.2.6.2 Shenyang J-8II ACT (1990–1999)

Planned as successor to the In-Flight Simulator JJ-6 BW-1 and J-8 ACT test demonstrator, during the mid-1990s a version of the basic fighter aircraft J-8II was equipped with canards above the engine intakes for decreasing aerodynamic stability and improving maneuverability as well as with a digital three-axis and quadruple redundant Fly-by-Wire system. It was designated as J-8II ACT (see Fig. 6.77). Two MIL standard 1553B data bus interfaces were provided for computer communications. A total of 49 test flights were performed during the period from the first flight on December 29, 1996 to the last flight on September 21, 1999. The J-8II ACT played a significant role in the development of next generation of fighter aircraft.

6.3 German Demonstrators

6.3.1 Dornier Do 27 DFBW and Percival Pembroke DFBW (1964–1972)

Johannes Tersteegen

6.3.1.1 Introduction

The basic feasibility of analog electrohydraulic control systems could be demonstrated in the early nineteen-sixties at the RAE, for example on Avro 707 C (see Sect. 6.2.1.2) and the Concorde. For emergencies, however, it was necessary to provide a mechanical back-up control system. The required reliability of these systems could be achieved only by redundancy, that is, through multiple parallel channels. The analog Fly-by-Wire systems encountered difficulties in monitoring the redundant channels by majority decision because of the inevitable drifts in signal transmission. Furthermore, the necessary resolution could not be achieved to connect control law processor signals.

These general difficulties in analog systems could be overcome using digital electrical systems. As the transmission in digital systems occurs incrementally in just two clearly identifiable states, the redundancy and majority decision is considerably simplified. Complex compensation and stabilization measures, necessary in analog electrical systems, are not required. It was anticipated that the digital system would provide high resolution in the future. The degree of resolution depends on the quantization of the actuation system outputs. At that time practical solutions were unknown, meeting the high technical and reliability demands for digital control systems. This was one of the main research focus at the DFVLR Institute of Flight Guidance under the leadership of *Karl-Heinrich Doetsch* and *Walter Metzdorff* in the sixties.

6.3.1.2 Digital Electrohydraulic Aircraft Control

The digital electrohydraulic flight controls consisted mainly of:

- Digital position transducers as A/D (analog to digital) converter,
- Digital signal processing and transmission system,
- Fault detection system,
- Digitally controlled electrohydraulic actuation system, and hydraulic power amplification.

Component Development

Electronic System

Besides the theoretical work on reliability, redundancy, error or fault detection and correction, the focus was initially on developing basic components for system operation. Therefore, it was necessary to develop devices such as mechanical-to-electrical transducers (encoders to convert the pilot control column movement into a digital electrical signal). Additionally, signal processing and transmission, as well as fault detection systems, had to be developed, as adequate components were not available on the market. Thus already in 1962, an incremental displacement sensor as analog-to-digital (A/D) converter emerged. This device generated the information content by incrementally adding steps. Later on, an encoder was developed that was based on the principle of the brush impulsing and magnetic core matrix switch to generate the binary information (see Fig. 6.78).

This new technology made it necessary to develop and implement new control electronics, consisting of sensor

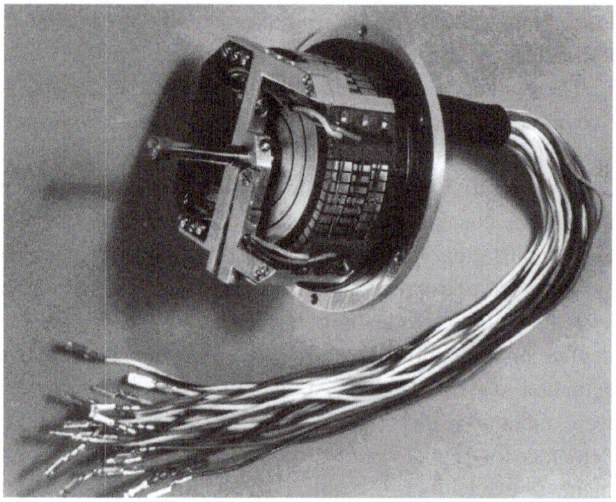

Fig. 6.78 Digital mechanical-electrical position transducer (Encoder)

clock generator, impulse storage and detection, monitoring electronics, and auto-diagnostic and failure detection with channel switching. As the assessment of the reliability of digital electronic components was uncertain at that time due to the lack of operational experience, a system design was implemented in the late sixties, so that system repair could be performed during the flight by manually changing the components (*in-flight repair*).

Hydraulic System

Quite soon it became apparent that a fully digital Fly-by-Wire control system required not only encoders and electronic components, but must also include the hydraulic actuation system. To perform research in the area of aeronautical hydraulics, it was necessary to establish an appropriate hydraulic laboratory. Due to lack of funds, hydraulic components were recovered from crashed and scrapped aircraft of the German Air Force during 1962. In addition, the German Air Force provided a hydraulic trainer of the F-86 Sabre. This trainer was used for familiarization in the field of aircraft hydraulics. Simultaneously, a first small hydraulic power unit was set up from these hydraulic components and an electrohydraulic actuator was assembled for testing purposes. Having done this, a complete digital electrohydraulic control chain could be set up for the first time and a very simple laboratory simulator with a single degree of freedom for the elevator surface was assembled.

Electrohydraulic Actuation

The research in electrohydraulic actuation systems included investigations about the optimal location of the analog to digital (AD) conversion. The research was focused on electrohydraulic switching valves, which served as a link between digital signal-processing electronics and the analog hydraulic power amplification. Thus, not only devices such as a digital torque motor emerged as a purely mechanical device (serial switching of pistons with binary weighted strokes), but also fluidic solutions and the pulse-modulated actuator systems. Fast-switching valves were required for these new developments, which were not available at that time and had to be developed.

Conrad R. Himmler, a former DVL employee and renowned expert from France, was asked to support the team in the design of airworthy electrohydraulic actuation systems for flight testing. According to his inputs and DFL specifications, electrohydraulic actuators with mechanical feedback were built by the company "Centre de Recherches, Hydrauliques et Électriques, Paris", of which he was the director. These actuators were then installed in the Do 27 (see Sect. 6.3.1.3) and later in the HFB 320 aircraft (see Chap. 7) [44].

Figure 6.79 shows the schematic structure of an electrohydraulic actuator. The actuator was designed with a mechanical feedback. The torque motor itself is not within the control loop in this type of feedback mode. However, the static characteristic of the overall drive deteriorates, as friction develops during the sampling. With the introduction of an additional "dither" (high-frequency signal with minimal amplitude), however, a satisfactory performance could be obtained. The mechanical feedback brought simplification into the system design compared to an electrical feedback, thus enhancing the system reliability. In the case of electrical supply failure, the actuator with mechanical feedback shows "fail neutral" behavior because the actuating piston moves to the safe center position. On the contrary, in systems with electrical feedback, the loss of electrical supply leads to "hardover" (piston moves to either end position). In the initial tests with the Do 27, an electrical D/A converter was used to generate analog torque motor input signals, whereas for subsequent flight tests with the Do 27 and HFB-320 (see Chap. 7) a digital electro-mechanical torque motor with a binary weighting of coil windings was developed. Applying appropriate voltages to this binary coil matrix, the required binary coded torque values in the motor were generated.

6.3.1.3 Test Aircraft Do 27

Do 27 (YA 913) available at that time at DFL provided a suitable platform to flight test the experimental unit of the digital FBW. The objective of flight testing was to investigate the influences of quantization and sampling in the digital controller on the controllability of the closed loop system (pilot inputs, signal processing, and aircraft) under real operational conditions. The Do 27 is a lightweight, single-engine STOL (short take-off and landing) multi-role aircraft, manufactured by Dornier, Germany. As a high-wing monoplane with four to six seats, it was primarily deployed by the German Federal Armed Forces for military purposes. After the first flight in 1956, over 600 aircraft were

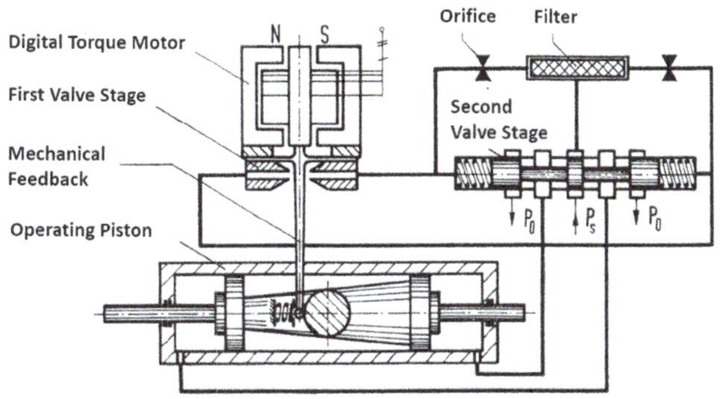

Fig. 6.79 Electrohydraulic actuator with digital torque motor as D/A converter and mechanical feedback

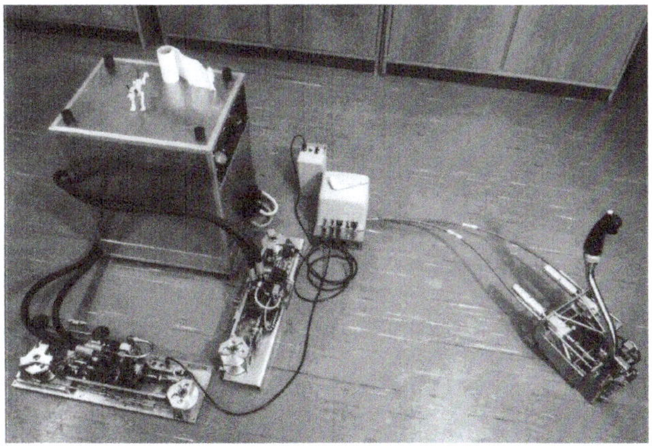

Fig. 6.81 Digital electrohydraulic aircraft control for Do 27, consisting of (from *right* to *left*): control unit, control electronics, electrohydraulic actuator for aileron and elevator, hydraulic system with switching mechanism

Fig. 6.80 Test aircraft Do 27

manufactured until 1965. Thus, it was the first German aircraft design which was produced on a large scale after the Second World War (see Fig. 6.80).

Hydraulic System

As the Do 27 was not equipped with a hydraulic power supply, due to the shortage of funds and unavailability of off-the-shelf products, the hydraulic power pack was built in-house. The constant hydraulic pressure of 3000 psi (210 bar) was generated by a self-regulating axial piston pump, driven by an electric motor that was powered by the onboard power supply. The initial experiments were performed with a fixed displacement pump. The maximum pressure in this system varied by about 435 psi (30 bar) and was controlled by a two-point pressure switch for the upper and lower limit values As already mentioned, this hydraulic system consisted essentially of components from discarded or scrapped F-86 Sabre aircraft of the German Federal Armed Forces. Furthermore, off-the-shelf industrial by-pass ball valves were utilized and new hydraulic filters with clogging indicators and by-pass valves fabricated for aeronautical purposes, too. Due to space limitations, the hydraulic system had to be mounted directly behind the pilot's seat and enclosed by a cabinet housing. The panel with the electrical controls and hydraulic monitoring devices was also part of this cabinet. Additionally, a "central control panel" could be cable-plugged to the cabinet for remote control and monitoring. The hydraulic circuits for ailerons and elevators were separated and connected by approved high-pressure hoses to the hydraulic system via shut-off ball valves. Due to the very compact design, the hydraulic system heated up significantly in a relatively short time, so additional ventilation of the interior became essential. Even so, the tolerable maximum hydraulic fluid temperature of about 70 °C posed severe limitations on the flight test duration.

The hydraulic system described here was used in the Do 27, Pembroke (YA 558) and also in HFB-320 (see Chap. 7) in DFL/DFVLR. Later, it was also incorporated in the DHC Beaver at the Technical University in Delft, The Netherlands, and for ground tests in the Technical University of Berlin and by the flight test department of DFL/DFVLR.

Installation of FBW System

After extensive ground tests in 1967, a two-axes-simplex system was installed for the elevator and aileron in the Do 27 (YA 913) (see Figs. 6.81 and 6.82).

FBW Control Unit

For the FBW operation, the usual mechanical control was removed and replaced by a control unit for the test pilot on the right. The control column movements of the test pilot were transmitted via Flexball Bowden cable to encoding gear units. An adjustable spring damper system for force feedback was mounted on the same lever to replicate the

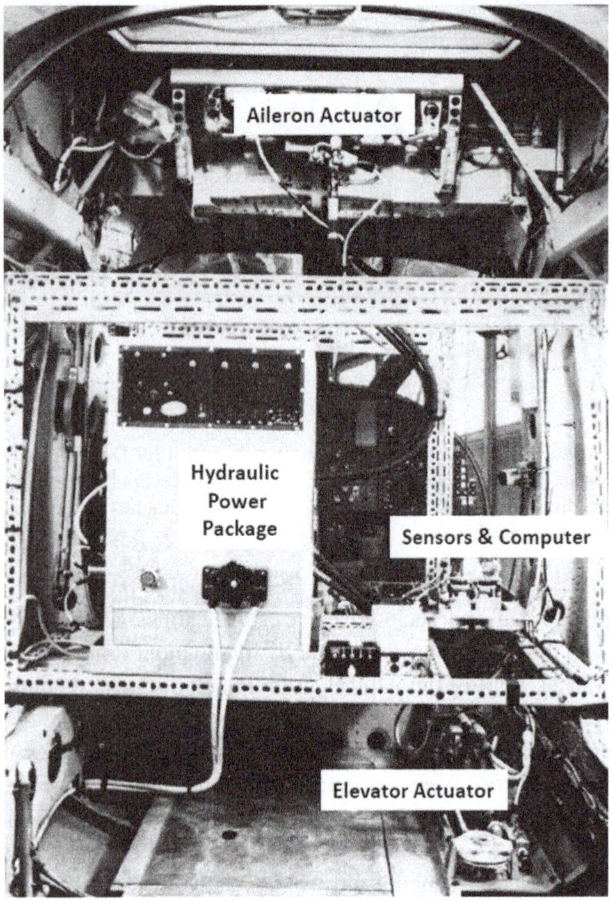

Fig. 6.82 Components of digital FBW control in the Do 27 Cabin

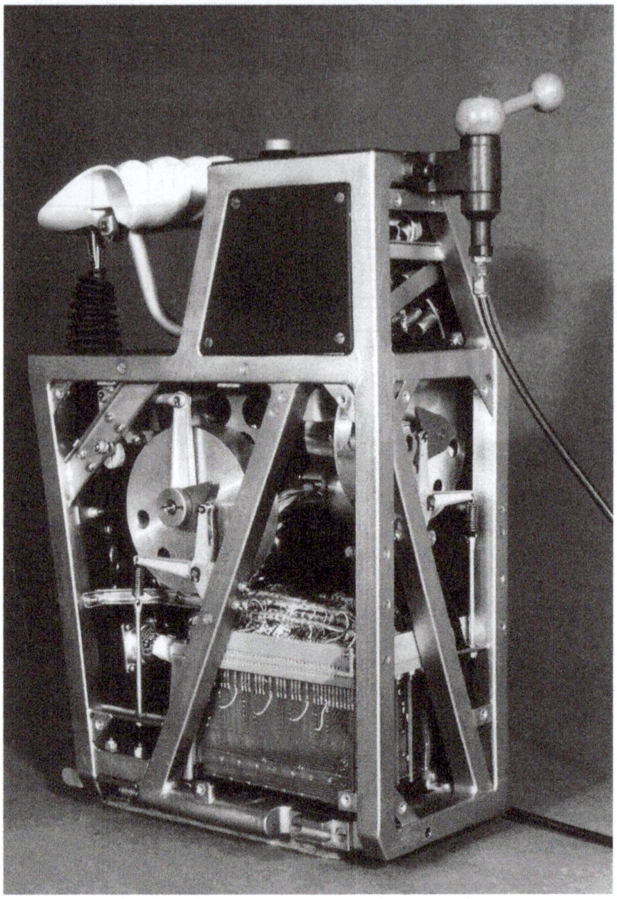

Fig. 6.83 Sidegrip with control force simulation in a casing with encoders and signal processing

usual feedback of the aerodynamic control forces. The encoding gear unit with the connected encoders and electronic components was located in a box mounted to a baseplate fixed to the aircraft structure.

Figure 6.83 shows an improved version of the control system for the test pilot, which was incorporated in the Do 27 as well as in the Pembroke. In this so-called "sidegrip" control device, the control information was tapped under the wrist via a gimbal ring. The range of sidestick operation covered ±45° in two degrees of freedom, namely for the pitch as well as for the roll axis. The term "Fliegen durch Handauflegen" ("Fly by manual contact") was introduced for control with a sidegrip, because the pilot's palm movements corresponded directly to the aircraft movements with the arm resting on an armrest.

A self-centering spring damping system with a specially shaped cam was used to generate force feedback for each axis. This design provides a stiff preloading in the neutral position, while deflected the spring had a softer characteristic. Coaxially with the cam disc, a hydraulic vane pump with adjustable flow characteristics worked as a damping device. The sidegrip, the encoder, and the electronic signal processing formed a compact unit in this design.

Electrohydraulic Actuation

The design of the electrohydraulic actuators has already been described in Sect. 6.3.1.2. The piston travel output of the electrohydraulic actuator systems, mounted on a baseplate, is connected via two rope loops and a switchable magnetic gear coupling onto the basic mechanical control of Do 27. Thus the experimental system could be disconnected from the aircraft's basic mechanical control system (see Fig. 6.84). Additionally, a shear pin device is integrated into each gear coupling (see Fig. 6.85). The shear pins are designed in a way that only a fixed maximum force of the actuators can be transferred to the control surface cables to avoid overloading.

Activation of the Experimental FBW Control

Before activation of the FBW control, a synchronization of the two control systems was essential to balance out the control travels of both the safety pilot command input devices as well as the test pilot systems. Matching was achieved when both the pilots brought their controls into the neutral position. The FBW control could be switched to the

Fig. 6.84 Elektrohydraulic actuator with magnetic gear coupling

basic control by the magnetic gear coupling, but only when both the pilots operated their command switches simultaneously.

Safety Concept

> "The safety concept was designed in such a way, that the safety pilot could take over control at any time with the aircraft's basic controls, for example, when critical operational problems occurred during test flights or when malfunctions happened in the experimental FBW system".

When the FBW system was activated and the test pilot on the right-hand side was in command, the safety pilot on the left side in the cockpit could still monitor the test pilot inputs at any time as he had a feedback of the surface movements on his control stick by the aircraft basic mechanical system. In the case of a failure of the normal disconnection procedure or in the case of a malfunction in the FBW control system, the safety pilot was able to overrun the shear pin and thereby disconnect the FBW test controls from the basic mechanical controls. The force required to break the shear pin was about 10% of the maximum load limit of the basic mechanical control.

Already in the normal disconnection procedure, activated by a switch on the control stick, the piston chambers of the electrohydraulic actuator were short-circuited via bypass valves. By this by-pass between both piston chambers, it was secured that mechanical control remained operational even in the very unlikely failure of the coupling device. The safety pilot had only to apply increased control force to override the experimental system.

This general principle was also adopted for the DLR test helicopter Bo 105-S3 which was equipped with a parallel digital FBW control system (see Chap. 8). Instead of using the shear pin and the magnetic gear coupling for safe disconnection of the experimental control system from the basic controls, the pistons of the electrohydraulic actuator were by-passed and thus easily driven by the mechanical controls.

Fig. 6.85 Coupling point between FBW- and mechanical control with integrated shear pin

Pressure switches in the hydraulic system detected either pressure excess or hydraulic hose ruptures and limit switches at the actuator detected a hard-over position. Each of these failures led to an automatic disconnection of the FBW system.

Flight Tests

The worldwide first experimental flight with a digital electrohydraulic primary aircraft control system was performed with the Do 27 (YA 913) on June 12, 1967 (see Fig. 6.86). During the Do 27 test and demonstration flights, dubbed DPFS ("Digitale Flugzeug Primär Steuerung"—Digital Aircraft Primary Control), measurements were carried out for standard maneuvers, in which the maximum of tolerable and noticeable time delay was determined by the pilot. The quantization, that is the number of steps per full displacement, appeared to be an important parameter for the system design. The test system enabled a change of the quantization during flight trials to evaluate the effect of different increment values. This could be considered only as a basic attempt because the transmission characteristic was

Fig. 6.86 Control inputs in Do 27 through a sidegrip inceptor by *Johannes Tersteegen*

Fig. 6.87 Test aircraft Percival Pembroke (YA 558)

Fig. 6.88 Components of digital FBW control in the Pembroke cabin (*right*, from front: digital computer, two hydraulic systems, inertial platform LN3, *left*, input console for digital computer, central operating desk for hydraulic system, triplex actuator drive)

influenced by the flexibility, friction and dead band present in many elements of Do 27 basic mechanical controls.

The available electrical energy, payload and space in the Do 27 cabin were insufficient for advanced investigations and measurements in flight. Nevertheless, based on the experience gained with the first digital FBW control system and the theoretical findings for the required reliability as well as the development of digital electronic components, another technically advanced system was developed for the test aircraft, Pembroke.

6.3.1.4 Test Demonstrator Pembroke

The Pembroke (YA 558) is a twin-engine, multipurpose, light transport aircraft produced by the British manufacturer Percival Aircraft Ltd. A total of 136 aircraft were manufactured in different versions from roughly 1952–1959. The Pembroke is a high-wing monoplane with main landing gear integrated into the two engine nacelles and has two 9-cylinder radial engines, each with 540 hp. With the Pembroke, an experimental aircraft was available with a considerably larger cabin for a higher payload of equipment and space for an additional test engineer (see Fig. 6.87).

In 1968, a triple-redundant digital electrohydraulic primary aircraft control system was installed in the Percival Pembroke (YA 558) to drive the elevator (see Fig. 6.88). The digital signal processing and transmission, as well as the fault detection system, consisted of a central unit with triple-redundant subsystems (see Fig. 6.89) [45]. Investigations of potential reliability models and the findings from reliability calculations led to these functional subsystems, in which all parts (mechanical and electronic) with random failure characteristic were pooled. The subsystems could be replaced manually after component failure and its detection by the inherent self-diagnostic system. All three parallel units were connected by a hinge mechanism via guides and safety gears so that all of them received the equal mechanical and electrical information.

The advanced test system was designed to enable inputs from a digital controller. The programmable digital controller consisted of a general-purpose digital computer Honeywell H316 and a flight data acquisition system [46].

A triple-redundant electrohydraulic system from H. M. Hobson Ltd., Wolverhampton, England was used for elevator deflection (see Fig. 6.90, [47]). Duplex torque motors as input D/A converter per channel were developed

Fig. 6.89 Manually replaceable, triplex electronic subsystem with functional units and fault detection; two subsystems removed

by the DFL. The aileron control was provided by a simplex actuator described in Sect. 6.3.2.1. In General, the coupling between the experimental system and the basic mechanical control of the Pembroke was similar to the design in the Do 27, using the same switching procedures and devices.

Initially, the flight tests were focused on safety and reliability issues, and the response characteristics to signal transmission. As already determined in ground tests, it was also confirmed during flight tests that each detected and repaired failure in the digital system during operation did not produce more than 0.3% discrepancy in the control signal.

Further flight tests had the objective to determine the effect of quantization on the controllability of the overall system with the pilot in the loop [45]. The influence on glide path accuracy in ILS approaches was investigated carefully. The analysis was carried out applying statistical methods. It was observed that for the most precisely quantization, that is, the time highest number of 127 steps, the aircraft controllability was nearly as good as with the basic mechanical control system. These results were confirmed by subjective evaluations of a number of test pilots. After installation of the above-mentioned programmable digital controller, further tests were performed by changing the flying qualities of the test aircraft. Results of these tests were included in on-going investigations in the field of digital flight control, in particular with regard to the experimental evaluation of new systems for a range of different flying qualities (see Chap. 7).

6.3.1.5 Epilogue

The first digital FBW landing took place in 1968, with the digital electrohydraulic primary aircraft control system integrated into the Pembroke aircraft.

On various occasions, the Do 27 and the Pembroke were flown also by guest pilots, scientists and even politicians. To

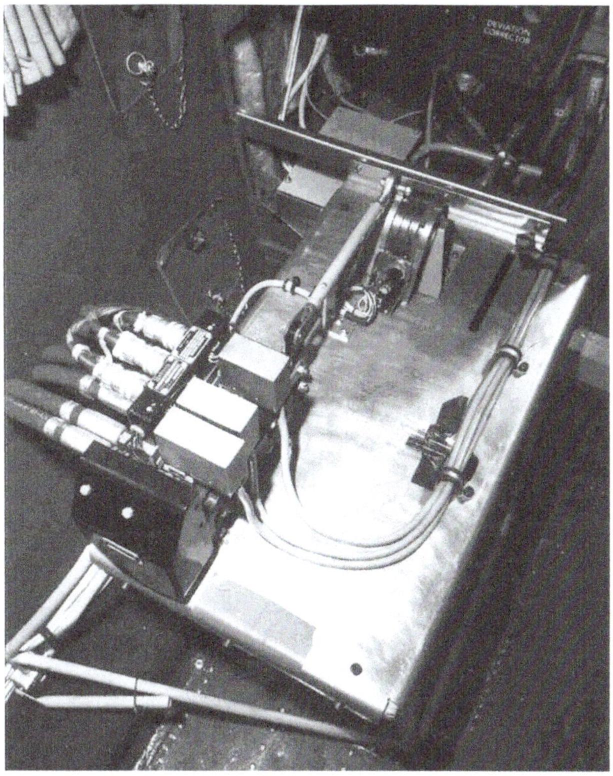

Fig. 6.90 Electrohydraulic triplex-actuation with magnetic gear coupling for elevators

mention just a few, these included the renowned expert in control engineering, *Winfried Oppelt* from Germany, former Chief Minister and sovereign of the German state of Lower Saxony, *Georg Diederichs*, as well as the first man on the moon, the astronaut *Neil Armstrong* (see Figs. 6.91 and 12.11). During his visit of the DFL, it was hard for *Neil Armstrong* to imagine how these few employees could get a digital FBW control system operational in a test aircraft.

After detailed theoretical system studies, the development of appropriate components and test systems and proof of their suitability, initially in the laboratory and then in flight test, it could be shown that the problems of reliability in electrohydraulic flight control systems arising due to digital signal processing were manageable. It was now required to demonstrate by endurance testing, that the theoretical approach of the reliability concept was virtually valid. For this purpose, the equipment developed for digital FBW flight controls was, after extensive adaptations, installed and successfully tested in the rudder control of the research vessel "Planet" of the Oceanographic Research Centre of the German Navy, providing a possibility of 24-h operations (see Fig. 6.92, [48]). The ship was built in 1967 by Norderwerft in Hamburg with a gross register tonnage of 1950 tons. It was decommissioned in 2004.

Fig. 6.91 After a demonstration flight with Pembroke (from left: test pilot "*Halu*" *Meyer*, FBW expert *K.-H. Doetsch* and Astronaut *Neil Armstrong*

6.3.2 Research Aircraft Dornier Do 28D Skyservant (1968–1992)

Gunther Schänzer

6.3.2.1 BGT Do 28D FBW (1968–1978)

The Bodenseegerätetechnik company (BGT) in Überlingen was the leading German developer and manufacturer of flight control systems in the nineteen-sixties and seventies. Among others, prototypes of all three German vertical takeoff vehicles (Do 31, VJ 101, VAK 191) were developed, manufactured and tested, in cooperation with the manufacturers Dornier, Entwicklungsring-Süd (MBB) and VFW Bremen [19]. At the behest of German Lufthansa (DLH), an autothrottle was designed, manufactured, and tested for speed control. It was installed on the Boeing 707 fleet of DLH and thereby fulfilled the requirements for reliable approaches under bad weather conditions according to CAT 2 [19]. These experiences enabled BGT to participate successfully in the call for tenders for flight controller for the European multirole combat aircraft Tornado and Airbus A300.

In order to test more efficiently flight control concepts and other equipment, BGT decided in 1967 to obtain a twin-engine experimental aircraft. This new aircraft was foreseen for various tasks and should possess flexible equipment, allow quick modifications and acquisition, and be cost effective in operation.

Fig. 6.92 Research vessel Planet (*Credit* Frank Behling)

A twin-engine aircraft was envisaged, capable for IFR operations, with a relatively large installation volume, allowed flexible modification possibilities and adequate payload, as well as forgiving flying qualities. Maximum airspeed, flight altitude, and range were of secondary importance. The Dornier Do 28D Skyservant largely fulfilled these requirements (Fig. 6.93). The box-shaped fuselage provided a large installation volume and sufficient headroom. A pressurized cabin was not essential because of the lower maximum altitude. Holes in the fuselage skin were easy to implement, for example for antenna installations, also in relation to the re-certification of the modified aircraft. The aircraft had a civilian certification for all required applications.

In addition to the basic aircraft controls, electromechanical actuators were installed for elevator, rudder, aileron, and thrust control. Furthermore, electric trim motors were provided for elevators, rudder and ailerons for automatic trim in order to compensate the high control column forces during asymmetric one-engine flights and flaring maneuvers. For reasons of flight safety, all actuators could be disconnected by the pilot through adjustable slipping clutches with a specific, calibrated slip moment limit. Additionally the clutches were disengageable through emergency switches on the control columns. In this case, the pilot took over a properly trimmed aircraft. The electromechanical actuators were directly connected to relevant control rods and moved with the control columns. The pilots were able to instantaneously follow the control column movements and thus monitor the system. With this simple and reliable safety concept, a civil certification of the modified aircraft was relatively easy to achieve.

Precise measurements of static and dynamic pressures could only be achieved to a limited extend due to strong propeller slipstream effects around the flow close to the

Fig. 6.93 Do 28D Skyservant with nose boom and flight log

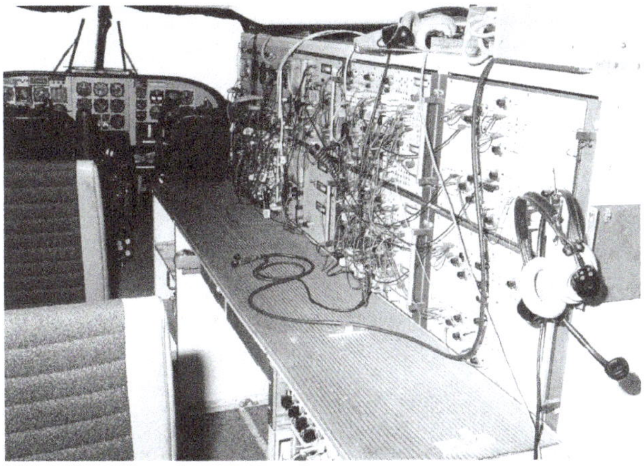

Fig. 6.94 Cabinet with electronics equipment and analog computer

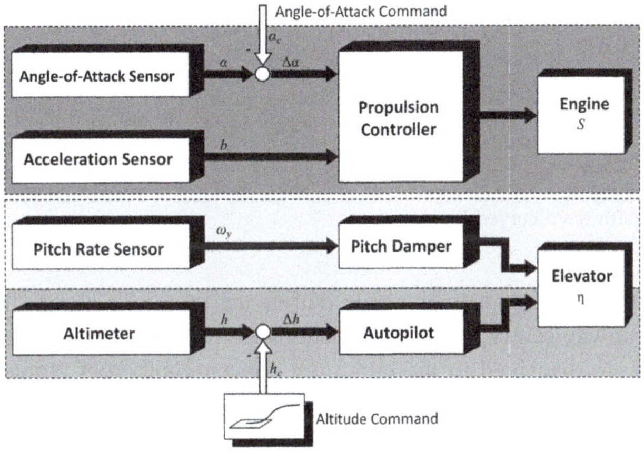

Fig. 6.95 Block diagram of a conventional flight controller (longitudinal motion)

aircraft. The basic version of the Do 28D had two adjacent *Prandl* tubes on top of the vertical tail. The accuracy of these *Prandl* tubes was sufficient for the normal flight operations, but not for experimental test operations. In the BGT test aircraft Do 28D, a nose boom including a precise *Prandl* tube was installed by Dornier before delivery, that aligned itself in the flow direction. With this flight log, a special air data sensor, precise measurements of the true airspeed vector could be made (see the lower part of Fig. 6.93).

To enable flight testing of different flight control laws more conveniently without major modifications, all kinds of sensors were pre-installed. A certified cabinet was provided to accommodate these sensors and an analog computer (see Fig. 6.94). The analog computer could access all sensors through appropriate signal converters.

Due to the relatively low weight as well as low maximum airspeed of Do 28D, powerful hydraulic actuators were not essential. The simple "*all electric aircraft*" proved to be more than adequate as an experimental aircraft. An analog multichannel ink recorder was used for data recording. In normal flight test operations, the experimental aircraft was flown by two pilots and accompanied by a flight test engineer.

Additional sensors required for the flight control and guidance system were (1) body-fixed angular rate gyros in all 3 axes, (2) body-fixed linear accelerometer along all 3 axes, (3) 3-axis inertial reference platform, (4) altimeter (barometric and RADAR), and (5) radio navigation receiver (VOR, DME, ILS) [49].

It was sufficient to address the first project with this set of basic equipment, namely "*Proof that an automatic trajectory tracking was possible with the military microwave landing system SETAC (SEctorTA Can and TACAN (Tactical Air Navigation) of the company SEL*" [50].

Initially, a conventional flight control system consisting of a damper, autopilot, and autothrottle (see Fig. 6.95) had to be realized on the analog computer. The aircraft characteristics and aerodynamic derivatives were determined with handbook methods and additional flight test data [49]. The flight controller was designed in a conventional manner and simulated and optimized in the laboratory together with the flight dynamic model of the aircraft using a ground-based flight simulator. Once the unexpectedly large thrust moment of the Do 28D was accounted for, the flight controller worked satisfactorily on the ground and in flight. With the standard instrument landing system (ILS), the aircraft flew

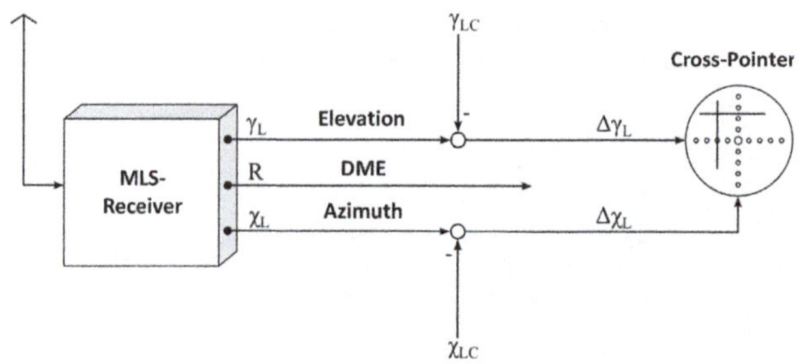

Fig. 6.96 Standard display for a microwave landing system (MLS)

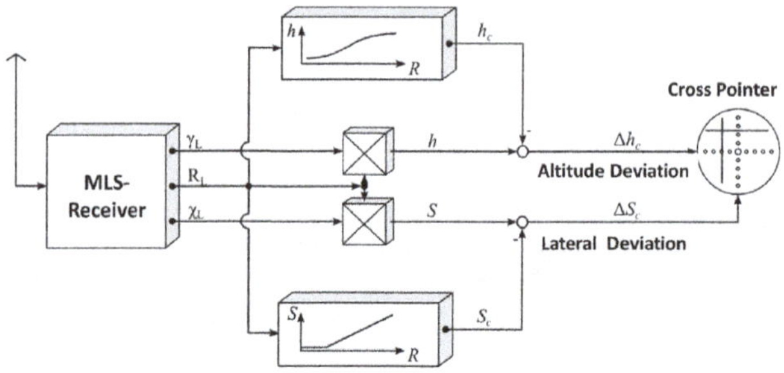

Fig. 6.97 MLS-receiver with trajectory tracking device to generate earth fixed curved flight path

with a 3° glide path angle during the approach along the runway center line. The deviations from the nominal values were displayed to the pilot on a conventional cross-pointer instrument (see Fig. 6.96) and could be corrected manually or automatically. For the SETAC system, the pilot selected the desired values manually using a knob on the SETAC-operating console.

In the automatic control mode, for example, the deviations were fed as command signals for course and altitude (see Fig. 6.96). In the SETAC mode, however, the approach paths could be arbitrarily chosen by the pilot within the permissible boundaries (see Fig. 6.96). If the centerline is specified as target azimuth and 3° as target glide path angle for the SETAC system, then the conditions for manual and automatic approaches are nearly the same in comparison to conventional ILS approaches (except for better signal quality and higher accuracy with SETAC). If the transmitted distance signal is used in the simple trajectory tracking device (see Fig. 6.97), even desired trajectories with elevation and azimuth curvatures could be generated with SETAC. This trajectory tracking device was initially assembled based on analog technology (operational amplifier and non-linear components). The extreme temperature fluctuations in the non-airconditioned fuselage of the Skyservant gave rise to unacceptable discontinuities (bends, jumps) in the target trajectory, which had to be followed by the precisely operating flight controller. An elaborate calibration of the analog trajectory tracking device was desirable before each flight.

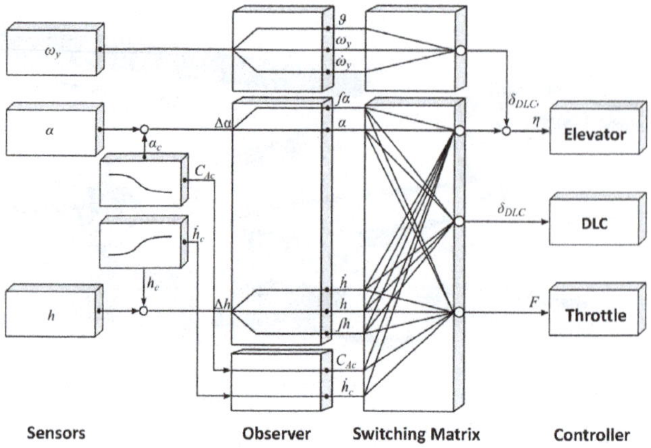

Fig. 6.98 Integrated flight control system

These problems could be resolved later on with an airworthy programmable digital computer, which was fast enough for calculating simple nonlinear functions.

The conventional flight control system with separate components of the damper, autopilot, and thrust controller (see Fig. 6.95) encountered operational limits when trajectories steeper than 4° and low airspeeds were reached. The large moment variations as a result of flight mechanics couplings were no longer manageable. These coupling effects had also to be emulated appropriately in the flight controller as well (see Fig. 6.98).

The full state feedback was a new method in the beginning of nineteen seventies and appeared promising to tackle the described flight control problems. The methods usually applied for designing a flight controller were based on transfer functions and eigenvalues. Given a large number of states and feedbacks, these standard methods were no longer practical.

The theory of optimal state feedback provided a theoretical approach, namely the cost function Q which was to be minimized to achieve optimal flight control behavior. After many iterations between flight testing and computer simulations, a special distribution of various factors contributing to the function Q (quality criterion) was arrived at: change in the target trajectory—3%, change of commanded angle of attack—3%, wind shear—10%, and atmospheric turbulence —84% [51]. The cost functional Q was composed of the four elements, namely, deviation from the target trajectory, deviation from the desired speed, low throttle activity, and passenger comfort. These elements had to be weighed against each other, as a simultaneous minimization was not possible. The question whether a speed deviation of 1 m/s or height deviation by 1 m is unfavorable needed to be answered before the optimization. It was obvious that the answer depended on the current aircraft flight condition, for example at what altitude the aircraft is flying. The working hypothesis that the energy variations due to an altitude error and due to velocity errors are equally uncomfortable, led to good results in flight operations [51]. A very high importance was assigned to the pilot evaluation of low throttle activity, acceptable accelerations at the pilot's station as well as low pitch and yaw motions. Under good visual conditions pitching and yawing motions were for the pilots more annoying than response due to weather conditions. This criterion played only a minor role under poor visibility conditions.

Under calm atmospheric conditions, all requirements could be fulfilled almost error-free using the coupled flight control system for pre-defined flight trajectory and speed commands. In this case, the cost function, Q, is almost zero. In order to account for the flight test results in the optimization of flight control laws, many feedbacks had to be selectively modified. It was possible to assess only three weighting factors through the four basic elements of the cost function Q. The squared average value of the error between the flight path and the aerodynamic flow conditions during flight testing was easy to gather, as well as the throttle activity. The assessment of the subjective pilot and passenger comfort had to be carried out together with the pilots. After introductory training, pilots and flight test engineers were well capable of proposing quantitative changes in the weighting factors of the cost function. The flight control laws were adjusted with these modified weighting factors in the flight simulator to minimize the cost function. These values were transferred to the analog computer of the experimental aircraft. Thereafter, a new flight test was started. This iteration converged quite fast. In the implementation of the coupled autopilot, those contributions of less than 1% to the total cost function value were ignored.

Figure 6.98 shows the realized control structure. The states not directly measured were estimated by an observer. Initially, only elevator, elevator trim, and throttle were considered as independent variables. The direct lift controller was added later. The open loop nonlinear control, which generates the required actuator movements for the command functions (flight path, airspeed), is not separately shown in Fig. 6.99. The thrust is a function of the commanded flight path angle and the glide ratio C_L/C_D dependent on the aircraft weight G. The elevator deflection and thus the pitching moment is dependent on the commanded lift coefficient C_L. Thrust changes generate a thrust moment that had to be compensated by means of the elevator.

The controller for the lateral motion consisted primarily of two angular rate gyros in the roll and yaw axes, a sideslip angle sensor (as part of the flight logs), as well as a body-fixed lateral accelerometer. These sensor signals were also fed to the ailerons and rudder. The lateral acceleration signal catered primarily for sideslip free flight during the banked turns. The necessary lift in a turning flight, which increases with the bank angle, was additionally fed to

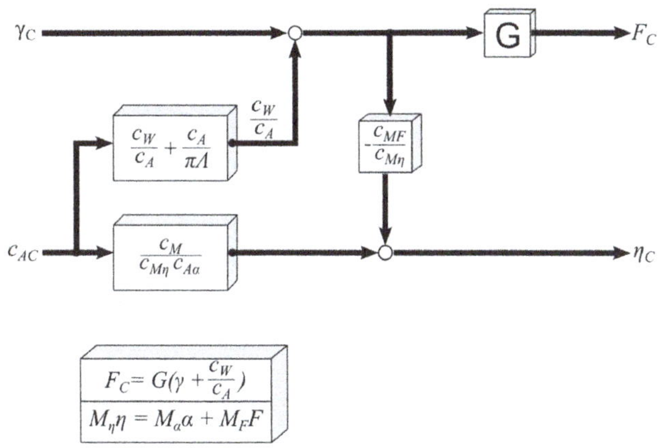

Fig. 6.99 Open loop control system for Fig. 6.92

longitudinal motion controller to precisely hold the altitude. Without going into further details, it can be stated that the flight controller emulates the inverse flight mechanical behavior of the airplane, at least as far the flight performance is concerned.

Steep approaches with flight path angle up to 8° on curved trajectories were flown in flight test with the integrated flight control system (see Figs. 6.98 and 6.99) and the trajectory tracking device (see Fig. 6.97). Figure 6.100 shows a flight test result. An inclination of about 2° is observed for intercepting the earth-fixed nominal flight path profile. This was followed by a curved transition to the 7° step approach. Before the final flare, the vertical speed had to be reduced corresponding to a standard flight path angle of 3° in a further curved profile at the right time. For the typical approach speed of only 70 kts in the STOL (*short takeoff and landing*) mode of the Skyservant, this guided flare is barely visible. The approach was completed with a conventional automatic landing. Suspected overshoots in trajectory and speed did not occur. During the approach as depicted in Fig. 6.100, a strong wind shear of maximum 7 kts/100 ft was encountered, which was fully compensated by the flight controller.

In STOL approaches, the aerodynamic flow condition must be precisely and reliably controlled. In the experimental aircraft three different methods were employed based on dynamic pressure (conventional), angle of attack α, and lift coefficient C_L.

The lift coefficient was determined with the aid of pressure ports on the upper side of the wing. Ports in the quarter-chord line outside the propeller slipstream influence are preferred. All three methods worked satisfactorily. Even nearly undamped oscillations of the Dornier flight log posed no problems to the integrated flight controller. The oscillation frequency of the flight log increases linearly with airspeed and can excite the nose boom, which may lead to structural oscillations.

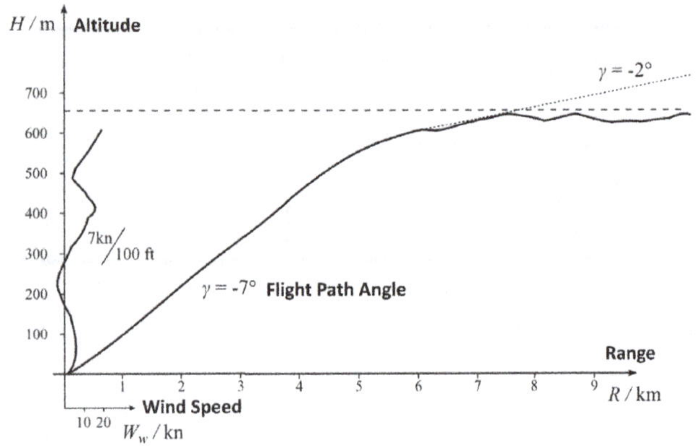

Fig. 6.100 Automatic steep approach along a curved flight path, including automatic landing

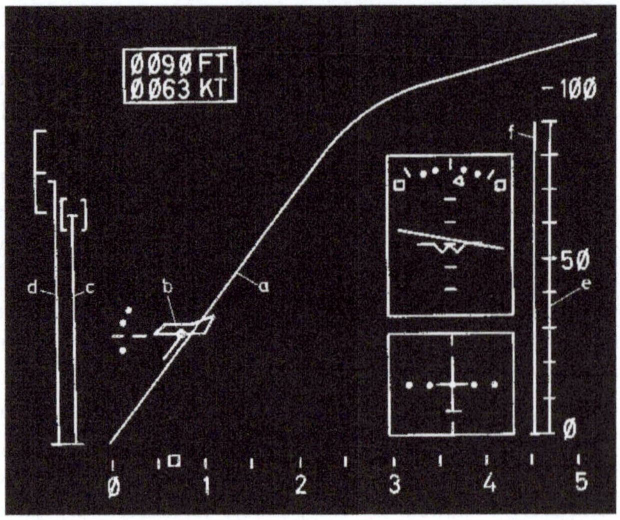

Fig. 6.101 Display of curved approach

A precise knowledge of the airspeed vector (aircraft speed relative to the moving atmosphere) is significant for various research tasks such as flight control law design, parameter identification, wind shear, wake vortices, optimal gliding. Hence, all relevant signal sources (standard air data sensors, flight log, static pressure, wing pressure) were optimally estimated in an air data computer.

Although all tasks related to STOL landing approaches based on the microwave landing system SETAC were successfully completed, two observed shortcomings in flight operations had to be eliminated:

1. The system operation during the transition from different flight phases, for example, from long-distance flight into the steep approach flight phase was inconvenient and was associated with increased pilot workload.
2. The information of deviations from the nominal values of trajectory and aerodynamic flow conditions was based on display concepts, which had proved successful only for conventional linear (non-curved) approaches with fixed nominal values, for example, the cross-pointer instrument for ILS approaches.

The important issue for the pilot was that he should have the right knowledge about the exact location of the aircraft on the flight path, and how to change the thrust to recover the aircraft on a curved trajectory. Of particular interest was the final flare and how the thrust should be increased at the proper time during the transition in order to reduce the high vertical speed. The delayed approaches for aircraft noise reduction, discussed during that period, posed no problems to the flight control system, but considerably bothered the pilot under poor visibility conditions. In cooperation with the DLR and the company VDO Aviation (today: Diehl Avionics), an electronic display including a symbol generator was developed, that satisfied the pilots and increased significantly their comfort in dynamically eventful approaches. The realization of such electronic displays and its symbol generators was in the early nineteen-seventies quite a challenge (see Fig. 6.101).

To improve the ease handling of the flight guidance system, both Do 28D control columns were equipped with sensors that recorded the control forces applied by the pilot and fed them to the integrated flight control system as nominal values. The control inputs in the roll axis were, as with other Fly-by-Wire projects, defined roll rate commands and additionally integrated to a commanded bank angle (roll attitude hold). If the pilot did not exert forces any more, the commanded bank angle was maintained and the aircraft flew a constant curve, provided the nominal values in the longitudinal motion were met. If the commanded roll angle was reduced to zero, the aircraft maintained its course, as in the case of an uncontrolled aircraft. To simplify the pilot operations, especially in turbulent atmosphere, the achieved course was fixed by the autopilot, as soon as the commanded roll angle fell below a predetermined tolerance value.

Similarly, the Fly-by-Wire function was realized for the longitudinal motion. Generally, the control column force signal was interpreted by the autopilot as pitch rate command which was integrated to a pitch attitude (pitch attitude hold). If the control force signal went to zero, the commanded pitch attitude remained unchanged. This is the standard method for *control wheel steering* (CWS) for highly maneuverable aircraft, which is equivalent to the C* method (see also Sect. 6.2.2.4).

In the case of an integrated flight control system, according to Fig. 6.98, however, the relevant control column force was interpreted as commanded vertical acceleration and integrated to yield a commanded vertical speed. Once the commanded vertical speed was achieved within specified tolerance, the altitude reached was hold thereafter. For compatible boundary conditions, a smooth transition to a spatially fixed nominal target trajectory could ensue in both commanded altitude as well as commanded vertical speed.

In the case of an integrated flight control system, the aerodynamic flight condition, for example, the angle of attack, was in addition to the flight path predefined, as the pitch attitude cannot be directly controlled by the pilot. The pilots needed to get used to this behavior. Another phenomenon was, on the other hand, accepted after a short familiarization phase. If the flight path was flattened in a descending flight (either automatically or by the pilot), the flight controller reduced accordingly the thrust (see Fig. 6.98). If the pilot commanded a nominal vertical speed change in the direction of the descent, he was pushing the control column. An increased vertical speed required a thrust reduction. Because of the large induced pitching moment of the Do 28D due to thrust changes, the pitch up moment had to be compensated From a control engineering point of view, this is a classical "non-minimum phase effect", which is considered generally difficult to control. This process is felt by the pilot as natural and hardly noticed because the unregulated aircraft during this maneuver behaves exactly as the controlled.

The topic of whether a direct lift control system (DLC) would be useful or not was interestingly and controversially discussed during the early nineteen seventies. In cooperation with Dornier, the landing flaps were equipped with a faster actuator to realize a DLC function [52]. In the normal mode, the flap continued to operate in the full range up to 52° and lock in three positions as before. Since the Do28D has had no hydraulic supply, the only feasible solution were electromechanical actuators. A powerful actuator was provided by Lear Siegler, resulting in an increased speed of flap movements from 2.3°/s to 15°/s. Due to flight safety reasons, the flap movement speeds were limited in the range of ±10° flaps. The resulting maximum change in the lift coefficient C_L was in the range of 0.6. With these values, it was possible to mostly suppress the gusts induced vertical accelerations. At a fixed position of 10°, the flaps generated primarily lift variations with secondary drag changes, and at 42° primarily drag changes with less lift changes (see also Sect. 9.2.5).

In the optimization of the integrated flight control system, a cost functional was used with and without the influence of DLC. With the aircraft parameters of the Do28D, a 40% reduction in the cost functional Q resulted with DLC compared to that without DLC (Fig. 6.102). This improvement was not negligible but occurred in areas which were commonly not directly considered in the design. The substantial improvement resulted with throttle activity. The flaps responded quickly to vertical and horizontal gusts disturbances, and the resulting drag changes made the higher frequency thrust changes unnecessary. The reduction of

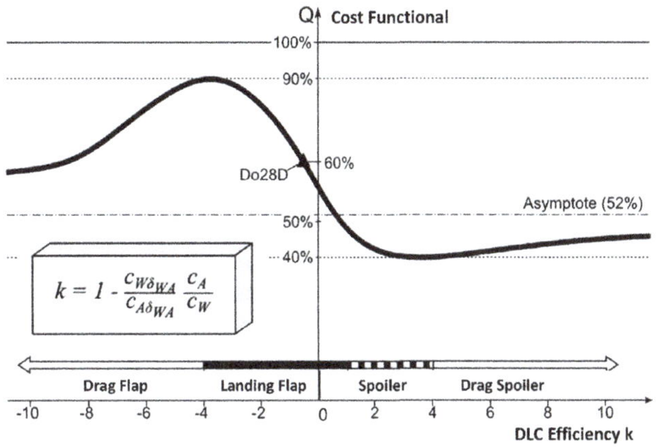

Fig. 6.102 Effect of direct-lift-control (DLC) on the normalized cost function Q of an optimized controller (Q = 100%: without DLC). Parameter k is a measure of the flap lift to drag ratio

unwanted pitch rates had the next largest influence on the cost functional. The drag variation in the DLC system of the Do28D was significantly more important than the lift variations. These effects could be formulated through the DLC effectiveness, which was defined as shown in Fig. 6.102, and varied in the range of −10 to +10 for a hypothetical aircraft [52]. The best results were provided by a spoiler (k ≈ 4) with a reduced cost functional of 40%. Spoiler deflection led to an increase in drag and at the same time reduction in lift. For a typical landing flap (k ≈ −3), the lift increased and at the same time, the drag increased slightly. In this case, the cost functional reduced to marginal 90%. Here the effort was not worthwhile. For very large values of k the cost functional approaches asymptotically to Q = 52%. For these values, the lift variation is small compared to drag variation. The behavior of spoilers (lift reduction with drag increase) was basically advantageous over lift flaps (lift increase with drag increase). Pure drag spoiler yielded very good results, which are usually easier to install and less critical during failures.

When comparing steep approaches with regard to the flight path accuracy, hardly any differences were noted with or without DLC. In contrast, additional DLC increased considerably the pilot and passenger comfort. Sufficient smooth thrust activities were the most important assessment criterion for a pilot. BGT had already learnt this lesson in testing their thrust controller in the Boeing 707 of DLH [19].

6.3.2.2 ITB Do 28D GPS-Based Automatic Landing (1980–1992)

The basic development of an integrated flight control system was demonstrated and completed successfully. Starting 1976 it became increasingly difficult for BGT to acquire sufficient contracts to maintain the aircraft and its team.

The Institute of Flight Mechanics of the Institute of Technical Braunschweig (ITB) had addressed the research topic of "wind and turbulence". The former BGT Do 28D

Skyservant well-equipped with sensors was highly suitable for measurement of the movement of air masses. Wind, wind shear, and turbulence were of interest to the flight mechanics engineers as well as for meteorologists. The increase of the wind measurement accuracy evolved as an attractive scientific topic. The ITB managed to acquire contracts to lease on daily basis the Skyservant from BGT and to gather wind and turbulence measurement data. A challenging scientific aim was to increase the measurement accuracy.

The objective of a wind speed measurement accuracy of 10 cm/s resulted in the requirement of a measurement accuracy of 10 cm/s for ground speed and airspeed as well. This specification necessitated a measurement accuracy of at least 0.15% of the Skyservant speed of about 70 m/s during a flight measurement campaign. Relatively quickly an accuracy of about 1 m/sec was achieved, which was for many anticipated tasks initially quite satisfactory. As the aircraft was increasingly deployed by the TU-BS with the financial support of the German Research Foundation (Deutsche Forschungsgemeinschaft—DFG), the ownership of the Do 28D Skyservant changed from BGT to the German State of Lower Saxony in 1980. The outdated analog computer became inadequate for the intended applications and was replaced by a highly complex digital computer system.

To increase the measurement accuracy, methods were available to formulate the measurement error in suitable error models. Good results were achieved with (stationary) Wiener filters, assuming the measurement errors to be time invariant and constant over the period of observation. This Wiener filter could also be realized by analog computers. Time-varying complementary filtering by means of Kalman filter could be implemented with the high-speed digital computer, and thus the measurement accuracy could be significantly improved. The Kalman filter had to be calculated in real time during the flight test. Although fast, due to the available finite computing power, the number of estimated error states had to be limited to those which were considered essential. Despite these practical limitations in flight tests, significant progress could be made with regard to the measurement accuracy.

The flight path measurement methodology was further developed towards precision positioning. With the support of German Federal Minister for Research and Technology (BMFT), an inertial navigation platform of the type Delco Caroussel 4 was procured and integrated. The development was continued, as satellite tracking with the Global Positioning System (GPS) for moving vehicles became operational from 1989. Once again with the support from BMFT, two highly accurate satellite positioning receivers were procured in 1987, which were developed by the Sercel company for geodetic tasks. Under these conditions, a positioning accuracy of less than a meter was achieved in real time using the differential method and GPS phase measurements. The high accuracy of differential GPS was very susceptible to interference due to only a few satellites being visible and the unusual low strength of the electromagnetic GPS signals. An excellent high precision positioning was realized through complementary filtering of precise but interference-prone GPS signals and inertial positioning signals with short-term high precision but long-term drifting characteristics. In the end, the inertial platform was calibrated with the support of satellite positioning when the GPS was reliably available.

The replacement of the previously employed microwave landing system (MLS), the locating method SETAC and the DLS (DME-based landing system) through a real-time differential GPS system with phase measurement proved comparatively easy. From the beginning of 1989, four GPS satellites in an appropriate geometric constellation were visible quite seldom for a period of ca. 30 min and flight testing of curved steep approach on the basis of GPS was made possible. During the symposium "Satellite Navigation Methods" of the German Society of Navigation (DGON), held in Braunschweig during July 1989, two approaches were publicly demonstrated, namely a conventional approach and a steep approach with curved flight path, each with the final automatic landing. Therewith the proof was furnished that satellite navigation has an application potential for the substitution of radio navigation methods. Since serious doubts were raised worldwide about this demonstration, the flight demonstrations were again repeated during the "First International Symposium on Real Time Differential Applications of the Global Positioning System" of the International Association of Institutes of Navigation in Braunschweig, during September 1991.

Eventually, with these investigations a base had been created for the European satellite navigation method "Galileo", currently being designed and tested. Over the following years, intensive studies including flight tests were carried out with the aim to prove the suitability of satellite positioning as a substitute for the conventional, reliable but inflexible radio navigation (VOR, DME, ILS, RADAR). The integrated flight control system was frequently involved in these flight tests, but it was rarely required to improve or pursue further research.

In addition to satellite-based precision positioning, the quality of the measurement of the air speed vector (aircraft speed relative to the moving atmosphere) was further pursued (see Fig. 6.102). Additional several sensors such as laser anemometer had the potential to increase further the measurement accuracy by further improving the complementary filter methods. For example, a number of measurement error elements have occurred in turning flights, which could be compensated to a large extent by assuming that the wind velocity components to be independent of the aircraft flight path. To calibrate the wind measurement system a turning flight was performed previously. It was possible to measure ground speed to 10 cm/s. The achieved high accuracy of the wind velocity component measurement excited the interest of meteorologists and led numerous research contracts and secured thereby funding for flight operations. The main client was the Alfred Wegener Institute for Polar and Marine Research (AWI). Since the Do 28D Skyservant of TU Braunschweig was not suitable for use in polar regions, the AWI management encouraged the Institute of Flight Guidance TU to set up a commercial company to implement the scientifically oriented flight measurement methods in the polar planes of the AWI. As a part of the technology transfer program, this company was established in 1985 under the name Aero Data Flight Measurement Technique, Ltd.

The precision positioning method developed by using the Skyservant prompted Aero Data to establish a calibration procedure for radio navigation systems. Based on the precision positioning method and the digital processing technology, the time and labor (including flight hours) for calibration of radio navigation systems could be significantly reduced. Aero Data developed and produced complete flight measurement systems (*Flight Calibration System*), which were delivered to the German Federal Air Traffic Control Authority (BFS) as well as to Swiss Control. Today Aero Data supplies about 70% of the world market with flight inspection systems (FIS).

The close cooperation with the AWI opened the possibility to take over one of its three aircraft, namely Polar 1, which is a Do 128-6 Skyservant with PTL engines. The necessary funding was provided by the BMFT, the German Federal state of Lower Saxony and resources of the Institute of Flight Guidance of the TU-BS. This aircraft was equipped with extensive sensors, a powerful digital real-time data processing system, and digital data recording. It was mainly used for precise wind measurements and precision navigation. An integrated flight control system was not installed for cost reasons. Since sufficient contracts could be acquired for two test aircraft, both the Do 28D and the Do 128-6 were operated jointly during 1986–1992 (Fig. 6.103).

As long as the German Armed Forces deployed Do 28D Skyservant aircraft maintenance was not a problem. Thereafter the Do 28D Skyservant had to be grounded after 24 years of successful flight test activities. The fully functional equipment could only partially be retrieved for other projects due to recertification limits. A re-registration for automatic landing with the Do128-6 appeared illusory. Thereby the CCV operation of the Do 28D Skyservant was terminated. The aircraft including complete equipment is displayed since 1992 in the Aviation Museum in Wernigerode (see Fig. 6.104).

Fig. 6.103 Do 28D (front) and Do 128-6 of TU Braunschweig

Fig. 6.104 Transport of Do 28D Skyservant to the Aviation Museum, Wernigerode

6.3.3 Dornier Do 128 TNT OLGA (1976–1980)

Bernd Krag and Horst Wünnenberg

6.3.3.1 Introduction

During the late nineteen sixties/early seventies, the NASA and the US aeronautical industry were engaged with a new generation of fighter aircraft, which was to have a whole range of unusual properties, such as high maneuverability, relaxed static stability, controlled flight at very high angles of attack, and artificial damping of structural vibrations while flying in turbulent air or close to ground (*Terrain Following*). This new generation of aircraft has been called "*Control Configured Vehicles*", abbreviated simply as CCV aircraft. Starting from the experience with Fly-by-Wire technology, and latest findings and methods in flight control technology, it was anticipated to arrive at aircraft configurations, which were optimized for a specific mission. Flying qualities were to be ensured through flight control augmentation.

In Germany too, this idea was pursued and appropriate projects were initiated in the industry and research organizations under the so-called ZKP program (Ziviles Komponenten Programm). Since flight testing of innovative controller functions was risky and just a computer simulation insufficiently conclusive, the Institute of Flight Mechanics of DFVLR chose a third way, which offered more realism with less risk, namely the "Dynamic Simulation in Wind Tunnel" test technique with controllable flying models in wind tunnels [53].

The primary objective of this work was not the development of new aircraft configurations, rather being mainly concerned with appropriate controller functions (*Control Laws*) for a futuristic, hypothetical Fly-by-Wire aircraft. With this new experimental technique, the complete chain from the sensor, control algorithm, and right up to the actuator could be realistically investigated and optimized. Furthermore, an interaction between several control functions and their impact on the flight behavior were of interest. In addition, answers were to be looked for to the question of to what extent the experimentally generated results can be applied to real aircraft.

One of these innovative control functions was focused on so-called "Gust Alleviation" to smooth the flight in the turbulent atmosphere. This feature was especially interesting for smaller transport aircraft. These aircraft fly at medium altitude with often pronounced turbulence. For this reason, the gust alleviation was also interesting for the Dornier company where the light transport aircraft Do 228 and Do 328 were under development. This led to launching the jointly planned and executed technology project called "Open Loop Gust Alleviation" (OLGA).

6.3.3.2 Dynamic Simulation in Wind Tunnel

NASA Research

Before discussing dynamic simulation aspects in wind tunnels, which were a precursor program for the German project OLGA, the extensive research programs of NACA and its successor organization NASA are briefly mentioned. Already at the beginning of the nineteen-forties, this organization had investigated the dynamic behavior of free-flying controllable aircraft models in wind tunnels. These models were held in a steady flight condition. A special free flight tunnel at NACA Langley was used for the experiments. Results from these experiments were correlated with those from flight tests.

This method proved to be extremely successful and was applied to almost all new aircraft projects right up to the nineteen fifties. The models became more and more complex with growing demands concerning control augmented stability with movable control surfaces and rate gyros. Further, the introduction of remote control made it possible to excite the model dynamics and to perform stability analyses. This technique was attractive for advanced configurations such as flying wings and aircraft with delta and swept wings.

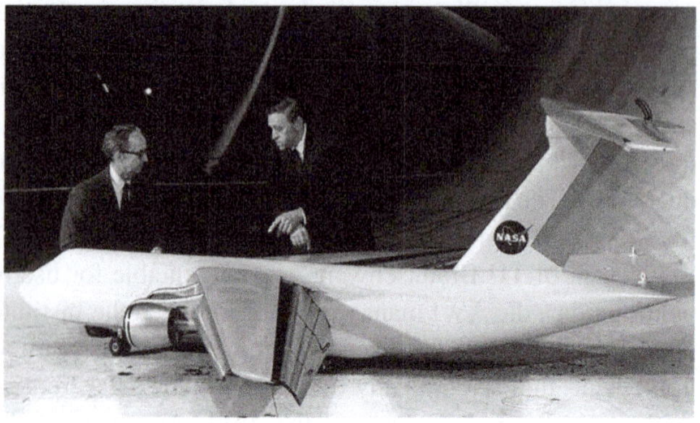

Fig. 6.105 *John P. Campbell*, left, inventor of the externally blown flap (EBF) concept, and *Gerald G. Kayten*, of NASA Headquarters, pose with a generic free-flight model of an STOL configuration at the Langley full-scale wind tunnel (*Credit* NASA)

Because of the heavy demand for high-lift investigations, the free flight technique was further developed for the large 30 × 60 ft and 40 × 80 ft full-scale tunnels at the NASA Langley and Ames Research Centers, providing more space to operate larger and more complex models. These models were powered, instrumented and could be controlled by an externally located pilot. Flight control laws could be implemented through an external computer via cables. Experiments with VTOL and tilt-wing aircraft were particularly successful to clarify stability problems during the transition phase from vertical to horizontal flight (Hawker P1127, LTV XC 142A). STOL transport aircraft configurations with externally blown flaps (EBF) were also investigated here for dynamic stability purposes (see Fig. 6.105). The development risk of several aircraft projects could be substantially reduced with this novel technique. A detailed review of the important role of dynamically scaled free-flight models in support of NASA's aerospace programs can be found in [54].

DLR Research Program BASE

Dynamic simulations in a wind tunnel with a controlled and flexible wind tunnel model served as precursor program to the German OLGA project. Therewith an experimental technique was to be tested and validation procedures (System Identification) were to be developed for mathematical models to describe the flight physics of the so-called BASE project, a German acronym for gust alleviation and elastic mode damping.

The construction of the facility for dynamic simulations was started in 1972. The setup was dimensioned for operation in the 3 m low-speed wind tunnel NWB at DFVLR in Braunschweig ("Normal Windkanal Braunschweig"). The core of the setup was a stable tubular steel frame, large enough to allow free stream air flow (see Fig. 6.106). The NWB department built a so-called "Gust Generator" in the wind tunnel nozzle. The gusts generator consisted of two deflectable flaps, which were driven by electrohydraulic servos. These flaps ensured up and down deflection of the air stream. Driven by a signal generator, the gust generator could generate both sinusoidal gusts and real stochastic turbulence. A flutter model of the so-called AVS aircraft (*Advanced V/STOL Strike Fighter*) of the former company MBB-UF in Ottobrunn was adopted by the Institute as a "flying aircraft". The flexible (flutter) model emerged from a cooperation between Boeing and the predecessor of the later MBB-UF company. The AVS program was an important transatlantic aeronautical technology program [55].

At the Institute of Flight Mechanics, the model was equipped with electrically driven control surfaces (ailerons, flaps, elevator). The model could move freely around the vertical axis. Limited pitch and roll degree of freedom and vertical (heave) motions on the rod were made possible through a longitudinal ball bearing. The instrumentation of the model included rate gyros, potentiometers, and accelerometers. A fuselage mounted vane was used to measure the angle of attack (see Fig. 6.107). Due to the limited space inside the model and for weight reasons, the power supply was provided from outside via an umbilical cable. The cables were also used for signal transmission to and from the model.

A complete aerodynamic dataset including the effectiveness of control surfaces was obtained from measurements in the NWB. Similarity laws (scaling laws) had to be considered to ensure the applicability of wind tunnel generated results to the actual aircraft. The reference angle of attack of

Fig. 6.106 Facility for dynamic simulation in wind tunnel with BASE-Model

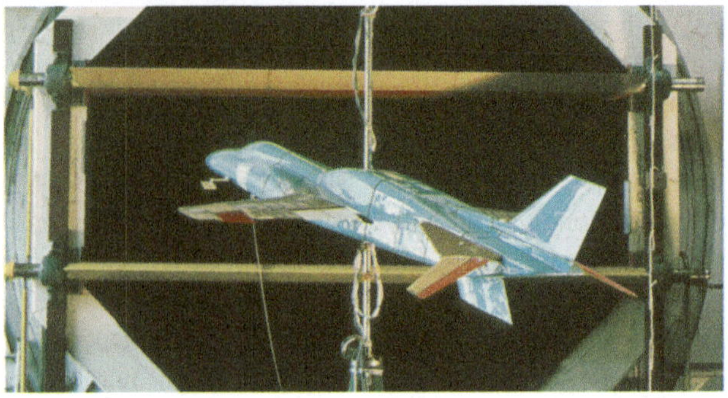

Fig. 6.107 BASE-Model in front of wind tunnel nozzle with gust generator

Fig. 6.108 BASE-Model with canards for damping of fuselage bending mode

the BASE model did not match with that of a free flight aircraft because the model was too heavy. As a consequence, a mechanism for "weight relief" was developed, which pulled the model with a constant force over a cable. A force sensor was mounted on this cable, whose signal was routed via a controller to a servomotor. This servomotor ensured that the force sensor measured constantly the same vertical force, independent of the model movement. All necessary signal processing were carried out outside the model.

The primary objective was to develop a gust alleviation system, with which the flight in the turbulent air could be smoothed. Sudden vertical gusts and their annoying vertical accelerations are perceived by passengers as particularly unpleasant. As such it was logical to measure this vertical acceleration and feedback the same to symmetrically moving flaps in an appropriate manner to produce a counterforce. But such a flight controller with acceleration feedback (*Feedback Control System*) responds also to accelerations generated by pilots through control commands.

For this reasons, the so-called "disturbance compensation" procedure was chosen to avoid the influence of a gust alleviator on the controllability of the model [56]. Taking into account the dynamics of the entire system from the sensor to actuator, the gust angle of attack and its rate of change were fed to the flaps and elevator. The aim was to minimize vertical accelerations caused by the gust along the body stations of the model. The exact knowledge of the gust angle of attack, being computed from the vane signal, was important and the phase margin in the vane signal resulting from its location far ahead of the wings had to be taken into account. This phase margin could be used advantageously to compensate for delays resulting from the servomotors.

Another controller for the damping of elastic vibrations (*Elastic Mode Control*) was tested with the BASE model. Thereby material fatigue of the wing and fuselage structures could be counteracted. The symmetric aileron actuators were fast enough to dampen the wing bending mode. The signals from miniature accelerometers on the wing tips were directly fed to the ailerons located close by with correct phase angles (*ILAF principle, Identically Location of Acceleration and Force Application*). Later, the BASE model was equipped with actively controlled canards on the nose (see Fig. 6.108). With this additional control surface, it was possible to dampen the first fuselage bending mode. Both controllers were successfully tested in 1974 and 1975 [57].

6.3.3.3 OLGA Project

In the nineteen seventies and eighties, a number of industrial technology programs were carried out with the Do 128 TNT Research aircraft (Tragflügel Neuer Technologie), a modified Do 28 (see Fig. 6.109). These programs were sponsored by the Ministry of Research and Technology (BMFT) and supported the preliminary development of the Do 228 and Do 328 regional aircraft.

One of these technological programs was devoted to improving passenger ride comfort of regional aircraft. This class of short haul aircraft usually fly in the lower and therefore more turbulent part of the atmosphere. The idea was to develop a gust alleviation system that measures the gust before it impinges the wing and uses symmetrically deflected ailerons as flaps to attenuate its effect on the wing and thereby the vertical response of the aircraft.

From investigations in the United States, it became apparent that mainly low frequencies around 0.3 Hz necessitated mostly the use the airsickness bags (see Fig. 6.110). Unfortunately, the frequency of the short period mode for this category of aircraft is also in this frequency range of 0.3–1.0 Hz, so that an excitation of this mode could be expected.

There was a whole series of questions that had to be answered first, before an expensive and elaborate flight test program could begin with the Do 128TNT research aircraft. The successful gust alleviation experiments at DFVLR with the BASE model provided a solid cooperative basis to embark with Dornier on a joint technology program dubbed

Fig. 6.109 Test aircraft Dornier Do 28 TNT (*Credit* Dornier)

Fig. 6.111 Wind tunnel model of TNT test aircraft in NWB (1976)

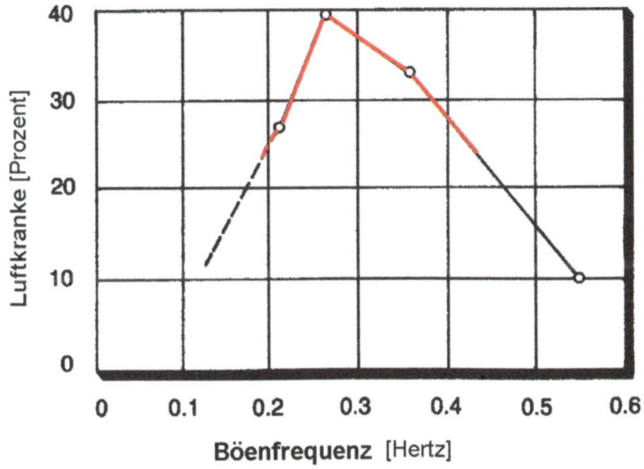

Fig. 6.110 Probability of becoming airsick as a function of the gust frequency (*Credit* Dornier)

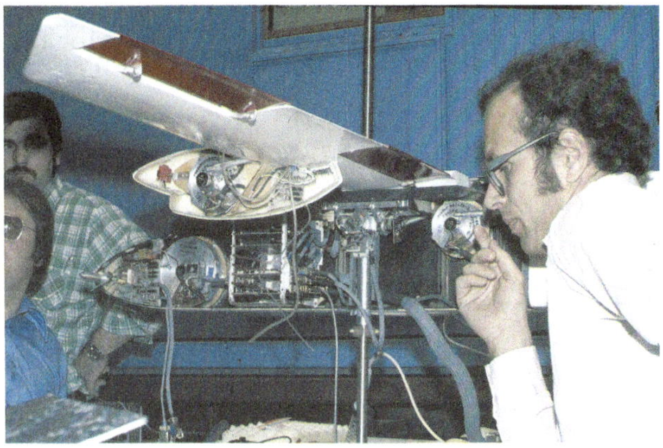

Fig. 6.112 Actuators and signal processing onboard the wind tunnel model (1976, on the right *Rüdiger Karmann*)

OLGA (Open Loop Gust Alleviation). The task of the DFVLR was twofold; first to demonstrate the functioning of the gust alleviation system in the dynamic simulation facility with a TNT model, and secondly to develop the necessary control algorithms, which could be realized later in a flight test program.

In the year 1976, a wind tunnel model of the Do 128TNT test vehicle was constructed for the first time at the DFVLR Institute of Flight Mechanics using the then innovative carbon fiber technology (mass 7 kg, see Fig. 6.111). A nose mounted vane provided measurements of the gust angle of attack. The ailerons deflected symmetrically served as lift-generating flaps and the all-moving tailplane was used to reduce pitching responses. The "flight" of the model in the wind tunnel could be adjusted using inboard-flaps.

As in the case of BASE model, all control surfaces were driven by electric servomotors. Besides the vanes, the measuring equipment included rate and attitude gyroscopes. An electronics box in the fuselage contained the power supply and signal processing unit (see Fig. 6.112).

Again, the TNT model had a weight-relieving controller establishing a trimmed, steady flight condition at the correct angle of attack. The steady aerodynamic parameters were determined during the first wind tunnel tests in 1977. As the knowledge of dynamic aerodynamic forces was important for the correct design of the gust alleviation system, special system identification techniques were employed to determine the aerodynamic parameters (see Chap. 3). In two further wind tunnel campaigns during 1978 and 1980, the Open-Loop Gust Alleviation system was optimized and the principle was successfully demonstrated (see Fig. 6.113) [58].

In this context, it is interesting to note that within a bilateral cooperation between DFVLR and ONERA, France, a second OLGA model was built and tested in the catapult free flight facility of ONERA. In this free-flight facility the model flew through a vertical wind field. Here too, proper functioning of the open-loop gust alleviators could be demonstrated.

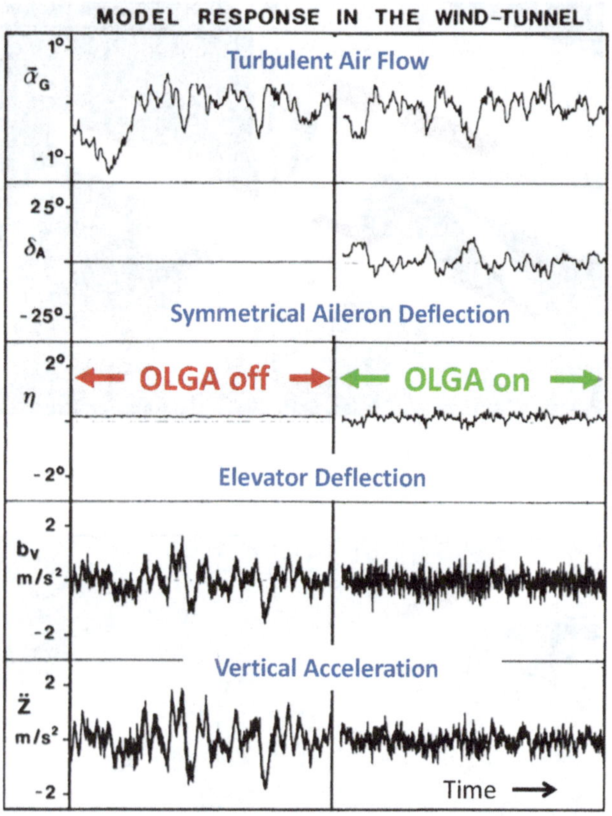

Fig. 6.113 Results of wind tunnel tests (1980)

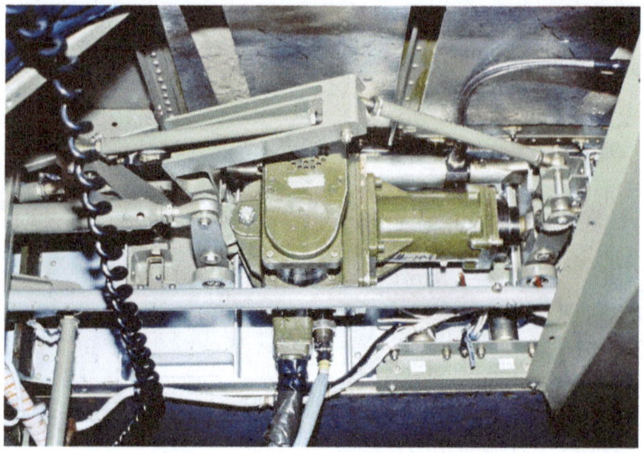

Fig. 6.114 Installation of electromechanical actuators of the aileron controls (1981) (*Credit* Bernd Krag)

This marked the beginning of a flight test program with the Do 128TNT research aircraft. For this purpose, the aircraft had to be modified to enable testing of the gust alleviation system. A technological challenge was to replace the mechanical aileron control via cables through rigid rods, otherwise, the higher-frequency components of the gust alleviation would be absorbed by the elastic cables. The function of symmetric aileron deflection had to be integrated into the normal aileron control with appropriate priority. In addition, the horizontal tailplane had to be operated by a servo to compensate for pitching. This was realized using a servomotor acting in series with the existing trim motor of the horizontal tailplane. A highly dynamic electromechanical actuator was selected for activation of the control surfaces because electrohydraulic actuators would have been too complex and expensive. A Cobalt-Samarium servomotor was chosen for this purpose. The motor operated a threaded rod, which was extended and retracted. The entire actuation system had a bandwidth of 1 Hz and was sufficient to meet the system requirements (see Fig. 6.114).

The angle of attack sensor from an Alpha-Jet served as a gust sensor. In addition, a corresponding digital signal processing and operator console had to be developed and integrated into the test vehicle (see Fig. 6.115). All additional installations in the test vehicle had to be certified for the flight.

The ideal conditions of the OLGA wind tunnel experiments could not be expected in the present case. Only certifiable hardware components could be installed. To tune the gust alleviation system a so-called "Hardware-in-the-Loop Simulation" was built. A computer simulation of the TNT test vehicle was coupled with the devices used for the gust

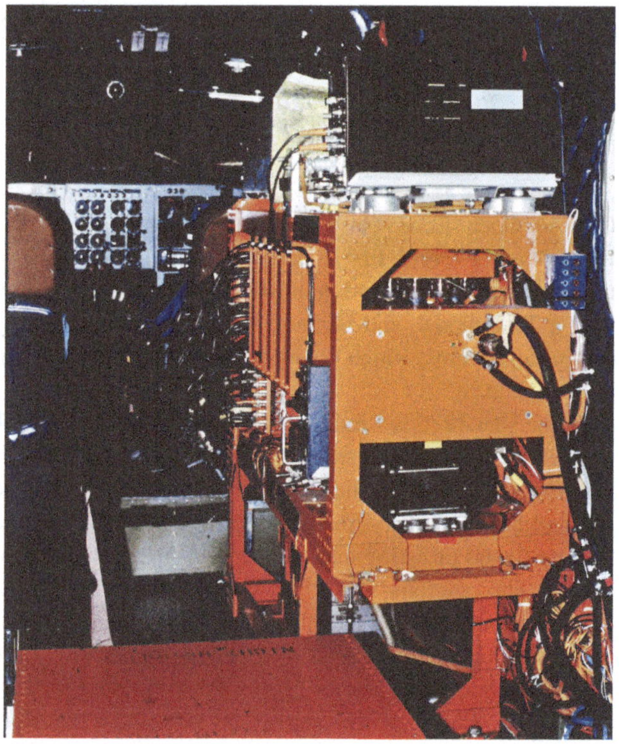

Fig. 6.115 Signal processing in the TNT test aircraft (*Credit* Bernd Krag)

alleviation. Prior to this, flight tests were performed in which the aircraft dynamic motion was excited through control pulses. Applying system identification methods a precise set of aerodynamic data could be obtained, which was required for the mathematical model of the test vehicle. In this simulation, the computational algorithms for the gust angle of attack could be validated and the influence of various hardware components on the gust alleviation effectiveness and on the flight safety could be investigated. Among others, it had to be ensured that in the case of a malfunction in the control system the experimental system could be safely turned off, without affecting the controllability of the aircraft. Thereafter, the entire actuation system was certified for flight operation.

The first few flights were used for approval and verification that the system engaged does not affect the flight behavior. The first flight with the activated gust alleviation system showed that, contrary to expectations only the wing bending mode of 3 Hz had been excited. This was caused by the excessive time delay between measured angle of attack and control surface deflections. By raising the computational cycle time from 20 to 40 Hz, and through the introduction of a notch filter to suppress the wing bending oscillation, the gust alleviation system showed its desired effect (see Fig. 6.116). The desired "Wow" effect, however, failed to appear due to the slow response of the actuators. The lower gust frequencies around the "spitting frequency" were alleviated well by the system, but Dornier test pilot *Dieter Thomas* objected that at higher frequencies he had a feeling of flying over "cobblestone" [59].

A corresponding adaptation of the system to production stage was aborted due to the involved additional costs and also due to the lack of interest of potential customers. And so the afflicted passengers had to wait for the latest generation of passenger jets to experience the benefits of such a system.

Within the OLGA technology program, a complete validation chain from wind tunnel test, system identification, Hardware-in-the-Loop Simulation, and right up to flight test was achieved for the first time. Although the cost of dynamic simulation in wind tunnel facility was significant, the investments paid off, because it provided the experimental evidence that a gust alleviation, based on the principle of open-loop control, actually worked in practice. Familiarization with these results justified the investment from industry to explore so far untested new technology. The OLGA technology program constituted an excellent symbiosis of complementary research at DLR and industrial development

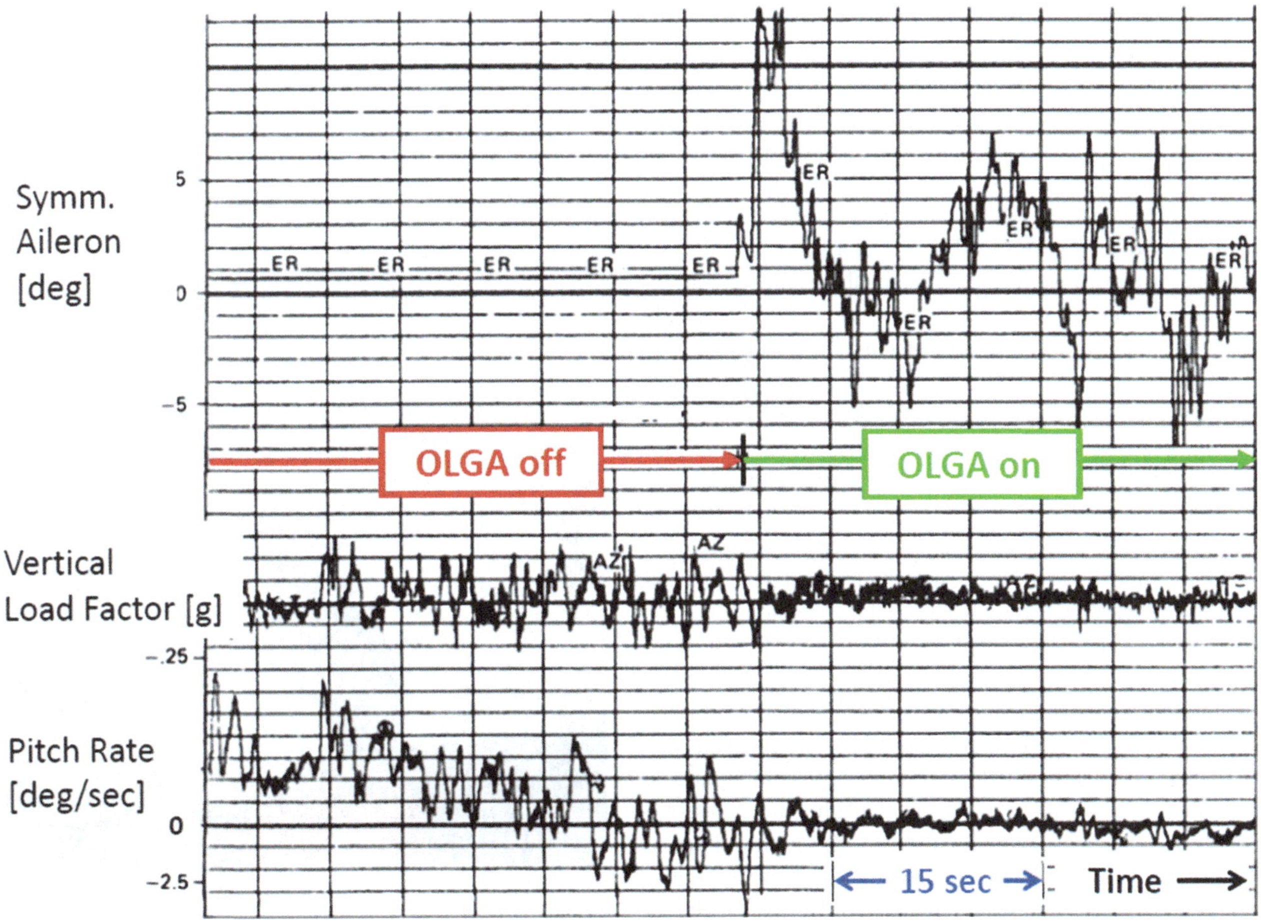

Fig. 6.116 Flight test result from TNT test aircraft (*Credit* Dornier)

Fig. 6.117 TNT test aircraft and TNT wind tunnel model (1982, in front, project manager *Bernd Krag*)

Fig. 6.118 Lockheed/MBB CCV F-104

(Dornier) (see Fig. 6.117). Since 2012, the Do 128 TNT OLGA wind tunnel model is displayed in the Dornier Museum at Friedrichshafen.

6.3.4 Lockheed F-104 CCV (1977–1984)

Hermann Hofer

In the year 1974, Messerschmitt Bölkow-Blohm (MBB) started an experimental program to test the principles of CCV technology (Control Configured Vehicles). The task included the development and testing of an advanced high-bandwidth, redundant Fly-by-Wire flight control system for stabilizing an aerodynamically unstable supersonic aircraft in the entire flight envelope.

Classical aircraft with inherent stability are designed such that the center of gravity and the wing lift generate a nose-down pitching moment. This behavior is required to restore the controllability of an aircraft by the pilot even after flow separation. The negative pitching moment, however, must be compensated by a corresponding elevator upward deflection resulting in a downward force. Thereby the overall aircraft lift is reduced, and the wing lift must be increased accordingly resulting in a significant aerodynamic drag increase.

In contrast, if the center of wing lift is located ahead of the center of gravity, the trim force generated by the horizontal tail plane now contributes directly to the overall lift. At the same time, the aerodynamic (induced) drag is reduced. However, such an aircraft loses its inherent stability and can only be safely flown as a CCV configuration.

Since some of the classical stability requirements of flight mechanics and structural dynamics have no longer to be considered with the integration of redundant Fly-by-Wire flight control systems, there is the possibility to design an aircraft exclusively from the viewpoint of performance (see also Sect. 6.1.2). Thereby significant improvements can be achieved compared to conventional configurations: (1) Reduction of the take-off weight by 10–20%, (2) Reduction in the wing area by up to 35% resulting in drag reductions and fuel savings, (3) Almost complete flutter suppression especially when flying with variable external loads, (4) Enormous increase in maneuverability and agility, and (5) Reduction of sensitivity to gusts at low altitudes.

A Lockheed F-104G was selected as a test vehicle (see Fig. 6.118). The shifting of the total lift ahead of the center of gravity was achieved by the integration of small auxiliary aerodynamic surfaces (so-called *Canards*) behind the cockpit above the engine intakes (see Fig. 6.119) and additional masses (ballast) in the aft fuselage. In the case of an emergency, the ballast mass could be jettisoned to restore aircraft controllability. The canard surfaces, structurally identical with F-104 horizontal stabilizer, were fixed. The conventional hydro-mechanical actuator was augmented through the integration of a quadruplex Fly-by-Wire flight control system. The main components of the FBW system were (1) air data sensors which were mounted fourfold underneath the fuselage behind the radome, (2) cockpit sensors for gathering pilot inputs, (3) a quadruplex central computer for implementing the control laws, (4) control loop electronics for the actuators, as well as (5) LAT triplex actuators (Hydraulic Duo Duplex) for the aileron, elevator and rudder. In addition, the angle of attack and angle of sideslip were measured through quadruple redundant vanes arranged diagonally underneath the front fuselage. In the Fly-by-Wire

Fig. 6.119 CCV F-104 canards

Fig. 6.120 CCV triplex actuator

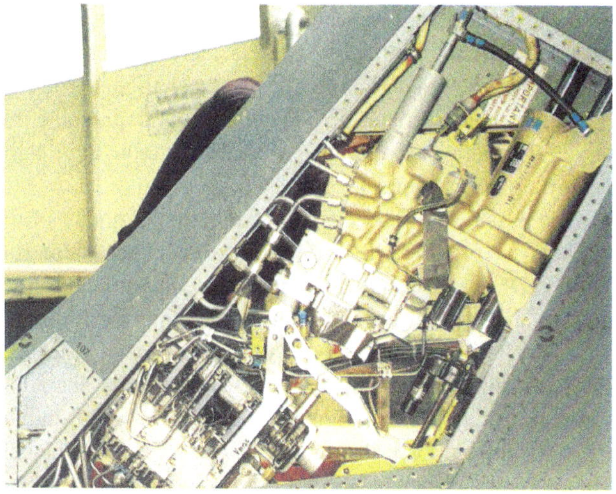

Fig. 6.121 CCV actuator integrated in the vertical fin

mode, the actuators were connected with the control linkage via hydraulically operated clutches. In this configuration, the aircraft was flown fully electronically using control commands calculated by the central computer.

Switching between the CCV mode and normal operation was carried out by the pilot activating a redundant hydraulically actuated clutch. Switching mechanism and triplex actuators were integrated into a single unit developed and manufactured by Liebherr-Aerotechnik (LAT) (see Fig. 6.120). The LAT triplex actuator consisted of three actuator modules operating on the common output lever with a pre-defined elasticity. The unit was engaged or separated from the control linkages to the boosters of the individual control surfaces through a redundant electrohydraulically activated clutch. Each actuator module consisted of a two-stage servovalve, bypass valve, linear ram and ram position sensors. Figure 6.121 shows the actuator installation for the rudder in the vertical fin.

Uncharted technological territories were covered in this program through the first-time application of an innovative monitoring concept for the actuators. An electronic model of the actuator and the servo valve were programmed in the central computer. The desired positions computed by the model were then compared in real time with the actually measured values. Exceeding the set limits is interpreted as an error and as a consequence, the affected module is switched off.

The overall functionality is still preserved through the remaining healthy modules. The principle of "electrical monitoring", replacing the hitherto complex and inflexible hydraulic comparators, allowed the construction of compact fault-tolerant actuators suitable for modern electronic flight controls. Due to the lack of space, the necessary actuator electronics required for loop closure and monitoring were

Fig. 6.122 F-104G CCV in flight test (*Credit* WTD-61)

housed in the central onboard computer. The first flight took place on November 20, 1980, with a CCV configuration of the F-104G, destabilized by canards. Until 1984, a total of 176 test flights were performed (see Fig. 6.122). With a 660 kg ballast in the tail section, an instability of nearly 20% of the mean aerodynamic chord was achieved with supersonic speeds up to Mach 1.3 or 650 knots [60]. The flight test results were of great benefit for the development of the Eurofighter Typhoon as well as for the two X-31 experimental programs (Sects. 6.3.6 and 6.3.8).

6.3.5 Dornier/Dassault AlphaJet DSFC (1982–1985)

Horst Wünnenberg

6.3.5.1 Statement of Task

During the early nineteen eighties, a military technology program was initiated at Dornier with the objective to increase maneuverability and operational flexibility of combat aircraft using new control techniques. Primarily concerned with air-ground missions the pilot must carry out in a very short time precise attitude and trajectory corrections, when he leaves his low-level cover flight through a "pop-up" maneuver, pulling up sharply and pursuing his target. Through this pop-up maneuver, the exposure to the enemy fire is minimized.

One possibility to accelerate the precise alignment of the airplane is provided by additional direct force controls (DFC). DFC is exerted directly at the aircraft's center of gravity and thereby instantaneously changing the trajectory without altering the attitude. The normal control surfaces (elevator, aileron, and rudder) generate moments about the center of gravity, resulting in attitude changes which induce the aerodynamic forces leading to desired trajectory changes. This classical maneuvering process takes longer but can be avoided through the use of direct force control.

These new control effectors are generally termed as Direct Lift Control (DLC), Direct Side Force Control (DSFC) and Drag Control (DC). They allow new operating modes such as those depicted graphically in Fig. 6.123.

Within a preliminary study program involving the Dornier/Dassault Alpha Jet (see Fig. 6.124) it could be demonstrated that using such direct force controls an appreciable potential is available for improving the flying qualities, thus simplifying controllability during difficult flight phases. Therefore, Dornier was assigned by the Federal Minister of Defense (FMoD) to carry out relevant studies to realize such a feature on an Alpha Jet prototype. The flight test phase was to be conducted in cooperation with the German Air Force Flight Test Center (WTD 61, formerly E-61) and AFB LG IV (*Richard Rosenberg*) of the Federal Office of Defense Technology and Procurement (BWB).

In the DFC design, the following features were to be incorporated: (1) Realization of DLC by maneuvering flaps on the wing, (2) Realization of DSFC by symmetrical deflection of additional split control surfaces on the pylons, and (3) Implementation of DC by simultaneous deflections of the additional split control surfaces on the pylons.

In addition, to compensate for gust disturbances und coupling effects a single-channel three-axis damper with an

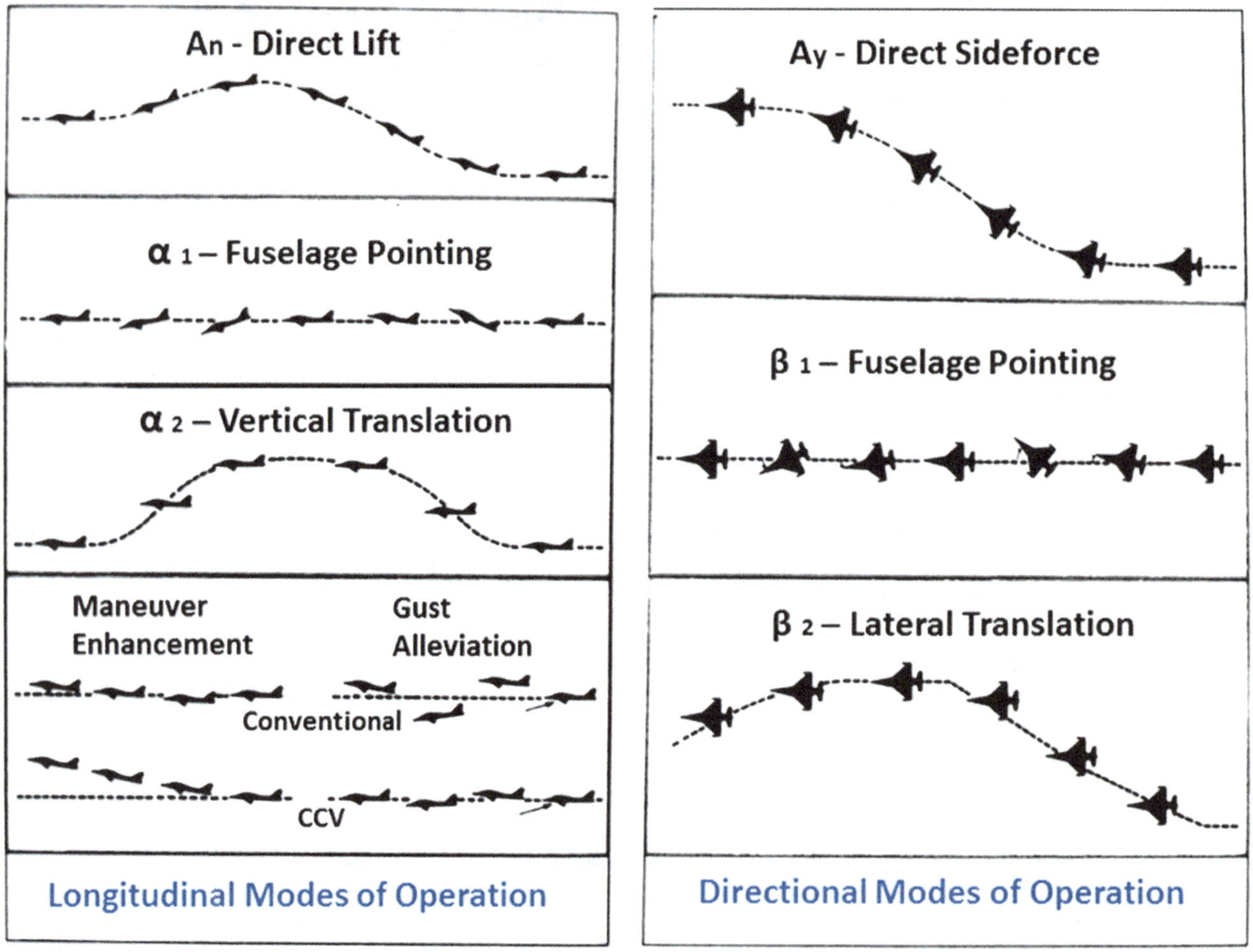

Fig. 6.123 DFC operating modes

Fig. 6.124 Dornier/Dassault Alpha Jet

attitude hold mode was conceptualized. In a further expansion phase with the German Aerospace Center (DLR), it was planned to use this demonstrator aircraft as host for a future in-flight simulator for research and pilot training purposes.

6.3.5.2 Program Implementation

The program was executed in several stages and the aforementioned individual aspects of DFC were accomplished in separate campaigns. Because of financial constraints, the originally planned integration of all individual aspects of the DFC in an experimental vehicle could not be realized. However, in a technology program spread over several years, a couple of aspects were investigated incorporating flight test and simulator studies involving test pilots from the German Air Force Flight Test Center (WTD-61). They were assessed with regard to their operational advantages in order to generate a reliable data base for the design of future fighter aircraft. In particular, they pertained to:

- Design and flight testing of the effectiveness of maneuvering flaps for the supercritical transonic wing (TST) of this special Alpha jet prototype (TST—*Transonischer Tragflügel*),
- Realization and flight testing of direct side force control (DSFC) and a drag control (DC) effectors through split flaps control on the rear part of the pylons,
- Simulator studies with the Dornier Alpha Jet development-simulator investigating possible operational modes of an integrated DFC system on Alpha Jet and their operational utility, and
- Concept study Alpha Jet In-Flight Research and Training Simulator (Project CASTOR, in cooperation with the DLR and WTD-61).

6.3.5.3 Results and Conclusions of Direct Lift Control

Direct Lift Control on Alpha Jet was to be implemented with maneuvering flaps, which were designed and implemented in the framework of a separate technology program on transonic wing design (TST). They consisted of leading-edge slats and trailing edge flaps. For cost reasons, the flaps were not continuously adjustable but were preset in

Fig. 6.125 TST test aircraft with deflected maneuvering slats and flaps

a fixed position on the ground. With this setting, the flight tests were performed in the desired flight envelope (see Fig. 6.125). The results provided the basis for the design of the planned direct lift control system.

The test results, in this case, promised a sufficient effectiveness for the intended application and were incorporated in the conceptual design of the DLC system for subsequent simulator investigations. A planned extension of this program with automatic flap control could not be realized.

Direct Side Force and Drag Control

The direct side force control system was investigated in a separate technology program. The necessary side control forces were generated through small, electrohydraulically driven split flaps, that were mounted on both sides at the rear of the pylons (see Fig. 6.126). When deflected asymmetrically, they generated lateral forces, whereas driven symmetrically they generated additional drag like an airbrake. The operating modes depicted in Fig. 6.127 were possible

Fig. 6.126 Electrohydraulically driven pylon split flaps

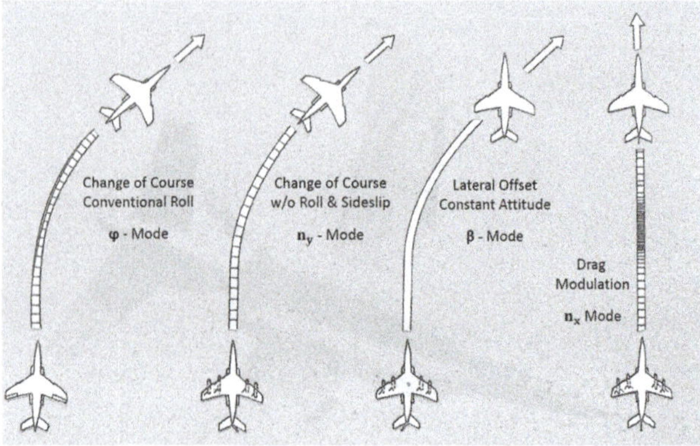

Fig. 6.127 Tested DSFC operating modes with pylon-split flaps

Fig. 6.128 DSFC test aircraft with deflected pylon split flaps

with this set-up and were selectable by the pilot. In addition, the resulting interference moments were compensated with the single-channel 3-axis damper with a switchable attitude hold mode. The damper effectiveness of was also investigated separately during the flight test phase.

The first flight took place on March 22, 1982. This flight test phase, in which WTD 61 participated, served initially the purpose of envelope opening and system optimization. The functionality of DFSC device was demonstrated in all operating modes (see Fig. 6.128). Above all, the operating mode "sideslip free course change" (ny mode) provided a noticeable improvement in the target tracking accuracy. The operating mode "drag control" (nx-mode) was also very effective resulting in decelerations up to 0.75 g and opened new tactical possibilities through steeper dives with constant speed and less intercept altitudes.

The program was continued with operational testing including shooting approaches. This campaign was organized by the WTD-61 and was carried out during 1984 on the shooting range at Meppen and with the multiple target-lamps-facility GRATE in Manching, which was developed by DLR (GRATE—*Ground Attack Test Equipment*, see also Sect. 12.3.2).

The results of this campaign showed the basic feasibility of a DSFC system using pylon split flaps. The results provided, however, surprising insight with respect of an operational advantage of such a device. A significant reduction of the pilot-target error was, indeed, identified while using the DSFC effectors, but similar positive results were achieved while using just the 3-axis controller, when sideslipping was allowed [61, 62].

Integrated DFC System

Once the basic feasibility of the essential components of a DFC system on the Alpha Jet prototype was established in the course of previous technology programs, an overall DFC system for the Alpha Jet was designed and tested. This included all DFC components, like direct lift control, direct side force control and drag control in conjunction with a command and stabilization controller.

The main objective of this study was to prepare a further flight test program to provide the basic data for the design of an appropriate system for air and ground missions with a future combat aircraft or for the improvement of the combat efficiency for the Alpha Jet.

In-Flight Simulator Proposal

This proposal was conceived as a joint task between Dornier, DLR, BWB LG IV and WTD-61. It included the development of an in-flight simulator based on the Alpha Jet with DFC components. For this purpose, a study under the acronym CASTOR (*Combat Aircraft for Training, Operations, and Research*) was submitted to the Federal Ministry of Defense (see also Sect. 11.2). Because of changed priorities in the procurement philosophy of Federal Ministry of Defense with regard to the European Eurofighter development, this proposal was, however, abandoned.

6.3.6 Rockwell/DASA X-31A EFM (1990–1995)

Peter Hamel

The objective of this first transatlantic X-research program X-31A EFM (*Enhanced Fighter Maneuverability*) was to prove that an aircraft can safely maneuver in the post-stall regime and to investigate what effects such post-stall maneuvers would have in an air-to-air combat mission. For this purpose, the construction of a "*Low Cost Demonstrator*" as delta canard configuration was envisaged, which had to be similar to the planned European combat aircraft, as much as possible.

The prerequisite for this was the development of a highly reliable digital Fly-by-Wire flight control system for the integration and practical testing of a thrust vector controller. The main emphasis of the program was on the flights far beyond maximum lift in the post-stall regime. In this flight regime, the lift and control surface effectiveness diminish, while the drag increases at the same time due to flow separation. Thrust vectoring has then the potential to compensate for the loss of aerodynamic control effectiveness at high angles of attack.

To save on the cost, many components and subsystems from other aircraft programs were used, which posed some problems during the system integration. Even minor geometrical asymmetries on the fuselage nose of the two testbeds resulted initially in different flight behavior in extreme flying conditions. These asymmetries were eliminated by mounting nose strakes. The vortex shedding at the forward fuselage at high angles of attack then occurred always symmetrically at the same point on both sides [63]. Thrust vectoring was provided by three heat-resistant graphite composite paddle-like vanes mounted on the airframe adjacent to the engine exhaust nozzle (developed by MBB/DASA and produced by SGL/SIGRI). These thrust-vectoring vanes could generate sufficiently strong combined pitching and yawing moments in the high-angle-of-attack flight regime. In addition, besides the trailing edge wing flaps and the rudder, the so-called canard control surfaces in the forward fuselage area were driven actively (see Fig. 6.129). In trimmed flight, these canard control surfaces floated freely in the wind, without generating any moment. Thereby the actual angle of attack during the spectacular flight maneuvers, for example at the Paris Air Show in 1995, could be recognized very well even by a layman.

The biggest challenge, however, was to check out the hardware and software of the innovative digital, integrated thrust vector control system, whereby MBB/DASA was responsible for the flight control system and could advantageously use the know-how of the digital flight control system gathered during the F-104 CCV program [64] (see Sect. 6.3.4). The hardware included a triplex system with a fourth computer without CPU, which took over the function of a digital arbitrator (*Digital "Tiebreaker"*) in the case of a fault. An analog or mechanical backup system was not provided from the very beginning. The first flight of the first X-31A prototype (S/N 164584) took place on October 11, 1990 from Palmdale flight test center in Rockwell California, USA, and that with the second prototype (S/N 164585) followed on January 19, 1991. The first flight with thrust vector control was conducted on February 14, 1991. 108 test flights were completed up to the end of 1991, achieving 52° angle of attack.

On January 20, 1992, the two aircraft were transferred to the NASA *Dryden* Flight Research Center (since 2014: *Armstrong* FRC). The first flight at NASA with the 2nd prototype was on April 23, 1992, with *Karl-Heinz Lang* as the pilot. An angle of attack of 70° and bank angle up to 45° was flown for the first time with the 1st prototype on September 18, 1992 (see Fig. 6.130). Modeling inaccuracies in flight physics at increasing angles of attack necessitated corrections in the controller software, which could be carried out expeditiously on the advanced digital computer system without any major complications. The Institute of Flight Mechanics of DLR at Braunschweig (since 1999: Institute of Flight Systems), through its methodological experience in the field of system identification, participated successfully in the determination and validation of flight physics models and introduced the so-called methodology of Single or Separate Surface Excitation (SSE) [65, 66]. In this process, all control surfaces were deflected individually, in parallel with the active flight controller, in order to eliminate the coupling between the individual control surfaces that resulted from the high degree of controller feedback. Only through such independent separate input signals was the determination of the individual, independent flight mechanical parameters possible.

The improvements achieved through the SSE method becomes apparent from Figs. 6.131 and 6.132, showing as an example the identification of the wing flap (elevator)

Fig. 6.129 Rockwell-MBB X-31 EFM S-N164584 (*Credit* NASA)

Fig. 6.130 X-31A EFM post-stall demonstration

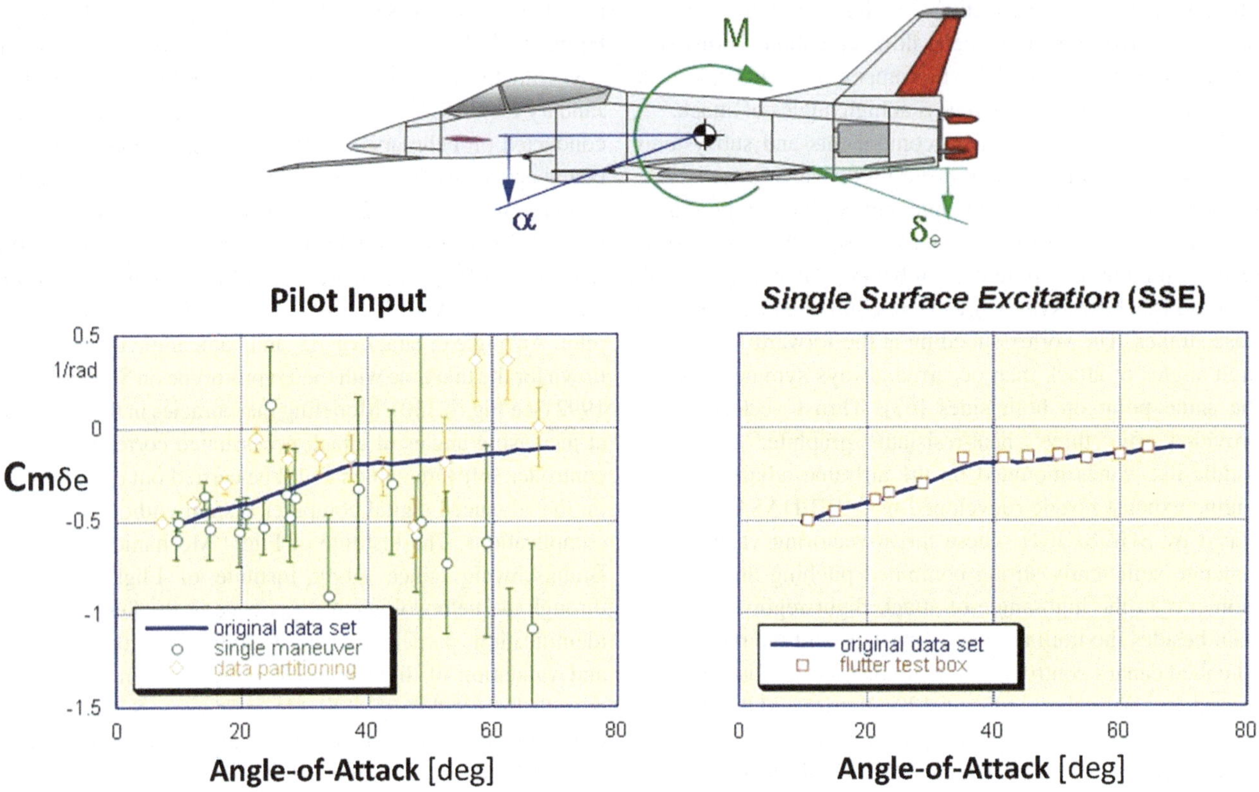

Fig. 6.131 X-31A identification of the wing flap effectiveness (*Credit* Detlef Rohlf)

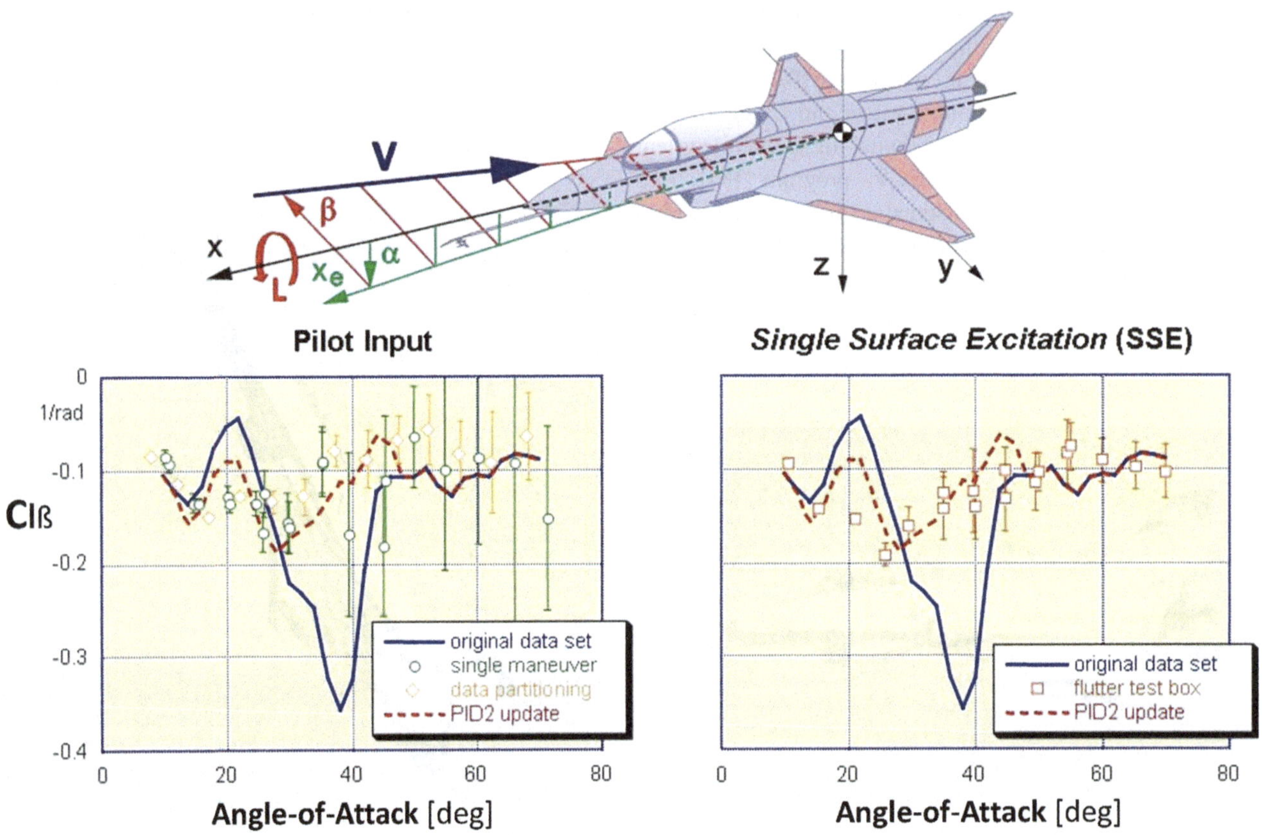

Fig. 6.132 X-31A identification of lateral stability parameter (dihedral effect) (*Credit* Detlef Rohlf)

effectiveness and of the dihedral effect respectively. The dihedral effect parameter $C_{l\beta}$, that is the rolling moment due to sideslip, is an important parameter for assessment of the lateral stability. In the case dihedral effect $C_{l\beta}$ the wind tunnel predicted large value between 30°–45° could not be confirmed. As another interesting result of the parameter estimation, the hysteresis in the unsteady aerodynamic lift force for large variations in the angle of attack could be clearly identified by combining a variety of different flight maneuvers (*Multi-Run Evaluation* of 29 pull-up/push-down maneuvers, see Fig. 6.133 [67]).

Rolling maneuvers about the velocity vector (*Velocity Vector Roll*) at 70° angle of attack were successfully performed for the first time on November 6, 1992. On April 29, 1993, a spectacular maneuver with a 180° change of heading (*J-turn*) at 70° angle of attack was finally flown with the second prototype. This so-called "Herbst maneuver" was named after the German development engineer and post stall protagonist *Wolfgang Herbst*.

After the successful completion of the post-stall maneuverability flight tests, tactical flight testing in many air combats, primarily against the F-18, as well as F-14, F-15 and F16 were undertaken. This tactical testing was even more successful than the many previous ground-based manned simulations.

Furthermore, in the year 1994, the feasibility of flying tailless, for example without fin and rudder control by exclusive thrust vector stabilization, was successfully flight-tested. Through special software, gradual destabilization of the lateral motion was achieved with the help of the rudder and thus a quasi-tailless configuration was simulated (see Fig. 6.134). Both at supersonic speeds (Mach number = 1.2) as well as in at low speeds, the effectiveness of the thrust vector control and stabilization system could be

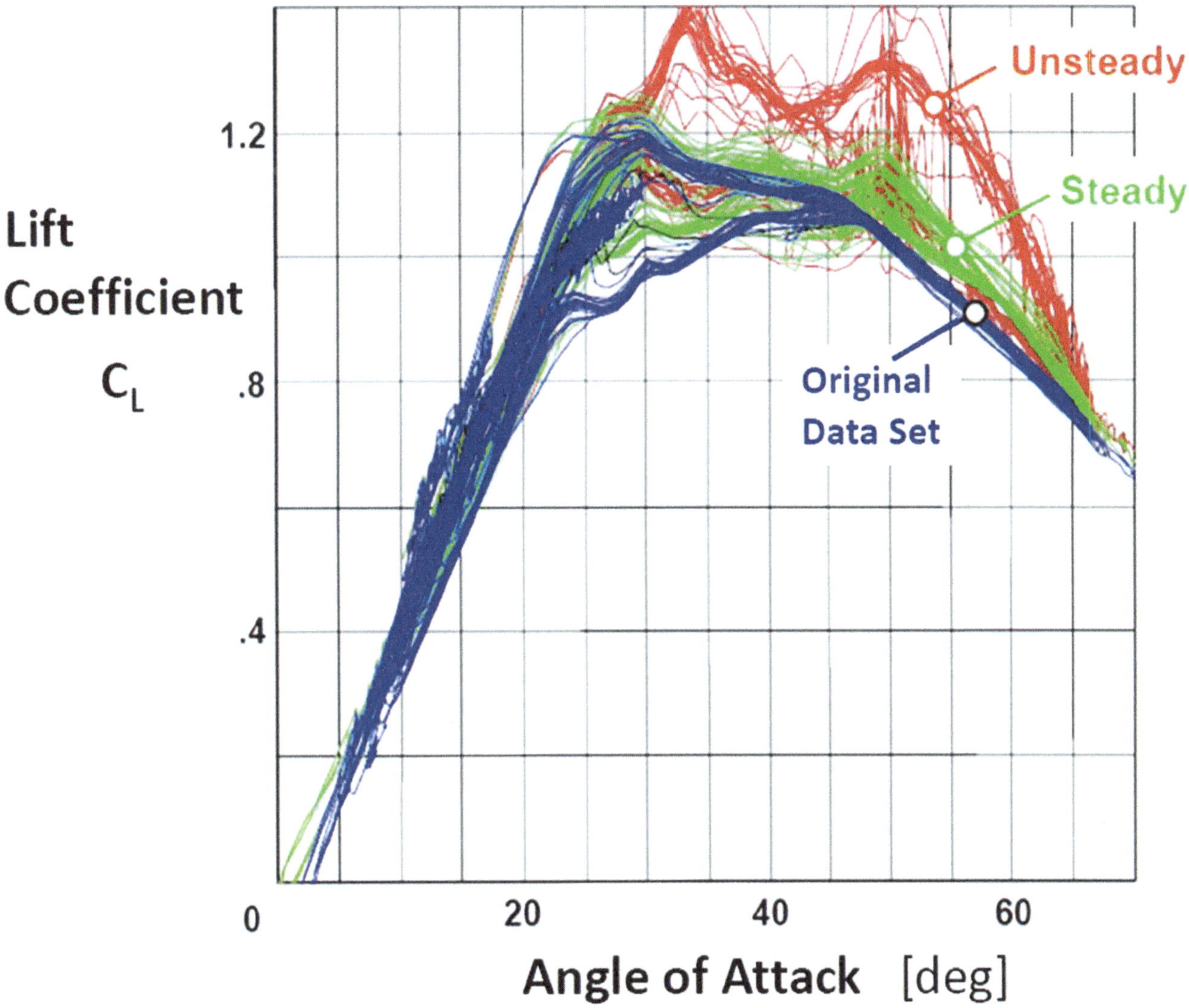

Fig. 6.133 X-31A identification of lift hysteresis (*Credit* Detlef Rohlf)

Fig. 6.134 X-31A quasi-tailless with extended airbrakes

Fig. 6.135 Romania commemorates X-31 (1994)

demonstrated. Here the experimental technique GRATE/ATLAS, developed at the DLR Institute of Flight Systems at Braunschweig, to evaluate precision flying qualities, proved to be highly successful [68] (see also Sect. 12.3.2). It is obvious that a tailless aircraft configurations would reduce the aircraft drag, and weight as well as the RADAR-detectability.

On January 19, 1995, after performing successful test maneuvers the first prototype crashed while returning to the base, whereby the German pilot *Karl-Heinz Lang* of the German Air Force Flight Test Center 61 ejected safely and parachuted to the ground. The icing on the non-heated air data sensors was found to be the cause of the accident (see Sect. 6.1.1 and Fig. 6.1).

After a quick inventory check, however, it was decided to revise the flight control system and to demonstrate the unique aircraft as planned to the professional world and the general public at the Paris Airshow in Le Bourget during June 1995. Thereby the second prototype completed successfully spectacular maneuvers, and earned the title "*Star of the Paris Air Show*".

Within the EFM (*Enhanced Fighter Maneuverability*) program and the Quasi-Tailless tests, the two X-31 aircraft absolved a total of 559 test flights, and further 21 flights in Europe for the preparation and implementation of the flight demonstrations at the Paris Airshow 1995. The European sovereign state Romania even drew attention on the X-31 by bringing out a special postal stamp (see Fig. 6.135) with a subtle indication of its famous scientist *Henri Coanda* (1886–1972), to which the principle of jet deflection is attributed ("*Coanda* effect").

Three years later, the X-31 was to be revived within the framework of the so-called VECTOR program (see Sect. 6.3.8).

6.3.7 VFW 614 ATD (1996–2000)

Hartmut Griem and Hermann Hofer

6.3.7.1 Overall System

Another test demonstrator, namely the ATD (*Advanced Technologies Demonstrator*) is worth mentioning. At the national level, the German aeronautical industry had focused on getting ready to face challenges in the field of digital Fly-by-Wire flight control for new generations of civil aircraft. The projected aim was to gather experience in modern flight control systems technologies. For this purpose, during the mid-nineties, a research project "*Electronic Flight Control System*" (EFCS) was initiated, which was financially supported by the German Federal Ministry for Research and Technology (BMFT).

The objective of the EFCS project was development, qualification and flight testing of a fully electronic flight control system taking into account all certification aspects of an operational program. The project was carried out in a cluster under the leadership of EADS Airbus in Bremen, with the participation of Bodenseewerk-Gerätetechnik (BGT), meanwhile Diehl Aerospace), LAT (Liebherr Aerotechnik), and German Aerospace Center in Braunschweig (DLR).

The new EFCS system comprised of (1) replacement of control columns by sidesticks for both pilots, (2) sensor system with adequate redundancy for air and inertial data (3 times angle of attack, 3 times ADIRS (*air data and inertial reference system*), (3) new flight control computer (*primary flight control unit*-PFCU), (4) new actuation systems of different technologies for elevator, aileron, rudder and spoilers, (5) upgrading the power system to provide sufficient redundancy and level of performance, including a newly developed *hydraulic power package* (HPP) in the tail,

Fig. 6.136 Second "first flight"

and (6) supplement the cockpit displays for EFCS related elements, including installation of a primary flight display.

For this purpose, EADS Airbus brought the VFW 614, S/N G15, obtained from Air Alsace after 3 years of operation, to an airworthy status. It was not a simple task because the machine had been used as a spare part dummy for ATTAS, and for training of student. The second "first flight" of G15 (see Fig. 6.136) took place on February 28, 1996, with *Dietmar Sengespeik* as the pilot, *Stefan Seydel* as copilot and *Bodo Knorr* as flight test engineer. Thereby the aircraft was transferred to the Airbus plant in Bremen for conversion.

The workshare of EADS Airbus consisted primarily in the integration of a newly developed Fly-by-Wire system, including development the control laws. These were similar to the proven flying qualities of Airbus A320 type aircraft in order to support a family concept and adapted to the flight dynamics of VFW 614. The technological focus was on automation of the flight control law development process using a model-based approach. EADS Airbus was responsible for carrying out all Fly-by-Wire system related aircraft modifications. In addition, EADS Airbus had to take care of all related tasks, which were necessary for conversion of the test vehicle and carrying out the test program. These included weight and energy balance, structural integrity, check compliance of load limits, evidence of freedom from flutter (including a complete ground vibration test), as well as proof of reliability and safety, generation of all documents for certification of the overall system for flight clearance, maintenance, and aircraft operation. Finally, an emergency exit was installed and all onboard test equipment installed and operated. This involved flight simulators and system test rigs (see Fig. 6.137), as well as design, construction and integration of onboard equipment for acquisition, recording, and display of experimental data, and airborne and ground-based telemetry system (see Fig. 6.138). Finally, the responsibility of implementation and analysis of laboratory, ground and flight tests, and certifications rested on EADS Airbus.

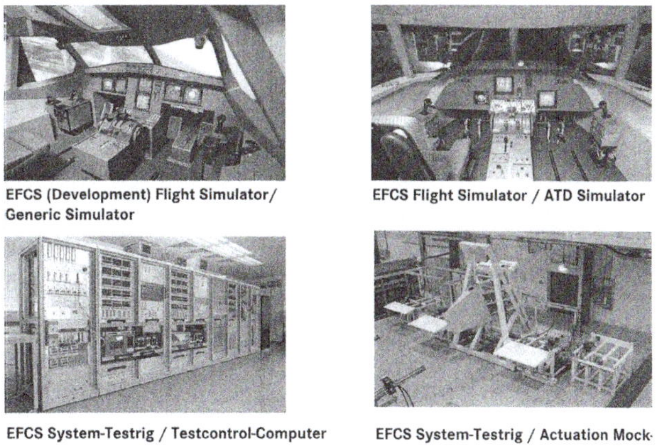

Fig. 6.137 Simulators and test rigs

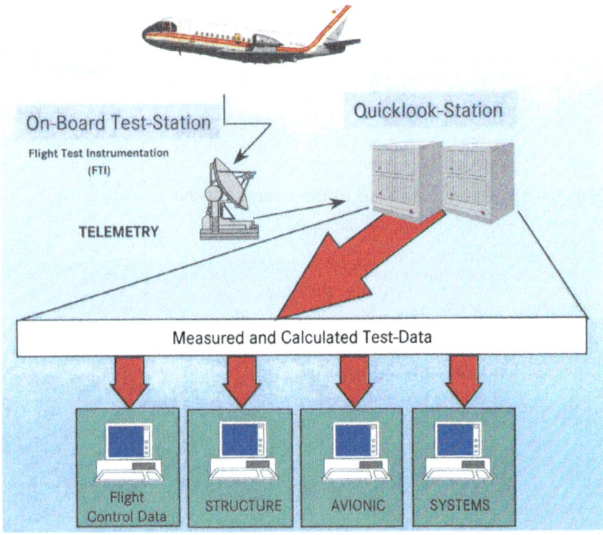

Fig. 6.138 Measurement and telemetry facilities

6.3.7.2 Primary Flight Control Unit (PFCU)

The contribution of BGT was the central flight control computer used in PFCU (primary flight control unit). It was the heart of the system. It generates the actuator commands for elevator, aileron, rudder and spoilers from the pilot inputs according to the control laws, using measured air, inertial and radar data. The PFCU interface in the overall system and the redundancy structure are shown in Figs. 6.139 and 6.140 respectively. Therein ISM/OSM were the input/output signal management modules that monitor the integrity of the input/output signals. RM is the redundancy management module that monitors the switching to alternative control channels, and RA is the radio altitude provided by the radar altimeter. ADIRS (Integrated air data inertial reference system) supplies air data (airspeed, angle of attack and altitude) and inertial reference (position and attitude) information to the pilots' electronic flight instrument system displays as

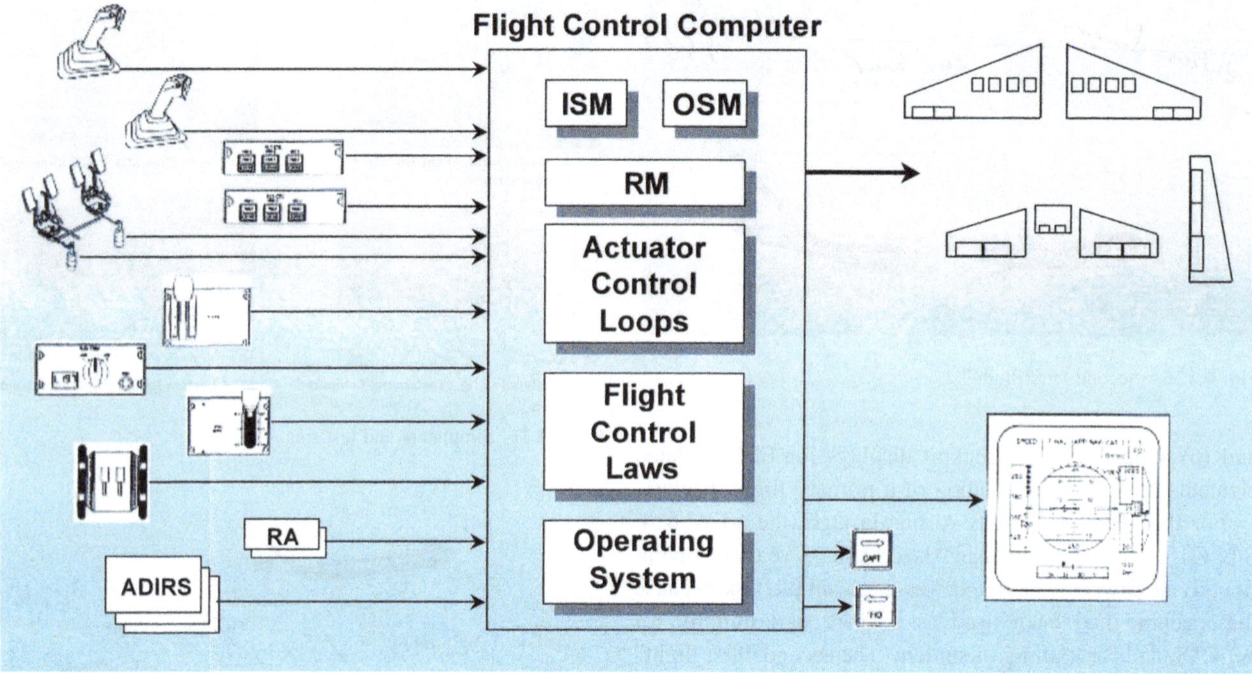

Fig. 6.139 PFCU interface in the overall system

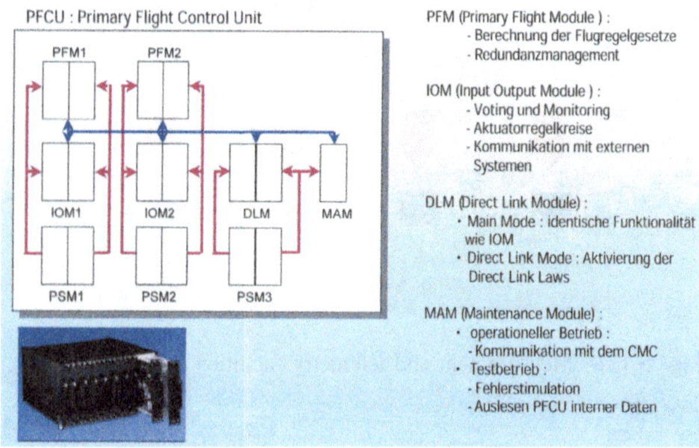

Fig. 6.140 PFCU redundancy structure

well as other aircraft systems such as the engines, autopilot, aircraft flight control system.

It is distinguished between two modes of control laws, namely (1) *Direct Laws*, in which control surface deflections are generated proportional to pilot input and (2) more comfortable *Normal Laws*, in which the pilot input is interpreted as a command to change the flight state. For example, roll rate proportional to the pilot lateral inputs at all airspeeds and load factor proportional to pitch input. The PFCU took over the redundancy and fault management of the entire control system.

Since the first flight of the A320 on February 22, 1987, Fly-by-Wire technology had become the state of the art for flight controls in civilian aircraft, including sidesticks as a substitute for the classical control devices. The certification regulations according to JAR/FAR/CS 25.1309 demand that any error combination with catastrophic effect must be extremely improbable. Quantitatively, 10^{-9} failures per flight hour are interpreted as "extremely unlikely". For flight control systems, JAR/FAR/CS 25.671 demands furthermore that no single failure may lead to a catastrophic effect. This qualitative requirement complements the quantitative requirement. The flight control system of Airbus A320 meets the above requirements and the EFCS of the ATD too.

These demands could be fulfilled only through a system with sufficient redundancy. Due to a high number of system states in a complex digital system, it cannot be ruled out that in spite of careful validation tests a design error remain undiscovered. In the case of an identical structure of redundant channels, these so-called generic errors would not be detected by a fault monitoring system based on the principle of majority rule, because all channels are then equally erroneous. Therefore, they lead to false commands and may lead to control surface "hardovers" and possibly to a loss of aircraft. Even such an error should not result in the aircraft loss. In the present case this aspect was taken into account by a dissimilar redundant structure, that is, redundant channels were pairwise structured differently. The dissimilarity was achieved by using different processors (Transputer T805/T400 and Power PC MPC 505) with different memory architectures in redundant channels.

If a disparity occurs between the two channels of the channel pair PFM 1, then this channel pair passivates itself. Nevertheless, in order to maintain control capability after this error, a second dissimilar pair PFM 2 is installed, which will be activated by the previous error. The increase in safety is, however, at the cost of reduction in the availability because of the larger number of components used and leads to a methodical as well as disciplinary extremely demanding process for both the design teams and for the certification authority.

PFCU was the first modular avionics cabinet for safety critical applications. It led to reduced number of external interfaces within the flight control and to the environment compared to the conventional construction. As a result, it led to a reduction in the development and unit costs (for example cabling), as well as breakdowns and maintenance costs. An integration of additional functions was more readily possible due to the modular concept, for example, autopilot functions.

The BGT tasks included several aspects such as (1) conceptualization of hardware and software for PFCU, (2) development and construction of the agreed number of devices, (3) implementation of the software, including control laws delivered by EADS Airbus and actuator electronics delivered by LAT, (4) test devices for functionality, (5) proof of compatibility of all the required load conditions, (6) proof of reliability, (7) support and guidance during aircraft ground tests, and (8) flight tests and implementation of all changes that may accrue based on the flight test results.

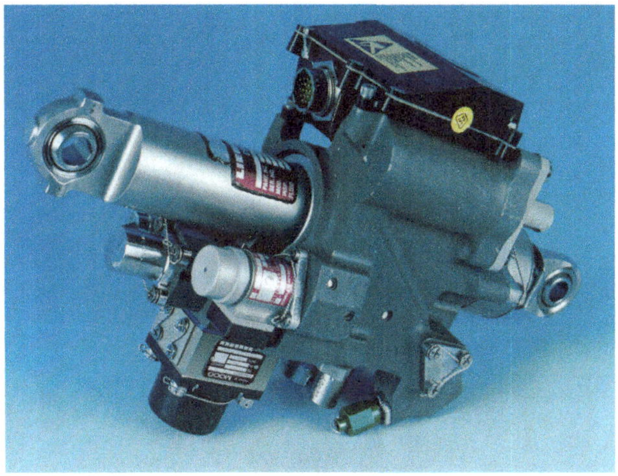

Fig. 6.141 Aileron actuation system

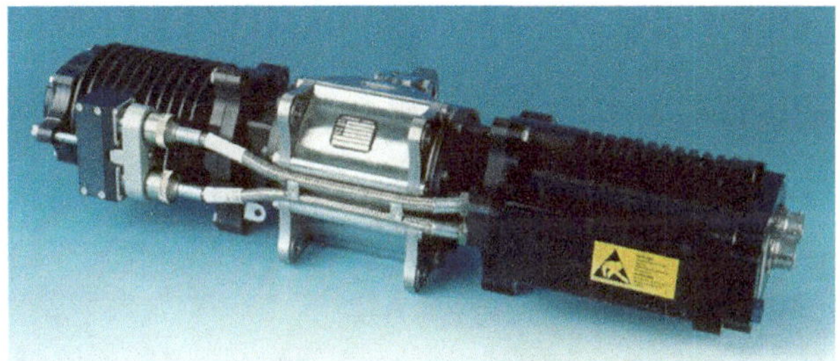

Fig. 6.142 Spoiler actuation system

6.3.7.3 Actuators and Hydraulic Power Supply

The work share of LAT comprised of the development, production and certification of actuator systems for the three primary axes using different actuator technologies. While linear electrohydraulic drives in parallel arrangement were envisaged for ailerons, electro-mechanical rotary actuators with integrated signal and power electronics were used for spoilers. All actuators for roll control were designed as *smart units* with integrated electronics. The elevator and rudder retained conventional servo actuators, powered by a central hydraulic system. Their control loop electronics was housed in a separate cabinet of PFCU, in the form of individual IMA modules. The smart actuators of the roll control communicated with the PFCU directly through an ARINC 429 data bus. A comparison of different solutions was thus possible for signal transmission and type of energy supply under the same operating conditions.

Two major development steps were undertaken pertaining to the actuator technology. One, the control loop and monitoring function were integrated into the actuators for the aileron (see Fig. 6.141) and spoiler drives (see Fig. 6.142). The technology employed herein is called "smart" and retains the connection to the higher-level computer through a serial data bus. Furthermore, an electro-mechanical rotary drive, which can perform all spoiler functions, was developed as a substitute for conventional hydraulically driven spoiler.

For the lateral and vertical control (see Figs. 6.143 and 6.144), the mechanical structure of the actuators was essentially further developed, such that a higher functional density in the hydraulic valve block and extended access possibilities in the switching functions could be realized. The elevator actuators were provided with an additional centering function, which permitted zeroing of the servo valve by means of mechanical controls in the case of electrical failures. The third hydraulic supply system required for redundancy was provided by electrically driven power packs (see Fig. 6.145). It was needed only for elevator and rudder actuators located in the tail and provided a new solution of Power-by-Wire in the form of a decentralized, local system. All newly developed components of PFCU, actuator systems, and HPP had to undergo a rigorous qualification test program to the prove suitability of devices for flight operations.

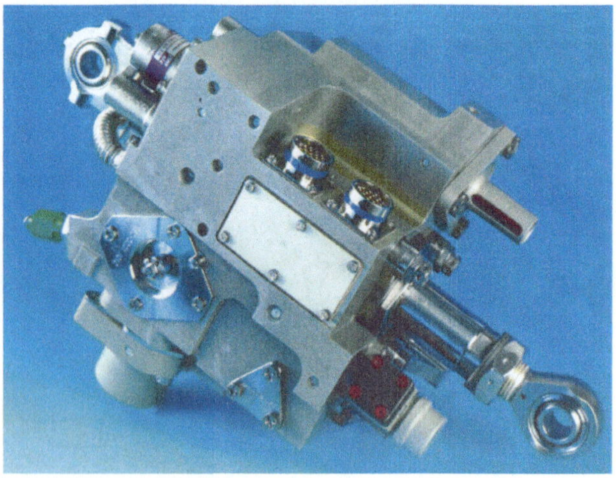

Fig. 6.143 Rudder actuation system

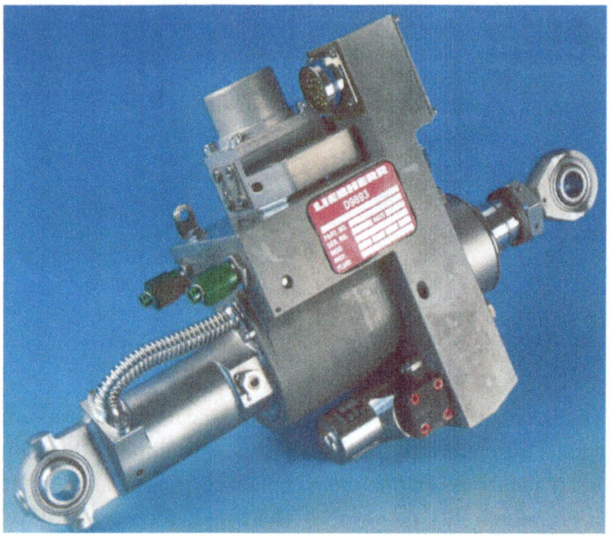

Fig. 6.144 Elevator actuation system

6.3.7.4 Ground and Flight Tests

Accompanied by Wagner music and in the presence of the German Federal Minister for Research and Technology, the converted aircraft was rolled out from the Hangar through of a cloud of steam on August 10, 1998 (see Figs. 6.146 and 6.147). After a series of ground checkouts, the third "first flight" took place on August 13, 1999, now as ATD with the registration number D-ASAX, with test pilot *Gert Rainer Selleske* and *Axel Widmann* and the flight test engineer *Uwe Krome*.

The test bed was thoroughly tested in 12 test flights until October 13, 2000. Figure 6.148 shows the aircraft after landing from one of the following flights with the first-flight crew, accompanied by the flight test engineer *Martin Läpple*.

Fig. 6.145 Hydraulic power package

Fig. 6.146 Roll out

6.3.7.5 Results

The new control system proved to be very promising in the ground tests. The system could be checked for all states under the various conditions in a tight and extensive flight test program. The errors found were rectified and system or equipment changes were carried out where necessary. A preliminary airworthiness certification was granted on August 16, 1999 by the German Federal Aviation Authority. In spite of the short duration of flight testing, the flight controls were tested throughout the entire flight envelope and configuration in normal as well as direct law, including extreme conditions such as engine failures and stalls. Controllability and handling qualities with the new sidesticks

Fig. 6.147 Roll out with German federal minister *Bernd Neumann* (on the stairs) and ATD program manager *Hubertus Schmidtlein* (3rd from right)

Fig. 6.148 Return after successful flight test (from left, *G. R. Selleske, Martin Läpple, Uwe Krome, Klaus Dietrich Flade)*

were assessed as good in both control law modes. The EFCS system performed as specified, including reconfigurations in error stimulations, the correct interpretation of monitoring thresholds, mode switching, the precise functioning of protections, such as angle of attack, airspeed, and load factor limitations. The new EFCS system, as well as the entire aircraft, proved robust and reliable. None of the stimulated failures had any unacceptable impacts on system design or functional integrity and could be eliminated easily without major difficulties.

6.3.7.6 Impact on Future Projects

About 100 employees from the participating companies have further qualified themselves by contributing to this challenging task and could contribute to future Airbus products through newly acquired experiences in the development of Fly-by-Wire flight control systems. The safety-critical issue of secondary flight controls could be strongly anchored at the national level through the ATD program. This was essential for the "*High Lift Center*" in Bremen, which emerged from the creative achievements in the fields of aerodynamics, structural design, as well as wing assembly, and then became the systems engineering center for the secondary flight controls. As a consequence, EADS Airbus Bremen took over the primary responsibility for the hydraulic systems for all Airbus aircraft.

The data acquisition, transmission, display, and analysis system for the ATD program was designed using mostly commercially available components. It was highly successful in terms of functionality, flexibility, and range of applications at moderate cost. A few important parts of this concept were adopted in the ATRA (*Advanced Technologies Testing Aircraft*) of DLR. Primarily, ATD served as an efficient and proven concept to take over the responsibility for the wing test benches of the Airbus family at Bremen site, namely the Airbus high lift center. Meanwhile, test benches for A380, A400M and A350 have been built and put in operation, an impressive demonstration of top-class technology (see Fig. 6.149) that could not have been realized without the ATD/EFCS program.

BGT (today: Diehl Aerospace), which was in advance envisaged as the supplier of SFCC (*secondary flight control computer*) for the Airbus family, could consolidate its leading position for the future Airbus family members through their

Fig. 6.149 Wing system test bench

achievements in the ATD program. Thus, BGT could prevail in supplying SFCC for A380, A400M, and A350. Electronics circuits developed by BGT for PFCU are implemented in SFCC of Airbus A340-500/600. An improved scaled-down software for dissimilarity, derived from the PFCU development, is implemented in A330/A340 SFCC redesign.

The technologies developed by LAT for the ATD program are incorporated now in numerous applications in primary and secondary flight control systems, including Airbus A380. LAT provided for this aircraft the rotary actuators for slat actuation as well as four spoiler actuators, all with integrated electrical and hydraulic supply to better survive malfunction in power supply.

As a spinoff of the ATD program, LAT was awarded a contract to deliver the Fly-by-Wire flight control system for the Russian regional aircraft Superjet S-100. Designed and built by the Sukhoi, the Superjet S-100 is a twin-engine, single-aisle, short and medium haul aircraft for 95 passengers. Amid the tough international competition, Sukhoi chose the LAT proposal, which was substantially based on the ATD technology. The development and production contract included the entire primary control system, the high-lift system, the control devices, and all system computers including flight software as an integrated package. The full electronic flight control system contains sidesticks and rudder pedals as cockpit control devices, the control loop and power electronics for all electrohydraulic and electromechanical actuators as well as flight control units for computation of control laws. All electronics are multiple redundant and are designed using the dissimilarity principle to protect against generic design problems. A mechanical back-up could be dispensed due to these system properties and the high degree of redundancy. The Superjet Sukhoi S-100 (first flight May 19, 2008) received European type certification from EASA in February 2012. Thereby the way was paved for its commercial service not only in Russia but also worldwide.

A first step towards "more electric aircraft" was taken through the technologies tested with ATD, a topic whose importance is perceivable in the Boeing Dreamliner. The next generation of aircraft starting with the regional area will be fitted completely with Fly-by-Wire and Power-by-Wire flight control systems. Electro-mechanical and electrohydro-static actuators with decentralized signal and power electronics will successively replace the hydraulic and pneumatic power supply systems. LAT successfully participated in Airbus A 380 and A 400 M with this new generation of actuators. Based on the smart wing concept of ATD, LAT has also developed a standardized *remote electronic unit* and brought it to production maturity. This technique is intended for flight control systems of the next generation, where increasingly digital data buses are used as communication medium.

The application of the latest production methods and the use of highly integrated microprocessors allowed the miniaturization of electronics to such an extent that they become an integral part of the actuator. The benefits are the increased system aptitude of the device, and at the same time lower costs and improved reliability. Participation in the various national aeronautics research programs with significant own resources, together with a long-term company strategy, helped LAT to establish itself as a so-called global player for complete aircraft systems (*nose-to-tail*), among others in the field of integrated flight control systems.

The cooperation with DLR in Braunschweig (Institute of Flight Systems) once again proved to be successful through the exchange of information and experience between research and industry, as in the past within the ATTAS project. DLR contributed to the ATD program by providing measurement and telemetry systems, 10 scientists by creating sponsorship jobs in Bremen and Braunschweig, and by assigning test pilots. Furthermore, it also provided an important contribution to the ATD success through pre-testing of the ATD-flight control laws on the In-Flight Simulator ATTAS under the so-called SAFIR program (see Sect. 9.2.6).

In summary, the two programs ATTAS (Chap. 9) and ATD in combination with the corresponding helicopter test demonstrators ATTHeS (Chap. 8) and FHS (Chap. 10) had stimulated not insignificantly the development of internationally competitive products in the German aeronautical and military industry, and thereby represents a good investment. It also needs to be emphasized that these programs provided ample chances to acquire and retain system capability, which is an essential prerequisite to survive in the present days of global competition.

6.3.8 Boeing/EADS X-31 VECTOR (1998–2003)

Peter Hamel

The highly successful post stall demonstration flights described in Sect. 6.3.6 provided the foundation for the German-American revival of the remaining second X-31 prototype (S/N 164585). Here, the goal was to utilize thrust vector technologies as an integral part of the digital flight control system for fully automatic and extremely short landings at low approach speeds (ESTOL—*Extreme Short Take-Off and Landing*). That was one of the issues because a reduction in the approach speed and thereby in the kinetic energy would preserve the structure and lead to a reduction in wear and tear, also during aircraft carrier landings. It was also the starting point of investigations pertaining to post stall flying at MBB during the beginning of the nineteen seventies. This concept tested with the X-31 is currently being applied to the X-47B, an unmanned technology demonstrator capable of launching from and landing on a seaborne aircraft carrier.

Fig. 6.150 Boeing-EADS X-31 VECTOR graphite composite thrust vector deflectors (*Credit* www.456fis.org)

Fig. 6.151 X-31 VECTOR in landing approach (*Credit* www.456fis.org)

In late 1998, Boeing took over the conversion of the test demonstrator. This included retaining the X-31A EFM thrust vector control system, with the three heat-resistant graphite composite paddle-like vanes mounted on the airframe adjacent to the engine exhaust nozzle (see Fig. 6.150). On the tip of the fuselage nose, adopted from the Eurofighter, a high-precision and miniaturized air data system FADS (*Flush Air Data System*) was installed for precise measurement of high angles of attack and sideslip (NordMicro). This necessitated the relocation of the noseboom mounted hitherto underneath the fuselage to a location on the upper side. The data from FADS were recorded, whereas the flight control system was, however, further provided with data from the hitherto ADS (*Air Data System*) installed on the nose boom and fuselage.

In addition, a GPS-based navigation system (IntegriNautics) was installed, which provided an extremely accurate aircraft position. A new autopilot was developed for automatic landings at high angles of attack. The X-31 was still equipped with the triplex FBW flight control system and an additional tiebreaker computer. Analog or mechanical emergency systems (back-up) were not available.

The so-called VECTOR-Program (*Vectoring, Extremely Short Takeoff and belonged Landing, Control and Tailless Operation Research*), was organized jointly by the US Naval Air Systems Command (NAVAIR) and the German Federal Office of Defense Technology and Procurement (BWB). The integrated flight test team consisted of members from these two organization, and those from Boeing, EADS, and DLR Institutes of Flight Systems (up to 1999: Institute of Flight Mechanics) and Robotics and Mechatronics. While accompanying the flight testing, the DLR Institute of Flight Systems had developed and validated a global flight dynamics model valid over the entire operation envelope applying system identification methods [69, 70].

On February 24, 2001, the X-31 took off for its "second" first flight (see Fig. 6.151). Up to the end of April 2001, initial flight tests were carried out, primarily to check the functionality of the systems, and also to familiarize the pilot with this unusual research aircraft and FBW technology demonstrator.

The modified X-31 completed its final flight on April 29, 2003 at the Patuxent River Naval Air Station and landed fully automatically with an angle of attack of 24° (normal: about 10–12°) and a touchdown speed of just 220 km/h (normal: about 320 km/h). For this precision maneuver, a special highly improved GPS (*Global Positioning System*), the so-called IBLS (*Integrity Beacon Landing System*) combined with 3 inertial platforms was necessary, which provided an accuracy in the range of a few centimeter. From Fig. 6.152 it is apparent that the de-rotation (*Pitch Down*)

Fig. 6.152 X-31 VECTOR ESTOL landing (*Credit* Navy Mil)

Fig. 6.153 US-German X-31 VECTOR-Team (Deutsches Museum Oberschleißheim)

required for the touchdown takes place at the very last moment, just about 60 cm above the ground [71–73].

Since August 2003, this extraordinary aircraft is displayed at the branch office of the Deutsches Museum in Oberschleißheim, Germany, in good company with many other interesting aircraft such as the DLR In-Flight Simulator ATTAS (see Fig. 6.153).

6.3.9 Stemme S15 LAPAZ (2007–2013)

Robert Luckner

Motivated by the growing market for airborne reconnaissance and surveillance tasks, the German aircraft manufacturer Stemme AG (Strausberg) investigated the advantage of FBW technology for its single-pilot, single-engine, high-performance utility aircraft Stemme S15 (now called Ecarys ES15). This light aircraft with a high-aspect ratio and a take-off mass of about 1.1 ton, that can carry payloads of 100–300 kg, represents a cost-effective and efficient solution for such tasks. The S15 was specifically designed for commercial applications and certified according to FAR Part 23 (wingspan 18 m, max. cruising speed 270 km/h, endurance 8 h can be increased by installation of ferry tanks).

In missions where the pilot has to fly the aircraft while simultaneously operating the payload, an Automatic Flight Control System (AFCS) is needed or may be compulsory to support the pilot. In extremely difficult, long endurance or dangerous missions, Stemme's objective was to operate the S15 as an optionally piloted vehicle by replacing the pilot with the AFCS and ultimately to have an unmanned aircraft, where the cockpit space becomes available for payload. This required an AFCS with full control authority and high reliability.

The development, certification and production of such a complex and safety-critical system and its software according to the certification specifications EASA CS-23 or FAR Part 23 at competitive cost was a major challenge. This challenge was addressed by the LAPAZ technology project, in which an AFCS was developed for the S15, see Fig. 6.154 [74]. The acronym LAPAZ stands for "Luft-Arbeits-Plattform für die Allgemeine Zivilluftfahrt" (*Air utility platform for the General Civil Aviation*).

The LAPAZ project had three partners: Stemme as the coordinator, Institute of Aircraft Systems (ILS), University of Stuttgart, and Department of Flight Mechanics, Flight Control and Aeroelasticity (FMRA), Technical University of Berlin. Stemme provided the aircraft and was responsible for the integration of the AFCS as well as for the execution of

Fig. 6.154 Stemme S15-LAPAZ in flight

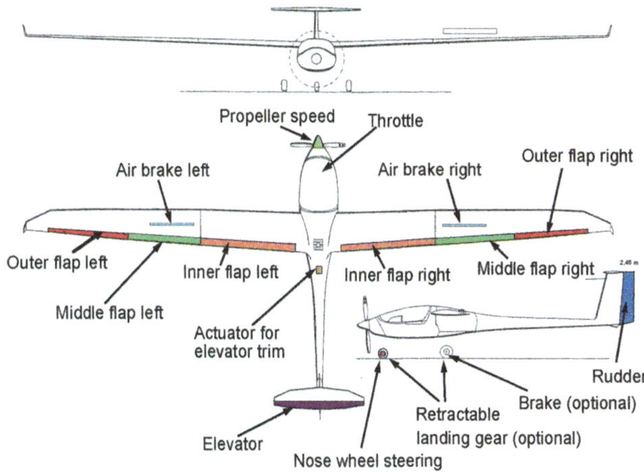

Fig. 6.155 S15 flight control surfaces

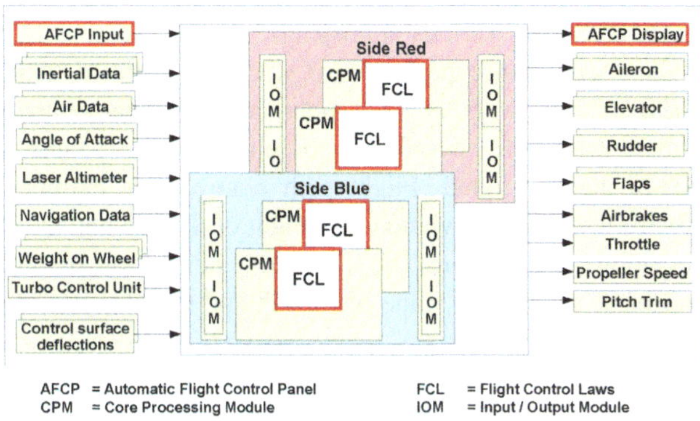

Fig. 6.156 Integration of FCL into the AFCS

hardware-in-the-loop (HIL) simulator tests and flight tests. ILS developed the fault-tolerant Fly-by-Wire flight control computer platform for the AFCS, including all redundancy mechanisms. FMRA developed the flight control laws, the flight mechanical simulation model, the HIL flight simulator, the pilot's interface as well as a specially designed development process for certification of the flight control law functions and software.

The definition of a modular and scalable system architecture, the use of modern standard electronic flight control components, and the definition of a stream-lined system development process for the AFCS functions, hardware and software were the key elements for success. The LAPAZ project has successfully demonstrated and validated this development process as well as the AFCS architecture and functions.

Interestingly, the heads of the university institutes, *Reinhard Reichel* and *Robert Luckner,* who had participated as principal engineers for Diehl BGT and for EADS Airbus respectively in the development of the Electronic Flight Control System (EFCS) of the VFW 614 ATD (see Sect. 6.3.7), contributed significantly to the overall success of this project.

The mechanical flight control system links the pilot's control devices to control surfaces by rods and cables, see Fig. 6.155. The AFCS commands are mechanically added by means of electromechanical actuators. In case of a failure, the AFCS redundancy management opens the affected clutch or clutches. In this way any failed actuator can be isolated. A fast decoupling function (FaD) was installed as a safety measure that instantly disconnects all actuators from the primary controls of the aircraft by switching off the electrical power. The pilot can use the FaD to take over control whenever he wants.

The AFCS was designed for high-precision flight path control during airborne measurements and surveillance tasks with a lateral and vertical flight path accuracy of a few meters. Also, improved attitude stabilization for certain payload sensors was provided for flights in atmospheric turbulence.

The flight control laws (FCL) were developed with a model-based approach, where the functions were immediately tested within a ground-based flight simulation. The flight control law source code was automatically generated.

The FCL were integrated into the Flexible Avionics Platform (see Fig. 6.156) developed by the ILS (see Sect. 6.3.10).

On March 22, 2012, a major project milestone was achieved with the first automatic landing at the airfield Neuhardenberg, Germany using a GPS with enhanced positioning accuracy by satellite-based augmentation (EGNOS). Two different automatic landing modes were implemented: 1) a motor landing mode with airbrakes in a fixed position using thrust and elevator for approach path control and 2) a glider-like landing mode with thrust in idle that uses dynamic airbrake and elevator deflections to control the glide path. A complete automatic mission from takeoff to touch-down with roll-out was flown at Strausberg on November 23, 2012. During the flight test campaign

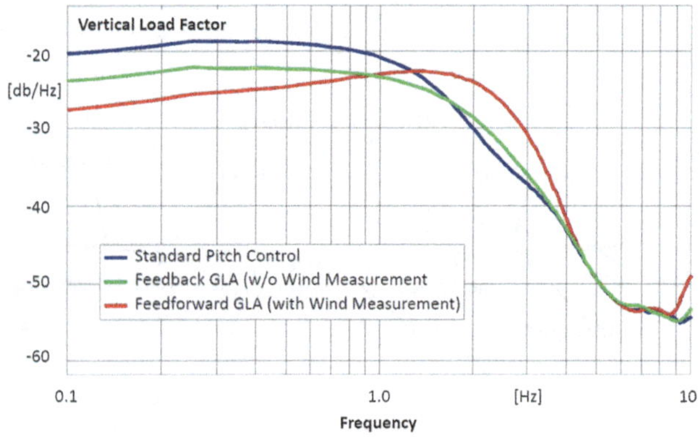

Fig. 6.157 Vertical load factor spectra

Fig. 6.158 DA42 MNG FBW research aircraft

more than 100 successful takeoffs and landings were performed from different airfields, on grass and concrete runways, and under various atmospheric conditions.

The surveillance performance was demonstrated taking aerial photos with a fixed-mounted and non-stabilized camera while the aircraft had to follow a scan pattern in 700 m above ground. Flight path control was extremely precise with mean deviations in height of 0.11 m and lateral deviations of 0.187 m. The excellent stabilization of pitch and bank angle (rms value of 0.202° and 0.082°) provided superb photo quality.

Figure 6.157 illustrates the influence of three different control laws of the LAPAZ AFCS on the vertical load factor spectra during flight in turbulence: standard pitch control (blue), feedback gust load alleviation (GLA) (green), and feedforward GLA (red). The GLA functions reduced the load factor by 3 and 6 dB respectively below frequencies of 1.5 Hz (9.5 rad/s); above 1.5 Hz the load factor increases as predicted.

6.3.10 DA42 MNG FBW Research Aircraft (Since 2008)

Lars Peter

In 2008, development of a versatile research aircraft platform incorporating a Fly-by-Wire system was initiated by the Institute of Flight System Dynamics (FSD) of Technical University of Munich (TUM) with the participation of multiple companies and institutions and supported by the Bavarian ministry of economic affairs. After a thorough assessment of available aircraft, the DA42 MNG (Multi Purpose Platform New Generation) by Diamond Aircraft Industries was selected as the basic aircraft [75]. The driving factors for this choice were its unique restricted type certificate, available exterior nose and belly payload pods. It also provided for a dedicated 2.8 kW power supply for experimental equipment. The FBW research platform was successfully flown for the first time in September 2015 (Fig. 6.158).

High-quality reference sensors such as multiple GNSS receivers, multiple air data booms, a navigation grade inertial navigation system, laser and radar altimeters and data links were installed as a part of specially developed versatile flight test instrumentation (FTI). The unique feature of the aircraft is, however, the access to every axis of the flight control (elevator, aileron, rudder, throttles, and trim surfaces) through an experimental FBW system for in-flight control and guidance experiments. During the design and development of the experimental control system, the highest requirement was to ensure safe operation of the aircraft under all circumstances. As in the case of all small General Aviation aircraft, the DA42 features a purely mechanical flight control system of pushrods for elevator and aileron control, cables to the rudder and Bowden cables for trim surfaces. The installed multi-stage safety system allows testing of experimental components and functional algorithms at an early developmental stage by ensuring a safe disconnect of the digital flight control system and reversion to manual controls following faults in the experimental system.

The experimental FBW control system is based on electromechanical rotary actuators backdriving the existing mechanical controls, including EECU (Electronic Engine Control Unit) throttle levers. The primary flight control actuators were specifically designed for the DA42. They are fully reverse-operable, feature a dual redundant motor winding, redundant power and control electronics, and a single dual-stage planetary gear system with very low backlash. While the output torque and speed of the actuators can be limited via the motor control electronics, these parameters can also be influenced physically through MIL-COTS DC-DC converters. Replaceable analog trimming resistors enable to limit actuator rate (voltage) and torque (maximum current) for each flight, resulting in variable performance ranging from slow autopilot like dynamics to full authority control with surface rates in excess of 60°/s.

An adjustable safety release clutch can also independently limit the maximum transmission torque without slip and functions as a backup disconnect mechanism in case of mechanical failures in the drive system. Adjustable drive internal hard stops can limit actuator travel during high bandwidth operation to protect against high authority actuator runaway scenarios.

Primary means to engage and disengage the Fly-by-Wire control system are electromechanical clutches, designed for safety critical applications and able to open under all load conditions when electric power is interrupted. Multiple independent and software-free means to disconnect power to the clutch (and actuators) are available to all crew members, including existing AP (Autopilot) Disconnect buttons and emergency stop switches. The closing of the clutch is armed electro-mechanically by the crew and executed by a safety computer dubbed "*ENSURE*" if engagement conditions are met. Based upon a qualifiable aerospace data concentrator unit, *ENSURE* continuously monitors operational engagement criteria, actuator operating parameters, Flight Control Computer (FCC) operations, and predefined aircraft envelope limits such as g-loads and will automatically cut system power in reaction to violations.

For flights utilizing the experimental FBW control system, the aircraft is typically operated by a three-person crew. The pilot in command, located in the front left seat, serves as safety pilot and controls the aircraft using the mechanical center stick and control system whenever the experimental control system is not active or needs to be disconnected. A test or evaluation pilot occupies the front right seat, where the mechanical control inceptor has been removed. For pilot-in-the-loop flight test scenarios, he can use either a removable passive spring-damper control stick to command inputs into the FCC or utilize an active center stick with freely programmable dynamic control loading functions. Besides these stick inputs, a removable experimental display combined with an autopilot mode control panel can be taken aboard and connected either in the front or rear of the cabin. Both pilots can engage or disconnect the FBW system using a small cockpit control panel and monitor system operations via LED indications. Additional experiment related information such as controller modes, trajectories or sensor and flight guidance information can be displayed on a smaller glare shield mounted display or on the cockpit Multi-Function Display (MFD).

A flight test engineer (FTE) station is located on the left rear seat, next to a 19-inch equipment rack that replaces the fourth cabin seat (Fig. 6.159). Besides controlling power distribution to all flight test systems, configuring and monitoring sensors and other experimental equipment, the FTE can set detailed configurations of the FBW control system, such as selecting an individual aircraft control axis that shall be engaged, configuring DC-trim resistors to preselect FBW authority or select active motor windings using an extended operation panel. The physical control panel also provides monitoring and status lamp indications for all safety relevant operating conditions. A much more detailed access to all control system related information and all other sensor and aircraft data is provided through a permanently mounted tablet computer connected to the FTI. Only if all engagement criteria are met and the FBW control system operates correctly, the FTE enables its operation by the pilots.

As a part of the overall DA42 MNG research aircraft, a hardware-in-the-loop (HIL) simulation, test-rigs, and a ground-based full cockpit simulator were set up to test new software. After installation in the aircraft, all functions and control system elements will be evaluated in aircraft-in-the-loop-simulations as final verification before flight tests (Fig. 6.160).

Commissioning of the research aircraft and its Fly-by-Wire control system capabilities started in 2015, after nearly six-year technical development and installation phase. Initial in-flight verification and validation of the control system components and its control authority were performed using a direct-law-control software. Confidence in system stability and all disconnect mechanisms was built up due to very extensive testing via aircraft-in-the-loop simulations and some ground taxi tests, and direct law flight tests revealed no issues. Flight controller engagement was performed stepwise, by first verifying the inner loop controller performance with preprogrammed command signals and

Fig. 6.159 Rear cabin view of FTE workstation and flight control system electronics

Fig. 6.160 Aircraft-in-the-Loop simulation with Ground Control Station

then adding more advanced outer loop functionalities such as autopilot and trajectory controllers.

Due to available pretesting capabilities with high fidelity desktop flight dynamics simulations as well as full system tests in HILS and AILS, developed primary flight control algorithms performed flawless from the very beginning, not requiring retuning of gains. Initial controller flight tests were performed over four flights on two consecutive days, with successful autopilot demonstration and an initial flight with full remote control from a ground station on the very same day. Already completed and ongoing flight campaigns centered around the development and demonstration of technologies for optionally piloted vehicles, as showcased on the 2015 Paris Airshow together with Diamond Aircraft, including further remote control functionalities and automated waypoint flights with STANAG compliant interfaces.

The research aircraft has been shown to be particularly useful in the demonstration of flight control technologies for unmanned aircraft, for example, remote controlled operations via data link from a ground control station. Even fully automated flights including automatic take-off, mission segments, and automatic landing could be performed without requiring segregated airspaces and closed airports, due to having a safety pilot on board that can always monitor system performance and take control if required. Overall Fly-by-Wire flight time accrued between 2015 and 2016 was around 50 h.

The DA42 MNG FBW research aircraft has been selected as a demonstration platform to develop methods of compliance to support certification of advanced flight controls in General Aviation for the US Federal Aviation Administration. System technologies and controllers designed for the DA42 MNG were successfully transferred into an electric single-seat aircraft and an experimental digital autopilot of a 19 seat commuter aircraft and reputation gained allowed to acquire new ongoing development contracts from UAV and new Part 25 commercial air transport manufacturers.

6.3.11 DIAMOND DA42—FlySmart-FBW23 (2012–2015)

Reinhard Reichel

The prospective demand for point-to-point traffic employing small aircraft is offering significant growth potential. The associated appropriate new type of small aircraft is expected to be of category CS23 (Class I + II) and to be characterized as a so-called "*Easily Piloted Aircraft*" (EPV), meaning an aircraft providing high safety in severe flight conditions, for example, IMC (Instrument Meteorological Conditions), even if controlled by low level instructed and inexperienced pilots. This requires a flight control system providing a fully automatic/autonomous flight (a/a-mode) from takeoff to landing. The pilot can take over control at any time within the safe flight or route envelope. If the aircraft is close to exceeding these safe limits, the flight control system intervenes to avoid exceeding the safe envelope and if this is not successful, the flight control system automatically returns back to the a/a-mode ensuring a safe flight along a safe route to the destination airport. While the flight envelope protection keeps the aircraft from getting into an unsafe flight condition, the route envelope protection keeps the aircraft from flying into terrain, into restricted respectively controlled air space without clearance or into bad weather areas. It also protects the aircraft against air traffic collision or simply against running out of fuel.

In this approach, the a/a-mode represents the basic control mode whereas the manual control mode represents an "on top mode"—a change of paradigm. This requires a full authority FBW system providing full reliability of the a/a-mode for the complete mission. It is not intended for these systems to allow any degradation back to a so-called "Direct Law" or even back to a mechanical backup due to failures. Therefore, the FBW system must ensure not to "lose" the a/a-mode with a probability higher than 10^{-7} (1 h mission) and Design Assurance Level (DAL) B (the regulations covering these safety features are still under work by the authorities) [76].

The design and development of a FBW system providing the a/a-mode of an EVP is a real challenge because (1) it represents a functionally highly integrated system, (2) no degradation of the a/a-mode is accepted even in cases of failures resulting in a complex redundant Fly-by-Wire architecture, (3) highly integrated and complex redundant FBW systems are associated with the high development and qualification cost, and (4) The FBW system shall be applied to lower cost category aircraft of category CS23.

To meet these challenges, the Institute of Aircraft Systems (ILS) of the University of Stuttgart developed an avionics approach providing the potential to develop, qualify and manufacture complex, highly integrated FBW systems at low cost. The approach called "Flexible Avionics Platform" is characterized as follows:

- **Flexible Platform**: Any FBW system is built up as an instance (a concrete system) of the Flexible Platform [77].
- **Distributed Architecture**: The FBW system architecture is a distributed one based on generic hardware modules communicating with each other exclusively via a redundant network (dual FlexRay arrangement).
- **Platform-Management/Middleware**: The management of a FBW system, that is, the communication, failure detection, consolidation, and consensus generation (replica control) within redundant modules, between non redundant and redundant modules, redundant sensors, redundant and non-redundant actuators, is strictly segregated from any cybernetic function, for example, control law, called System-Function. The management of the FBW system (Platform-Management) is realized as middleware. Any module of a FBW system must be provided with an instance of the Platform-Management.
- **Simplex system minded Design of Control Laws**: Any System-Function is embedded in a virtually non-redundant, non-distributed FBW architecture. Consequently, System-Functions (control laws) can be designed in a "simplex system minded way".
- **Platform-Management/Basic Services**: Any instance of the Platform-Management is generated by selecting suitable Basic-Services (generic software services), specializing each Basic-Service instance, establishing data- and control-coupling of all Basic-Service instances.
- **Axx-Process**: The generation of all Platform-Management instances is performed automatically by means of a Tool-Suite executing the Axx-Process [78]. The systems engineer starts the design building a high level system characteristics model by means of a GUI (Graphical User Interface). In a next step, the Tool-Suite analyses this model and transfers it into a more detailed model representing the software high level design. In a consecutive step, this model is analyzed again and transferred into the most detailed model representing the software design result, which is transferred in a final step into the source code of a Platform-Management instance.
- **AAA-Process**: The models generated by the Tool-Suite contain all system and software characteristics of the considered FBW system. Therefore, after design all models are analyzed stepwise again in order to automatically generate the Systems Requirements Document, the Software Requirements Documents and the Software Design Document in a readable format (xAx-Process). Based on these requirements, the corresponding test cases or test scripts are generated automatically (xxA-Process). The development of the xAx-/xxA-Process is still in the works. The combination of the Axx-, xAx- and xxA-Process results in the AAA-Process, which is **A**utomatic design and code instantiation, **A**utomatic generation of requirements- and design-documents, **A**utomatic verification.

A big step towards the development of an EPV was feasible through the research projects SAFAR (EU) and FlySmart (LuFo). At the end of SAFAR, a Diamond Aircraft DA42 was equipped with a FBW system designed by means of the Flexible Platform with basic System-Functions implemented (see Fig. 6.161). Within the FlySmart program, the DA42 FBW system was upgraded and all System-Functions implemented, which were necessary for an automatic flight including takeoff and landing (ATOL). Most of the verification activities necessary to achieve the Permit-to-Fly were performed at the HIL (hardware-in-the-loop) test rig in cooperation with Aviotech GmbH. The System-Functions were developed by the IFR (Institute of Flight Mechanics and Flight Control of the University of Stuttgart). The flight tests in FlySmart with the FBW system started in July 2015 in Wiener Neustadt and those with automatic takeoff and landing in September 2015 (see Fig. 6.162). The resulting

Fig. 6.161 FlySmart FCS integration

Fig. 6.162 FlySmart Diamond DA42

performance was considered to be satisfying, with the average lateral and vertical deviations relative to the planned flight path recorded to be less than 4 and 2 m respectively. The aircraft for these programs was provided, modified, and operated by Diamond Aircraft Industries GmbH. Other partners with significant contributions to both projects were Airbus DS Airborne Solutions GmbH (Coordination) and SET GmbH (Hardware).

References

1. Hamel, P.: Challenges of Flight Systems—A View from the Past to the Future. In German, Invited Paper, TUHH, 19 Apr 2010
2. Croft, J.: Problematic Probe—Inaccurate Airspeed Readings Due to Ice Crystals Prompted Unnecessary Dives on a United Flight, AW&ST, pp. 32–33, May 23–June 5, 2016
3. http://www.historynet.com/lawrence-sperry-autopilotinventor-and-aviation-innovator.htm
4. Sutherland, J.P.: Fly-by-Wire Flight Control Systems, Joint Meeting of Flight Mechanics and Guidance and Control Panels of AGARD, Oslo, Norway, 3 Sept 1968
5. Tomayoko, J.E.: Helmut Hoelzer's fully electronic analog computer. Annals of the History of Computing **7**(3), 227–240 (1985)
6. Bollay, W.: Aerodynamic stability and automatic control. Annual Wright Brothers Lecture, December (1950)
7. Milliken, W. Jr.: Dynamic stability and control research. In: Proceedings, 3rd Anglo-American Aeronautical Conference, Brighton (1951)
8. Schmitt, V. R., Morris, J. W., Jenney G. D.: Fly-by-Wire—A Historical and Design Perspective. SAE International (1998)
9. Tomayko, J. E.: Computers take Flight: a history of NASA's pioneering digital fly-by-wire project. NASA SP 2000–4224 (2000)
10. McRuer, D., Graham, D.: A flight control century: triumphs of the systems approach. AIAA J. Guidance Control Dyn. **27**(2), 161–173 (2003)
11. Hamel, P.: The Birth of Sweepback—related research at LFA-Germany. AIAA J. Aircraft **42**(4), 01–813 (2005)
12. Meyer, H.U. (Hrsg.): Die Pfeilflügelentwicklung in Deutschland bis 1945, Die Deutsche Luftfahrt, Band 33, Bernard & Graefe Verlag, ISBN 3-7637-6130-6 (2006)
13. White, R.J.: Investigation of lateral dynamic stability in the XB-47 airplane. J. Aeronaut. Sci. **17**(3), 133–149 (1950)
14. Campagna, P.: Storms of Controversy: The Secret Avro Arrow Files Revealed. Dundurn Press (2010)
15. Holloway, R. B. et al.: Introduction of CCV Technology Into Airplane Design, Aircraft Design Integration and Optimization, AGARD CP-147, Paper 24, October 1973
16. Morris, A.: MDO—A tool for the future. Aerogram **9**(4), 19–24 (1999)
17. https://www.federalregister.gov/documents/2016/12/08/2016-29431/special-conditions-bell-helicopter-textron-inc-bhti-model-525-helicopters-interaction-of-systems-and#h-14
18. Tomayko, J. E.: Blind Faith: The United States Air Force and the Development of Fly-by-Wire Technology. In: Neufeld, J. et al. (ed.) Technology and the Air Force—A Retrospective Assessment, CreateSpace Independent Publishing Platform, pp. 163–185 (2012)
19. Reerink, H.P. (ed.): BGT—Die Geschichte eines Hochtechnologie-Unternehmens. Diehl BGT Defence GmbH & Co. KG, Überlingen (2006)
20. Schubert H. (Ed.): Die Deutschen Senkrechtstart-Flugzeuge EWR VJ 101C, Dornier Do 31 und VFW VAK 191B – Historische Nachlese und Bewertung, DGLR-Bericht 2000–01, Bonn (2000)
21. Campbell, L.: Project Morpheus—Flying a Test Bed for Future Landers, 11 July 2014. https://www.rocketstem.org/2014/07/11/project-morpheus-flying-test-bed-future-landers/
22. Matranga, G. J. et al.: Unconventional, Contrary, and Ugly: The Lunar Landing Research Vehicle. NASA SP-2004-4535 (2005)
23. Anon.: EWR VJ 101 – Alle Varianten, Teil 1, Klassiker der Luftfahrt, Flug Revue-aerokurier. http://www.klassiker-derluftfahrt.de. 23 May 2013
24. Smyth, R.: Vertikal startendes und landendes Aufklärungs- und Kampfflugzeug VAK191B Technologie-Träger, DGLR-Vortrag, Hamburg, 22 June 2007
25. Morgan, M.: Keeping the pilot happy—the contribution of research and development. Aeronaut. J. **75**, 630–647 (1971)
26. Hunt, G.H.: The evolution of fly-by-wire control techniques in the UK. Aeronaut. J. **83**(821), 165–174 (1979)
27. Anon.: RAE Electric Hunter, FLIGHT International, pp. 1010–1011, June 28, 1973
28. Warwick, G.: Fly-by-Wire Jaguar, FLIGHT International, p. 816, 12 Sept 1981
29. Moxon, J.: One-Eleven Trail Blazers, Flight International, pp. 32–36. 7 Sept 1985
30. Mcloughlin, A.: Smart fly by light, lucas aerospace flight control systems. Presentations at the Institution of Electrical Engineers (IEE) and British Computer Society (1994)
31. Landis K. H. et al.: Advanced flight control technology achievements at Boeing helicopters. In: Tischler M. B. (ed.): Advances in Flight Control. CRC Press, pp. 103–141 (1996)
32. Picirillo, A. C.: Fly-by-Wire: making the electric jet. In: NASA's Contributions to Aeronautics. NASA SP 2010-570 **1**, 631–733 (2010)
33. Landis, K.H., et al.: Advanced flight control technology achievements at Boeing helicopters. Int. J. Control **59**(1), 263–290 (1994)
34. Terry, J. L. et al.: The Advanced Digital-Optical Control System (ADOCS) User Demonstration Program, USAAVSCOM TM-89-D-2, September 1989
35. Norton, W. J.: Advanced Electromechanical Actuation System (EMAS) Flight Test, 4950th Test Wing FTR-86-4, June 1986
36. Linch, N.: Dream Machine—Thrust Vectoring in the F-16, AcesAero.com (1995)
37. Jensen, S. C., Jenney G. D., Raymond, B., Dawson, D.: Flight test experience with an electromechanical actuator on the F-18 systems research aircraft. NASA TM-2000-209015 (2000)
38. Boris, S.: Flying Testbeds and In-Flight Simulators—An Efficient Instrument for Joint Research within European Projects. EU-Russia Aeronautics Research Cooperation Workshop, Moscow, Russia, 20–21 Oct 2003
39. Belyakov, R.A., Marmain, J.: MiG—Fifty Years of Secret Aircraft Design. Naval Institute Press, Annapolis, Maryland (1994)
40. Gordon, Y., Gunston, B.: Soviet X-Planes. Midland Publishing (2000)
41. Gordon, Y., Komissarov, D.: Soviet and Russian Testbed Aircraft. Hikoki Publication Ltd. (2011)
42. Chatrenet, D.: Use of ground-based simulators and in-flight simulation for the development of the A320 flight control system. In: Flight Research at Ames—Fifty Years of Development and Validation of Aeronautical Technology, NASA SP-1998-3300, Paper 19 (1991)
43. Kubo, Y. et al.: The development of FBW flight control system and flight management system for ATIC BK117 experimental helicopter. In: American Helicopter Society 57th Annual Forum, Washington, DC, 9–11 May 2001

44. Metzdorff, W.: Elektro-hydraulische Stellglieder für die digitale Steuerungs- und Regelungstechnik, VDI-Bericht Nr. 107 (1966)
45. Metzdorff, W.: Neuere Arbeiten zur Flugzeug-Primärsteuerung am Institut für Flugführung der DFL, DFL-Mitteilungen, Heft 8, Seiten 393–404 (1968)
46. Metzdorff, W.: Zuverlässigkeit digitaler elektronischer Flugzeug-Primärsteuersysteme, DLR-FB 73–85 (1973)
47. Hobson, Engineering Proposal: Triplex Electro-hydraulic Actuator. Report No. ED/357/2/EP-1, 25. Wolverhampton, September (1964)
48. Tersteegen, J.: Einbau einer digitalen elektrohydraulischen Flugzeug-Primärsteuerung in das Forschungsschiff "Planet" zur Langzeit-erprobung, DFVLR-IB 082-72/15, April 1972
49. Hartmann, U.: Übertragungsverhalten des STOL-Flugzeuges Do 28D Skyservant, BGT TB 000D350/68 (1968)
50. Böhret, H., Schänzer, G.: The influence of microwave landing systems on guidance and control. In: AGARD Guidance and Control Panel Symposium on Night and All Weather Guidance and Control Systems for Fixed-Wing Aircraft, Cheltenham Spa, 3–6 May 1976
51. Schänzer, G.: Auslegung von Autopiloten durch Zustandsrückführung, Regelungstechnik, 26. Jahrgang, Heft 3, Seiten 75–104 (1978)
52. Schänzer, G.: Direct Lift Control for Flight Control and Gust Alleviation. AGARD Guidance and Control Panel Symposium on Guidance and Control Design Considerations for Low Altitudes and Terminal Flight, Dayton, Ohio, October 1977
53. Krag, B., Subke, H.: Flugversuche im Windkanal zur Entwicklung und Erprobung eines Böenenabminderungssystems. DFVLR Nachrichten, Heft 7, Seiten 561–562, Juli 1974
54. Chambers, J.: Modeling flight—the role of dynamically scaled free-flight models in support of NASA's aerospace programs. NASA SP 2009–575 (2010)
55. Heumann, G. W.: Das AVS-Projekt. Deutsch-amerikanisches Mehrzweck-VSTOL-Kampfflugzeug. FLUGREVUE 1967/4, S. 14–16
56. Krag, B., Subke, H.: Dynamic simulation in wind tunnels, part II. In: 46th Meeting Flight Mechanics Panel on Flight/Ground Testing Facilities Correlation, Valloire, June 9–13, 1975, AGARD CP 187, Paper 5 (1975)
57. Hoffmann, G.: Stabilisierung, Böenkom-pensation und Schwingungsdämpfung am elastischen, beweglichen Flugzeugmodell im Windkanal. Deutsche Luft- und Raumfahrt, DLR-FB 76–44 (1976)
58. Krag, B., Wünnenberg, H.: OLGA, a gust alleviation system for improvement of passenger comfort of general aviation aircraft. Vortrag auf dem 12. ICAS-Congress in München, Oktober (1980)
59. Böhret, H., Krag, B., Skudridakis, J.: OLGA—open loop gust alleviation system AGARD CP 384. Paper 13 (1984)
60. Löbert, G.: CCV Reglergestützte Flugzeuge, MBB-UF-Bericht, S. 84–91
61. Esch, P., Wünnenberg, H.: Direct side force and drag control with the aid of pylon split flaps. AGARD CP-262, Paper 10 (1979)
62. Jacob, D., Welte, D., Wünnenberg, H.: Experimental Flight Test Programs for improving Combat Aircraft Manoeuvrability by Manoeuvre Flaps and Pylon Split Flaps, AGARD CP-319, Paper 10, October 1981
63. Pletschacher, P.: DASA/Rockwell X-31A: Auf dem Absatz kehrt, Aerospace 1/95, S. 12-15 (1995)
64. Beh, H., Hofinger, G.: X-31A Control Law Design. AGARD CP-548, Paper 13, March 1994
65. Hamel, P., Jategaonkar, R.: Evolution of flight vehicle system identification. AIAA J. Aircraft **33**(1), 9–28 (1996)
66. Weiss, S., Friehmelt, H., Plaetschke, E. Rohlf, D.: X-31A system identification using single surface excitation at high angles of attack. AIAA Paper 95-3436, August 1995
67. Rohlf, D.: Direct Update of a Global Simulation Model with Increments via System Identification, RTO-MP-11. Paper No. 28, March 1999
68. Bosworth, J. T., Stoliker, P. C.: The X-31A Quasi-Tailless Flight Test Reults, NASA Technical Paper 3624 (1996)
69. Rohlf, D., Brieger, O., Grohs, T.: X-31 VECTOR System Identification—Approach and Results, AIAA-2004-4830 (2004)
70. Rohlf, D.: Global model approach for X-31 VECTOR system identification. J. Aircraft **42**(1) (2005)
71. Huber, P.: X-31 VECTOR Technologieprogramm – Der Ritt auf dem Strahl, Hochschule für Angewandte Wissenschaften Hamburg, 13 March 2003
72. Ross, H., Robinson, M.: X-31: 20 Jahre erfolgreiche internationale Zusammenarbeit zwischen Deutschland und den USA, DGLR-JT2003-053 (2003)
73. Grohs, T., Fischer, B., Heinzinger, O., Brieger, O.: X-31 VECTOR —ESTOL to the ground—flight test results and lessons learned. AIAA-2004-5029, August 2004
74. Luckner, R., Dalldorff, L., Reichel R.: A utility aircraft for remote sensing missions with a high-precision automatic flight control system. IEEE international conference on aerospace electronics and remote sensing technology 2014 (ICARES), Yogyakarta, Indonesia (2014)
75. http://www.fsd.mw.tum.de
76. Anon.: F3061, Standard Specification for Systems and Equipment in Small Aircraft. ASTM Committee F44 on General Aviation Aircraft, current edition approved August 1, 2016. Published September 2016
77. Goerke, S., Riebeling, R., Kraus, F., Reichel, R.: Flexible platform approach for fly-by-wire systems. IEEE/AIAA 32nd digital avionics systems conference (DASC), pp. 2C5-1–2C5-16, 25–29 Sept 2016
78. Mueller, P., Belschner, T., Lehmann, M., Reichel, R.: AAA-process—A New Approach to Affordable Fly-by-Wire Systems for CS23 Aircraft, DGLR Congress 2016, Braunschweig, Germany (2016)

Author Biography

Peter G. Hamel was the Director of the Institute of Flight Mechanics/Flight Systems of the German Aerospace Center (DLR/DFVLR) (1971–2001). He received his Dipl.-Ing. and Dr.-Ing. degrees in Aerospace Engineering from the Technical University of Braunschweig in 1963 and 1968, and his SM degree from M.I.T. in 1965. From 1970 to 1971 he was Section Head of Aeronautical Systems at Messerschmitt-Bölkow-Blohm in Hamburg. Since 1995 he is an Honorary Professor at the Technical University of Braunschweig and a founding member of three collaborative research centers at the University. He was Chairman of the National Working Group on Helicopter Technology (AKH) (1986–1994) and the appraiser for the National Aviation Research Program (LuFo) until today. He was also the Manager of DLR's Rotorcraft Technology Research Program and the German Coordinator for the former AGARD Flight Mechanics/Vehicle Integration (FMP/FVP) Panel. He is a member of the German Society for Aeronautics and Astronautics (DGLR) and of the American Helicopter Society (AHS), and a Fellow of AIAA. He is the recipient of the AGARD 1993 Scientific Achievement Award, of AGARD/RTO von Kármán Medal 1998, of AHS Dr. A. von Klemin Award 2001, and of the prestigious DGLR Ludwig-Prandtl-Ring 2007.

7 In-Flight Simulator HFB 320 FLISI

Knut Wilhelm

Assisted by Volkmar Adam, Dietrich Hanke and Hans-Heinz Lange

K. Wilhelm (✉)
Braunschweig, Germany
e-mail: uk.wilhelm@t-online.de

P.G. Hamel (ed.), *In-Flight Simulators and Fly-by-Wire/Light Demonstrators*,
DOI 10.1007/978-3-319-53997-3_7

7.1 Project Kickoff

During the early nineteen sixties, variable stability aircraft were developed and used intensively, especially in the USA, to investigate flying qualities. Based on the positive experiences with these aircraft, during the mid-nineteen sixties the need was recognized for the development of such a device in Germany too. Accordingly, the Institute for Flight Mechanics of the DVL (German Aeronautical Test Establishment) in Oberpfaffenhofen (Head: *Gerhard Brüning*) prepared a research program, with the objective to establish "a universally usable in-flight simulator with variable flying qualities and programmable control characteristics". The initial preparatory work was performed on a Piaggio P 149D aircraft, see Fig. 7.1, within the framework of a research contract from the German Federal Ministry of Defense (FMoD) (Project leader *A. Pietraß*). The Piaggio P 149D was a 4-seater training, aerobatic and liaison aircraft and a total of 194 pieces were manufactured at the Focke-Wulf company in Bremen under license agreement.

Initially, the concept of a variable flying qualities aircraft was implemented only for the elevator control and tested in flight during 1967 [1–4]. For this purpose, the control column on the right-hand side was separated from the elevator and replaced through a control stick with an electrically driven actuator. The stick force was provided by a simple spring. The elevator command from the evaluation pilot was modified by the analog computer such that different characteristics in the pitch response due to elevator deflection could be created.

The elevator was driven by an electrical servomotor developed by the German Bodenseewerk Gerätetechnik company (BGT). The servomotor was connected to the basic elevator control system over a push-rod through an electromagnetic clutch. The elevator could be controlled such that the pitch response of Piaggio followed the flight characteristics of a different aircraft being simulated. Eight different pitch response characteristics were pre-programmed and could be selected in flight and assessed by test pilots.

Fig. 7.1 Variable stability aircraft Piaggio P 149 D of DVL

The main variables of aircraft motion and of the control system were recorded using a 20-channel onboard measurement system. The data were also transmitted to a ground station using a telemetry system. The safety pilot in the left seat in the cockpit could at any time open the clutch by pressing a switch on his control column, and thereby detach the servomotor and take over the aircraft control. As an additional security measure, an acceleration switch was installed which operated the clutch automatically on exceeding adjustable maximum load factor. The DVL Piaggio P 149D was at disposal until about 1970 as the first test aircraft for flying qualities investigations at DVL in Oberpfaffenhofen.

While looking for a suitable test vehicle to continue this work, the German FMoD authorized the DVL to acquire up to December 23, 1968 and operate in future an aircraft of the type HFB 320 Hansa Jet, together with additional data acquisition systems and ground support equipment at the most favorable conditions from the aircraft manufacturer Hamburger Flugzeugbau (HFB).

On December 18, 1968, the contract for sale was signed between HFB and the German Aeronautical Research and Test Establishment (DFVLR) to deliver the HFB 320. The purchase price of the HFB 320 was 2.6 million DM plus 11% VAT.

7.1.1 The HFB 320

The HFB 320 Hansa was a twin jet engine business aircraft manufactured by Hamburger Flugzeugbau GmbH (HFB, later renamed Messerschmitt-Bölkow-Blohm-Unternehmens-bereich Hamburg—MBB-UH). The first flight of HFB 320 V1 took place on April 21, 1964. A total of two test versions and 45 series aircraft were built, the last in the year 1980. The HFB 320 was a mid-wing aircraft with a T-tail and two jet engines mounted on the rear fuselage. It was characterized by 15° forward sweep wings (see Fig. 7.2).

Fig. 7.2 HFB 320 Hansa S1

The mechanical flight control system consisted of rotating shafts with gears and push-rods without hydraulic power. The rudder and aileron trimming was carried out electrically, the trimming of the horizontal stabilizer, on the other hand, was done hydraulically. Landing flaps, leading edge slats, and air brakes were actuated hydraulically.

The first production aircraft with the serial number 1021 and the registration D-CARA had its first flight on February 2, 1966. After the crash of the first prototype V1 during a stall test flight in May 1965 at Torrejon (Spain), as a replacement the first serial aircraft S1 (1021) was fitted with a data acquisition system and deployed in the flight test program. Upon completion of the testing and type certification, the HFB 320 S1was purchased by DVL for research purposes and modified later to an in-flight simulator in the DFVLR research center at Braunschweig. The HFB 320 S1 played an essential role in aeronautical research programs on the development and testing of new technologies and processes. Especially emphasized were the flight test programs carried out within the framework of Zukunfttechnik Luft (ZTL, Future Technology in Aeronautics) and within the cooperation with the Federal Ministry of Defense (FMoD) and the US Air Force.

Within the framework of a medium-term program on variable stability test aircraft VVS, (Versuchsträger Variabler Stabilität), the suitability of this aircraft type was attested and published in December 1969 jointly by DFVLR, Dornier GmbH, Messerschmitt-Bölkow-Blohm GmbH, as follows [5]:

> The in-flight simulator HFB 320 Hansa should serve as research aircraft to enable experimental work in the fields of flight simulation, flying qualities, and flight control. The interests of the flight test center (E-61) of the German Air Force are also to be particularly accounted for during system conception.
>
> The aircraft HFB 320 meets particularly well the combined requirements. The aircraft size offers enough reserves in installation space, load capacity, and power supply. Still, the aircraft is small enough so that structural elasticity do not raise serious problems. Being a jet aircraft, it is aerodynamically "clean", which simplifies the simulation of targeted properties. Considering its configuration, it is particularly suitable for the simulation of horizontal takeoff combat and transport aircraft. The conversion of an aircraft designed and built in Germany promises to be less involved and economical than that of a foreign vehicle of similar size.
>
> Powerful onboard system, which can be checked separately, makes the in-flight simulator HFB 320 flexible. This includes an electrically driven high-performance actuation system, a programmable on-board computer, a programmable controller simulation system and a PCM measurement system with 125 channels.
>
> The in-flight simulator HFB 320 Hansa is absolutely suitable for addressing most of the problems mentioned in Sect. 7.2. According to the stipulated research tasks, it should, however, be mainly used for basic research. They would be oriented towards the national projects being dealt with, currently MRCA.

7.1.2 The Project In-Flight Simulator

On May 25, 1969, the HFB 320 S1 was flown from Hamburg to DVL Oberpfaffenhofen by the pilots *Gerd Puhlmann* (German Federal Aviation Authority) and *V. Wilkens* (DVL) (see Figs. 7.3 and 7.4). In the same year, the conversion of the aircraft to an in-flight simulator FLISI (Fliegender Simulator) began as part of a contract from the FMoD [6–8].

The first conversion pertained to an electrically driven thrust and flap actuation system, that is, the modification of landing flap to direct lift control (DLC) flaps. MBB-UH, who had developed the HFB 320, was commissioned to develop the thrust actuation system and Messerschmitt-Boelkow-Blohm-Unternehmensbereich Flugzeuge (MBB-UF) in Ottobrunn with the electric flap actuation system. The integration the systems was to take place at MBB-UH.

Fig. 7.3 HFB 320 Hansa S1 with test pilots *Puhlmann* and *Wilkens* in Oberpfaffenhofen (May 1969)

Fig. 7.4 HFB 320 Hansa S1 at DVL in Oberpfaffenhofen

Due to the merger of research organizations DFL, DVL and AVA to DFVLR in 1969, research activities using aircraft as a flying vehicle were concentrated in Braunschweig. As a consequence, the operations pertaining to in-flight simulation were transferred to Braunschweig. As desired by the FMoD (RüFo4, *MinR v. Halem*), the research activities of in-flight simulation were passed on to the Institute of Flight Mechanics (Head: *Peter Hamel*) Braunschweig in April 1971. In the following years, further application-oriented research with the in-flight simulator HFB 320 FLISI was, therefore, pursued in Braunschweig and evolved into a special research domain. *Dietrich Hanke* from the Department of Aircraft Flight Mechanics (Head: *Knut Wilhelm*) was named the project leader for further development of the in-flight simulator HFB 320 FLISI. On February 22, 1972, FLISI was flown from Oberpfaffenhofen to Braunschweig by the pilots *G. Puhlmann* and *A. Brünner* (DFVLR, Oberpfaffenhofen). Three days later FLISI was flown by *G. Puhlmann* and *Hans-Peter Joenck* (DFVLR Braunschweig) to Hamburg for the integration of the developed thrust and flap system at MBB-UH.

Due to insufficient funds, important equipment components had to be installed step by step after prolonged financial negotiations with industry and government authorities. Frequently, financially affordable simple and suboptimal solutions had to be sought. To facilitate research activities, many systems such as hydraulic power system, measurement system, stick force simulation, operator console, and others subsystems had to be developed and integrated by DFVLR. Even a self-developed analog computer had to be used in the first flight tests. The priority was to redefine the overall system and to address the operational and certification issues [9].

Initially, it was intended to provide autopilot actuation systems for the three primary controls (elevator, ailerons, and rudder) in the HFB 320. Frequency response measurements in Hamburg showed, however, that the performance capabilities of these actuation systems were inadequate because they were too slow and not powerful enough. While, as a part of research on digital electro-hydraulic flight control (DEHS) [10], electro-hydraulic actuators for all primary control axes were planned for installation in the HFB 320 aircraft by the Institute of Flight Guidance, it was decided to use the same systems for the in-flight simulation too. Thereby the cost and development time could be reduced.

For the research institutes in Braunschweig, it was the first jet-powered experimental aircraft, with *H.-P. Joenck* as the only trained pilot in Braunschweig. As such help had to be sought from pilots *A. Brünner* from Oberpfaffenhofen and captain *G. Puhlmann* from the German Federal Aviation Authority. *H.-P. Joenck* had received the type rating for the HFB 320 in the year 1972. Later followed the pilots *H.-L. Meyer* (1973), who succeeded *Hermann Bieger* to take over the Flight Test Center at DFVLR Braunschweig in 1973, see Fig. 7.5, and in the year 1977 *Wolfgang Beduhn*.

Fig. 7.5 The noseboom extended in the corridor (from *left*: *Hermann Bieger*, *"HaLu" Meyer*)

Once the HFB 320 was completely equipped as an in-flight simulator, the test vehicle was intensively used by DFVLR and the German aeronautical industry during the next 6 years, that is from 1977 to 1983.

7.2 HFB 320 FLISI System Development and Installation

7.2.1 Introduction

In-flight simulation is a special simulation technique with the aim to provide the visual and motion cues as well as to perform the flight tasks under realistic conditions. During the early nineteen seventies, the technology for visual and motion simulation for ground-based simulators was still very limited. Pilots became quite often sick, because visual and motion cues were not correctly reproduced. In particular, for the development of flying qualities criteria, it was important to minimize the influences of the lack of vision and motion cues in the pilot assessments to arrive at reliable results. The solution provided by in-flight simulation consists of generating the aircraft motion by a host aircraft, and then operate the aerodynamic control surfaces using control techniques such that the host aircraft now behaves like the target aircraft, that is, like an aircraft being simulated. Once this is achieved, the visual and motion cues are perfectly correlated and the flight tasks are realistically represented. The aircraft and controllers represent the visual and motion system analogous to the ground-based simulator. All other functions correspond to those of a ground-based simulation. For the realization of this concept, a host aircraft is required, in which a freely programmable electric (Fly-by-Wire) flight control system is available in parallel to the basic controls. Furthermore, independent control surfaces are required to control all 6 degrees of freedom. This implies direct lift control (control of vertical motion) and a lateral force control (control of lateral translational motion) besides the usual aircraft control surfaces.

The in-flight simulation has to meet the following demands that affect the flying qualities (see Fig. 7.6), namely (1) vision, motion, accelerations, (2) aircraft dynamics, (3) dynamics of display systems, (4) dynamics of control systems (force/displacement characteristics), and (5) operating units. Besides the specific hardware equipment, realization of an in-flight simulator is also associated with a very sophisticated control engineering task, namely designing a control system capable of adapting the 6 degrees of freedom of an aircraft to match the commanded dynamics of another vehicle (see Chap. 3).

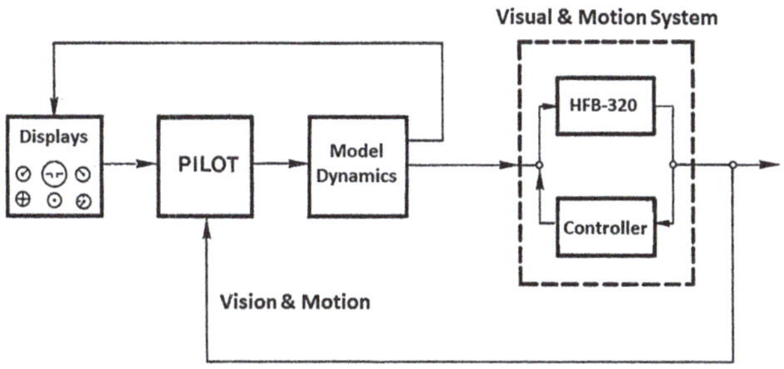

Fig. 7.6 Vision and motion information in an airborne simulation

Figure 7.7 provides an overview of the experimental equipment. These tasks were carried out during a period that spread from 1972 to 1977. The most important system units are elaborated in the flowing subsections. In addition, the attention was focused on setting up those basic facilities which were essential prerequisites for in-flight simulation, such as ground-based simulation, model following controller.

7.2.2 Safety Concept

The safety of the test flight is of paramount importance and had to be guaranteed at any time. To achieve this, the following general requirements had be met: (1) separation of basic and experimental systems, (2) test equipment malfunctions should not affect the basic system, (3) basic control as a backup for the safety pilot, (4) fast switching off of the test system by safety pilot, (5) clear display of the operating and failure status (6) protection against ground collision, possibly through sufficient minimum height, and (7) known flight sequence and trained takeover procedures. These safety requirements were fulfilled through separation of all the test systems and their power supplies from the basic

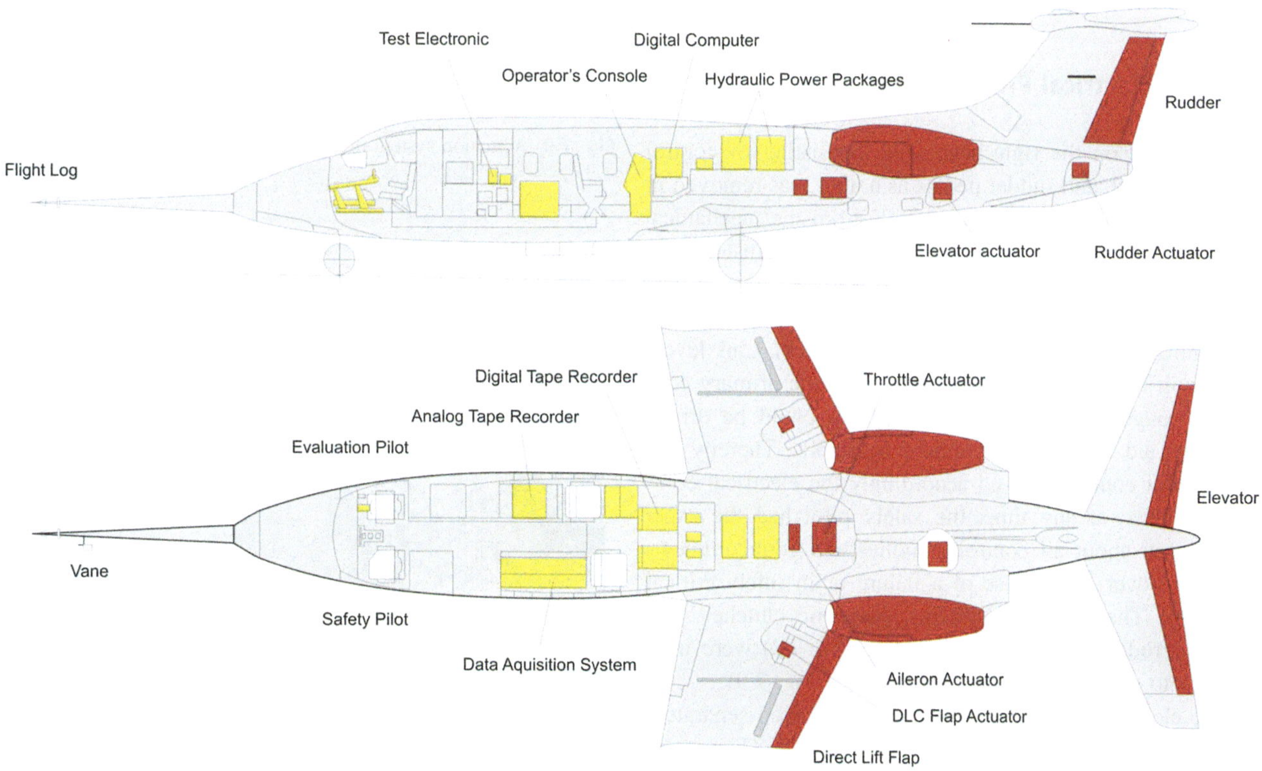

Fig. 7.7 Overview of experimental installation

aircraft systems. The rapid transition from test mode to the normal operation was made possible for the safety pilot (left seat) at all times. For this purpose, the controllers of the evaluation pilot (right seat) were disconnected from the host aircraft control system. In the flight test mode, the aircraft was flown electrically from this side.

The primary electronic flight control system was designed in parallel with the mechanical back-up control system and connected to the same through link levers, magnetic clutches, and overload pins. The magnetic clutches could be operated via the mode control unit. The experimental system could also be deactivated by the safety pilot by pressing an "EMERGENCY OFF" button on his control column. Simultaneously, with the control command to open the magnetic clutches, the actuation systems were switched into a passive state, so that the safety pilot could move basic controls even if the magnetic clutch is blocked. However, in such a case he had to exert an increased control force.

Additionally, a fast switching off could be achieved by counteracting the controller movements. Applied forces sheared the overload pins and disconnected the test system. A spring separated the elements of the coupling device. At the same time, a switch is operated and thereby the magnetic clutch is disengaged. Separation of the systems was displayed to the pilot.

In order to provide safety against ground collision, the flight altitude was limited to a minimum of 500 ft above ground in experimental modes.

7.2.3 Electrical Primary Control

The controls on the right-hand side, consisting of control column, wheel, and rudder pedals as a unit, was removed in order to detach the electrical controls from the mechanical basic controls for elevator, rudder, and ailerons. Instead, the same unit with potentiometers for measuring the control movements and with mechanical springs for reproduction of the control forces was installed (see Sect. 7.2.8). Link levers were built into the mechanical rods of all the primary controls, through which the control surfaces could be driven with the aid of electro-hydraulic actuators. Thereby the safety pilot controls moved accordingly the electrical commands. Thus, at any time, the safety pilot had the exact information about all control activities and could immediately take over in the case of a failure or in safety critical situations. The actuation systems, magnetic clutches, and coupling rods with overload pins (see Fig. 7.8) were integrated into the HFB 320 during 1972 (see Fig. 7.9).

The electro-hydraulic actuators were powered separately by hydraulic pumps that were driven electrically. The pump system was installed in two closed aluminum casings in the

Fig. 7.8 Two electro-hydraulic actuator systems on a board

Fig. 7.9 Actuation system installed in the tail section

Fig. 7.10 Cabin view: two hydraulic actuator power packs in luggage compartment behind magnetic tape recorder, onboard computer H316 and inertial platform LN3

luggage compartment (see Fig. 7.10). The electro-hydraulic primary controls were commissioned in 1973.

7.2.4 Direct Lift Control (DLC)

The landing flap system of the HFB 320 was modified for direct lift control applications. This work was financed within the ZTL program "variable flying qualities aircraft". The development and construction were carried out by the MBB-UF in Ottobrunn [11]. Completion and acceptance testing took place at MBB-UH on September 21, 1971. The installation in the HFB 320 was then performed by MBB-UH (see also Sect. 7.2.5). Figure 7.11 shows a system overview of the flap actuation system.

Electrical control of the flap control shaft was provided by two identical electro-mechanical servo systems, one each for the right and left wing, which were installed directly in the rotational control shafts close to the hydraulic flap servos. Without excitation of servos, they would rotate with flap actuation through the basic system. The drive, synchronization, and system monitoring with an interface to the modes operating unit BBG (Betriebsarten-Bediengerät) was done in an electronics box (see Sect. 7.2.11).

For the DLC operation, the landing flap setting speed was increased from 2.5°/s to 10°/s in order to meet the in-flight simulation requirements. For this purpose, the hydraulic throttle was adjusted by an electrically actuated bypass valve. The range of landing flap deflection was up to 40°.

7.2.5 Thrust Actuation System

The overall thrust actuation system is illustrated in Fig. 7.12. For the electrical control of the engines, a parallel connection to the servomotors was created near the engines, which was carried over flexball cables. The servomotors were manufactured by the Labinal company.

A gearbox with separating clutch (magnetic tooth and friction clutch) was installed ahead of the servomotor to switch on and off the system. As the engine pressure ratio (EPR) corresponds directly to the thrust generated by the engine, it was commanded as electric input for the engines, instead of the fuel control unit (FCU) position. The measured EPR from the basic aircraft system was available as feedback variable. The friction clutch was rated such that the clutch could be pushed over by the basic throttle levers, even in the event of magnetic gear malfunction. During the system acceptance tests, however, it turned out that the pilot forces were too high to override the blocked system. To solve this problem, the throttle levers were extremely elongated (three times long) to enable the required torque (see

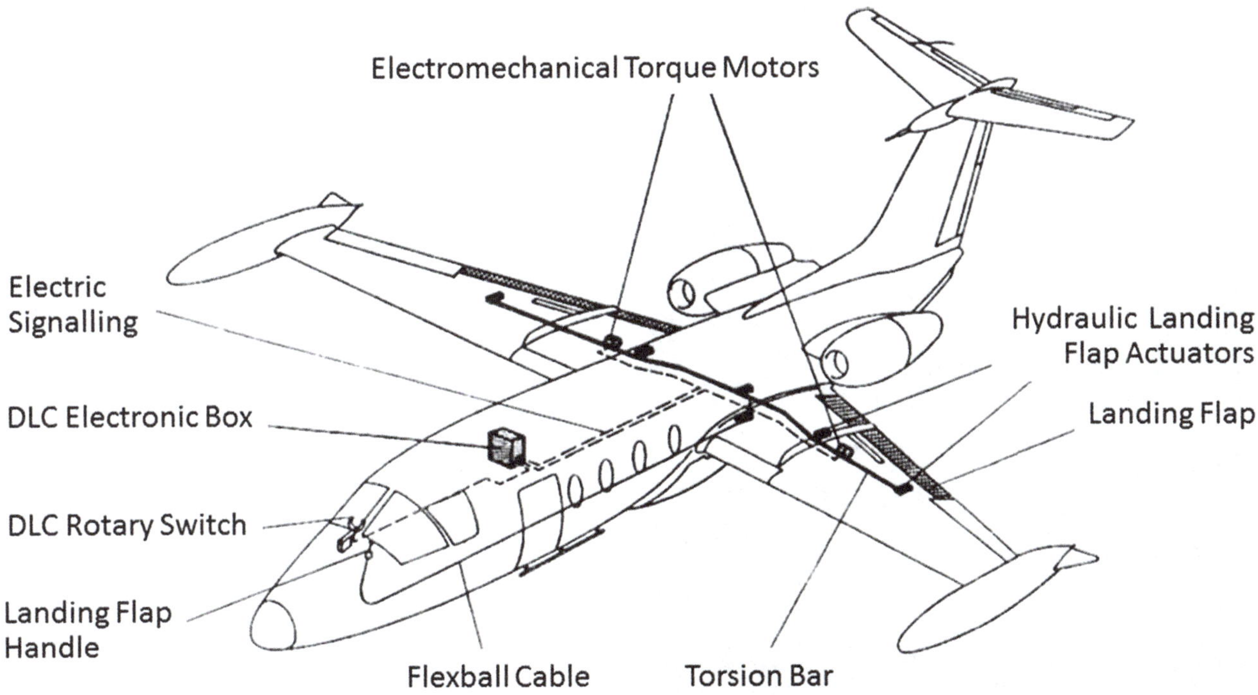

Fig. 7.11 Overview of landing flap actuation system

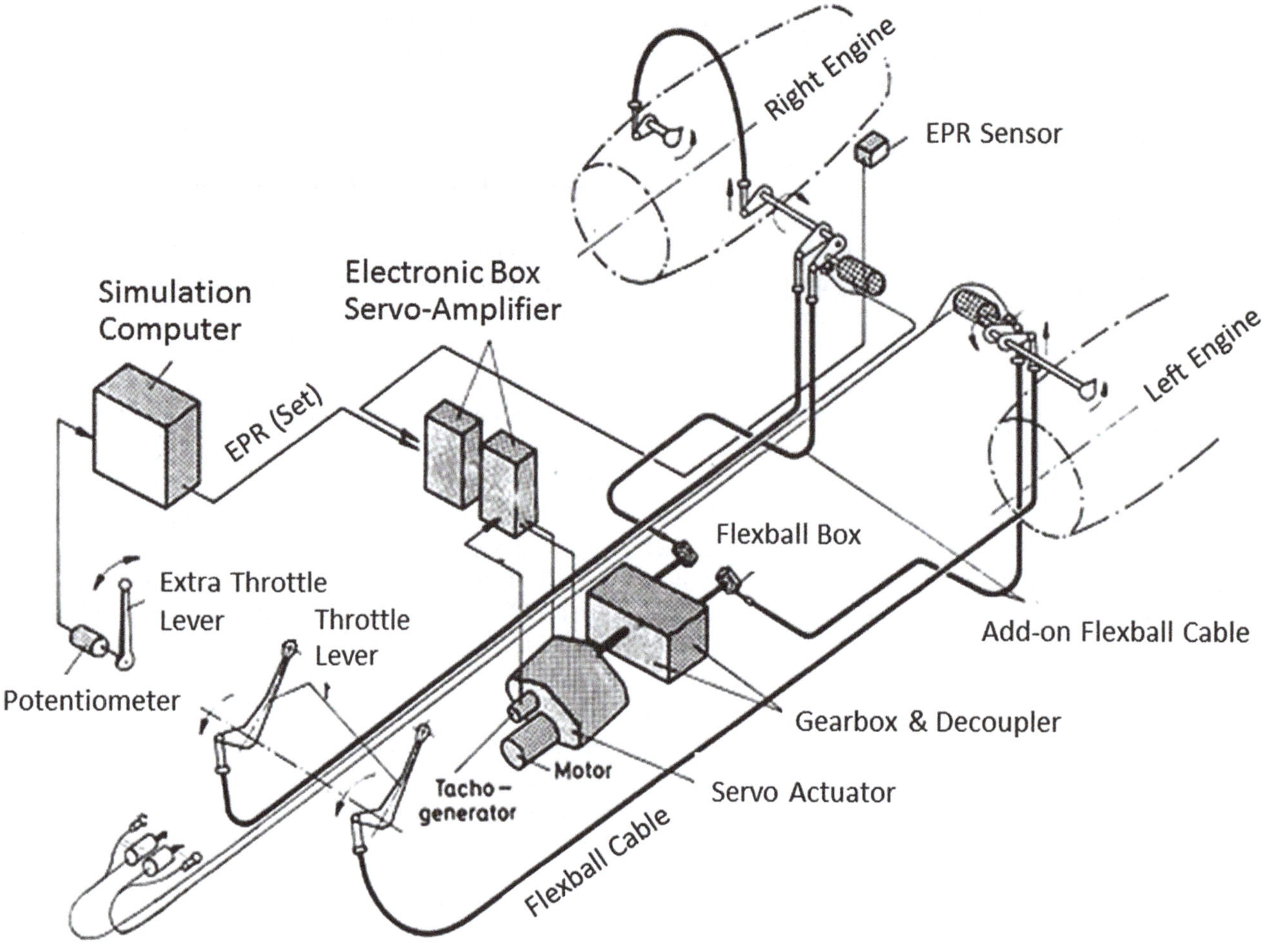

Fig. 7.12 Overview of thrust actuation system

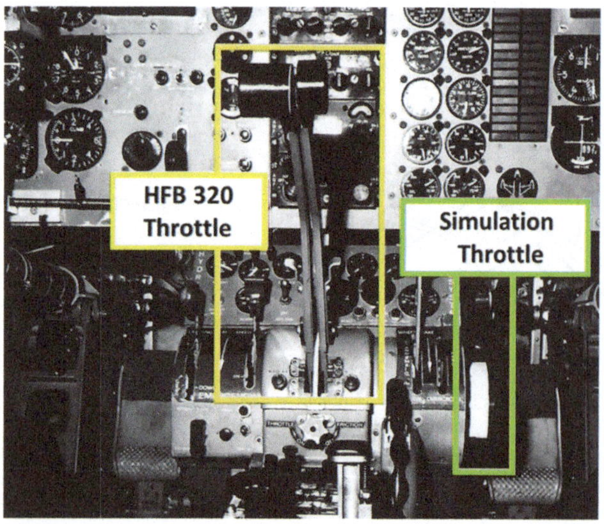

Fig. 7.13 Modification of throttle levers (simulation throttle lever on *right*)

Fig. 7.13). For the in-flight simulation, an additional thrust lever was incorporated in the cockpit, through which the test pilot could command a simulated engine model.

During the tests, it was observed that the engines could be drawn below the flight idle position, due to the flexibility in the flexball cables from the servo to the engine. To prevent this, a magnetically operable stop was introduced, which could be activated by the safety pilots for the test. Its position was displayed in the cockpit.

The acceptance tests of the installation of the electrical flap, the electrical thrust, and the electro-hydraulic actuation system, as well as the additional electrical power supply system, took place in the years of 1972 and 1973 at MBB-UH [12–14].

7.2.6 Auto-Trim System

An automatic elevator trim is an important function that is required in the in-flight simulation mode. Trim changes during the simulation phase in the model can lead to steady elevator deflections of the host aircraft. In the event of an emergency takeover by the safety pilot, the elevator would abruptly fall back to the old trim condition and thus cause a jump in the control column of the safety pilot and a hefty aircraft motion. The auto-trim system trims the aircraft automatically by adjusting the elevator hinge moment to zero. Thus, the transition back to the host aircraft is sufficiently smooth.

The auto-trim system was developed and integrated into the HFB 320 FLISI by the DFVLR [15, 16]. For this purpose, the electric trim system of the aircraft was modified. The elevator hinge moment or rod force was measured by a piezoelectric sensor mounted on the push rod from the elevator to the electro-hydraulic actuator. It was fed to the host trim coupler through a low pass filter that supplied corresponding trim pulses to the trim motor till the rod force was reduced to zero.

The value of the hinge moment was displayed on the cockpit instrument panel of the safety pilot, so that the proper functioning could be monitored. For safety reasons, a maximum manageable mistrim was defined, which was determined from flight tests. The auto-trim system was shut down automatically, if the mistrim exceeded a pre-defined duration.

7.2.7 Spoiler DLC

In the framework of a cooperative flight test program with MBB-UH on the application of DLC on commercial airplanes, it was decided to investigate the application of wing spoilers instead of the landing flaps for direct lift control. In the case of spoilers the lift increase leads to a reduction in drag whereas in the case of landing flaps a drag increase is induced. Under the ZKP program "Flight Guidance", MBB-UH was commissioned with the development and installation of an appropriate system. The basic spoiler surfaces were enlarged to increase the effectiveness. Wind tunnel tests were conducted at DVFLR in Braunschweig to determine the spoiler efficiency [17]. The system was flight tested in 1977 (see Fig. 7.14, [18]). Unfortunately it turned out that the spoilers produced unacceptable vibrations, although they were perforated and as such no further attempts to use them for the intended purpose were undertaken.

Fig. 7.14 HFB 320 with DLC spoilers

7.2.8 Pilot's Control Unit and Model Trimming

A complete HFB 320 control station was purchased from MBB-UH and converted to an experimental device for in-flight simulation applications [19]. As all control parts, such as control column, wheel, and pedals, could be easily mounted on a chassis, it was relatively easy to install adjustable spring packets for control forces and potentiometers to measure the control movements. Figure 7.15 shows the overall system. For the investigations pertaining to DLC, a thumbwheel was mounted on the control wheel to command directly the DLC flaps. Thereby the flight trajectory could be controlled directly (see Fig. 7.16) without

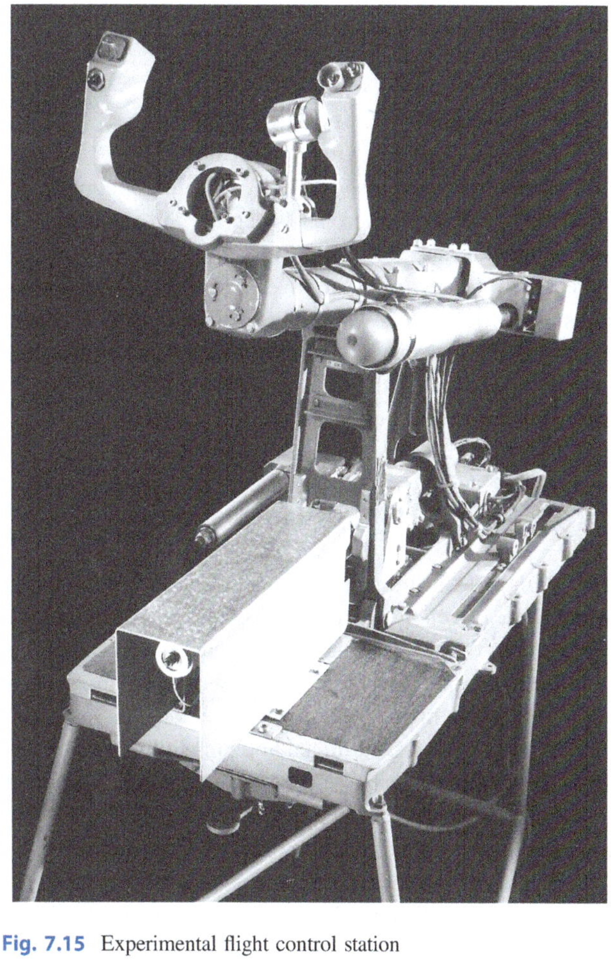

Fig. 7.15 Experimental flight control station

Fig. 7.16 Thumb wheel controller for direct lift control (DLC)

pitch commands. To trim the model equations in the onboard computer, a box with trim potentiometers was developed for all three control axes and installed in the cockpit for operation by the test pilot.

7.2.9 Side Arm Operating Unit

For several investigations on Fly-by-Wire flight control (see Sect. 7.3) a side arm controller (sidegrip) unit was used. The sidegrip controller for pitch, roll, and yaw control was attached to the test pilot side (see Fig. 7.17). The sidegrip units were developed within the research project "digital electrical flight control". The two sided handle was integrated on the seat for the left and right hand. It was used for the first time in 1973 on the HFB 320 [20].

7.2.10 Flight Test Instrumentation

The flight test instrumentation included cockpit displays, which were driven by the onboard computer. They displayed the motion variables of the model which were computed onboard. For this purpose the following display systems were integrated in the instrument panel on the right-hand side (see Fig. 7.18): (1) artificial horizon with ILS and *Flight Director* display (*Primary Flight Display*), (2) cross-pointer instrument (test specific, see also later Fig. 7.41), (3) trim indicator, (4) engine pressure ratio (EPR) indicator and (5) coincidence indicator of actuator and surface positions. In the middle, at the top of the instrument panels, a slot was provided for an autopilot control unit.

Fig. 7.17 Sidegrip operating unit for pitch and roll control ("*HaLu*" *Meyer* in cockpit)

7.2.11 Mode Control Unit BBG

The mode control unit (BBG) was a key element of the in-flight simulator. The system was designed, developed, and integrated by DFVLR in 1972 [8].

The BBG provided the interface between the pilot, flight test engineer, and the overall system. It consisted of a switch-board (processing logic signals) and the control buttons with button lights indicating the status of the control equipment. The individual actuation systems and subsystems

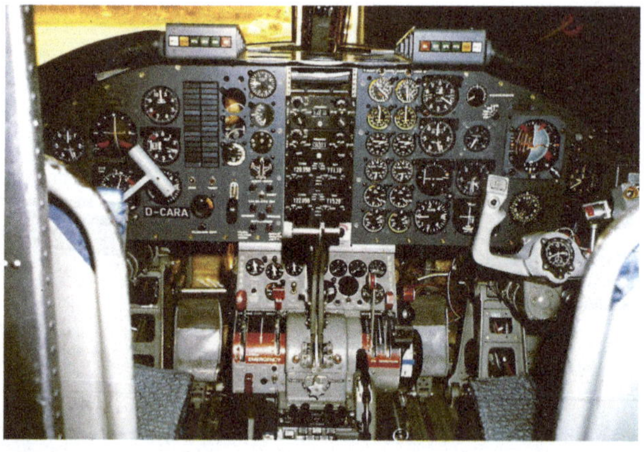

Fig. 7.18 Instrument panel and test pilot station on the *right* with primary flight display

Fig. 7.19 Operator console with mode control

could be selected via push buttons (indicator WHITE). After selection, error signals were checked. If errors were not encountered, the system indicated a readiness to activate the actuation systems (indicator YELLOW). Altogether, the control systems were provided for elevator (HRS), ailerons (QRS), rudder (SRS), flaps (LKS), and thrust (SS). Furthermore, the subsystems could be activated for elevator trim (HRT), control column (PPS), and onboard computer (BR). When errors occurred, a warning lamp was activated and switching on the corresponding system was prevented. In the other case with no system errors, all selected systems could be switched on by pressing the OPERATION button (indicator GREEN). The systems could also be operated and switched on individually. On encountering errors during the operation, the state was displayed in the cockpit by a red, flashing light. Normally, the unit was operated from the cockpit. For ground tests, it could also be operated from the cabin. The BBG electronic unit with an operating panel was located in the cabin (see Fig. 7.19), and another panel for the pilots was located in the middle console in the cockpit.

In addition, a row of lamps was provided for each pilot on the upper edge of the instrument panels, which clearly displayed the system state. The display indicated which side of the control mode was active (Basic or FBW), the transition state to and from the simulation, and a warning in the case of failures (see Fig. 7.18).

7.2.12 Onboard Computer

Onboard Analog Computer

After taking over the HFB 320 FLISI by DFVLR in Braunschweig, flight tests were carried out with the delivered actuation systems (electrically controlled landing flaps and engine thrust).

Analog computer modules were used to control the flap as a function of elevator deflection and for speed control. The computer modules were installed on the left side of the control panel in the cabin, next to the mode control panel (see Fig. 7.20). The control computation was carried out via connectors, as it was common in analog computing.

Onboard Digital Computer

During 1973 a Honeywell H316 digital computer was installed in the HFB 320 and tested under the "digital electric hydraulic flight control system (DEHS)" program. It became available for in-flight simulation purposes. The rack mounted H316 was a commercial 16-bit minicomputer, that was specifically developed for data processing applications by Honeywell in 1969. The computer was installed on *shock mounts* in the cabin on right-hand side (see Fig. 7.21).

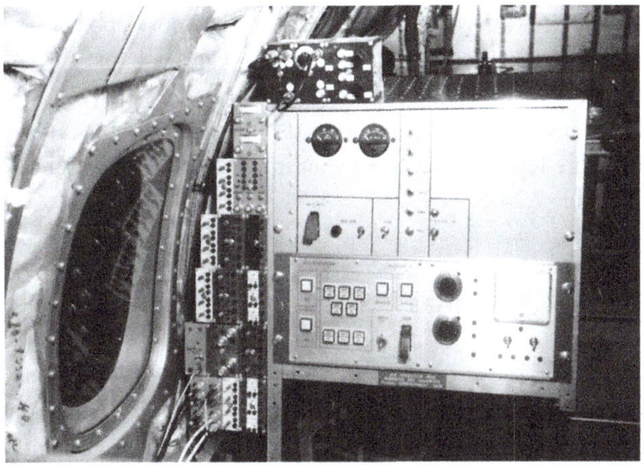

Fig. 7.20 Analog computer components (on the *left* close to operator console)

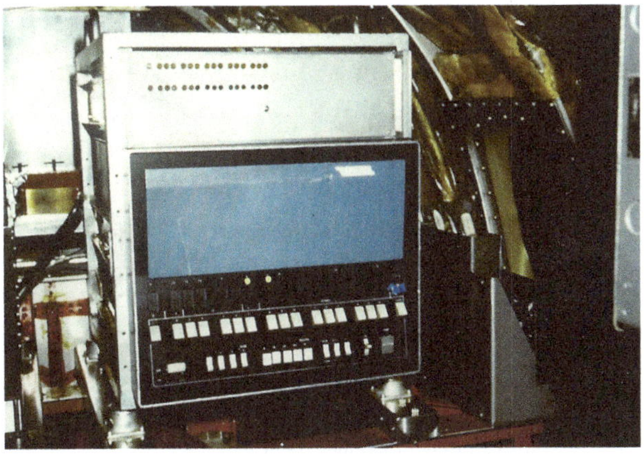

Fig. 7.21 Digital onboard computer Honeywell H316

7.2.13 Data Acquisition and Sensor System

A large number of measured flight state and system variables was necessary for in-flight simulation purposes. The acquisition of these data was carried out by separate experimental sensor systems. The following variables were measured: (1) control surface deflections by potentiometers, (2) actuator positions by potentiometers, (3) computer generated positioning commands, (4) system state variables as logical signals provided by the system, (5) attitude angles using a 3-axis gyro, (6) angular rates by a 3-axis rate gyro, (7) linear accelerations using 3-axis accelerometer, (8) navigation and ILS data, (9) true and indicated airspeeds, and (10) angle of attack and angle of sideslip.

A Dornier flight log was used to measure the air-flow directions, that are angles of attack and sideslip. The flight log also provided the measurement of true airspeed (TAS) by means of a small propeller (see Fig. 7.22). The flight log was very sensitive to rain and icing, and could not be used under adverse weather conditions. Therefore, an additional angle of attack vane was mounted on the nose boom (see Fig. 7.23). Later, the angle of attack vane was lost through flutter, due to frequency neighborhood between the vane dynamics and the noseboom oscillation (10 Hz). It was therefore replaced later on by an improved Dornier flight log (see Fig. 7.24).

Fig. 7.23 Flight log with angle of attack vane

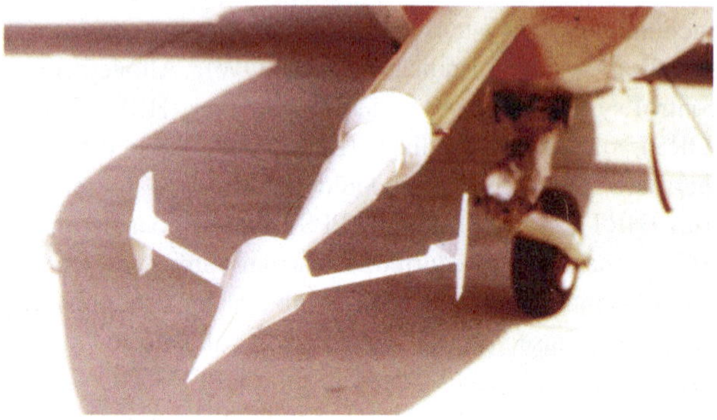

Fig. 7.24 Improved version of Dornier flight log

7.2.14 Antennas

In addition to the antennas for navigation, landing approach, and communication with air traffic control, a telemetry antenna was installed for transmitting measured data and for communication. A special antenna of the microwave landing system TALAR (Tactical Landing Approach RADAR) was installed in the fuselage in the aircraft nose which was used for flight testing of steep noise abatement approaches (see Sect. 7.3.4).

Fig. 7.22 Noseboom with Dornier flight log

7.2.15 Data Recording

The data were recorded onboard in digital form using an analog magnetic tape (Ampex) (see Fig. 7.25). The magnetic tape served also to load the programs on the onboard computer. The recording included all measurements as well as system and program specific data.

The sampling rate varied depending on the frequency content of the measured variables. The master clock rate was 100 ms. That was the highest possible rate. Slowly varying data were recorded at a multiple of the basic clock rate. A total of 79 measurements and 77 system parameters were digitally recorded onboard. 18 measured variables could be transmitted via telemetry to the ground station and recorded there in analog form.

Fig. 7.25 Onboard magnetic tape recorder

7.2.16 Control Concept for In-Flight Simulation

Besides the special hardware equipment (Fly-by-Wire), the control concept and the digital real-time program are crucial for in-flight simulations [21]. Moreover, program and data interfaces must be provided such that the user functions could be implemented easily and safely. Furthermore, their realization and flight testing by an operating team should be possible more or less routinely.

The following requirements were imposed on the control concept: (1) simulation of the aircraft dynamics in 6 degrees of freedom in real-time, (2) unified control law structure for all applications, (3) simple and easily understandable optimization criteria, (4) parameter insensitivity, and (5) digital controller. The sequence of generating the control system software program consisted of modeling, control law design with optimization, check-out with a nonlinear ground-based simulation, up to in-flight evaluation.

The approach used for the in-flight simulation development was based on the model description, control law design with Riccati optimization, testing of all functions in the ground-based simulator, and a final testing under real flight conditions. The controller design program provided the matrix coefficients for the model and for the forward and feedback loops of the controller on punched cards, which were read by a hybrid computer EAI Pacer 600. The simulation of the HFB 320 dynamics on the Pacer 600 was nonlinear, including non-linear effects of the actuation systems, such as backlash and hysteresis. The controller was re-optimized accounting for the nonlinearities. After that the program was implemented and compiled in the programming language of the onboard computer H316 (FORTRAN IV and Assembler) and finally transferred to the onboard computer by means of magnetic tape.

For verification of the control accuracy, step and doublet input signals were provided. The model and host aircraft responses were transmitted via telemetry, so that changes were possible even during the flight.

To fulfill the aforementioned demands, a model following controller was chosen with an explicit model representation and full state vector feedback (multivariable controller). The advantage of the explicit model following is that the controller can be optimized independent of the model characteristics and does not have to be changed for various models of different aircraft that may be tested. The actuator outputs, as well as the aircraft and model responses, were used as state variables. An integral term in the controller caused the stationary deviations to converge to zero.

7.2.17 Controller Optimization

A computational method for optimization was developed that involved on one hand the Riccati optimization with squared integral criteria and on the other hand permitted shifting the poles of the closed loop system to the desired locations applying a pole placement method [22]. The optimization was performed assuming a linear control system. The controller was subsequently re-optimized in conjunction with the HFB 320 nonlinear real-time ground-based simulation and ultimately adjusted based on flight test results.

7.2.18 Development of Real-Time Onboard Computer Programs

The programming of the model following controller was prescribed by the Honeywell H316 onboard computer. It was connected to the data acquisition and actuation systems through special interfaces. The H316 was a 16-bit word length processor with a core memory of 8 KB. A floating point arithmetic operation was not supported. All arithmetic operations had to be performed with the basic functions, that is with fixed point arithmetic, because programmed floating-point arithmetic could not meet the requirements of computational speed.

Specific program requirements resulted from the low computational power of the four-year-old and already outdated computer. In order to meet the computational time requirements, the complete model and controller calculation were performed in a 16-bit, scaled number representation.

For the calculation of model equations, double precision representation in 32-bit had to be adopted. Special limitations were also posed by the accuracy of the analog/digital converters, which worked with 12-bit resolution. That means a division of the signal by 1:2048, positive and negative.

The program sequence was controlled by interrupts generated with a real-time clock (RTC). This allowed to comply strictly with the time frames for the model and in particular the discrete controller program.

The execution of various program parts can be described as follows: the data acquisition and the fly-by-wire program were executed on every 2nd pulse of the master clock running at 10 ms clock. The actual controller program accepted every 100 ms a new data set and processed these in 26 ms to a new set of control variables which were outputted at a defined time point. Once the computations of controller program were completed, the remaining time was utilized to drive the displays in the cockpit as well as to generate automatic control commands, such as steps and doublets. In addition, to minimize the nonlinear effect in the control systems, special functions were programmed for hysteresis and backlash compensations.

7.2.19 Controller Performance in Flight Test

The assessment of the model following quality was carried out by comparison of responses from model and host aircraft excited by step and doublet inputs. A typical example of the achieved controller quality is given in Fig. 7.26 which compares the airspeed, pitch attitude and pitch rate responses between the model (large Airbus-type transport aircraft) and the HFB 320 host aircraft due to step function inputs. The time histories are in good agreement. The slight deviations are indicated by the white areas shown.

7.2.20 Ground-Based Simulations

With the start of the project work in 1972, a 100 V vacuum tube EAI analog computer was employed for simulation investigations. This machine was replaced in 1972 through a hybrid computer EAI PACER 600, see Fig. 7.27. The hybrid computer consists of an analog and a digital computer where all potentiometers of the analog computer were set by the digital system. The Pacer 600 was used for nonlinear

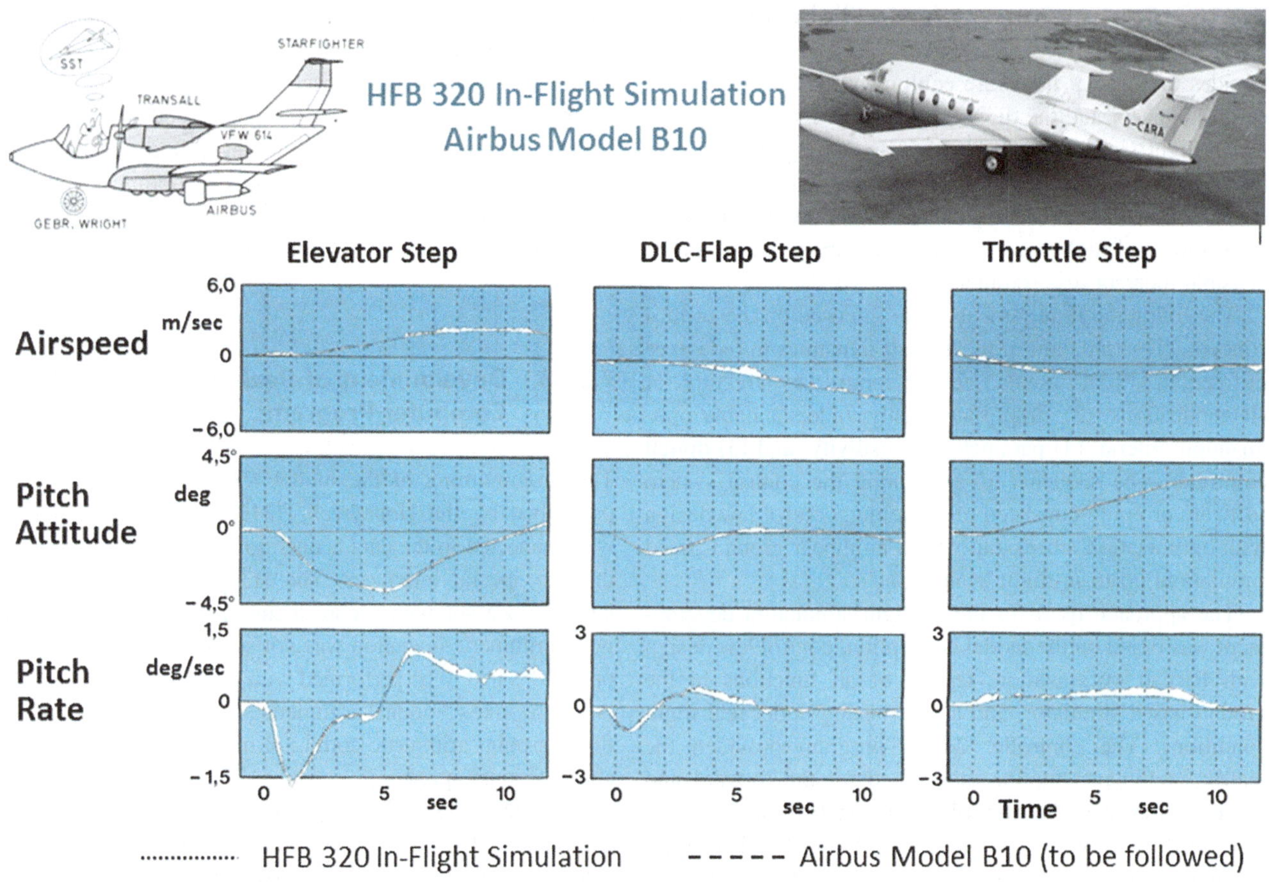

Fig. 7.26 Step response quality control of HFB 320 in-flight simulation

Fig. 7.27 Analog computer Pacer 100 for the real-time simulation of HFB 320

simulations as well as for program developments for the Honeywell onboard computer H316, exploiting the advantages of both types of computers.

The advantage of analog computing was its real-time simulation capability, that is, the equations of motion were computed in real time. The disadvantage was the difficult representation of complex aerodynamic functions.

In the digital simulation the accuracy of the controlled system must be better than that of the controller itself, that is, of the onboard computer program. This required a reduction of the computational cycle time for the simulation and also improved functional representations. A powerful computer, most modern at that time, namely AD10 from Applied Dynamics, was used to calculate the aerodynamic functions in shortest time.

On the AD10, the cycle time for the simulation of HFB 320 equations of motion was 2 ms, while the real-time process on the onboard computer was running with 50–100 ms time frame. Further, it was also important to reproduce correctly the time frames for the analog/digital conversion of the sensor signals.

7.2.21 Data Analysis

At the time of taking over HFB 320 in 1971, suitable analysis tools for mass data processing and for flight test data evaluation were not available at DFVLR. Based on the requirements specification, an appropriate program for the applications to in-flight simulation and assessment of flying qualities was acquired by the American visiting scientist *John McCracken* in a joint work project with the DFVLR computer center [23]. The program included the following functions: (1) conversion of the measured data in Ampex format to Siemens computer format, (2) calibration, (3) quick look and time history plots, (4) quick look printout, (5) data analysis (mean, variance, correlation, power spectrum, cross spectrum, frequency response, coherence, probability distribution), and (6) printout of data analysis. This program provided the basis for the development of a dialog oriented data analysis tool DIVA (Dialog Orientierte Versuchsdaten Auswertung), which was employed later also successfully in DFVLR flight test campaigns with ATTHeS and ATTAS in-flight simulators (see Chaps. 8 and 9).

7.2.22 Telemetry

The flight tests were monitored via a telemetry system. 18 measured signals were transmitted to the ground station, where they were recorded in real time on two 8 channel ink jet recorders. Via audio link it was possible to communicate with the flight test engineer or the pilot and thereby to guide the test procedure from ground. The entire board communication between the pilot and air traffic control could be listened to. The data reception took place over a large dish antenna with automatic tracking.

7.2.23 Parameter Estimation

Accurate knowledge of the dynamic behavior of aircraft and of the actuating system characteristics was an important prerequisite for an optimal controller design (see also Sect. 7.3.2). The simulation model developed for the HFB 320 was initially based on wind tunnel data, but it was improved continuously from 1971 by applying the modern methods of parameter estimation from flight test data. For this purpose, dedicated flight tests were carried out, in which the aircraft was excited dynamically in all axes at different flight conditions by onboard computer generated control input signals (steps, doublet). The gathered measurement data were used to determine the model parameters with the support of special parameter estimation programs (see also Sect. 7.3.2) [24, 25].

7.2.24 Operation and Certification

The maintenance and operation of the HFB 320 FLISI aircraft were carried out by the Flight Test Department in Braunschweig. An aircraft technician was always on board during the flight tests to support the pilots. Test equipment installation in the aircraft cabin was approved by DFVLR inspectors and partially directly by the German Federal Aviation Agency.

7.3 Application Examples and Results

7.3.1 Flight Test Overview

Until decommissioning of HFB 320 in May 1984, it was continuously used for flying qualities investigations and for testing of new technologies and methodologies. The most important utilization programs which were carried out from 1972 to 1984 are chronologically summarized in Table 7.1. A few selected projects are briefly elaborated hereafter to illustrate the HFB 320 utility.

7.3.2 The First Flight Test Program: Noise Abatement Approaches (1972–1973)

The HFB 320 was available for flight tests by the end of 1972 after functional testing and the certification of the actuation systems for landing flaps, engine thrust, and the operating unit as well as certain measurement components by the German Federal Aviation Authority, (see Fig. 7.28). Already in December 1972, the installed equipment (see Fig. 7.29) enabled a first flight test program, which was aimed at the improvement of flight path control by the application of direct lift control (DLC) during steep noise abatement approach trajectories.

For direct lift control operation, an electrical coupling was integrated between the elevator and DLC-flaps through the onboard analog computer, so that the flaps were actuated simultaneously whenever the elevator was commanded by the pilot. As a result, up-lift or down-lift could be produced without aircraft rotation (see Fig. 7.30). An increasing degree of coupling implied an increasing effect of direct lift. Figure 7.31 shows the basic difference of a flight path change during the approach between a conventional control and a control with DLC. In the case of elevator control, a non-minimum phase behavior can be observed in the aircraft time response, which delays the desired flight path change. This effect is caused by the initially generated down-lift of the elevator. These disadvantages were overcome by introducing direct lift control. Sluggish aircraft behavior in conjunction with higher descent rates is perceived by the pilots as unacceptable. To reduce the pilot workload, fast and precise flight path control behavior is required, especially just before landing. The aim of the flight tests was to demonstrate that the flight path control during steep approaches could be improved through direct lift control [26].

After extensive simulation investigations in ground-based simulations during 1971/1972, a number of ILS approaches were carried out at Hannover airport to determine the optimum setting for elevator-flaps ratios. The assessment of the direct lift control was done as a part of the noise-reducing steep approaches. To reduce the noise level during the landing approach, the approach angle was increased in order to achieve a greater distance between the noise source and noise sensitive areas. The noise abatement flight path profile consisted of two segments. The first segment was a steep 6° approach and the second segment of a conventional 2.5° ILS

Table 7.1 HFB 320 flight test statistics

Applications	Period	Participants
Functional testing LKS, SS and BBG	1972	DFVLR, MBB
Noise abatement approaches with DLC	1972–1973	DFVLR
Digital FBW control with failure detection	1973–1974	DFVLR
Handling Qualities with DLC	1974–1975	DFVLR, USAF
Flight path tracking, pilot modeling with DLC	1975	DFVLR, NLR
Automatically flown steep approaches	1976	DFVLR
Decelerated approach (*Autflap/Autothrottle*)	1976	DFVLR, NLR
In-flight simulation with digital model following	1977	ZKP
Spoiler-DLC testing and parameter estimation	1977	ZKP
A310 (B10) in-flight simulation with spoiler-DLC	1978	ZKP
Integrated flight guidance system (IFGS)	1977–1979	DFVLR, ZKP-partner
Pitch command control with attitude hold and DLC	1979–1980	DFVLR
Flight guidance system in onboard computer MUC161	1979	DFVLR, ZKP partner
Reduced static stability	1981–1983	DFVLR, USAF
4D guidance in TMA	1982–1983	DFVLR
Time delays in electrical flight control systems	1983	DFVLR, USAF
Pilot and flight test engineer training	1984	DFVLR, IPTN

Fig. 7.28 HFB 320 ready for DLC flight tests (1972/73)

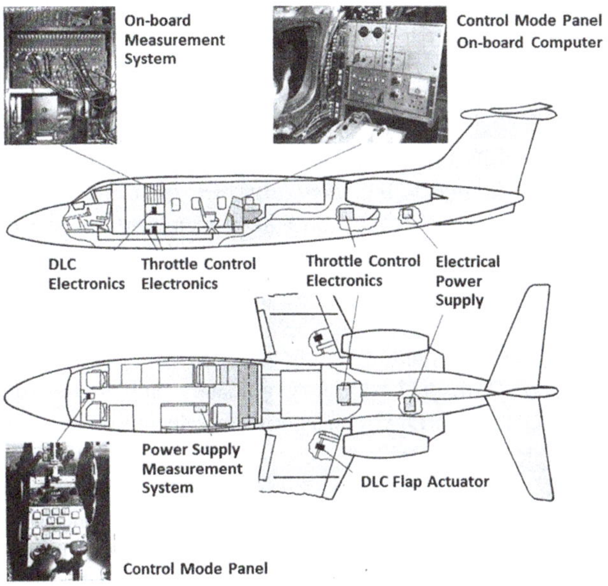

Fig. 7.29 HFB 320 FLISI equipment status 1972

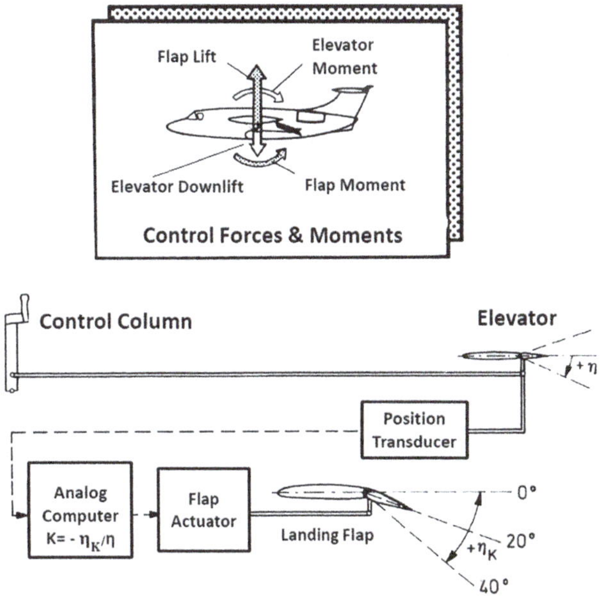

Fig. 7.30 Direct lift control concept

approach (see Fig. 7.32). Direct lift control was activated during the 6° segment, and was switched off after the transition to the 2.5° final approach for safety reasons. The approach profile had been calculated in advance and was displayed to the pilots as nominal target profile.

Noise measurements were carried out at 8 locations under the approach path. A precision radar system was used to measure the aircraft position. The noise measurements showed a significant reduction in noise propagation during steep approaches compared to those with a conventional 2.5° slope (see Fig. 7.33) [27].

7.3.3 Digital Fly-by-Wire Control with Failure Detection (1973–1974)

Basic research on a highly reliable digital FBW control technology began during the mid-nineteen-sixties. After the experience with the digital FBW test systems in Do 27, Percival Pembroke and on the research vessel Planet (see Sect. 6.3.1), a three-axis primary control system with "fail-safe" characteristics was successfully built in 1972, which was to be tested on HFB 320. In this context, the research was focused on design and implementation of

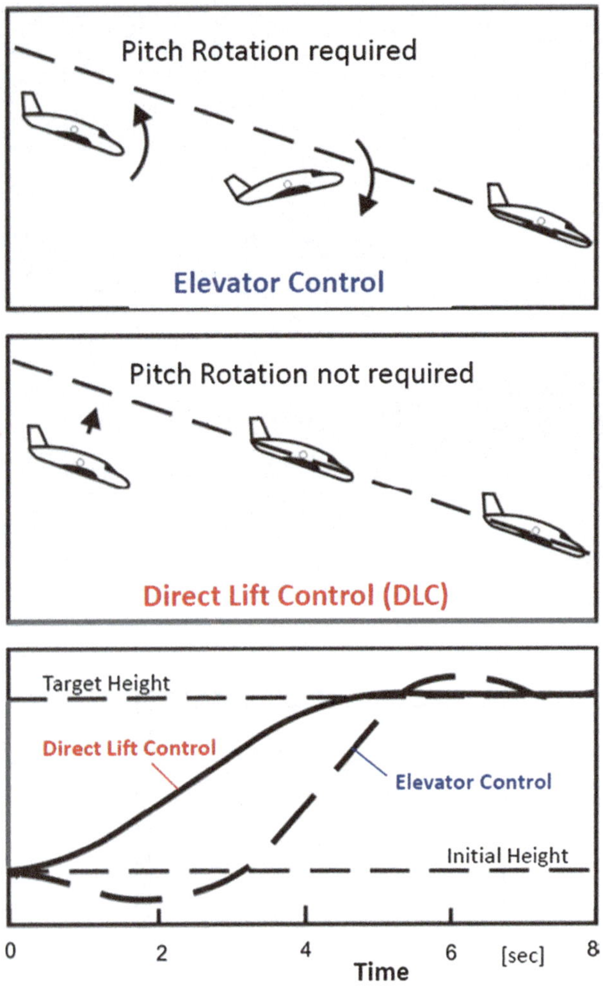

Fig. 7.31 Trajectory variations with and without DLC

digital flight control system, in particular for flight path guidance for the HFB 320.

Besides the electrical Fly-by-Wire control for elevator, ailerons, and rudder in duplex-stand-by arrangement, a modified copilot seat was installed with two sidegrip operating units with armrests (twin sidegrips, see Fig. 7.34). The two sidegrips were mechanically coupled to each other and had three rotational degrees of freedom, so that elevator, ailerons, and rudder could be operated 1:1 directly.

The electrical signal transmission between the sidegrip and electro-hydraulic actuator was equipped with a failure detection system, with which malfunctions and failure could be detected in those components which were temporarily not involved in the signal generation (so-called "sleeping errors") [28].

Suitability of the twin sidegrip control inceptor system as a substitute for the conventional control wheel as well as the influence of sampling time and quantization on the guidance accuracy were primarily investigated in flight tests. A typical pilot task consisted of, for example, holding precisely the altitude. For this purpose, the quantization of elevator deflection angle was varied and the accuracy of the altitude hold at different airspeeds was measured. It was found that the accuracy of altitude held deteriorated for a quantization of 0.3° for an airspeed of 200 kts, whereas at 260 kts the degradation is apparent even for smaller quantization of 0.1°.

An artificial counterforce in the right armrest simulated the control surface loads. Evaluation of the test flights also elucidated simultaneously the importance of counterforces in the sidegrip. Soft pinning down of the sidegrip control led to a lower mean value of the altitude deviations and small scatter in particular. Thereby a first conclusion could be drawn for the comparison between force-stick (extremely hard bound) and displacement-stick (not restrained by counterforce). Furthermore, the failure detection was to be demonstrated in flight operations. However, a permanent failure in electronic or electro-mechanical components for the A/D conversion did not occur during the test flights with a total of about 30 h of flight time.

7.3.4 Handling Qualities with DLC Applications (1974–1975)

Another flight test program pertaining to DLC was performed during the summer of 1974. The investigations were carried out under the *Memorandum of Understanding* (MOU) with the US Air Force in cooperation with the US Air Force Flight Dynamics Laboratory in Dayton, Ohio [29] (see also Sect. 12.3.2). The aim of the study was to investigate the influence of the degree of pitch and vertical (heave) motion coupling during the 6° steep approach on the accuracy of manual trajectory guidance and on the pilot assessment. Two DLC concepts were selected for this purpose (see Fig. 7.35).

The first concept consisted of coupling elevator and landing flap, whereby different ratios of flap and elevator led to varying degree of heave and pitching motion. In the second concept, a direct path control was adopted. In this case, just the heave motion was generated by activating the flaps, whereby the pitch attitude was held constant by a controller. The ratio of the elevator to flap control deflection was varied here. Figure 7.36 shows the principles of flight path control in the form of aircraft response to a step command.

The 6° nominal trajectory was realized by a microwave landing system TALAR IV (Tactical Landing Approach RADAR) with variable glide path capability (see Fig. 7.37). This device was provided by the US Air Force under the MOU. For this purposes, the TALAR receiver was installed in the HFB 320 nose (see Figs. 7.38 and 7.39).

A total of 54 approaches were performed. The steep approach procedures using the different DLC configurations

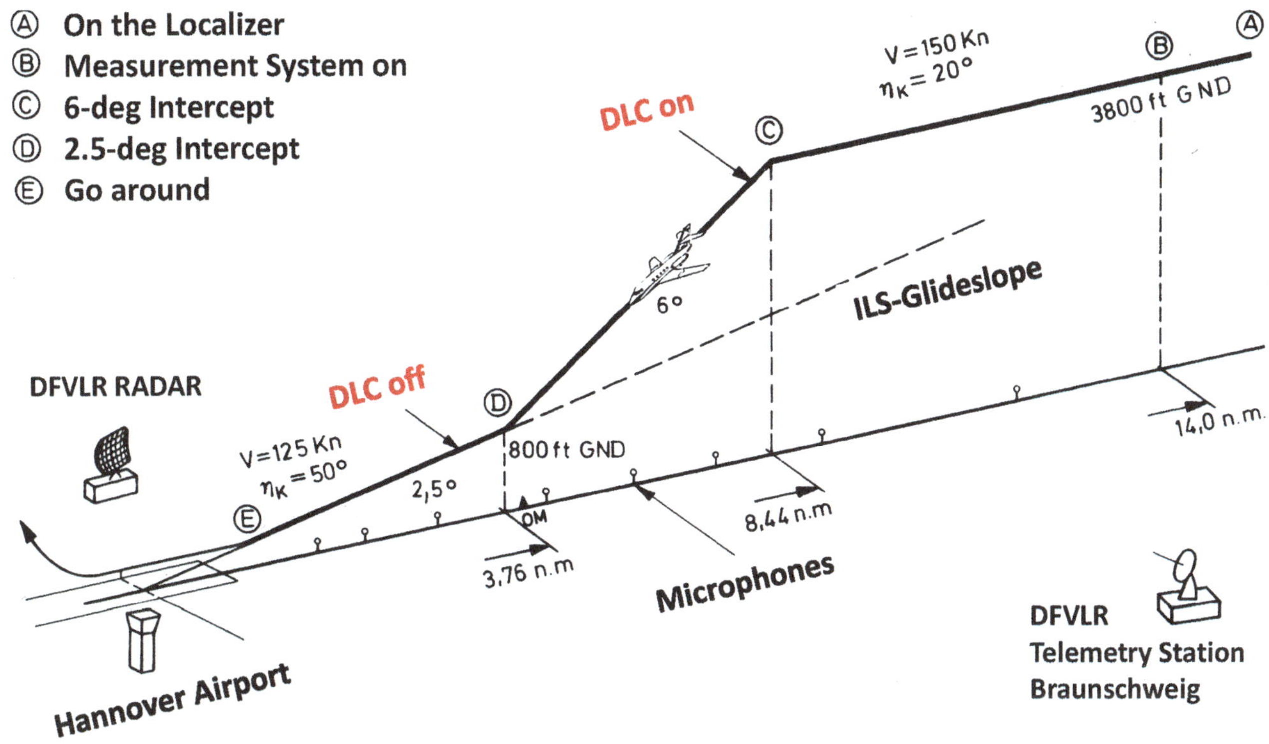

Fig. 7.32 6° steep approach profile and noise measurement stations

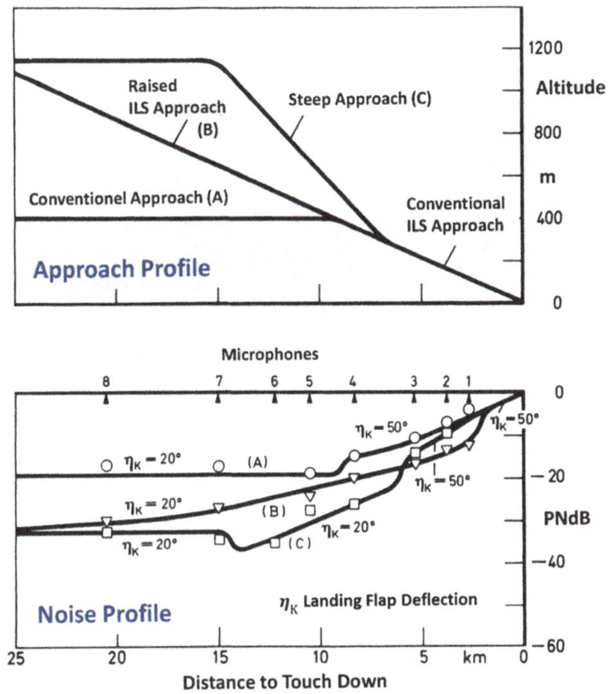

Fig. 7.33 Noise reduction potential during conventional and steep approach

Fig. 7.34 Test pilot *Hans-Peter Joenck* operating twin sidegrips

showed that the manual flight path control could be improved through DLC. This applies both to the coupled case as well as for the case of constant pitch attitude. Problems were, however, encountered in holding the speed manually. The flaps as direct lift device produced a strong coupling between heave and speed response. This effect could not be decoupled manually by the pilot. The problems of maintaining speed masked the improvements due to DLC [30–33].

In continuation of this flight test program, yet another flight test program addressing this subject was, therefore, carried out in 1975 [34]. During these flight tests, the speed was kept constant by a thrust controller. Approaches with a glideslope of 2.5° and 4° were flown. The 2.5° approach was

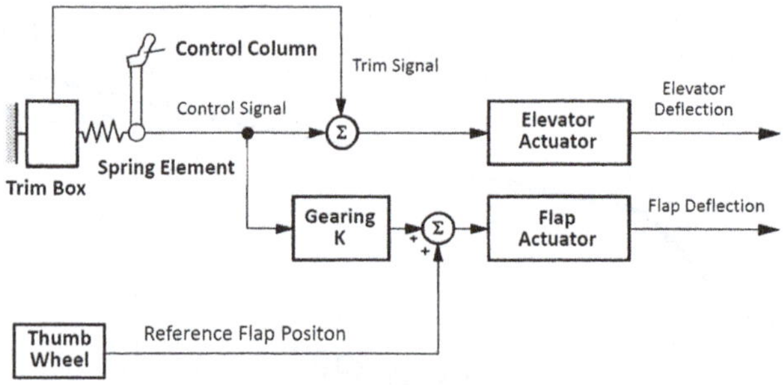

Fig. 7.35 Block diagram of DLC control

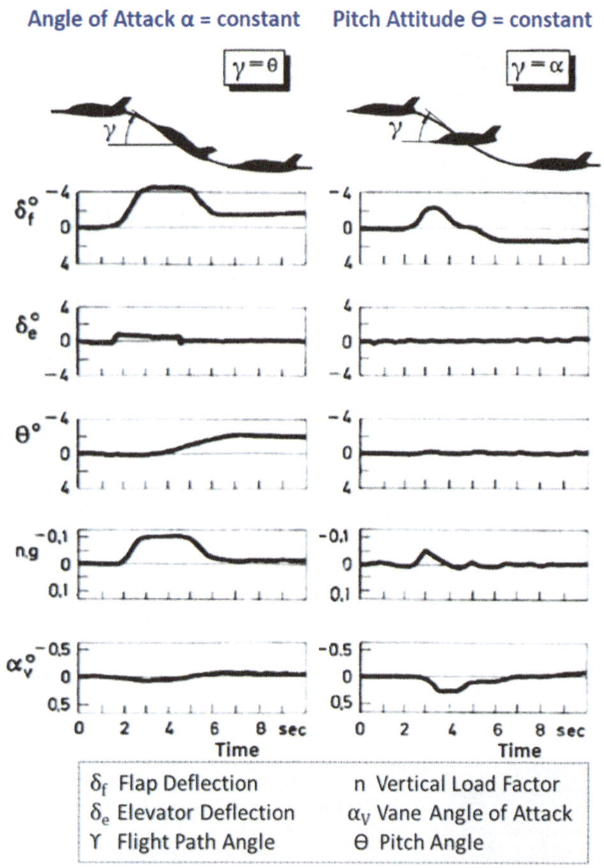

Fig. 7.36 Flight path control with DLC (constant angle of attack/pitch attitude)

carried out with the conventional ILS system, and the 4° approach with the TALAR system. A total of 41 approaches were performed.

The results demonstrated that a significant improvement in maintaining the flight path can be achieved through increased control sensitivity while using DLC in combination with pitch attitude control, in particular, compared to the conventional elevator control. The pitch attitude control was assessed by the pilots to be very pleasant. Through the use of a thrust controller, the problems of maintaining speed were not encountered anymore (see Fig. 7.40).

Fig. 7.37 HFB 320 in approach over TALAR station

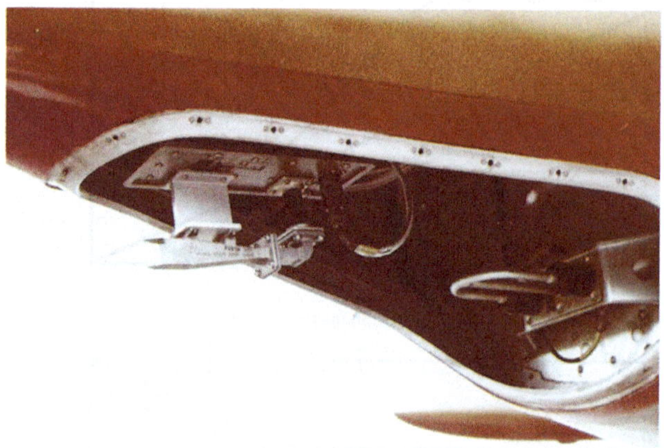

Fig. 7.38 TALAR antenna installation

7.3.5 Trajectory Tracking Experiments with Direct Lift Control (1975)

This flight test program was conducted under the cooperation with the NLR (Netherlands Aerospace Center) in the field of transport aircraft flying qualities. The objective was to identify a pilot model from flight test data for flight control applications for flight path tracking tasks [35]. Of particular interest was the influence of direct lift control.

Fig. 7.39 TALAR radom

Fig. 7.41 Cross-pointer instrument for flight tests

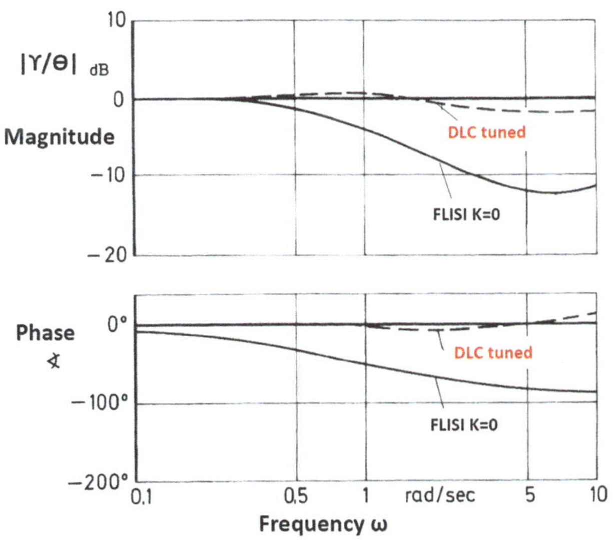

Fig. 7.40 Transfer function of flight path angle to pitch angle (γ/θ) without (K = 0) and tuned DLC

For this purpose, varying flight path commands (*forcing functions*), with a predefined frequency content was displayed to the test pilots in a horizontal flight, which he had to follow through manual control. The deviations between the desired path angle and the measured flight path angle were displayed on a cross-pointer instrument to the pilot (horizontal bar), which he had to manually regulate to zero (see Fig. 7.41).

The forcing function was played from an analog magnetic tape. The measured test data were stored on the same magnetic tape. The procedures to determine the pilot transfer function in the frequency domain were developed by NLR and were applied here. This task was carried out by two pilots for various ratios of DLC flap/elevator deflection, (*blended DLC*) with automatic speed hold (*auto throttle*). The flight test result confirmed that the pilot behaved analogous to the known *crossover model* (see Fig. 7.42).

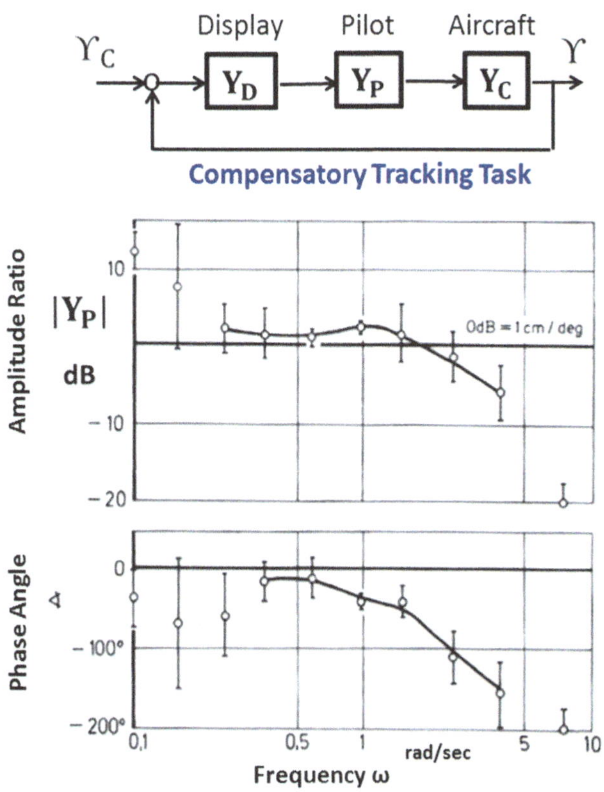

Fig. 7.42 Example of estimated pilot transfer function

The DLC configuration with a flap/elevator ratio of 7.5 showed the highest bandwidth and best stability with the smallest deviations from the desired path.

7.3.6 Automatic Steep Approaches with DLS

A prototype of the DLS (DME *Based Landing System, DME —Distance Measuring Equipment*) was tested and evaluated at Braunschweig airport during the years 1975–1977. DLS was a German proposal of a new approach and landing system submitted to the International Civil Aviation Organization ICAO.

In-depth investigations pertaining to command control for the pitch and roll axes in a ground-based moving cockpit simulator preceded these flight tests (see Fig. 7.43) [36]. An electronic steep landing display was installed for the first time in the instrument panel. Approaches with 3°, 4°, and 5° glide slopes were carried out both in the ground-based simulator as well as in flight tests with HFB 320.

To precisely guide the aircraft along a predetermined nominal path required correspondingly precise information about the current airplane position. The hybrid navigation system used the data from an inertial platform LN3, a barometric altimeter and a VOR/DME system to calculate noise-free, long-term stable position data in north, south and vertical directions with the onboard computer H316. A preliminary version of an experimental autopilot system was also programmed on the onboard computer, which included some typical autopilot functions such as preselection of altitude, course, and speed as well as automatic landing, besides the default command controller for the pitch and roll axes [37, 38].

For safety reasons, testing of digital autopilot system was performed only at height 2000 ft above ground level, because the HFB 320 experimental system had no redundancy. In the case of a malfunction of the experimental system, the pilots had then sufficient clearance to switch back to the mechanical basic controller without stress.

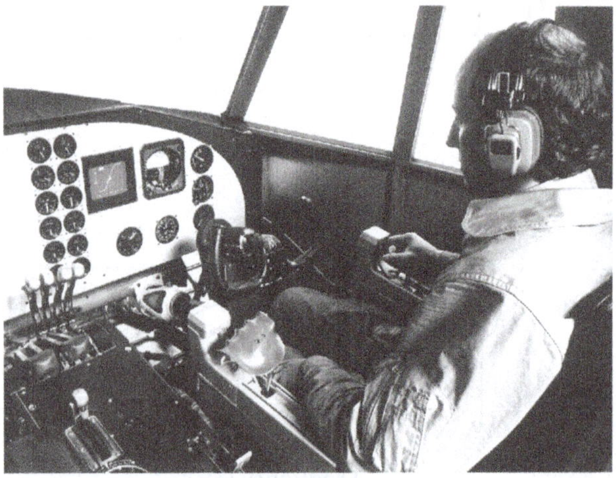

Fig. 7.43 Ground-based simulator investigations of different command control concepts for steep approaches (*Volkmar Adam* in cockpit)

The fully automated approaches to the runway began at an altitude of 4000 ft, about 10 NM northwest of DLS station, which was set up at Braunschweig airport, and ended with the flight over the DLS station at 2000 ft. Up to the starting point of the approach, the HFB 320 was flown by the pilots with the autopilot modes of altitude, course, and speed. Then the AUTOLAND system was switched on and the aircraft turned the course to the south. The airspeed was reduced to 140 kts and flaps were set to 40°. For a lateral offset of 3 km from the approach baseline, the aircraft turned to an interception path that intersected the extended runway centerline at 25°. The vertical reference trajectory, consisting of parabolic transitions and a linear descent segment, was calculated by the onboard computer depending on the flight condition at the time of switching on AUTOLAND. Depending on the chosen glideslope (3°, 4° or 5°), the descent began, therefore, sooner or later.

This first flight test phase with the digital integrated flight control system was completed quite successfully and clearly demonstrated the benefits of a digital system, namely the simple implementation of even complex calculations for navigation and flight control purposes.

7.3.7 Automatic Deceleration of Approach Speed (1976)

This flight test program was conducted together with Dutch NLR (National Aerospace Laboratory). The primary aim of the program was to determine whether noise reduction can be achieved through a continuous reduction of approach speed and thus of the engine thrust as well as by shorter approaches. Secondly, the aim was also to determine the pilot acceptance of such approach procedures, using an automatic continuous flap setting function (*autoflap*) with simultaneous speed hold, either automatically (*autothrottle*) or even manually.

A total of 37 landing approaches were performed at the Hannover airport and were flown and assessed by two pilots. For data analysis, the mean values and the standard deviations of localizer and ILS were used and correlated to the pilot assessments. The functions for *autoflap* and *autothrottle* were programmed on the onboard computer and the HFB 320 was flown from the right hand side in the fly-by-wire mode.

The aircraft was decelerated from 165 kts (flap 10°) to 125 kts (flap 40°). A cross pointer instrument was used to display the flap position (vertical arrow). The horizontal pointer represented the command display for manual speed hold with fast/slow for the speed deviations from the target speed (see Fig. 7.41). The result of the tests was that the method is flyable very well with the chosen deceleration gradient of 0.5 kts/s and the selected final height of 700 ft. Both methods with and without automatic speed hold yielded almost identical results, whereby the workload was higher, but acceptable, during manual speed control [39].

7.3.8 A310 (B10) in Flight Simulation with DLC (1977–1978)

This flight test program was carried out as a joint venture between MBB-UH, BGT, DFVLR, and Lufthansa as a part of the ZKP program "flight control" of the BMFT. The investigations pertained to a developmental configuration of Airbus, B10, which was later referred to as A310. The program consisted of two parts. The objective of the first part (1977) was to develop an in-flight simulation for the B10 configuration and to demonstrate the simulation quality in flight (see also Fig. 7.26). For this purpose, the B10 longitudinal and lateral-directional dynamics in the approach configuration had to be modeled first [40]. Then the model following controller had to be designed and verified in the ground-based simulation [41]. Finally, the created flight test software had to be implemented on the onboard computer and the entire system checked in flight tests [42].

The second sub-task carried out in 1978 was focused on the study of various DLC configurations. Since a direct lift control implies additional costs and weight, the aim was to determine: (1) achievable improvements with DLC compared to B10 without DLC, (2) effect of different DLC parameters on the performance of the pilot-aircraft system as well as on the pilot assessment, (3) flying qualities criteria to be applied while using DLC in the control system design. The investigated DLC concept consisted of simultaneous activation of elevator and wing spoilers for a pitch axis control command. The concept constituted of spoiler control in response to the pilot control input via a so-called washout term.

Besides the B10 without DLC, four DLC configurations were investigated, which were distinguished through coupling ratio between elevator and spoiler deflection as well as by the wash-out time constant of the spoiler actuator. A total of 63 ILS approaches were performed by two test pilots. In these tests, it was required to keep the glide path and localizer deviation within ±0.5 Dot (±0.015°) and the speed within ±5 knots. The assessment of the investigated configuration was carried out by pilots according to a specific workload scale, ranging from 0 (no load) to 10 (high load). Furthermore, the *Cooper-Harper* rating scale was used (see Fig. 2.6), which reflects a correlation between the performance achieved and the pilot workload.

The flight test results showed a significant degradation of flying qualities with increasing DLC function K

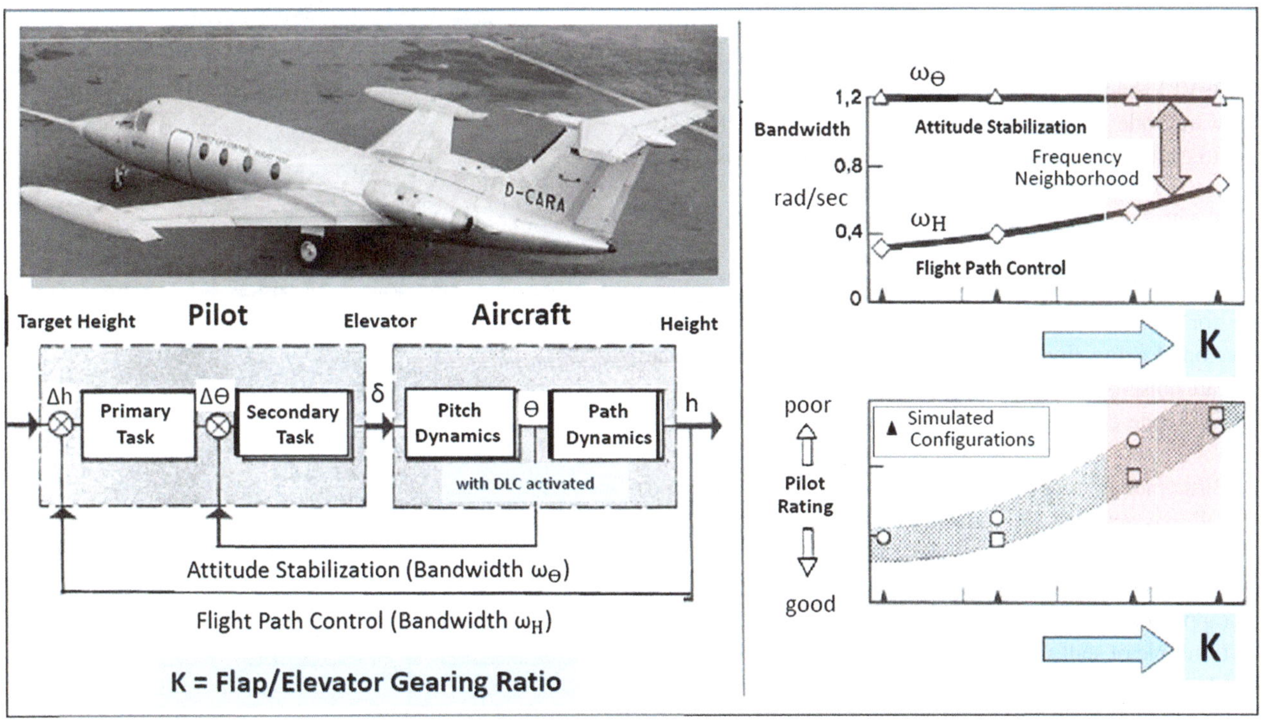

Fig. 7.44 Flying qualities criterion for flight path control with DLC (K = DLC flap/elevator gearing ratio)

(flap/elevator gearing ratio, see Fig. 7.44) [43, 44]. Based on this analysis, a new flying qualities criterion could be defined for aircraft with DLC [45, 46]. It is based on the frequency separation between the pitch attitude dynamics (inner loop) and the flight path dynamics (outer loop). This frequency separation is needed in order not to confuse with the pilot flight path control strategy. The frequency separation can be described by the phase relationship of the flight path to pitch attitude in the range of the short period frequency. It turned out that the frequency separation reduces with the increasing DLC application, so that the use of DLC can lead to flying qualities problems. Later attempts with a pitch rate command system (1979) showed that the benefits of DLC can be exploited optimally only with such a command control system.

7.3.9 Integrated Flight Guidance System (1977–1979)

The development of more and more powerful digital computers in the 1970s made it clear, that the next generation of transport aircraft would be equipped with digital autopilot/autothrottle systems instead of analog systems. To enable the German industry to gain early experience with the development of digital avionic systems, the German Ministry of Research and Technology (BMFT) sponsored a program called "ZKP". DFVLR and its industrial partners BGT, MBB-UH and VFW participated in this program with the objective to develop an *Integrated Flight Guidance System* (IFGS) suitable for a new transport aircraft of Airbus Type B10 and to evaluate it in flight on the HFB 320 aircraft.

DFVLR had the task to design the IFGS, to develop the software code for the onboard computer Honeywell H316 and to evaluate the IFGS in ground-based simulations and in flight trials. BGT developed a digital mode controller for the IFGS called ADB (Autonomes Digitales Bediengerät). VFW developed MUC161, a small avionics computer to run the IFGS software, as well as a 4D-Guidance function applicable for the time accurate arrival at an airport. MBB-UH developed a function "Delayed Flap Approach" to reduce aircraft noise on the ground.

The mode control concept was structured hierarchically:

- Basic modes were command control modes for pitch and roll axis actuated via deflection of the control wheel.
- Next level of modes comprised ALT ACQ (altitude acquire), HDG ACQ (heading acquire), VOR NAV (approach of VOR radial), CAS ACQ (CAS acquire), VS (multiple of stall speed), VX (steepest climb), VY (fastest climb).
- The highest level of modes included AUTOLAND, GA (Go Around), 3D-NAV (approaching a series of waypoints) and 4D-NAV (time accurate guidance to an approach gate).

Figure 7.45 shows the modified instrument panel with ADI (attitude director indicator) and HSI (horizontal situation indicator) in front of the test pilot seat with the ADB mounted on the glare shield.

In accordance with the hierarchical structure of the flight control system, dimensioning of the controller consisted of a step-by-step development within a coupled multivariable feedback system. All essential state data and command inputs were used to calculate deflections of rudder, aileron, elevator and throttle.

A control loop structure was chosen, which allowed designing the system eigenmodes separately from the determination of the command feedforward loop and the disturbance compensating loops (Fig. 7.46). Well damped eigenvalues were achieved for all flight conditions and for every conceivable combination of IFGS modes. This also provided an improved flight comfort in rough air.

A 'command model' generated continuous guidance commands both for the feedback 'controller' as well as for 'dynamic open loop control' of elevator and throttle. Aircraft configuration changes, such as extending/retracting of

Fig. 7.45 IFGS experimental cockpit equipment

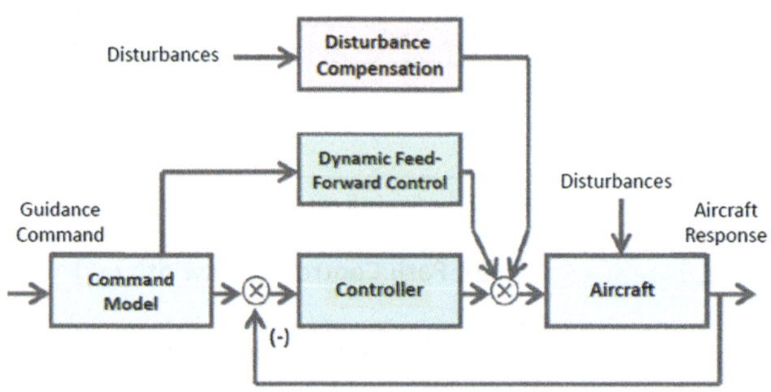

Fig. 7.46 Flight guidance command architecture

landing gear and setting of landing flaps, were taken into account by 'disturbance compensation', which was derived from the flight mechanical model of the aircraft.

High guidance accuracy was strongly dependent on the precision of the control surface deflections courface deflections commanded by the dynamic feedforward open loop control signal. On the one hand, the command model had not to exceed the physical flight limitations of the aircraft. On the other hand, all knowledge of the aircraft dynamic behavior during stationary and instationary flight phases had to be taken into account for the design of the feedforward loop. This was especially important for the throttle settings, because low throttle activity is highly desirable, otherwise applying thrust is the only means to increase the total energy of the aircraft.

Since the IFGS provided combinations of control modes for simultaneous control of altitude/vertical speed and airspeed, it was redesigned in 1978 for the first time as a total energy control system [47, 48]. Thrust was applied to control the total energy, whilst the exchange of potential energy and kinetic energy was compensated by the elevator.

The IFGS was extensively tested in the laboratory in conjunction with an elaborated nonlinear simulation of the HFB 320 aircraft system. Finally, all control modes and nearly all combinations of control modes were tested in 66 flights with a total of 95 flight hours. The airspeed ranged from 130 to 290 kts, flap settings from 0° to 40°.

Some results regarding total energy control can be summarized as follows:

- Climbs and descents were easy to fly and showed only minor speed deviations (less than 2 kts, as long as the throttle limits were not exceeded) and an adequate throttle settings.
- Stepwise extension of landing flaps from 0°, 10°, 20°, and 40° showed only minor speed deviations (less than 2 kts) and altitude deviations (less than 15 ft) and adequate throttle settings.
- Acceleration from 200 to 290 kts led to only minor deviations in altitude (less than 15 ft).

Further results concerning the other control modes can be found in references [49, 50].

7.3.10 Command Control with Attitude Hold and Direct Lift Control (1979–1980)

The aim of this flight test program was to investigate the influence of a controller-assisted command control in pitch and roll axes with automatic attitude hold. Such a system is today standard in Airbus aircraft, whereby vertical acceleration is commanded instead of the pitch rate. Furthermore, the influence of a direct lift control was studied in connection with the command control, as well as the impact of automatic course hold and a wings level control in the roll axis (*Wing Leveler*) with bank angles smaller than 2°.

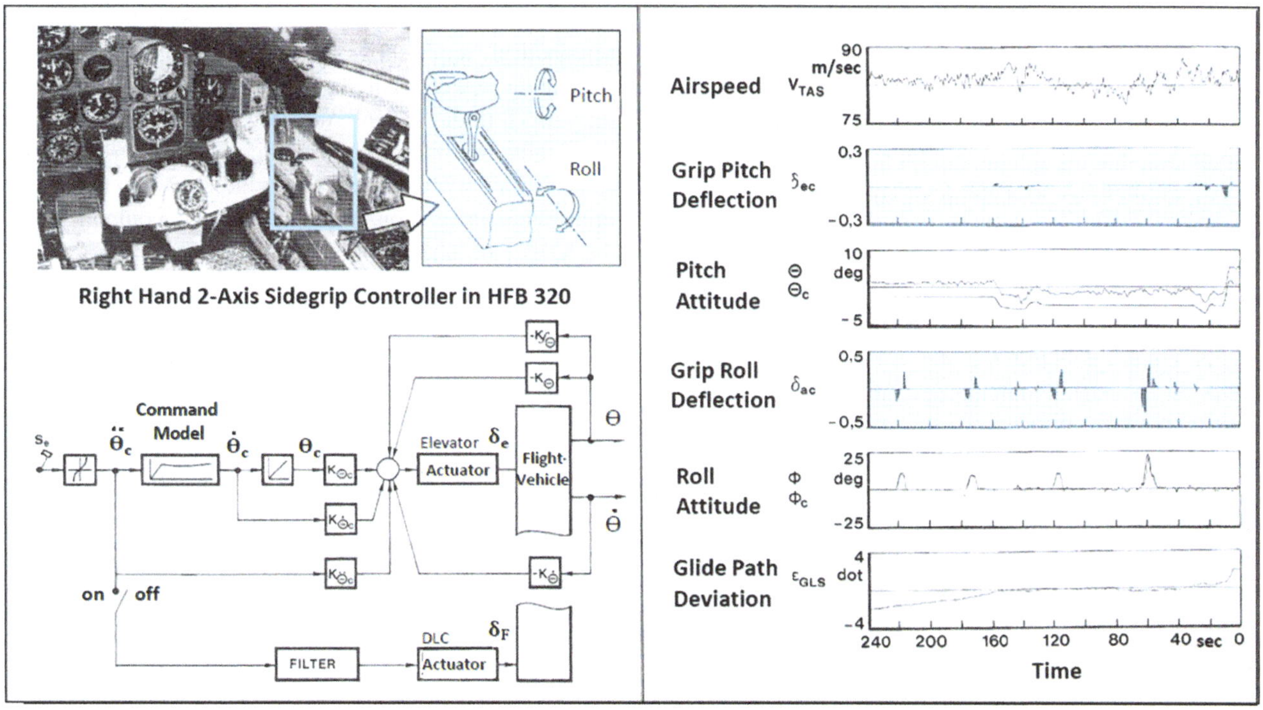

Fig. 7.47 Augmented flight control system with sidegrip

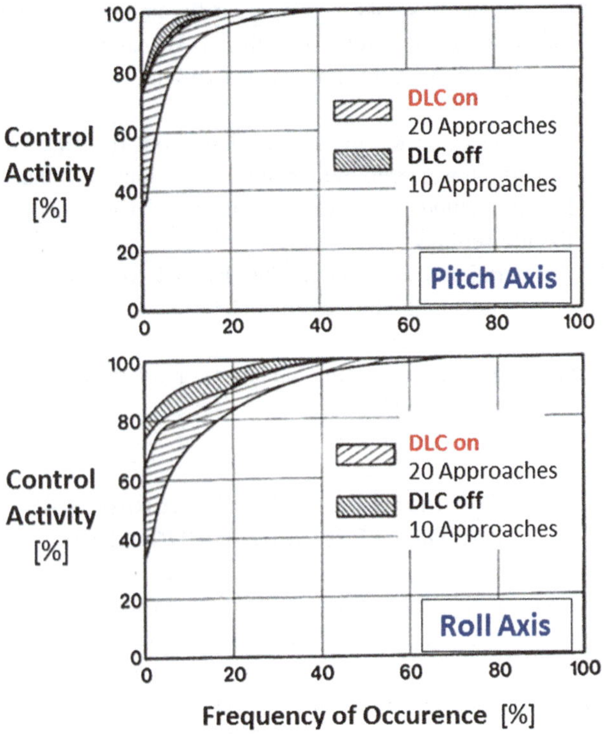

Fig. 7.48 Frequency of occurrences of sidegrip deflections for approaches with and without DLC

Fig. 7.49 Flight guidance computer MUC 161 and computer interface in HFB 320 cabin

HFB 320 as in-flight simulator was used for these flight tests. The flying qualities were assessed by two pilots for 85 landing approaches according to a *Cooper-Harper* rating scale (see Fig. 2.6). The system block diagram is shown in Fig. 7.47 for the pitch axis. For the pilot control, a 2-axis sidegrip control for the right hand was utilized. The time history plot of a typical approach is shown in the same figure. The control activity is pulse shaped and very low, because the aircraft flies stable following a flight path change and all disturbances are automatically compensated.

The results show that the pilot control technique in the present case differs significantly from that of the conventional control technique [51–53]. The pilot reacts more in the sense of open-loop control, because all disturbances are controlled by the system. The sidegrip as an input device was accepted quite fast and did not cause any problems. Behavior was judged of flying an overall system than just an aircraft. The DLC influence improved the flight path control and reduced the pilot activity as seen in Fig. 7.48. The automatic course hold and the *Wing Leveler* were assessed to be advantageous for cruise, however, not for landing approach. Small corrections in the roll attitude were necessary here, which were prevented by the *Wing Leveler*.

7.3.11 Flight Testing of a MUC161 Flight Guidance System (1979)

A compact digital computer MUC161 was built by VFW for aeronautical applications, which also provided enough computing power for flight control applications. The company BGT (today: Diehl Aerospace) had developed a corresponding interface for this purpose. MUC161 computer and interface were integrated into the HFB 320 (Fig. 7.49). Preparatory integration testing of both the components and the ADB took place in a laboratory.

After the flight testing of the Fortran computer programs of the AFCS on Honeywell computer H316 was largely completed, a considerable part of computer programs for the autopilot system was newly written in Assembler language by VFW and BGT. This affected all autopilot operating modes tested until then, as well as an approach procedure developed by MBB-UH, in which the flaps were computer-assisted continuously extended with a simultaneous reduction in airspeed (*delayed flap approach*). The computer programs for hybrid navigation ran thereby still with the Honeywell computer H316. Through this distribution, the sampling time of the controller could be reduced from 200 to 100 ms. All autopilot modes programmed in MUC161 were successfully demonstrated in 16 flights with 19 h of flight time.

7.3.12 Reduced Static Stability (1981–1983)

To increase the flight performance, transport aircraft are increasingly designed to fly over a wide range with a reduced longitudinal static stability. When the probability of a failure of the artificial *stability augmentation system* (SAS) is too high, then the pilot must be able to fly the aircraft manually after a controller failure [54]. This implies that the aircraft must have acceptable degraded flying qualities. Within this problem complex, extensive test programs were performed by MBB-UH and DFVLR as a part of ZKP task ACTTA (Active Control Technology of Transport Aircraft), investigating the influence of reduced longitudinal stability on the handling qualities. Thereby the aspects pertaining to a certain minimum stability after the stabilization system failure were addressed for safely performing climb and cruise flights, as well as safe landings.

Initially, basic experiments were performed in an A300 training simulator of German Lufthansa at Frankfurt. The results were then verified through extensive flight tests with the in-flight simulator HFB 320 FLISI. The experiments could be carried out with the in-flight simulator without safety risk, because the unstable aircraft response was artificially introduced. On the loss of control, the safety pilot could take over the control by pressing a button and bringing the aircraft back in a stable state. A total of 44 missions with 181 approaches were flown by 4 pilots (3 pilots from DFVLR, 1 from MBB-UH).

The simulation task consisted of ILS approaches at Hannover airport using "raw" ILS data. It was required to hold the glide path and speed as accurately as possible. The test pilots flew the entire mission (4 approaches) with the model dynamics, whereby the aircraft was guided by ATC. The approach altitude was 2500 ft with a go-around at 500 ft GND for safety reasons. The localizer and glide slope display was adapted for the test pilots in the onboard computer, such that at level-off the conditions of touch down on ground were displayed.

The evaluation pilots were required to describe after each flight the flying qualities according to a catalog with assessment queries. After performing a complete mission, the pilot had to assess the specific configuration flown with regard to control behavior, tendency to PIO (Pilot Induced Oscillations), etc. based on a *pilot comment card*. Simultaneously the evaluation pilots provided *Cooper-Harper* ratings.

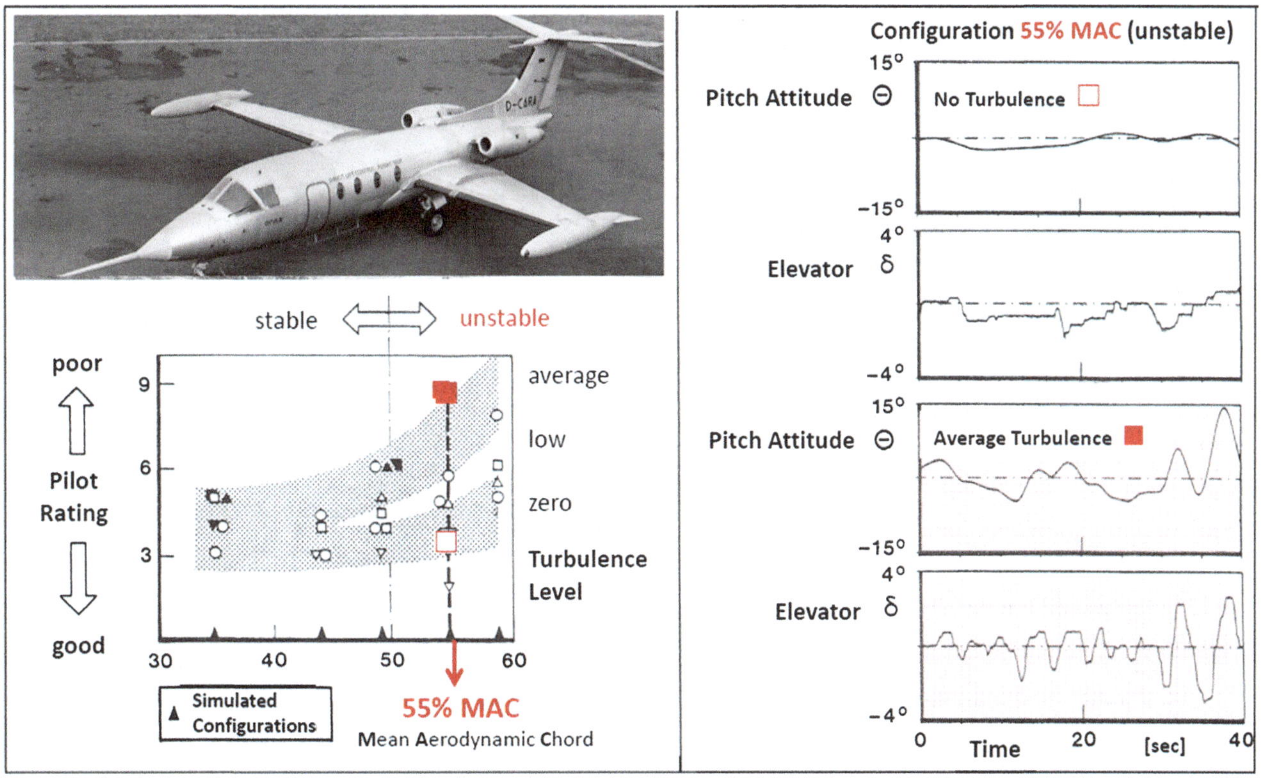

Fig. 7.50 Influence of reduced stability % MAC (mean aerodynamic chord) on handling qualities

Figure 7.50 elucidates a result of this experimental program [55–59]. It shows the influence of aft displacement of the center of gravity and thereby of reduced static stability on the pilot assessment according to the *Cooper-Harper* scale for a large transport aircraft (Fig. 2.6). It turned out that the pilot assessment is quite substantially affected by the turbulence level. For low turbulence, even the unstable configurations (55% MAC—Mean Aerodynamic Chord) are flyable. With stronger turbulence, the control limit is still close to the neutral point (indifferent behavior). The poor assessment for a center of gravity of 59% MAC (time to double 4.77 s) implies controllability is not acceptable anymore. Continuously increasing control activity in the pitch axis for the unstable configuration can be observed from the time history plots, implying that the pilot is no longer in a position to control the aircraft.

7.3.13 Time Delays in FBW Flight Control Systems (1983–1984)

The growing complexity of digital electrical flight control system generates non negligible dynamics and time delays, that have a tangible influence on the mission effectiveness and flying qualities. To investigate this effect, a flight test program was carried out with the in-flight simulator HFB 320 with the following objectives: (1) evaluate the impact of time delays in flight control system on the flying qualities during landing approach of a commercial aircraft, (2) determine maximum allowable values for the time delays in the pitch and roll control, and (3) compare the results with existing flying qualities criteria. For these investigations, the flying qualities of HFB 320 were not changed. However, time delays were introduced in the control system and its effect on controllability was assessed. Two different types of controls behavior were evaluated.

The first type, called basic control, consisted of the conventional control system of HFB 320, wherein the pilot control devices were mechanically connected with the aerodynamic control surfaces. This control constituted the basis control, in which time delays were not encountered.

In the second type of control, called fly-by-wire control, time delays arise through required computing time, though signal conversions, or through time delays in the actuating system, etc., which was of the order of 150 ms in the case of HFB 320. Additional time delays were implemented in the onboard computer.

Results of this flight test program are presented in Fig. 7.51. They show the deterioration of pilot judgments in the longitudinal and lateral-directional motion with

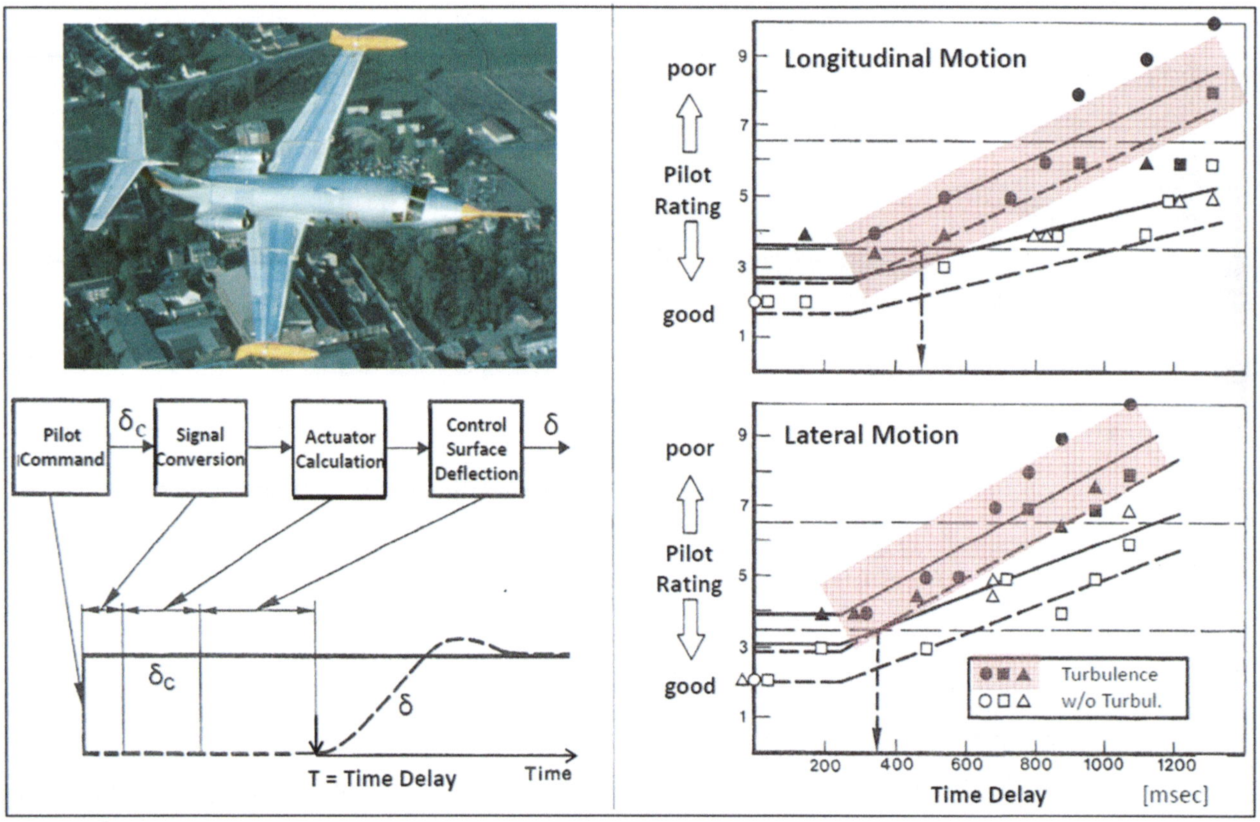

Fig. 7.51 Influence of system time delays on handling qualities

increasing time delays in the corresponding transmission chain from pilot input to the elevator or aileron control surface deflection. The time delays in elevator and aileron controls were varied from 0 to 1300 ms and implemented on the onboard computer. The individual configurations were flown under different atmospheric conditions (turbulence) by different pilots and were rated according to the *Cooper-Harper* rating scale.

The results showed that for aircraft in the category of HFB 320, time delays in the lateral-directional motion were regarded more critically than those in the longitudinal motion for the landing approach tasks [60]. Under moderate turbulence levels, time delays of the order of 300 ms in the pitch axis were tolerated before an appreciable degradation of Level 1 flying qualities (very good—satisfactory) was noticed. The time delays should not be greater than 250 ms in the roll axis. Time delays of up to 800 ms were tolerated by the pilots in the roll control system for a Level 2 (acceptable), whereas those in the pitch axis could be even larger.

7.3.14 Pilot and Flight Test Engineer Training (1984)

The final program with the HFB 320 FLISI consisted of its utilization as a variable stability flight demonstrator in a training campaign for IPTN (Industri Pesawat Terbang Nusantara, today: Indonesian Aerospace) in Bandung, Indonesia. During 1983–1984, IPTN was supported by DFVLR in the flight test certification phase of the Indonesian-Spanish joint commuter aircraft CN 235. DFVLR had the overall responsibility for the development and construction of a mobile flight test instrumentation system FTIS (*Flight Test Instrumentation System*) for the CN 235 certification testing in Indonesia. This included a "*Container City*" with 12 air-conditioned standard containers and an On-Board Data Acquisition System (OBDAS). Since 1984 the FTIS plant was successfully and independently operated in Bandung by IPTN [61]. Parallel to these activities, the CN 235 test personnel was introduced in Braunschweig to the FTIS design, and in the theory and practice of flight testing.

From June 11–14, 1983, the in-flight simulator HFB 320 FLISI was deployed as an ideal training device for the practical familiarization of Indonesian pilots and flight test engineers. The Indonesian pilots *Cpt. Mursanto*, *Cpt. Somersono* and *Cpt. Supriadi* were trained in the flight test techniques required for the specific purpose at hand. Instrument approaches were performed with the simulated CN-235 model, whereby the task for the pilots consisted of minimizing the deviations from the displayed target trajectory.

Fig. 7.52 HFB 320 and its successor VFW 614 G13

Model parameters were modified between the approaches in such a way that it resulted in different simulated CN 235 responses. The pilots were required to assess the resulting different controllability.

7.4 The End of Project

During the last few years of the HFB 320 in-flight simulator project, it became more and more difficult to have the required resources and experts available to maintain the intensive and smooth operation of HFB 320 as an in-flight simulator. In particular, the development and implementation of a new in-flight simulator ATTAS (see Chap. 9) based on a VFW 614 required concentrated participation of experts from 1981 onwards, who were until then available for the HFB 320 (see Fig. 7.52).

After winding up the utilization as an in-flight simulator, investigations and test maneuvers for Space experiments under Zero-G, that is under weightlessness, conditions were carried out with the HFB 320 during the fall season of 1983. With manual inputs, the weightlessness conditions could be created for about 22–23 s during the parabolic flight path.

Upon reaching zero-G, after a few seconds, the oil pressure warning on the engines (GE CJ 610-5) was activated, because even the oil floated in the oil tanks and the air was sucked through the oil pump exhaust. After consultation, engine manufacturer General Electric approved harmlessness for about 30 s with inadequate oil pressure.

Unfortunately, no further tests were performed for tasks related to Space missions. The HFB 320 FLISI was decommissioned on May 25, 1984, and deregistered on July 3, 1984 at the German Federal Aviation Authority (LBA). On June 12, 1986 at 13:20 h, the HFB 320 was transported with an army helicopter CH 53 from Braunschweig to the former Hamburger Flugzeugbau facility at Hamburg-Finkenwerder (see Fig. 7.53). The HFB 320 FLISI aircraft was externally restored back to the original condition of a business-type aircraft. The aircraft is now in Finkenwerder on the factory premises of Airbus Hamburg (see Fig. 7.54).

Fig. 7.53 HFB 320 FLISI transport with CH 53

Fig. 7.54 HFB 320 FLISI (D-CARA) at Airbus factory premise (*Credit* Torben Guse)

References

1. Brünner, A., Kuper, H.-J.: Umrüstung einer Piaggio P 149 zu einem Flugzeug mit Modell gesteuertem Übertragungsverhalten, DVL TV 11/67 (1967)
2. Pietraß, A.: Bestimmung von Trägheits- und Reibungsmomenten des Piaggio-Landeklappen-systems aus Einschwingvorgängen, DVL TV 6/68 (1968)
3. Pietraß, A.: Konzept der Umrüstung eines Flugzeuges zu einem Fliegenden Simulator, DVL, Technischer Vermerk 1/69 (1969)
4. Pietraß, A., Uhrmeister, B.: Umrüstung eines Flugzeugs zu einem fliegenden Simulator, DVL-Nachrichten, Heft 38, Seiten 405-408, Januar 1969
5. N. N.: Versuchsträger Variabler Stabilität (VVS), Mittelfristiges Programm, 10. Dez. 1969
6. Brüning, G.: Simulation, an introduction and survey, AGARD CP-79-70, Paper 1, January 1971
7. Pietraß, A.: Fliegender Simulator HFB 320 Hansa, Stand und Planung, DFVLR IB 9/70 (1970)
8. Schott, J.: Fliegender Simulator HFB 320 HANSA, Betriebsartensteuerung; Entwurf eines Bediengerätes, DFVLR Institut für Dynamik der Flugsysteme, ARNO 013-123/15 (1971)
9. N. N.: DFVLR Jahresbericht 1971, Institut für Flugmechanik, Seiten 101-116, 1971. Primärsteuerung am Institut für Flugführung der DFL, DFL-Mitteilungen, Heft 8 (1968)
10. Metzdorff, W.: Neuere Arbeiten zur Flugzeug Primärsteuerung am Institut für Flugführung der DFL, DFL-Mitteilungen, Heft 8 (1968)
11. Bestehorn, P.: Arbeiten zur Umrüstung des HFB 320 HANSA JET – Werk Nr. 1021 – zu einem fliegenden Forschungssimulator für Fluggerät mit wählbar veränderlichen Flugeigenschaften -3. Phase: Landeklappen-system, Anlage 1 zu UH-05-72 (1972)
12. N. N.: DFVLR Jahresbericht 1972, Institut für Flugmechanik, S. 106–107 (1972)
13. Hanke, D.: Erprobung der direkten Auftriebssteuerung beim Versuchsflugzeug DFVLR-HFB 320, DFVLR Nachrichten, Heft 7, S. 297-300 (1972)
14. Hanke, D.: Nachweis der Funktionstüchtigkeit des elektrischen Schubstellsystems der HFB 320 S1 im Fluge und Übergangsverhalten des elektrischen Schubstellsystems im Fluge, DFVLR Institut für Flugmechanik, ARNO 081-0147-F-3 (1973)
15. Hanke, D.: Systemspezifikation Höhenrudertrimmsystem für HFB 320 S1, ARNO 081-0417-E-2, DFVLR, Institut für Flugmechanik (1972)
16. Gartung, B., Schreier, A.: Automatisches Höhenrudertrimmsystem-Entwicklung und Konstruktion der Elekronik-Box, DFVLR IB 154-77/31 (1977)
17. Amtsberg, J.: Dreikomponentenmessungen an modifizierten Spoilern der HFB 320, DFVLR IB 157-76 C 11 (1976)
18. Anders, H., Hanke, D., Meyer, H.-L.: Flugerprobung des Spoilerstellsystems (SPS) der HFB 320 S1 im Rahmen des Forschungs-vorhabens Flugführung, Ergebnisbericht MBB-UH TN-HE 213-12/77 (1977)
19. Müller, M.: Steuerkraft-Simulationsstand für HFB 320 Hansa, DFVLR IB 154-74/34 (1974)
20. N. N.: DFVLR Jahresbericht 1973, Institut für Flugführung, S. 89-91 (1973)
21. Hanke, D., Lange, H.-H., Henschel, F.: Regelungstechnische Aspekte der In-Flight Simulation, 87. Wehrtechnische Symposium Regelungstechnik-Automation, Bundesakademie für Wehrverwaltung und Wehrtechnik, Mannheim, DFVLR IB 154-77/38 (1977)
22. Henschel, F.: The Determination of Flight-Tracking Controllers using Solheim's Pole Shifting Method, ZfW 1/1977
23. McCracken, J, Hempel, H.-D., Klingner, C.: Data Processing-Handling Qualities Investigation Project, DFVLR IB 154-75/20 (1975)
24. Rix, O.: Systemidentifizierung des DFVLR-Forschungsflugzeugs HFB 320 in der Längsbewegung mit besonderer Berücksichtigung der Steuerflächenwirksamkeiten, Institut für Flugmechanik, DFVLR-Mitt. 79-16, Nov. 1979
25. Rix, O., Hanke, D.: In-flight measured characteristics of combined flap-spoiler direct lift controls. In: AGARD Fluid Dynamics Panel Symposium on 'Aerodynamic Characteristics of Controls', Neapel, Italien, AGARD-CP 262, Paper 16 (1979)
26. Hanke, D., Lange, H.-H.: Flugmechanische Probleme beim Landeanflug mit direkter Auftriebssteuerung am Beispiel der HFB 320 Hansa, DGLR/DGON Symposium Neue Anflugverfahren, Düsseldorf, DGLR 73-024 (1973)
27. Hamel, P., Dahlen, H.W.: Erprobung lärmmindernder Anflugverfahren mit dem DFVLR-Forschungsflugzeug HFB 320, DFVLR-Nachr., Heft 10, Seiten 413-414, July 1973
28. Onken, R., Joenck, H.-P., Tacke, L., Gottschlich, M.: Digital fly-by-wire control system with self diagnosing failure detection. AGARD CP 173, Paper 22, Sept 1973
29. Wilhelm, K., Moorhouse, D. J. (USAFFDL): Summary of a joint program of research into aircraft flight control concepts, AFWAL TR-83-3057/DFVLR IB 111-83/21 (1983)
30. Hanke, D., Lange, H.-H.: Flugmechanische Probleme beim Landeanflug mit direkter Auftriebssteuerung am Beispiel der HFB 320 Hansa, DLR-Mitt. 74-28 (1974)
31. Hamel, P., Wilhelm, K., Hanke, D., Lange, H.-H.: Steep approach flight test results of a business -type aircraft with direct lift control, AGARD CP-160, Paper No. 21 (1974)

32. Lange, H.-H., Hanke, D.: Verbesserung der Flugbahnsteuerung durch direkte Auftriebs-steuerung (DLC), ZTL-Aufgabe 1974, Ber. f. Auftraggeber DFVLR- IB 154-74/41 (1974)
33. Wilhelm, K., Hanke, D.: Direct Lift Control From a Flight Mechanics Standpoint, Short Course "STOL Aircraft and the Community", Tullahoma/USA, Aachen/Deutschland (1974)
34. Hanke, D., Lange, H.-H., Henschel, F., Wilhelm, K.: Verbesserung der Flugbahn-steuerung durch direkte Auftriebs-steuerung, ZTL-Aufgabe 1975, Ber. f. Auftraggeber DFVLR IB 154-75/44 (1975)
35. van Gool, M.F.C., Hanke, D., Lange, H.-H.: Flight Path Angle Tracking Experiments in the DFVLR HFB 320 Equipped with Direct Lift Control, DFVLR IB-154-75/32 (1975)
36. Onken, R., Adam, V., Dierke, R.: The use of a flight simulator in the synthesis and evaluation of new command control concepts, AGARD CP 198, Paper 4 (1976)
37. Stuckenberg, N.: Entwurf eines Vorgaberegelsystems mit proportional-integralem Regler für den Steilanflug des Flugzeugs HFB320 Hansa nach der Theorie des optimalen Zustandsreglers, DFVLR FB 75-19 (1975)
38. Adam, V., Onken, R.: Evaluation of a new flight path control concept, 10. ICAS Kongress, Paper No. 76-56 (1976)
39. Hofmann, C.F.G., Lange, H.-H., Hanke, D.: In-Flight Investigation of Decelerating Approaches, DFVLR IB-154-76/10 (1976)
40. Schafranek, D.: Theoretische Untersuchung über den Einfluss der Spoilerdynamik auf die Fliegbarkeit eines Flugzeugs mit direkter Auftriebssteuerung, DFVLR FB 80-07 (1979)
41. Lange, H.-H.: Flugerprobung des Modellfolgereglers für die HFB 320 zur Simulation des Airbus A 310 im Fluge, DFVLR Mitt. 79-13 (1979)
42. Hanke, D., Lange, H.-H., Henschel, F.: Simulation der dynamischen Eigenschaften der B10 mit dem ‚In-Flight Simulator' DFVLR-HFB 320, Meilensteinbericht, ZKP-Aufgabe Flugführung, DFVLR IB 154-77/17 (1977)
43. Hanke, D.: Fliegbarkeitsuntersuchungen von DLC-Konfigurationen für den Airbus A 310 im Landeanflug mit dem In-Flight Simulator DFVLR-HFB 320, DFVLR FB 79-18 (1979)
44. Hanke, D., Lange, H.-H.: In-flight handling qualities investigations of various longitudinal short term dynamics and direct lift control combinations for flight path tracking using DFVLR HFB 320 variable stability aircraft, AGARD CP 260, Paper 21 (1978)
45. Hanke, D.: In-flight handling qualities investigations of A 310 airbus DLC-configurations on landing approach, using the DFVLR-HFB 320 in-flight-simulator (variable stability aircraft), ESA-TT-630, Translation of DFVLR FB 79-18 (1980)
46. Lange, H.-H., Hanke, D.: Use of DFVLR In-flight simulator HFB 320 Hansa for handling qualities investigations. In: 12th Congress of the International Council of the Aeronautical Science (ICAS), Munich, ICAS-80-9.4 (1980)
47. Adam, V.: Die Regelung von Fahrt und Höhe in einem integrierten Flugführungssystem, DGLR Fachausschuss-Sitzung, Hamburg (1978)
48. Adam, V., Leyendecker, H.: Erhöhung der Führungsgenauigkeit durch den Einsatz eines integrierten digitalen Flugführungssystems, DGLR/DGON Symposium "Fliegen im Flughafen-Nahbereich", Vortrag Nr. 79-043 (1979)
49. Adam, V.: Flight control modes for control of aerodynamic state parameters, 12. ICAS Congress, Paper No. 22.5 (1980)
50. Adam, V., Leyendecker, H.: Control law design for transport aircraft flight tasks, AGARDO graph 251, Paper 7 (1981)
51. Hanke, D., Wilhelm, K., Lange, H.-H.: Handling qualities aspects of CTOL aircraft with advanced flight controls, AGARD-CP-333, Paper 10, June 1982
52. Hanke, D.: Flying Qualities Experiments of Rate Command/Attitude Hold Systems in the HFB 320 In-flight Simulator, DFVLR FB 83-25 (1983)
53. Altenkirch, D.: Untersuchungen des Pilotenverhaltens bei Flugversuchen mit einem Rate-Command/Attitude-Hold Steuerungssystem, DFVLR FB 84-25 (1984)
54. Hanke, D., Schafranek, D.: Literaturstudie zum Thema "Fliegbarkeit von Flugzeugen mit reduzierter Längsstabilität", DFVLR IB 154-79/30 (1979)
55. Wilhelm, K., Schafranek, D. (DFVLR), Trosky, B. (MBB Hamburg): Flugeigenschafts-untersuchungen von Verkehrsflugzeugen mit reduzierter statischer Längsstabilität, DGLR Symposium "Leistungssteigerung von Flächen-flugzeugen", Frankfurt, November 1982
56. Wilhelm, K., Schafranek, D.: In-flight investigations of landing approach handling qualities of transport aircraft with relaxed static stability, preliminary analysis and results, DFVLR IB 111-83/24 (1983)
57. Wilhelm, K., Schafranek, D.: In-flight investigation of landing approach flying qualities of transport aircraft with relaxed static stability, DFVLR FB 84-11 (1984)
58. Wilhelm, K., Schafranek, D.: In-flight investigation of the influence of pitch damping and pitch control effectiveness on landing approach flying qualities for statically unstable transport aircraft, DFVLR FB 84-12 (1984)
59. Wilhelm, K., Schafranek, D.: Landing approach handling qualities of transport aircraft with relaxed static stability. J. Aircr. **23**(10), 756–762 (1986)
60. Wilhelm, K., Altenkirch, D.: In-flight investigation of the effects of time delay in control system on flying qualities in landing approach, DFVLR FB 84-35 (1984)
61. Hamel, P., Karmann, R., Klewe, H.-J., Freese, R.: Technical Assistance Agreement (TAA) between PT. Nurtanio and DFVLR on the development of a flight test instrumentation system (FTIS), DFVLR IB 111-84/16 (1984)

Author Biography

Knut Wilhelm was Professor at the Flight Mechanics and Flight Control Department of the Berlin Technical University (1993–2001). Prior to joining the University, he was a research scientist at the Institute of Flight Mechanics, DLR in Braunschweig (1966–1993), where he was head of Flight Mechanics of Fixed-Wing Aircraft department from 1972 to 1993. From 1979 to 1990 he was a project manager for the US-German Memorandum of Understanding (MoU) on Aircraft Flight Control Concepts. He received his Dipl.-Ing. degree (1966) and Dr.-Ing. degree (1968) in Mechanical/Aeronautical Engineering from the Technical University of Braunschweig. His research interests are flying qualities, flight dynamics, flight control, flight test, wind tunnel simulation, and in-flight simulation (fixed-wing aircraft).

8 Helicopter In-Flight Simulator Bo 105 ATTHeS

Bernd Gmelin

8.1 Introduction

The early phase of helicopter development, until about the end of Second World War, was characterized by the technical realization of individual components (for example, rotor, flight control, engine), by developing theoretical fundamentals (such as aerodynamics, rotor dynamics), and by the search for suitable configurations of the new flying device. The rapid further developments during the following decades led to higher speeds, improved maneuverability, enhanced efficiency, and at the same time to numerous ideas of exploiting the versatile applicability of this flying device. It became more difficult for pilots to fly the mostly unstable helicopter since the desire for better flying qualities was

B. Gmelin (✉)
Braunschweig, Germany
e-mail: bernd.gmelin@t-online.de

P.G. Hamel (ed.), *In-Flight Simulators and Fly-by-Wire/Light Demonstrators*,
DOI 10.1007/978-3-319-53997-3_8

often subordinated to the demands for higher flight performance and more complex missions. This was particularly evident during instrument flying under poor visual and adverse weather conditions.

One way out of this situation is the introduction of innovative flight control systems. This is only possible for helicopters if the mechanical/hydraulic control components are replaced by digital Fly-by-Wire (FBW) technology. FBW not only enables the use of stability and control augmentation, but, at the same time, it leads to reduced mechanical complexity, weight reduction, simplification of maintenance, and increased reliability. The introduction of Fly-by-Wire/Light (FBW/L) technology is a crucial step for a successful future of the helicopter because, by enabling the use of full authority stability and control augmentation, more precise flight maneuvers and flight envelope expansion can be realized with reduced pilot workload.

This chapter provides an account of modifications and equipping of the Bo 105-S3 helicopter and of the utilization of the in-flight simulator ATTHeS to address diversified aspects of FBW/L in helicopters.

8.2 History of Bo 105 (Serial Number 3)

The Bölkow Bo 105 is a light multi-role helicopter of the German manufacturer Messerschmitt-Bölkow-Blohm—MBB (then: Eurocopter, today: Airbus Helicopters). Its development began in 1961 and the first flight was on February 16, 1967 with the prototype V2 (see Fig. 8.1). This helicopter is deployed even today for governmental tasks, including the police, the military, civil defense and disaster control, as well as for air rescue in particular, and for various other tasks by civilian operators. For the first time the two gas turbine propulsion concept was introduced with Bo 105 in civilian helicopters of the 2-tons-class, and also a hingeless rotor head with fiberglass reinforced plastic (FRP) rotor blades. The 4-blade rotor "System Bölkow" features a rigid titanium rotor hub. The obligatory flapping and lagging motion of rotor blades, which is enabled in other helicopters through individual rotor hinges, is realized in this case through elastic FRP elements in the roots of the rotor blades. The rotor does not require any lead-lag damper. Altogether it consists of a substantially less number of components than the previous rotors. The construction enables high control power, quick control response and high effectiveness, and thereby a very good maneuverability of the helicopter.

Starting 1970, the Bo 105 was produced in different variants; a total of 1404 helicopters in Germany until 2001 [1]. The helicopter was also manufactured under license in Spain, the Philippines, Indonesia, and Canada. Altogether more than 1640 Bo 105 helicopters were built, of which many are in use even today.

The Bo 105-S3 was produced as Series A in July 1971 by MBB in Manching and registered as D-HEBV (see Fig. 8.2). Immediately after routine tests, the Bo 105 was exported to the USA and was operated by Boeing Vertol Co. with the registration N1149B. The cooperation between MBB and Boeing Vertol at that time had the following objectives: (1) to support type certification of Bo 105 by the FAA (Federal Aviation Administration) (April 1972) and (2) to

Fig. 8.1 First flight of Bo 105 helicopter (*Credit* Airbus Group)

Fig. 8.2 Bo 105-S3 D-HEBV (*Credit* Airbus Group, NA3T)

demonstrate the features of the hingeless rotor in the US and to recommend this technology for the project UTTAS (*Utility Tactical Transport Aircraft System*) of the US Army. The proposal of a prototype YUH-61A developed on this basis by Boeing Vertol was, however, defeated in the procurement process by the competitor YUH-60 of Sikorsky company, who received the development contract for the standard transport helicopter UH-60 of the US Army.

In July 1972, the Bo 105-S3 came back from the USA to MBB in Ottobrunn. After extensive modifications in controls and cockpit (see Sect. 8.3.1) and after upgrade to version C23 of the C series, the S3 flew from 1974 with 2 Allison 250-C20 engines and an all-up weight of 2.3 t. It was operated at MBB with the military registration 98 + 08 on behalf of the Federal Office of Defense Technology and Procurement, BWB (today: Federal Office of Bundeswehr Equipment, Information Technology, and In-Service Support—BAAINBw). As a part of the HSF (German: Hubschrauber-Schlechtwetter-Führung for helicopter adverse weather guidance) project, in the following years, the so-called variable stability helicopter was deployed for design and testing of flight control and guidance systems. In late 1980, BWB decided to sell the helicopter through the Federal Disposal Sales and Marketing Agency (VEBEG).

Based on the initiative of the Institute of Flight Mechanics and the Flight Test Facility, the German Aerospace Research Establishment—DFVLR (Today: German Aerospace Center—DLR)—acquired the helicopter, which finally arrived at Braunschweig in 1982 and was operated with the original registration D-HEBV based on a Permit to Fly of the Federal Aviation Office (Vorläufige Verkehrszulassung VVZ). In the subsequent years, the S3 was converted to the in-flight simulator Bo 105 ATTHeS (*Advanced Technologies Testing Helicopter System*) (see Fig. 8.3). ATTHeS accumulated over 1300 flight hours at DFVLR/DLR in numerous research and development programs (see Sect. 8.4). In a tragic accident due to a fatigue fracture in the tail rotor drive, the helicopter crashed on May 14, 1995, during a ferry flight near Stendal. Thereby the test pilot *Klaus Sanders* and the flight engineer *Jürgen Zimmer* were killed.

8.3 Modifications and Equipment

At the end of 1969, DFVLR, Dornier, and MBB submitted a memorandum "variable stability testbed (VVS), a mid-term program" to the German Federal Ministry of Defense [2]. Besides fixed-wing aircraft projects, realization of a variable stability helicopter based on the Bo 105 was proposed in this document. Theoretical investigation at DFVLR revealed that aircraft with vertical and/or short takeoff and landing capabilities (V/STOL) in low-speed regime could be simulated by a helicopter [3]. Due to its basic characteristics, the Bo 105 helicopter appeared to be particularly suitable for this task [4].

8.3.1 Control System

With the support of the Federal Ministry of Defense, the Bo 105-S3 helicopter was equipped in the years 1973/74 at MBB with a non-redundant electrical flight control system (Fly-by-Wire, FBW) with full authority for the main rotor and the tail rotor control. The test vehicle was designed for a 2-person crew: a simulation pilot and a safety pilot as the operator in command. The safety pilot sat in the rear left part of the cockpit and operated the helicopter through the mechanical control with hydraulic boosters almost similar to that in the production version. The simulation pilot sat centrally in the front part of the cockpit (see Fig. 8.4). His control inputs were converted into electrical signals and fed to the main and tail rotors through electrohydraulic actuators, together with additional signals from the control computer.

Fig. 8.3 Bo 105 ATTHeS

Fig. 8.4 ATTHeS pilots seating arrangement

The movements of the FBW actuators were mechanically fed back to the controls of the safety pilot, who was thereby always informed about the entire control input for the rotors and could monitor the control and assess their plausibility with regard to the planned flight task. The safety pilot could at any time switch off the FBW system or override the actuators and take over the helicopter command with the mechanical control system. The flight safety could be ensured this way even in the event of failure of important components of the non-redundant FBW system. In addition, an automatic safety system was installed, which monitored the bending moment of the main rotor mast and the lead-lag moments. Figure 8.5 shows the concept of the main rotor control modifications.

In the Introduction of the basic technical transcript of MBB [5], development engineer *Hans Derschmidt* describes the design principles as follows:

> The helicopter Bo 105 shall be converted to a test aircraft for VSTOL flight guidance and landing procedures. For this purpose, to increase the quality of simulation flights, the flying qualities and cockpit equipment shall be reproduced as accurately as possible, also for VSTOL aircraft which are likely to be considered. The helicopter has a crew of two pilots for simulation operation. The simulation pilot operates the helicopter with a control station that is equivalent of the aircraft being simulated, but connected only via a computer with the standard Bo 105 control ("Fly-by-Wire"). The onboard computer controls hydraulic actuators so that the Bo 105 motion corresponds to that of a pre-programmed model as best as possible. This deviation from Bo 105 flying qualities would be realized through a system of control simulation, in which the control inputs are converted into electrical signals. These signals and further signals, derived from sensors measuring the flight condition, would be converted by a special simulation computer and by a controller into Bo 105 helicopter control commands, which finally result in the modified flight behavior of the vehicle to be simulated. The differential equations of motion of the flight vehicle to be simulated are programmed in the simulation computer. The safety pilot, who is "program manager" at the same time, monitors the computer performance and can prevent critical flight conditions or limit the effect of controller malfunctions through direct intervention in the Bo 105 mechanical control. Therefore, computers and hydraulic actuators need not be redundant.

The first flight of the modified helicopter took place on July 16, 1974 in Ottobrunn and the successful first flight in FBW mode on August 22, 1974 [6].

The test helicopter could be flown in three modes:

1. Basic mode: The FBW system was turned off, only the safety pilot controlled the helicopter,
2. 1:1 FBW mode: The simulation pilot flew the host helicopter with full control authority, and
3. Simulation or VSS (Variable Stability System) mode: The simulation pilot steered the simulated helicopter with full control authority via the onboard computer.

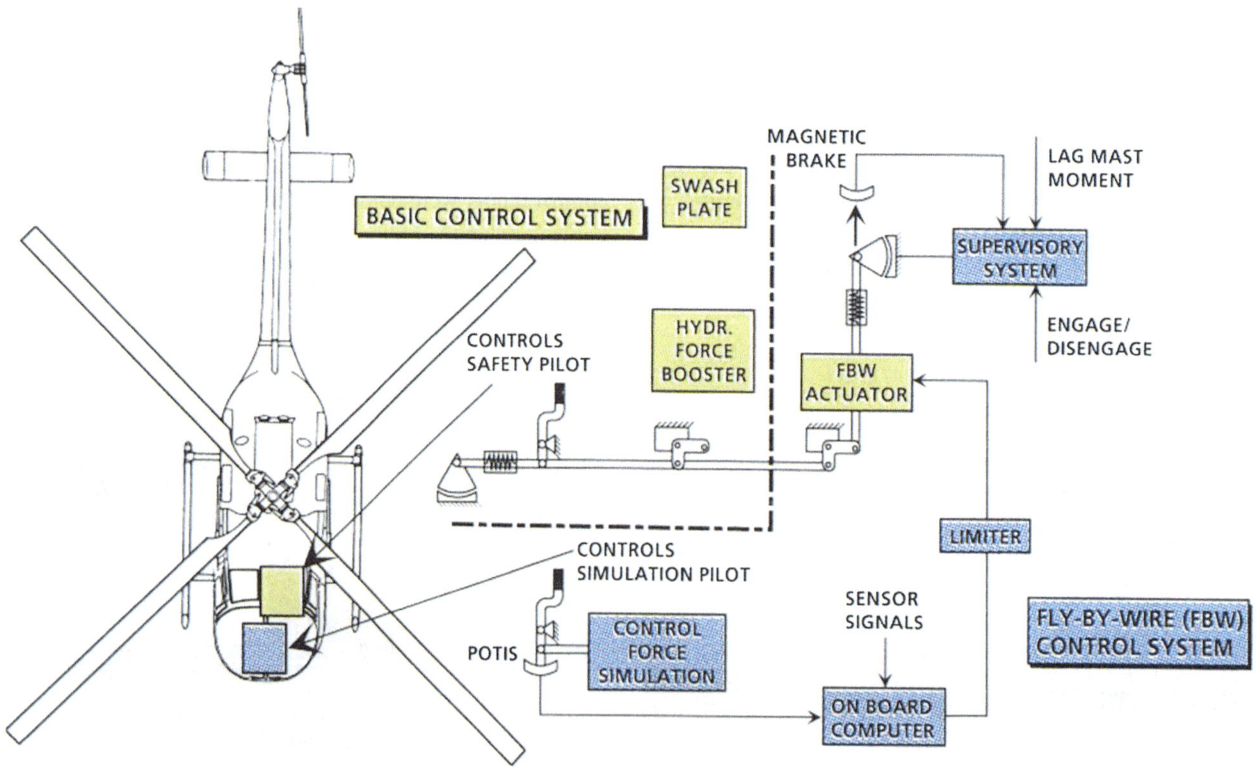

Fig. 8.5 Concept of main-rotor control modifications

In 1:1 FBW and simulation mode the flight envelope was limited to altitudes of at least 50 ft above ground when hovering and 100 ft in forward flight.

After the acquisition of the modified helicopter by DFVLR in 1982, the Bo 105, S/N 3 built in 1971 (Bo 105-S3) served as a host flying device for the helicopter in-flight simulator ATTHeS.

The next generation of civil and military helicopters should enable flight tasks with higher precision and maneuverability. These requirements were to be particularly considered in the development of ATTHeS. As the in-flight simulator capabilities depend highly on the dynamic performance of the basic flying vehicle, the high control effectiveness and fast response to control inputs of the hingeless rotor of Bo 105 were important prerequisites for the utilization of this helicopter as an in-flight simulator.

8.3.2 Tail Rotor Control with Optical Signal Transmission

From 1986, the project OPST1 (Optical Control, Phase 1) was pursued as part of a technology program of the Federal Ministry of Defense [7]. The FBW control system of the Bo 105-S3 tail rotor was replaced by an electro-optical flight control system (Fly-by-Light—FBL) and tested in flight jointly by MBB, LAT (Liebherr-Aero-Technik, today: Liebherr Aerospace) and DFVLR/DLR (see Fig. 8.6). On the basis of an existing duplex actuator of suitable dimensions, LAT designed an optically controlled "smart" actuator with integrated control electronics. The duo-duplex electronics could implement all functions in software and thereby could be changed without modifying the hardware. Besides the computations for the controller, the locally available computing power was utilized also for redundancy management and self-diagnosis purposes. Thereby the electro-magnetic compatibility was also improved and the amount of cabling considerably reduced compared to the central arrangement of actuator electronics. The smart actuator was developed and manufactured by LAT and could be driven by a triple redundant computer (see Sect. 8.4.3). The control variables generated in the triplex computer were converted into optical signals and transmitted via fiber optic cable to the electro-hydraulic actuator in the tail boom (see Fig. 8.7). The advantage of the optical signal transmission is its high immunity to electromagnetic interference, an important aspect for helicopter deployment, including those close to the ground and thereby in the vicinity of diverse transmitters [8, 9].

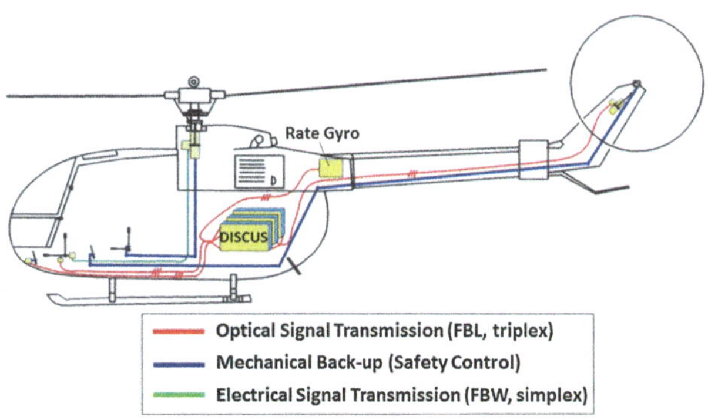

Fig. 8.6 Fly-by-Light (FBL) yaw axis control

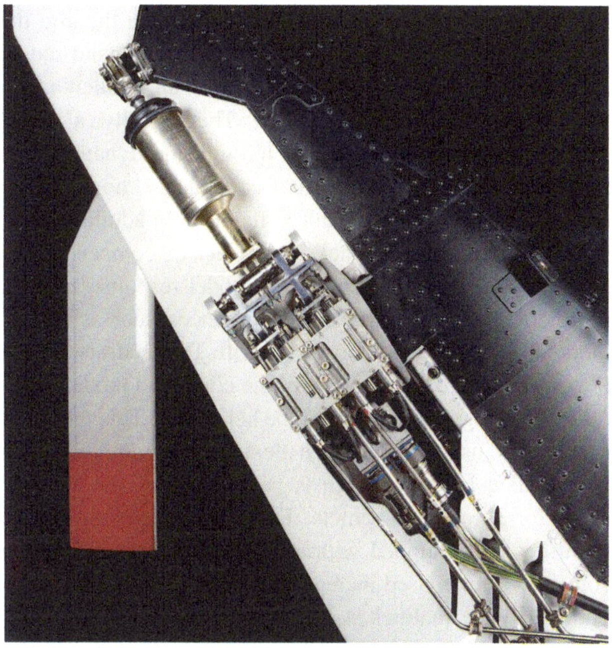

Fig. 8.7 Optically controlled actuator in tail boom

The system "smart actuator with optical transmission" was tested during 1988/1989 in conjunction with a yaw controller for heading hold (see Sect. 8.4.4) and then introduced successfully in several research programs [10].

8.3.3 Model Following Control System

The most promising and also most challenging approach to simulate the desired flying qualities, those that may differ from the host helicopter, is the development of a model following control system. In this process, the controller forces the host helicopter to follow the dynamic flight behavior of an explicit command model which is mathematically formulated in the onboard computer. The simulation pilot then flies a helicopter with properties of the mathematical model (see also Sect. 3.3).

The response of the command model due to pilot control inputs is calculated in real-time and fed to the control system. The dynamic feedforward controller, consisting of the known "inverted" model of the host helicopter, calculates the actuator inputs for the host helicopter with the aim that the host helicopter now responds like the command model. Figure 8.8 shows the basic structure of an explicit model following control system (MFCS). Theoretically, a perfect simulation in flight is achieved through the dynamic feed-forward. In the practical realization, however, a proportional-integral regulator is required in addition to the state feedback (PI feedback controller) to compensate the deviations between the command model responses and the host helicopter due to external disturbances, such as gusts or model inaccuracies and long-term drift. The feedforward and PI controllers are independent of the command model and of the current flight conditions of the helicopter. This method is particularly useful when high flexibility is required in the commanded models, which is particularly of great importance for a research vehicle. Even in modern operational flight devices equipped with a FBW/L control, this type of control system is used increasingly with the aim of realizing optimal flying qualities at different flight conditions [11].

The development of explicit model following control system began during 1983/84 in the *Vertical Motion Simulator* (VMS) at NASA Ames research center in the USA (see Fig. 8.9) as part of a joint transatlantic research program "US/German Memorandum of Understanding on Helicopter Flight Control" between US Army/NASA and DFVLR (see Sect. 12.3.3) [12]. The first simulator results showed a strong dependency of the model following quality on the command model dynamics. Improvements in the model bandwidth (the frequency range of the model response) led to higher demands on the controller. It was, therefore, necessary to account for the dynamic response of the actuators and their position and/or actuation rate limits in particular. The control system was developed for the helicopter Bo 105 and for UH-1H V/STOLAND (see Fig. 8.10) and tested in the simulator [13]. The results showed for both helicopters significant performance improvements at reduced pilot workload for specially defined dynamic flight maneuvers, namely fly over and around obstacles.

After implementation and evaluation of this concept in the variable stability research helicopter CH-47 of NASA (see Fig. 8.11) [14] (see also Sect. 5.2.3) and in the helicopter Bo 105 ATTHeS of DFVLR (see Fig. 8.12), it became apparent that further improvements in the original MFCS design were necessary. As a result of higher order effects, such as the rotor flapping motion dynamics as well as of the sensor filters and the time delays in the computers, large delays were observed between the control inputs and the helicopter reaction. It turned out that the efficiency and accuracy of the model following control system depend highly on the accuracy of the mathematical model of the host helicopter. The better the dynamics of the host helicopter and its systems are known, especially the short-time response, the more accurately the elements of the feedforward gain matrix can be calculated [15]. Additional factors are also of importance for the performance and the accuracy of the entire system, namely the dynamics of the pilot control devices, the shape of the pilot control inputs, the dynamics of actuators and sensors, and the processing of the electrical signals.

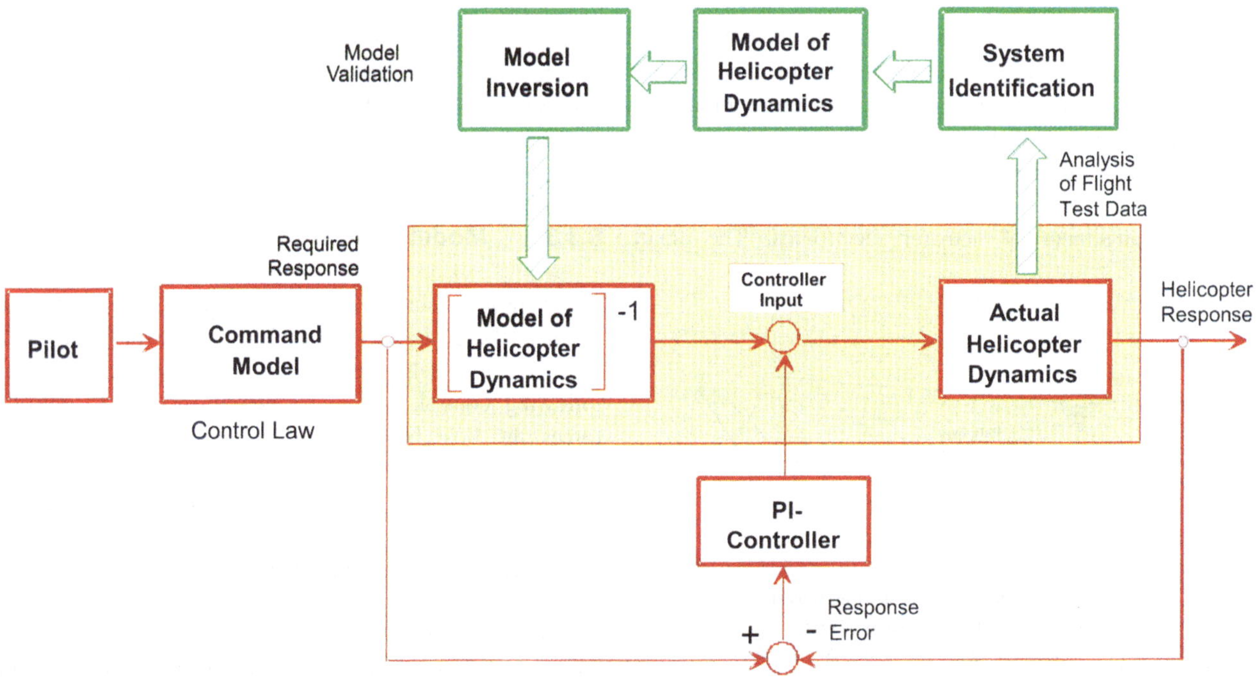

Fig. 8.8 Structure of an explicit model following control

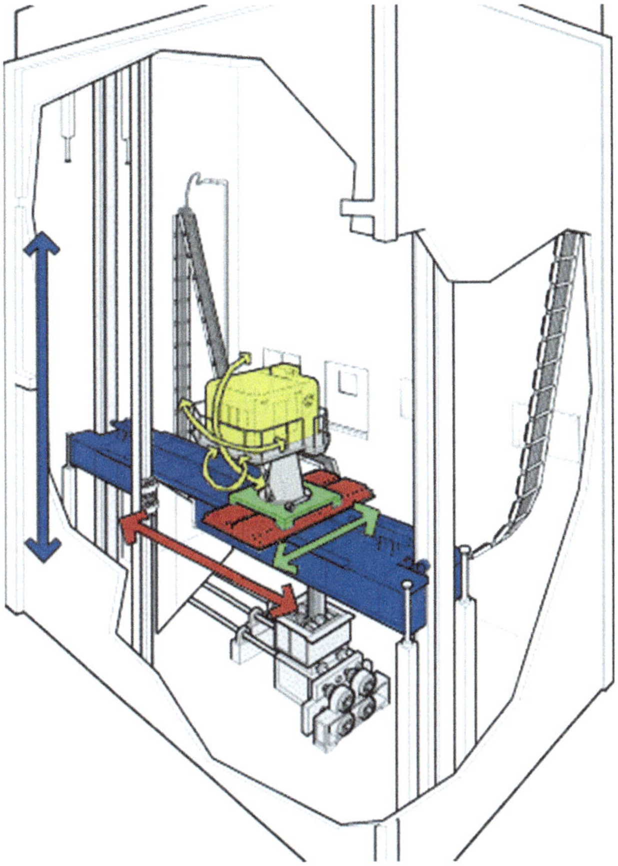

Fig. 8.9 Vertical motion simulator (VMS) of NASA (*Credit* NASA)

Fig. 8.11 Test helicopter CH-47B (NASA 737)

Fig. 8.12 NASA test pilot *Ron Gerdes* (*left*) and flight test expert *Ed Aiken* in front of Bo 105 ATTHeS (Manching, May 1984)

Fig. 8.10 Test helicopter UH-1H V/STOLAND (NASA 733)

The effect of unmodeled rotor dynamics is shown in Fig. 8.13. The three time-history plots on the top show for the in-flight simulator ATTHeS a comparison of the behavior demanded by the command model and the measured behavior in the roll axis due to lateral stick input for a 6 degrees of freedom rigid-body model of the host helicopter (Bo 105). The deviations between the desired and measured responses are significantly reduced by accounting for the rotor dynamics in the model of the host helicopter (8 degrees of freedom) and for the corresponding adaptation of the feedforward gain matrix as seen from the lower diagrams in the same figure [16].

After further improvements, for example, through powerful system identification methods to improve model fidelity using flight test data, an explicit model following control system was developed for the in-flight simulator ATTHeS that was tuned for airspeeds between 40 and 100 knots and for hovering [17–20].

The design of the model following control system was carried out substantially in four steps:

1. Definition of the mathematical model structure of the host aircraft including rotor dynamics at high frequencies,
2. Determination of the parameters of the defined model applying system identification methods or by simulation programs,
3. Determination of the feedforward structure through formal inversion of the defined model of the host flying vehicle, and
4. Definition of the feedback controller structure, optimization of the overall performance in simulation and confirmation in flight test.

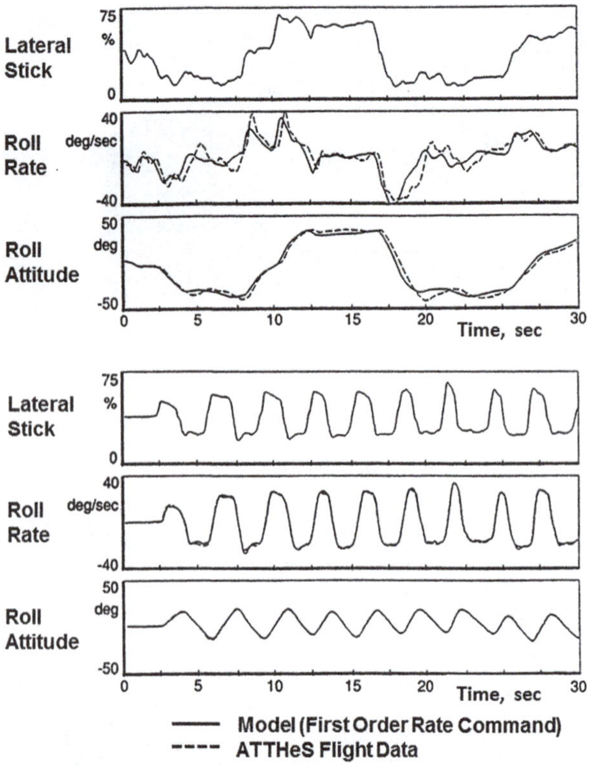

Fig. 8.13 Influence of rotor dynamic modeling on the simulation accuracy (*top*: 6° of freedom rigid-body model of Bo 105, *bottom*: 8° of freedom model accounting for rotor dynamics

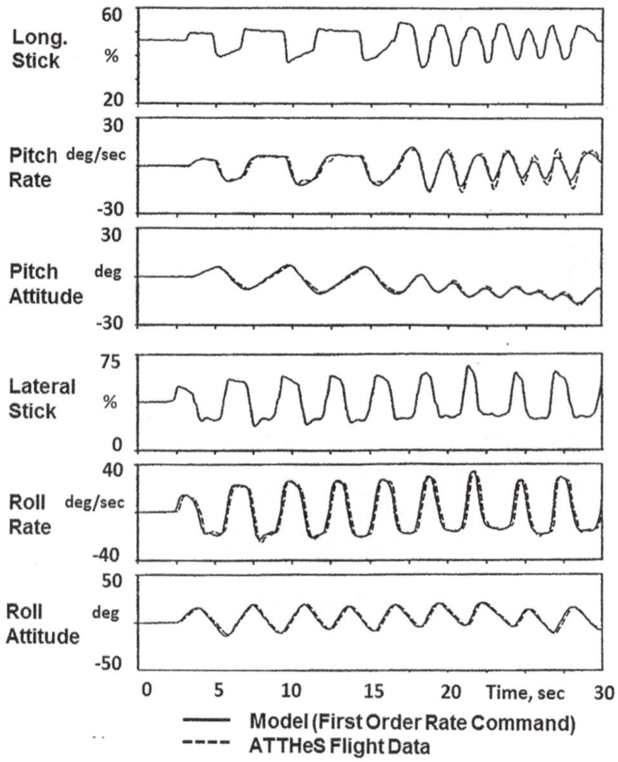

Fig. 8.14 ATTHeS model following quality

Poor model following quality results in differences between the actual behavior of in-flight simulator and the reaction demanded by the command model. The achieved model following quality is shown in Fig. 8.14; both flight-measured rotational rates, as well as the attitude angles in the roll and pitch axes, show satisfactory match with the model. The achieved decoupling between the roll and pitch axis compared to Bo 105 without control system is depicted in Fig. 8.15; here too the good controller performance is obvious [21].

To assess the required model following quality, a criterion was defined in the frequency domain that was based on the flight dynamics not perceived by pilots (unnoticeable dynamics). As an example, Fig. 8.16 shows the ratio of the roll rate of ATTHeS to the command model for dynamic pilot control inputs (transfer function). In an ideal model following case, the amplitude of the so-called error function will be 0 dB and the phase angle 0° over the entire frequency range. The boundary curves show ranges in which the pilot discerns significantly or does not notice respectively the differences in the flying qualities. For a good model following performance, the ratio of the roll rates has to be within the defined range for the unnoticeable dynamics. The controller based on the model taking into account the rotor dynamics fulfilled this requirement [22].

8.3.4 Onboard Computer and Measurement System

An onboard computer and a data acquisition system were constructed and installed in the helicopter to enable implementation of a flight controller for in-flight simulation. Considering the limitations of existing technology available during 1980 and subsequent years, the following requirements had to be accounted for: (1) available space in the helicopter is very limited, (2) software changes in control system had to be carried out on a so-called host computer on ground, (3) software modifications were to be introduced in flight only after verification in a system simulation on ground compatible to the onboard system, (4) controller tasks of the onboard system and the evaluation of the system performance must be clearly separated, and (5) flight tests are to be tracked and managed from a ground station. A block diagram of the onboard system is shown in Fig. 8.17. Two computers, which were "hardened" for the conditions during flight tests, were assigned to the data acquisition and the control/regulation tasks and permitted a largely independent data transfers of both the tasks. The control inputs by the simulation pilot and the necessary state variables for the control system were generated directly from the sensor signals with a sampling rate of 25 Hz. The total processing time for the command model and the control laws was 7 msec. The computer for data gathering was equipped with a 64-channel analog/digital converter, and all

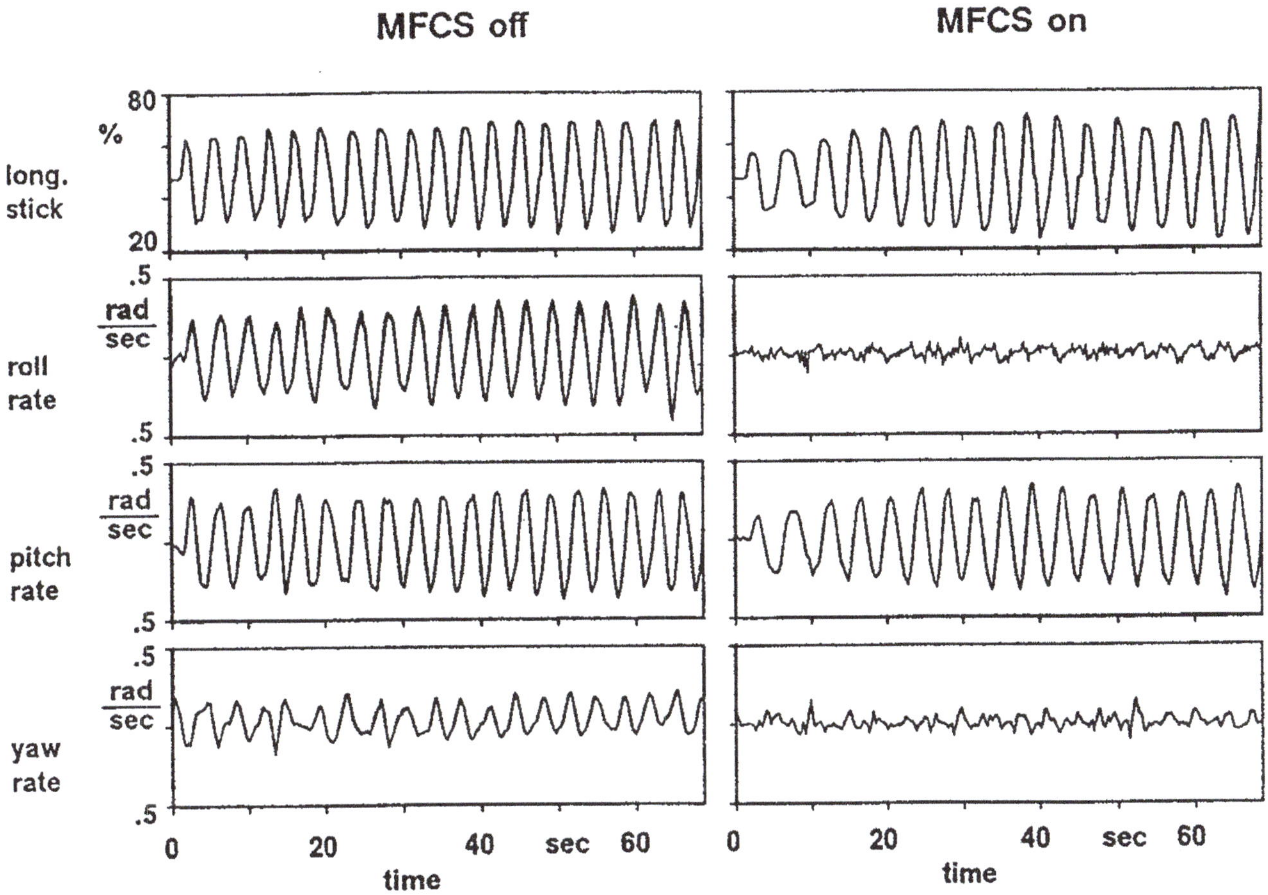

Fig. 8.15 ATTHeS decoupling between roll and pitch axis (*MFCS* model following controller)

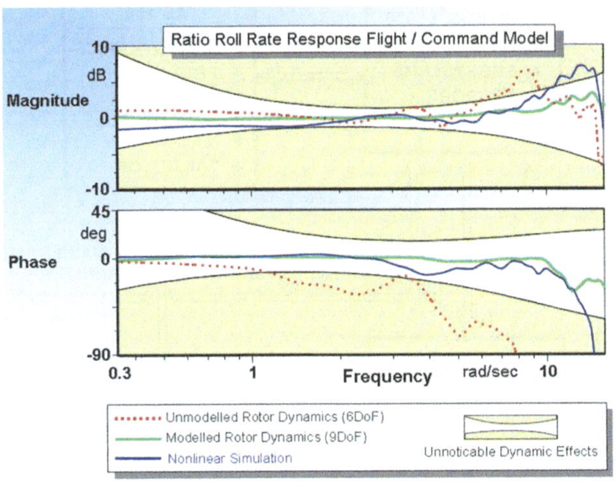

Fig. 8.16 Influence of modeling on the model following quality

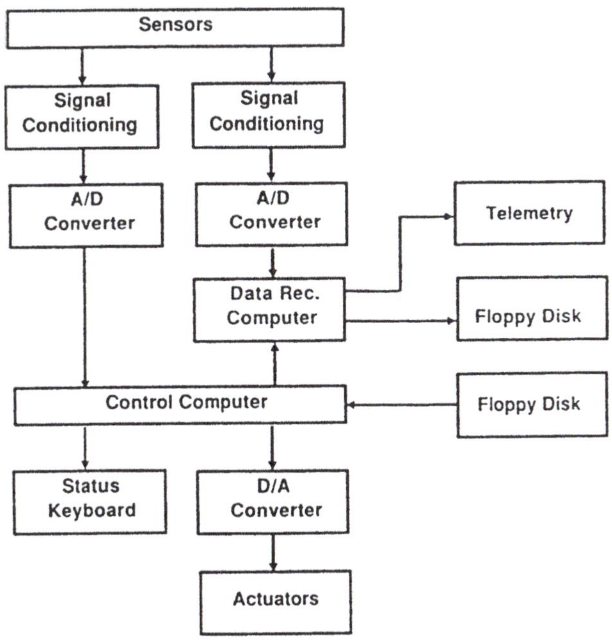

Fig. 8.17 Onboard computer system for data recording and flight control

sensor signals sampled at 100 Hz. The much higher sampling rate compared to the control computer was chosen to enable a more precise evaluation of overall system performance.

Furthermore, the so-called aliasing errors were thereby eliminated, which arise due to the digitization of higher frequencies at too low a sampling frequency. Both computers were connected through a memory with two inputs (*dual ported memory*), through which all data were gathered and recorded on an onboard floppy disk. Moreover, data were transmitted via telemetry and used in the ground station for quicklook. A computer compatible with the onboard control computer was available in the ground station on which all software modifications could be carried out and then transferred to the onboard computer using a floppy disk.

The software of the control computer consisted of two parts, the system level and the user level. The system software with real-time control, the input-output operations and pilot information was programmed in assembler programming language to minimize the computing time. The user software with the control laws, the command model and signal conditioning required for the controller was programmed in Fortran and C languages [23].

8.3.5 Ground-Based Simulator and Ground Station

Before the software could be used in flight, it had to be tested on the ground in real time to ensure the compatibility and to avoid extreme control amplitudes. A real-time simulator was, therefore, designed for the complete ATTHeS system, including the actuators and sensors, see Fig. 8.18. A nonlinear helicopter simulation program was implemented on a special simulation computer (Applied Dynamics International AD100), that computed all model elements in a cycle time of 2.5 msec. Thereby simulated signals of sensors and actuators were generated and used as input for a duplicate of the onboard control computer, on which the same software was run as that on the onboard computer.

The ground-based simulator was mainly used for functional testing of the software for the in-flight simulator. Therefore, it was adequate to use a simple cockpit with conventional helicopter controls and a display for the information to the pilot [24]. With this simulation on the ground, engineers and pilots could be trained to handle the in-flight simulator and could check the proper functioning of the software. After successful ground tests, flight testing was envisaged.

The limited space in the helicopter necessitated the construction of a ground station for the engineers as an integral part of ATTHeS system. The ground station contained the following facilities: (1) host computer for onboard systems, (2) PC for continuous recording of telemetry data and presentation of the helicopter position on a local map, (3) two PCs for quicklook representation of ten signals on each, (4) 3-dimensional visualization of the helicopter motion,

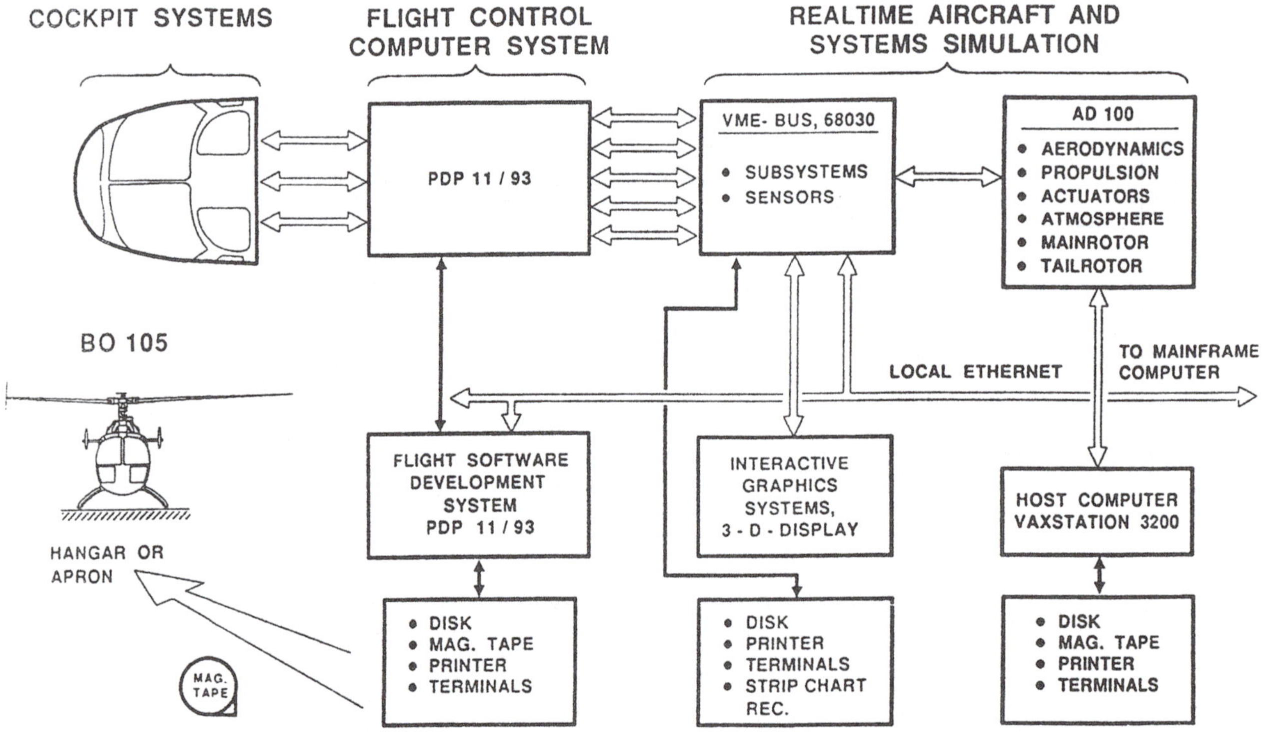

Fig. 8.18 Real time ground based simulation

calculated from the telemetry data, (5) terminal for real-time display of the helicopter position above ground from the laser tracking data, (6) TV monitor to display images from a camera mounted on the tracking antenna, and (7) computer for off-line analysis of flight data. The engineers were able to observe and guide the flight tests from the ground station. Furthermore, the mobile equipment facilitated flight tests at different places outside DLR [25].

The ATTHeS equipment and layout of the ground facilities was implemented at DLR in several successive stages from 1983. Corresponding to the achieved state of these facilities, different utilization programs were planned and carried out, which delivered the first important results, and were also fundamental for the further ATTHeS system expansion. An international symposium on in-flight simulation was held in July 1991 at DLR in Braunschweig, during which the experts shared their experiences, and various test aircraft and their applications were presented and discussed [26]. A special highlight during this period was the first flight over the Brocken in Harz mountains, shortly after the German reunification (Fig. 8.19).

Fig. 8.19 ATTHeS over Brocken in Harz mountains

8.4 Utilization Programs

8.4.1 Flying Qualities Investigations

8.4.1.1 Variation of Control Characteristics

After equipping the Bo 105-S3 with the necessary sensors, data processing, and digital computer for actuators control (see Sect. 8.3), first flights with digital control were conducted starting March 1983. Thereby the influence of the control characteristics on pilot workload was investigated during certain flight tasks. For example, the damping and control sensitivity in the roll axis were varied to assess the helicopter dynamic properties during flying close to the ground around obstacles (“*slalom task*”). Figure 8.20 shows results of this evaluation compared to previous studies [27]. Heretofore, appropriate data with high control sensitivity were not available for helicopters.

8.4.1.2 Contributions to Flying Qualities Criteria

One of the key objectives of flight mechanics research is to provide reliable data for the definition of flying qualities guidelines. The certification criteria for civil helicopters, for example, the documents CS-27 and CS-29 of EASA (*European Aviation Safety Agency*), contain general requirements pertaining to control and stability in order to ensure flight safety at all flight conditions. In the current guidelines for military helicopters, for example, ADS-33E-PRF of the US Army [28], detailed quantitative and qualitative criteria for the completion of a particular task or mission are defined in addition, the so-called mission-oriented flying qualities criteria. They also account for the integration of modern cockpit equipment and control systems, as well as flights under limited visibility conditions [29].

Development of new flying qualities guidelines was pursued by the US Army/NASA from roughly 1980 onwards, once the shortcomings of the flying qualities requirements

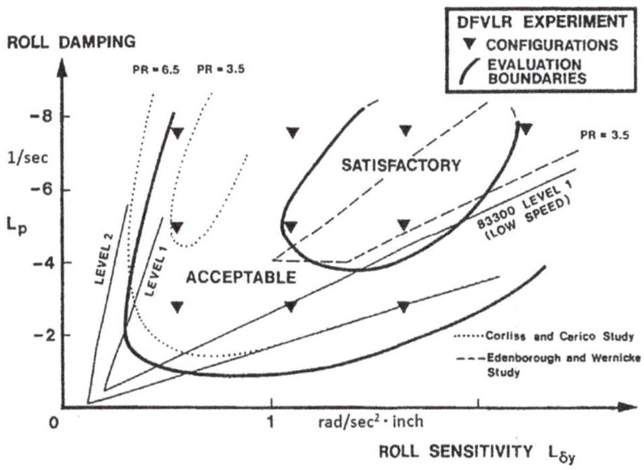

Fig. 8.20 Assessment of roll control sensitivity

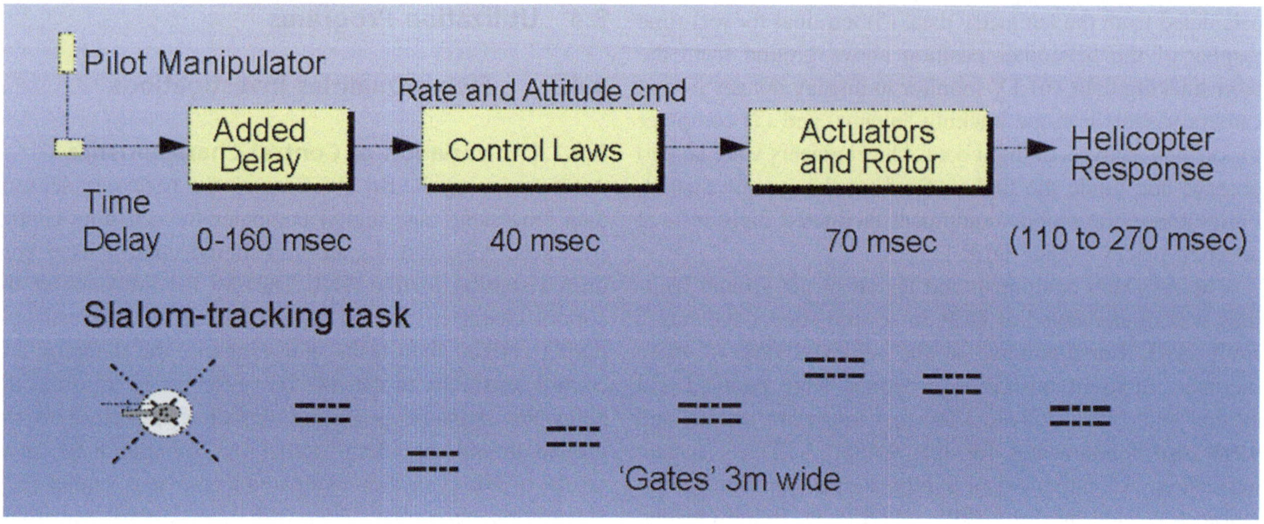

Fig. 8.21 Slalom tracking course

MIL-H8501, which were applied to military helicopters since 1952, were discussed sufficiently and established. An essential prerequisite for this purpose was the availability of an adequate, systematic and reliable data base. In order to generate this, collaboration was sought with other research organizations and institutions. The Flight Research Laboratory of National Research Council (NRC) in Canada (see also Sect. 5.3), the Royal Aircraft Establishment (RAE) in the UK and the Institute of Flight Mechanics of DFVLR supported the proposition in the subsequent years through flight test data and scientific contributions [30].

In collaboration with the US Army, systematic tests were carried out to verify and optimize the new flying qualities criteria. For this purpose, the ground-based simulator VMS (*Vertical Motion Simulator*) of NASA and the in-flight simulator ATTHeS were deployed complementarily (see also Sect. 12.3.3). As an outcome of this effort, important key data were generated and insights gained for the new *Aeronautical Design Standard* ADS-33. Flight tests with ATTHeS have significantly contributed to this joint venture [31].

Time delays appearing in helicopter response to pilot control inputs, for example, due to computational times in the control computer or by run-times in the control system, can lead to undesirable and dangerous couplings between the pilot and the helicopter (*Rotorcraft-Pilot-Coupling* - RPC) during difficult maneuvers [32]. To investigate these effects, a flight test was planned in which several pilots flew a slalom course defined by markings on the ground (see Fig. 8.21). The suitability of the helicopter characteristics for this particular task was rated based on the standardized *Cooper-Harper* scale. Along the marked course so-called tracking phases through the 3-meter-wide gates alternating with transition phases, which required rather lower frequency control inputs. The deviations from the prescribed flight path

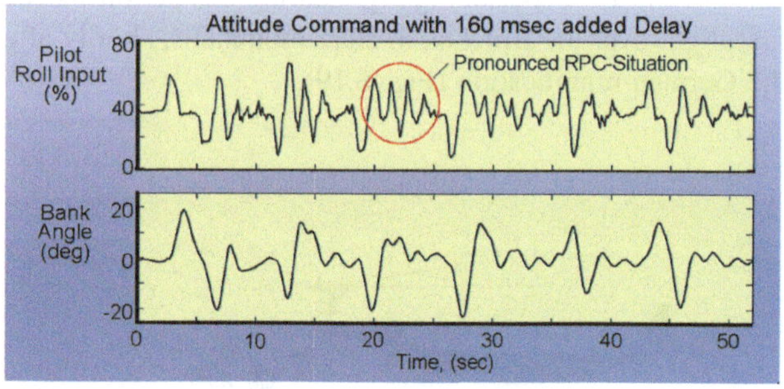

Fig. 8.22 Flight tests with additional time delay

were measured with a laser tracking system. Following each evaluation phase, the properties of ATTHeS were systematically varied. Additional delays in the helicopter response up to 160 msec could be inputted and different controller characteristics were programmed. Figure 8.22 shows the time histories of a test flight with an attitude command controller and an additional time delay of 160 msec. The individual phases can be clearly observed in the pilot control and in the roll attitude, also the strong dynamic in the control while flying through the gates due to RPC ("*pronounced RPC situation*"). The pilot commented on the flight as follows: "roll, pilot induced oscillation" and "very poor configuration". He assessed the flying qualities as objectionable with tolerable deficiencies, whereby adequate flight operation could be achieved only with considerable pilot compensation.

For such high precision tracking flight tasks, for example during landing, or during target tracking, two parameters are of prime importance, namely phase delay and bandwidth of the helicopter response to pilot control inputs. The correlation of both parameters could be determined from the systematic ATTHeS flight tests for this flight task. The DLR

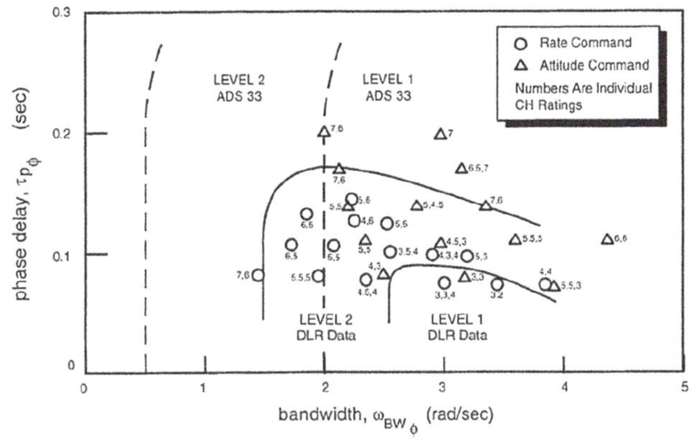

Fig. 8.23 Flight test data for new flying qualities criteria

data and the changed boundary curves between the flying qualities levels "satisfactory without improvement" (Level 1), "acceptable, warrant improvement" (Level 2), and "unacceptable, improvement required" are depicted in Fig. 8.23 [33]. The criteria in *Aeronautical Design Standard ADS-33* were modified based on these results.

Because of the specific properties of the Bo 105 and because of the flexible model following control system and the modular equipment, the in-flight simulator Bo 105 ATTHeS played a significant role in the formulation and verification of the new criteria, which are meanwhile applied worldwide to define and assess the mission-oriented flying qualities of military and also civil helicopters [34, 35].

8.4.2 In-Flight Simulation of Other Helicopter Types

The essential purpose of a simulator is manifold, for example, a verification of flying qualities of a helicopter in its design stage or after modifications, optimization of controller and equipment components of existing helicopter, and familiarization and training of pilots. For this purpose, often different simulators are successfully utilized, which are specifically customized for a particular task. In spite of high technical efforts, however, it is not always possible to reproduce realistically all factors in ground-based simulators, in particular the high-stress situation for a pilot in difficult flight phases. This aspect can be investigated more readily with in-flight simulators under realistic flight conditions. But, in fact, the more dangerous and highest stress scenarios at the edges of the flight envelope are best tested first in the ground-based simulator before they are finally evaluated in flight.

For the simulation of different helicopters with ATTHeS, a special linear mathematical model was developed, whose parameters were determined from generic simulation models or from flight tests, applying system identification methods. The nonlinear terms required for a continuous flight with large variations in flight conditions were explicitly programmed. These terms come from coordinated turns, gravitational terms, changes in the trajectory and Euler angles defining the helicopter position in the airspace. In addition, a 4-axis flight controller (*Stability Command Augmentation System*—SCAS) was programmed, with which the investigation of SCAS failures was possible.

As an example, the in-flight simulation of the helicopter Westland Lynx is presented. Due to the opposite direction of rotor rotation (clockwise as seen from the top), this helicopter has different coupling properties compared to the Bo 105 host helicopter of ATTHeS. Figure 8.24 shows an acceleration/deceleration maneuver at a constant altitude and constant heading. The speed was varied by changing the pitch attitude. The control inputs of the simulation pilot and the ATTHeS-actuator activity (= output of MFCS) show significant deviations, except for the collective control (graphs on the right). The reason for this deviation is attributed to opposite coupling of pitch rate due to collective input in the case of Lynx helicopter compared to the Bo 105. All the ATTHeS flight variables agreed well with those commanded of the Lynx model (graphs on the left-hand side). The pilots assessed the ATTHeS in-flight simulation as a representative of the Lynx helicopter [36].

Figure 8.25 depicts a case of control system failure in the simulated helicopter. In a right turn, after 10 sec in Fig. 8.25, the longitudinal SCAS malfunctions, and as a consequence the pitching motion is now unregulated and unstable, as is the case with most helicopters. After another 10 sec, the activity of the simulation pilot in the longitudinal control increases considerably, the Lynx helicopter starts to oscillate about the pitch axis. Even this situation could be replicated exactly with ATTHeS and was assessed by the pilots as representative of the real flight case.

8.4.3 Fault Tolerant Computer System (DISCUS)

The future demands on the spectrum of helicopter operations necessitate integrated digital flight guidance/flight control systems with full control authority. Such systems have to meet the same or higher demands on operational safety as in the case of conventional mechanical-hydraulic controls. Safety in this context implies fault tolerance, which can be achieved by multiple duplications with appropriate redundancy management for failure detection and elimination. In the DISCUS project (*Digital Self-healing Control for Upgraded Safety*), MBB and Liebherr-Aero-Technik, together with DLR, have developed systems with fault-tolerant features. These were implemented and flight tested with ATTHeS [37].

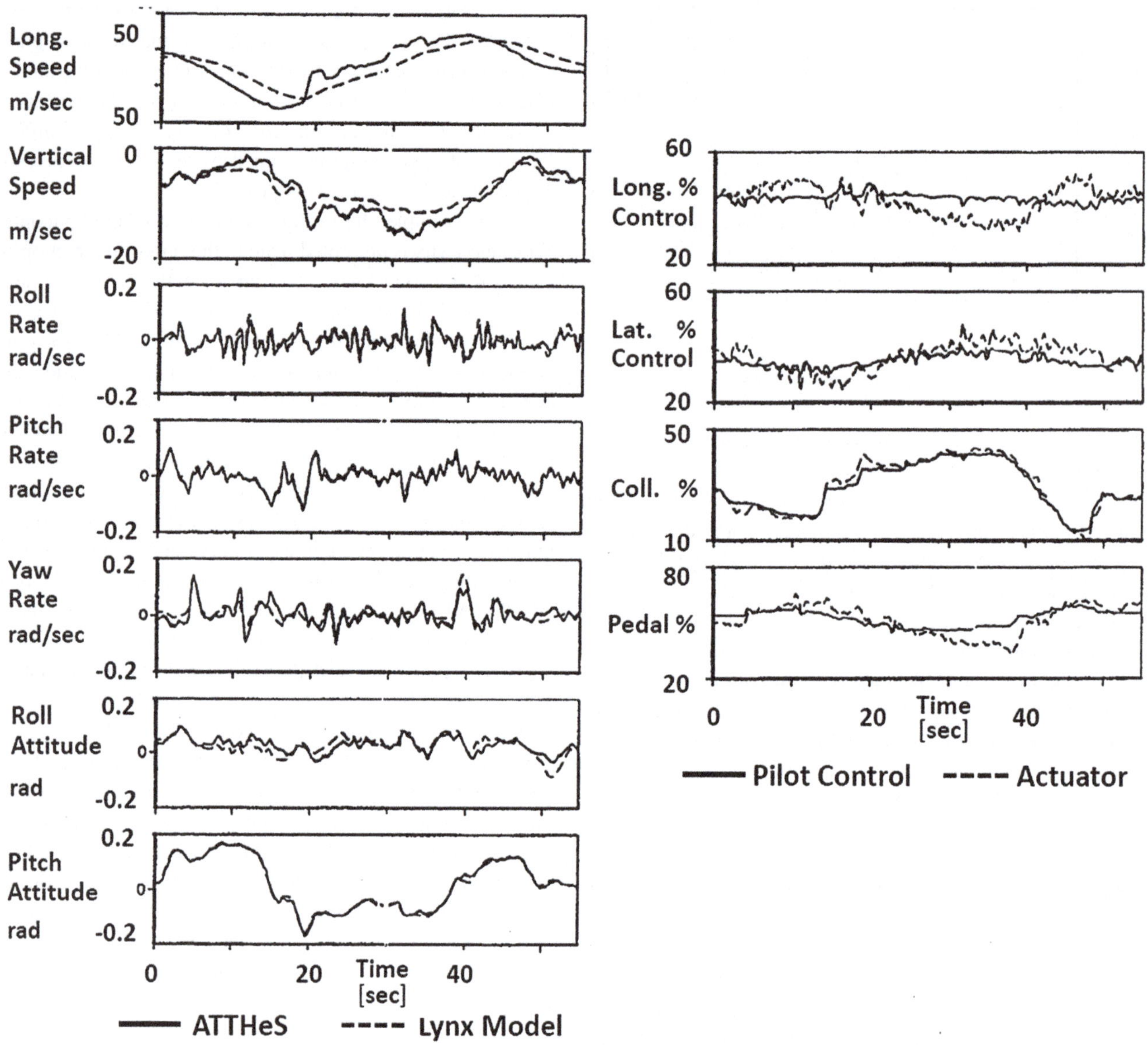

Fig. 8.24 ATTHeS simulation of Lynx helicopter

One of the essential building blocks of the system was the DISCUS computer, a fault-tolerant modular multiprocessor system (see Fig. 8.26). The ability of the computer to detect a first fault or failure, to isolate, to display and to overcome them (*One-Fail-Operational capacity*) could be accomplished by a triplex structure. Each computer channel was housed in its own casing, and the failure detection was ensured by a majority decision between the three identical channels (see Fig. 8.27). The data exchange between the parallel computer channels was by means of optical communication. Thereby the interferences due to other electromagnetic signals could be eliminated [37].

The realization of the flight control computer DISCUS featuring the required degree of redundancy and its testing in flight broke new ground at that time (before 1990). Thereby important insights could be obtained into the redundancy structure, fault tolerance and failure detection of flight critical systems [9].

8.4.4 Design of Flight Controllers

8.4.4.1 Yaw Controller for Pilot Workload Reduction

Using the tail rotor control modifications in the Project OPST1 (see Sect. 8.3.2) and the DISCUS computer (see Sect. 8.4.3), a yaw controller was developed, integrated and tested in ATTHeS jointly by MBB, LAT, and DLR. The objectives of the project were: (1) to improve the accuracy of helicopter control, (2) to reduce the interferences, for

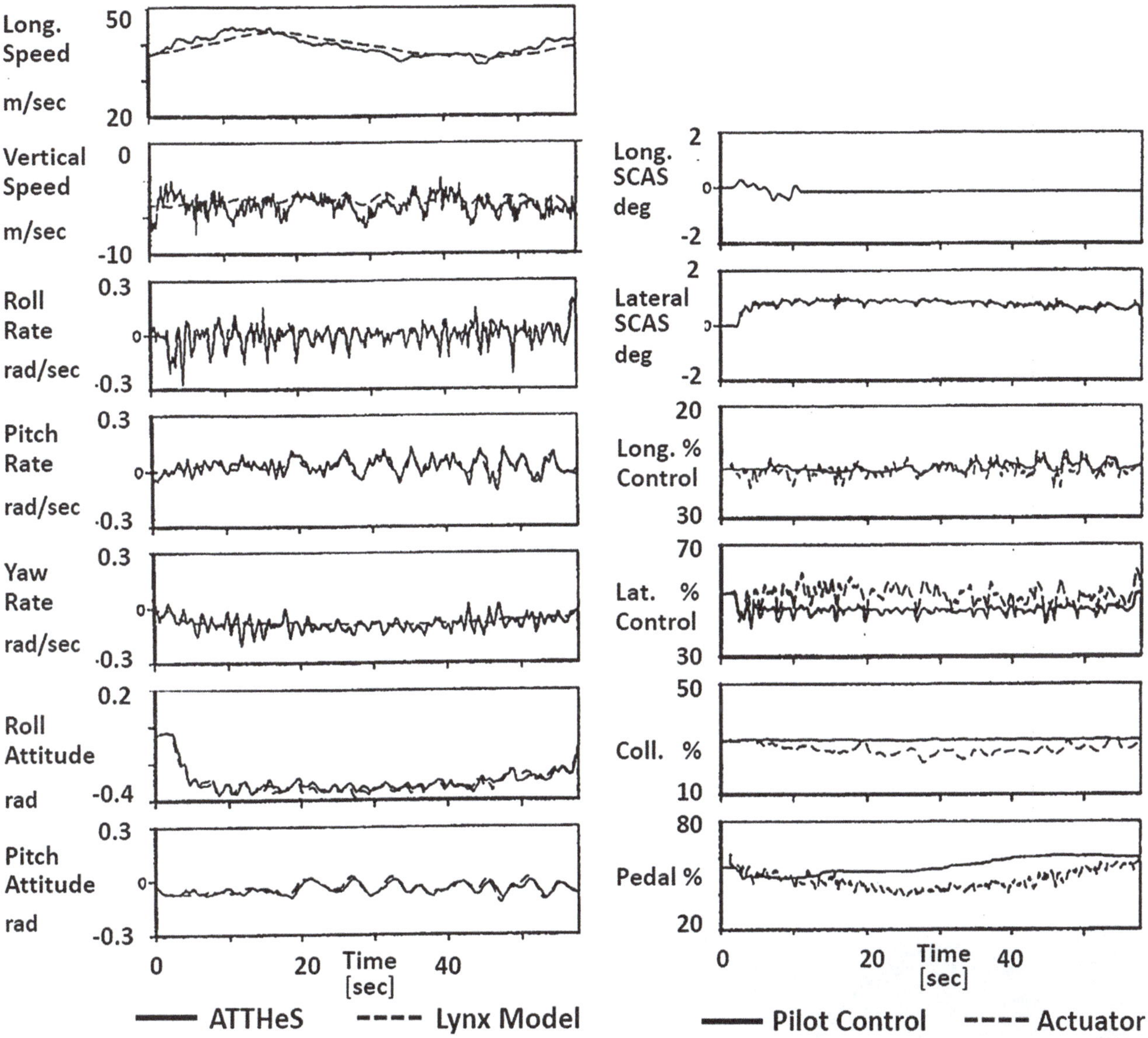

Fig. 8.25 SCAS failure in Lynx Model

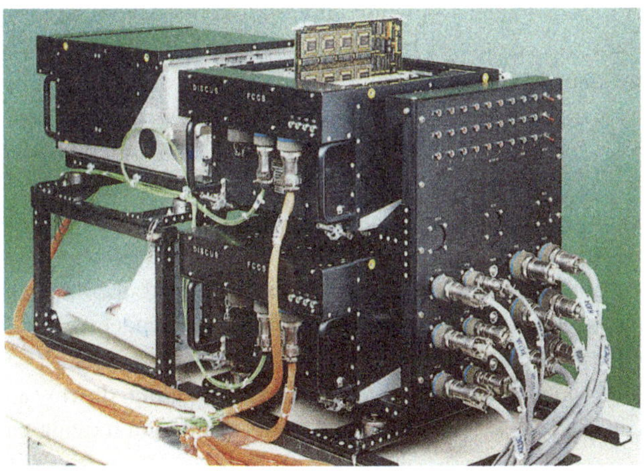

Fig. 8.26 Airworthy hardware of fault tolerant DISCUS computer

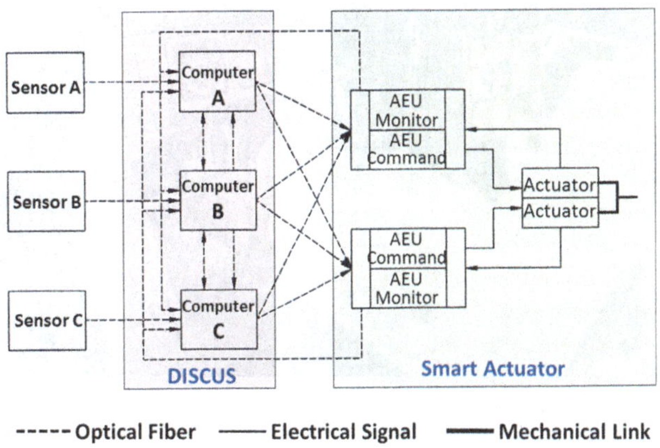

Fig. 8.27 Redundancy structure of DISCUS system

example, caused by strong gusts, and (3) to minimize the coupling between pilot collective control inputs and the helicopter yawing motion. Thereby the pilot could be significantly relieved, especially in difficult flight tasks and under bad weather conditions. The high actuation authority for the yaw controller, necessarily required to meet these objectives, demanded adequate fault tolerance of the entire yaw control system. According to the redundancy structure shown in Fig. 8.27, with the exception of the actuator and the hydraulic energy supply, all elements of the yaw control were tripled. They included the position encoders for actuation commands for the collective blade pitch angle on the main rotor, pedal position encoder for blade angle of the tail rotor, rate gyro to measure the yaw rate, control computer, and electric power supply.

The flights for testing and optimization of the systems took place during 1989 in Braunschweig. Several pilots evaluated the flying qualities in terms of the desired reduction of workload during difficult flight phases and maneuvers. The helicopter cockpit was equipped with electronic displays so that commands could be easily displayed, which the test pilot had to follow in accordance with the defined task (see Fig. 8.28).

The flight test data in Fig. 8.29, pertaining to a hover while maintaining the heading in the presence of crosswinds from left (wind 17 knots from 190°), clearly illustrate that the pilot required significant pedal control inputs to maintain the course with the unregulated helicopter. After switching on the yaw controller, pedal inputs were not required. The controller takes over the tasks of minimizing the course deviations and reducing the yaw rates resulting from disturbances. The targeted pilot workload reduction was confirmed by their comments [38].

Fig. 8.28 ATTHeS Cockpit with primary-flight-display and NAV-display

8.4.4.2 Autopilot Functions

Control systems with autopilot functions are increasingly installed in modern, especially the larger helicopters. For research in this field with ATTHeS, a corresponding navigation task was defined, namely to fly autonomously, at constant altitude and airspeed, a predefined course marked by fly-over-points.

The existing explicit command model, including speed command and attitude hold in forward flight, was extended by controllers for altitude, airspeed, and heading. A software module was programmed and integrated in the onboard computer, that computed the heading command as a function of the current and desired position and the actual wind using a wind estimator and GPS data. The fly-over-points and the resulting flight path in the vicinity of Braunschweig are illustrated in Fig. 8.30. The flight was carried out autonomously (with safety pilot) at constant altitude and airspeed, without pilot intervention, except pressing a button to initialize the system. The flight tests took place in 1993 [39].

8.4.4.3 Position Hold

The design of flight controllers for helicopters in hover or at low speeds is an essential prerequisite for expansion of flight operations, such as the police operation, rescue flights or offshore supply flights. A specific task is maintaining the helicopter position over a fixed or moving target in the presence of wind and gusts, such as over a ship deck or a lifeboat in stormy seas.

For these investigations, ATTHeS was equipped with an innovative measurement system for hovering over a target. A video camera and a computer for processing the optical information was used as an integrated sensor to determine the helicopter position relative to the target. The existing ATTHeS model following control system was modified for the requirements of position hold. In a helicopter in hover, the longitudinal and lateral accelerations can be controlled by variations in the pitch and roll attitude. Additional terms were necessary in the MFCS equations for position hold, which formulated the relations between the commanded pitch and roll attitude changes, the helicopter position relative to the target, and the corresponding speeds.

Before the first flight tests for position hold, the hardware and software were integrated and tested under real-time conditions in the ground-based simulator (see Sect. 8.3.5). The video camera on the helicopter pointing downwards captured the target, which was represented by a moving car.

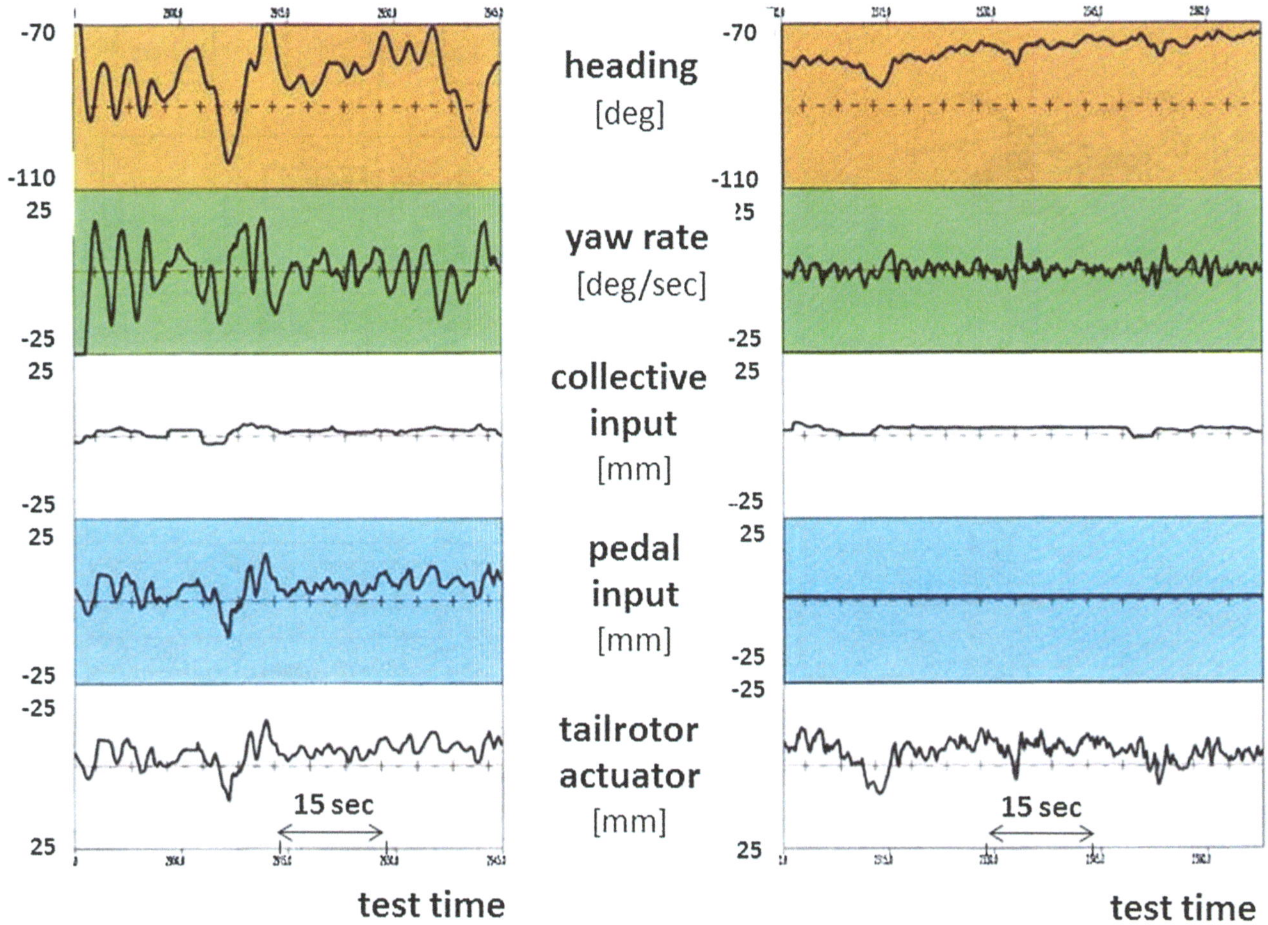

Fig. 8.29 Course hold in hover during wind from left; without controller on *left side*, and with controller on *right*

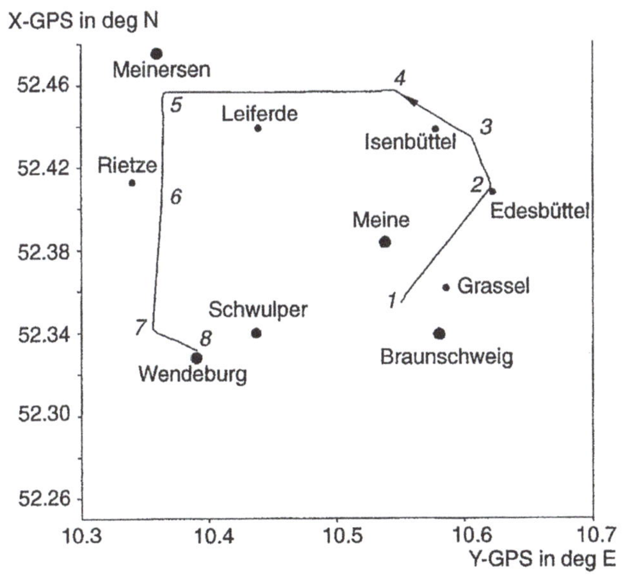

Fig. 8.30 Autonomous flight path control (fly-over points and track)

The camera signals were analyzed in the onboard computer, which then provided the control computer with position data of the vehicle. The control system steered the "observing" helicopter so that it flew directly over the target and followed the target movements at constant height and constant heading, without any pilot intervention (see Fig. 8.31). The flight tests took place in March 1994, with a public demonstration on March 16, 1994, partly under stormy weather conditions (15 knots of wind with gusts up to 30 knots, see Fig. 8.32). The vehicle on the ground drove in circles of about 40 m radius at a constant speed. Once the pilot had activated the system, ATTHeS followed autonomously the target with a standard deviation of not more than 1.6 m. Figure 8.33 shows the time histories of the x and y position coordinates and the pilot controls dx and dy, which were not operated manually during the test flight [40]. Corresponding flight tests with autonomous flying were unknown until then, and as such, the demonstration with ATTHeS attracted attention worldwide.

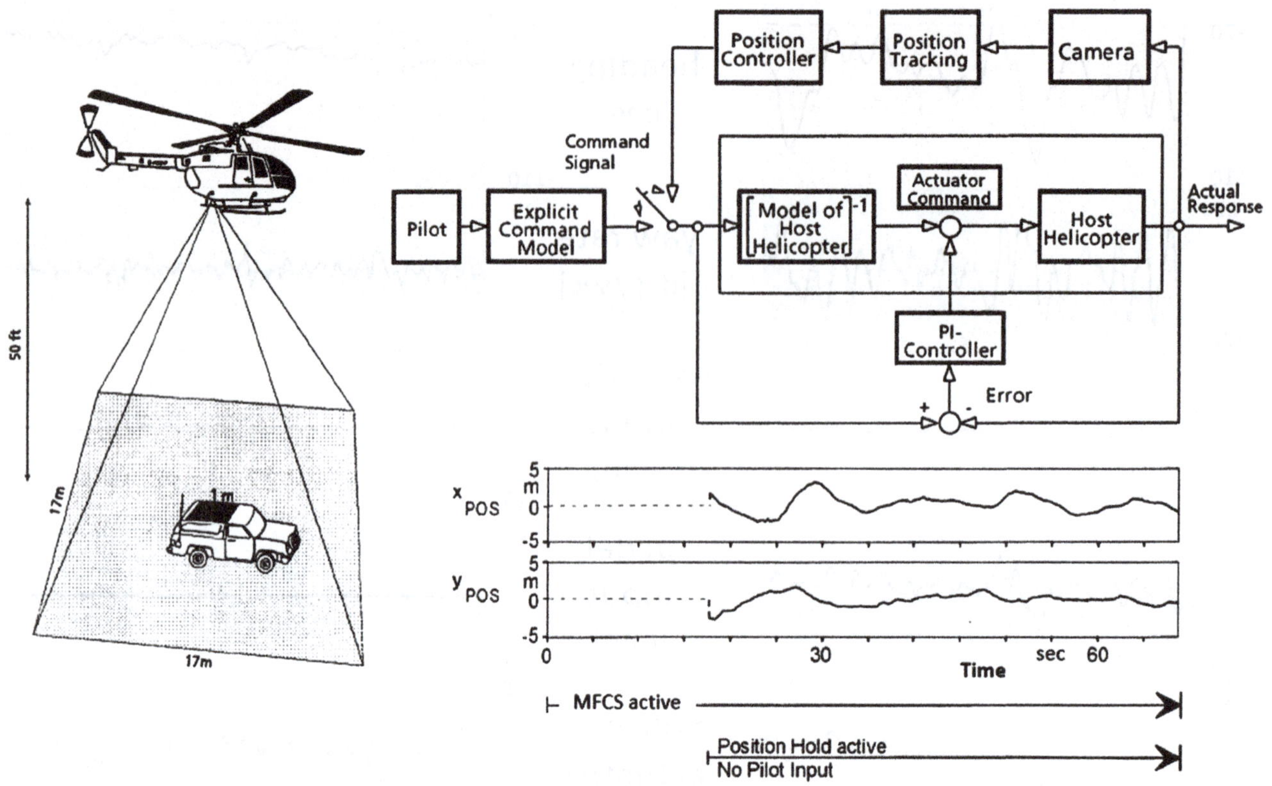

Fig. 8.31 Test flight for position hold (experimental arrangement)

8.4.5 Test Pilot Training

In cooperation with the English Empire Test Pilots' School (ETPS), ATTHeS was deployed for test pilot training since 1990, and also together with the French Ecole du Personnel Navigant d'Essais et de Réception (EPNER) since 1992 (see Fig. 8.34).

In both the cases, the prospective test pilots flew ATTHeS in various configurations without stabilization, for example, with different damping control sensitivity, and time delays. Furthermore, different controller functions and control couplings were tried out and finally also critical flight conditions with RPC effects (Rotorcraft Pilot Coupling) [32]. Following the description in Chap. 3, the aim of the training was to evaluate the configurations and to identify and avoid critical situations. Because of the ATTHeS flexibility, it was feasible for trainee pilots to test a wide range of possible flying qualities in a short time. Figure 8.35 shows a group of ETPS test pilots together with DLR staff members. These training courses were conducted in Braunschweig as well as in Boscombe Down, England, for ETPS, and in Braunschweig for EPNER.

8.4.6 Statistics of Utilization

The projects described in the foregoing demonstrate the flexibility and versatility of the in-flight simulator Bo 105-S3 ATTHeS. In addition, there are a variety of different projects that would not have been possible without ATTHeS [41].

As part of the European project ACT (*Active Control Technology*), flight tests with ATTHeS were performed in 1994 with an active sidestick, which was installed in place of the conventional controls. Various flight characteristics were investigated during pre-defined flight tasks.

In cooperation with the French research organization ONERA, flight tests with ATTHeS were carried out in 1995 to investigate the flying qualities in hover. The flight task consisted of tracking a moving vehicle as precisely as pos-

Fig. 8.32 Autonomous position hold in hover (*Credit* Braunschweiger Zeitung)

sible in a sideward flight. This task was to be accomplished by the test pilots with different preprogrammed helicopter flying qualities [42].

Altogether, ATTHeS was deployed at DLR from 1982 to 1995 for more than 1300 flight hours. Figure 8.36 shows the statistics of utilization and provides an overview of the numerous modifications and additions to the equipment, as well as of the various utilization and demonstration programs. The work carried out at DLR with ATTHeS and the unique flight test data are documented in numerous scientific publications of the DLR and of the partners, and were highly appreciated and utilized by the international research institutions and industry. Figure 8.37 shows a group of visitors at the DLR in June 1993 with senior representatives from ministries and industry, rightmost *Klaus Sanders*, a highly experienced safety pilot and engineer for many years at DLR, who provided significant contributions to the projects and flew as a reliable safety pilot in many flight tests with Bo 105 ATTHeS (see also Sect. 8.1).

Due to the age and other restrictions of the Bo 105-S3, and long before the ATTHeS crashed on May 14, 1995, DLR had initiated a new in-flight simulator [43, 44]. In the following years, this initiative evolved into the project ACT-FHS (*Active Control Technology Demonstrator – Flying Helicopter Simulator*) as a joint venture of the German Federal Ministry of Defense, the DLR and the industry (Eurocopter Germany, Liebherr Aerospace). This Flying Simulator was realized starting from 1996 based on a EC 135 helicopter and is since 2002 available at DLR for European research organizations and industry as a unique flying device for technology testing and demonstration (see Chap. 10).

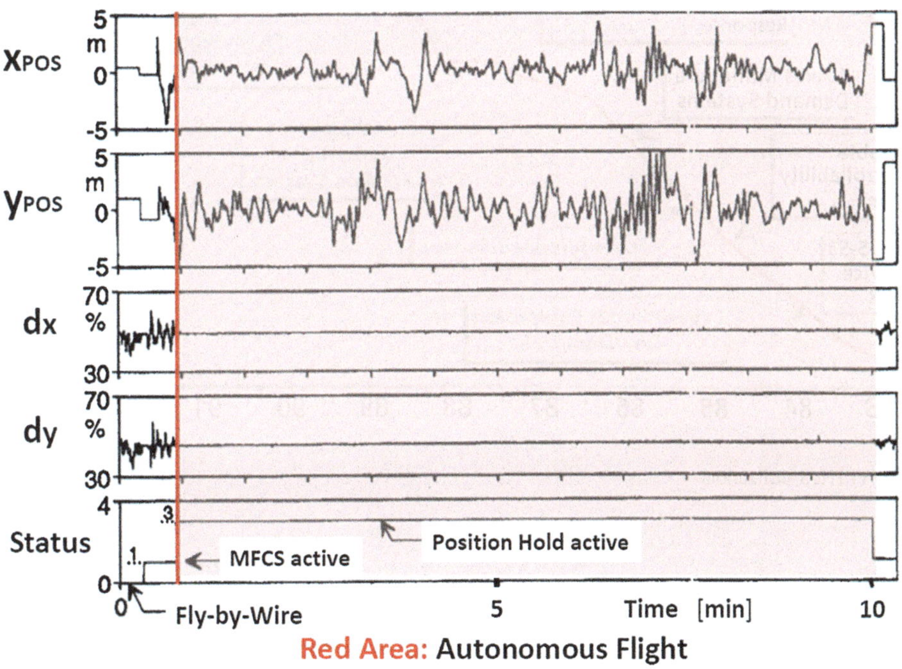

Fig. 8.33 Test flight for position hold (measurement protocol)

Fig. 8.34 Training of future EPNER pilots

Fig. 8.35 Training course with ATTHeS for ETPS pilots

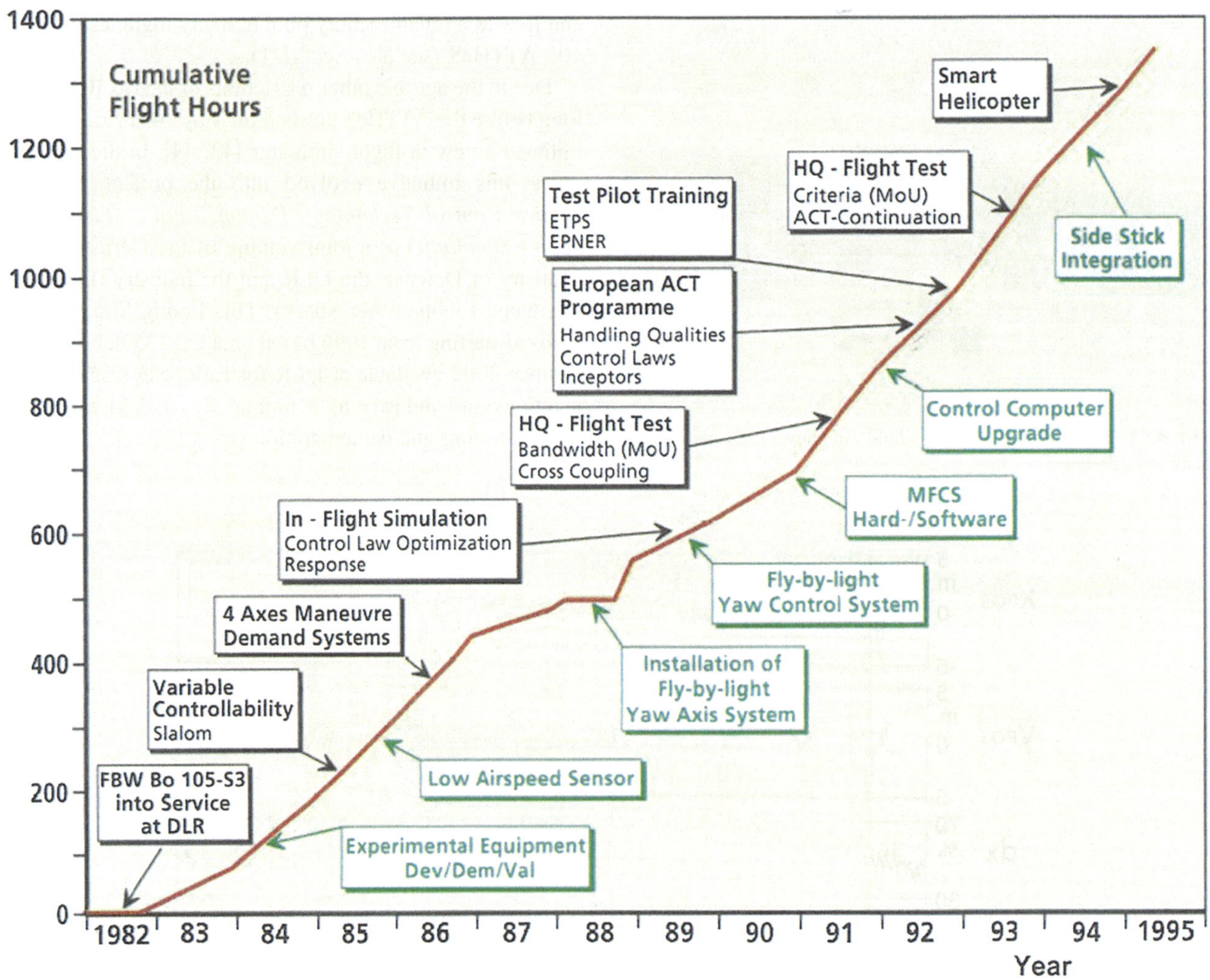

Fig. 8.36 Statistics of ATTHeS utilization

Fig. 8.37 Group of visitors at DLR (from *left*: *E. Eckert, B. Gmelin, H. Huber, F. Thomas, K. Schymanietz, Ives Richard, P. Hamel, H.-J. Pausder, L. Müller, Ph. Roesch, M. Rössing, J. Kaletka, K. Sanders*)

References

1. von Gersdorff, K., Knobling, K., (Ed.): Hubschrauber und Tragschrauber. Bernard & Graefe Verlag München (1982)
2. DFVLR e.V., Dornier GmbH, MBB GmbH: Versuchsträger Variabler Stabilität—Mittelfristiges Programm—Memorandum für das Bundesministerium der Verteidigung, Dec 1969
3. Marchand, M., Wilhelm, K.: Flugmechanische Untersuchungen über die Einsetzbarkeit eines Hubschraubers zur Simulation von V/STOL-Flugzeugen, DLR-FB, pp. 75–59, Braunschweig (1975)
4. Janik, K., Schmitt, H.: Hubschrauber BO 105 als Versuchsträger variabler Stabilität—Aufgabenstellung und Realisierungsmöglichkeiten, Symposium über Versuchsträger variabler Stabilität, Manching 12. Juni 1970, MBB-Bericht MBB-DU 52-70-Ö (1970)
5. Derschmidt, H.: Steuerung und Cockpit für BO 105 als Flugversuchsträger, MBB Technische Niederschrift TN D125-2/71 (1971)
6. Attlfellner, S., Rade, M.: Bo 105 in-flight simulator for flight control and guidance systems. In: First European Rotorcraft and Powered Lift Aircraft Forum, Southampton England (1975)
7. Formica, B., König, H., Bender, K., Mansfeld, G.: OPST1—an digital optical tail rotor control system. In: 14th European Rotorcraft Forum, Milano Italy, Sept 1988
8. Mansfeld, G., Bender, K., Holle, K.: DISCUS—Ein fehlertolerantes Fbw/Fbl-Experimentalsystem. DLR-Nachrichten **61**, 6–11 (1990)
9. Hofer, H.: OPST1/ DISCUS, Informationen zu Projektanteilen der Firma Liebherr Aerospace Lindenberg GmbH (2013)
10. König, H., Stock, M., Zeller, S.: OPST1—an optical yaw control system for high performance helicopters. In: 45th Annual Forum and Technology Display of the American Helicopter Society, Boston USA, May 1989
11. Tischler, M.B., (Ed.): Advances in Aircraft Flight Control. Taylor & Francis (1996)
12. Key, D.L., Gmelin, B.: A Decade of Cooperation. American Helicopter Society, September/October, Vertiflite (1989)
13. Bouwer, G., Hilbert, K.B.: A Piloted Investigation of Decoupling Helicopters by Using a Model Following Control System, 40th Annual Forum of the American Helicopter Society, Paper No. A-84-40-07-4000, Arlington Virginia USA, May 1984
14. Hilbert, K.B., Lebacqz, J.V., Hindson, W.S.: Flight Investigation of a Model-Following Control System for Rotorcraft, AIAA 3rd Flight Testing Conference, Las Vegas Nevada, Apr 1986
15. Kaletka, J., von Grünhagen, W.: Identification of mathematical derivative models for the design of a model following control system. In: 45th Annual Forum & Technology Display of the American Helicopter Society, Boston USA, May 1989
16. von Grünhagen, W., Kaletka, J.: Application of system identification for rotorcraft control system design, 4. In: Braunschweiger Aerospace Symposium "Sicherheit im Luftverkehr", Technische Universität Braunschweig, Sept 1994

17. Pausder, H.-J., Bouwer, G., von Grünhagen, W.: A highly maneuverable helicopter in-flight simulator—aspects of realization. In: 14th European Rotorcraft Forum, Milano Italy, Sept 1988
18. Pausder, H.-J., Bouwer, G., von Grünhagen, W.: ATTHeS in-flight simulator for flying qualities research. In: In-Flight Simulation for the 90's, International Symposium, DLR Braunschweig, DGLR Conference Proceedings, Paper 15, Braunschweig (1991)
19. Bouwer, G., Pausder, H.-J., von Grünhagen, W.: ATTHeS, a helicopter in-flight simulator with high bandwidth capability. In: 48th Annual Forum and Technology Display of the American Helicopter Society, Washington DC USA, June 1992
20. Huber, H., Hamel, P.: Helicopter flight control—state of the art and future directions. In: 19th European Rotorcraft Forum, Cernobbio Italy, Sept 1993
21. von Grünhagen, W., Bouwer, G., Pausder, H.-J., Henschel, F., Kaletka, J.: A high bandwidth control system for a helicopter in-flight simulator. Int. J. Control **59**(1) (1994)
22. Hamel, P.: Advances in aerodynamic modeling for flight simulation and control design. GAMM-Mitteilungen **2000**(1–2), 7–50 (2000)
23. Gmelin, B., Bouwer, G., Hummes, D.: DFVLR Helicopter In-Flight Simulator for Flying Quality Research, Vertica, Pergamon Press, Vol. 10, No. 1 (1986)
24. Gmelin, B., Bouwer, G.: Development and Application of Helicopter In-Flight Simulation at DLR, La Recherche Aérospatiale No. 1, pp. 59–72 (1995)
25. Pausder, H.-J., Bouwer, G., von Grünhagen, W., Holland, R.: Helicopter In-Flight Simulator ATTHeS—A Multipurpose Testbed and its Utilization, AIAA/AHS Flight Simulation Technologies Conference, Hilton Head Island SC USA, Aug 1992
26. Hamel, P., (Ed.): In-Flight Simulation for the 90's, International Symposium, DLR Braunschweig, DGLR Conference Proceedings, pp. 91–05 (1991)
27. Pausder, H.-J.: A study of roll response required in a low altitude slalom task. In: 11th European Rotorcraft Forum, London UK, September 1985
28. Anon.: ADS-33E-PRF, Aeronautical Design Standard, Performance Specification, Handling Qualities Requirements for Military Rotorcraft. United States Army Aviation and Missile Command, Aviation Engineering Directorate, Redstone Arsenal, Alabama, 21 Mar 2000
29. Gmelin, B., Pausder, H.-J., Hamel, P.: Mission-Oriented Flying Qualities Criteria for Helicopter Design via In-Flight Simulation, AGARD-CP-423, Paper 4 (1987)
30. Pausder, H.-J., Bouwer, G.: Recent results of in-flight simulation for helicopter ACT research. In: 15th European Rotorcraft Forum, Amsterdam The Netherlands, Sept 1989
31. Ockier, C.J., Pausder, H.-J.: Experiences with ADS-33 Helicopter Specification Testing and Contributions to Refinement Research, AGARD-CP-560, Paper 4 (1995)
32. Hamel, P.: Rotorcraft—Pilot Coupling, A Critical Issue for Highly Augmented Helicopters?, AGARD-CP-592, Paper 21 (1997)
33. Pausder, H.-J., Blanken, C.L.: Investigation of the effects of bandwidth and time delay on helicopter roll-axis handling qualities. In: 18th European Rotorcraft Forum, Avignon France, Sept 1992
34. Padfield, G.D.: Rotorcraft handling qualities engineering—managing the tension between safety and performance, AHS Alexander Nikolsky lecture. In: 68th Annual Forum and Technology Display of the American Helicopter Society, Fort Worth Texas USA, May 2012
35. Bouwer, G., von Grünhagen, W., Pausder, H.-J.: Model Following Control for Tailoring Handling Qualities—ACT Experience with ATTHeS, AGARD-CP-560, Paper 11 (1995)
36. Bouwer, G., von Grünhagen, W.: LYNX Helicopter In-Flight Simulation with ATTHeS, DLR-IB 111–92/47, Braunschweig (1992)
37. Bender, K., Leyendecker, H.: Untersuchungen zu einem Fehlertoleranten Hubschrauber-Flugregelungssystem mit Abgestufter Redundanz, DLR-Mitteilung, pp. 97–01. Braunschweig (1997)
38. Mansfeld, G., Bender, K., Tersteegen, J.: Entwicklung und Flugerprobung eines fehlertoleranten Fly-by-Light Gierregelungs-systems. DGLR-Fachausschuß-Sitzung Flugeigenschaften. Flugregelung, Braunschweig (1991)
39. Hamers, M.: Vollautomatische Navigation mit ATTHeS, DLR-IB 111-94/16 (1994)
40. Bouwer, G., Oertel, C.-H., von Grünhagen, W.: Autonomous helicopter hover positioning by optical tracking. Aeronaut J, **99** (984) (1995)
41. Bouwer, G., Pausder, H.-J.: Fliegende Simulation mit Hubschraubern. In: Luftfahrtforschung in Deutschland. Bernard & Graefe Verlag Bonn (2001)
42. Bouwer, G., Taghizad, A., Mödden, H.: ONERA and DLR Cooperation on the Smart Helicopter Concept—Handling Qualities Data Base for Hover and Low Speed Flight, AGARD-CP-592, Paper 11 (1997)
43. Anon.: HESTOR—Entwicklung eines Fliegenden Simulators für Hubschrauber, Rahmenvorschlag, MBB/DFVLR, June 1984
44. Gmelin, B., Kissel, G., Mansfeld, G.: Key Technologies for a New Flying Helicopter Simulator. In: In-Flight Simulation for the 90's, International Symposium, DLR Braunschweig, DGLR Conference Proceedings, Paper 7 (1991)

Author Biography

Bernd Gmelin was a research scientist and head of the Rotorcraft Department at the Institute of Flight Mechanics of DLR at Braunschweig (1973–1999). From 1999 to 2006 he headed the DLR Helicopter Program under the German/French DLR/ONERA Common Rotorcraft Research Program. Prior to joining DLR, he was a researcher at the Institute for Rotorcraft in Stuttgart (1968–1973). He received his Dipl.-Ing. degree in Aerospace Engineering from the University of Stuttgart (1968). His primary interests are mathematical modeling, helicopter simulation in the wind tunnel, control systems, flying qualities, helicopter in-flight simulation. He is a member of the German Society for Aeronautics and Astronautics (DGLR) and of the American Helicopter Society (AHS). He is the recipient of the AHS Agusta International Fellowship Award (1996), DLR/ONERA Team Award (2006), and AHS Fellow Award (2008).

In-Flight Simulator VFW 614 ATTAS

9

Dietrich Hanke and Klaus-Uwe Hahn

With contributions from 14 further coauthors

VFW 614 ATTAS *(Credit: WTD-61)*

The original version of this chapter was revised: Author biography has been corrected. The erratum to this chapter is available at 10.1007/978-3-319-53997-3_14

D. Hanke (✉) · K.-U. Hahn
Braunschweig, Germany
e-mail: dietrich.hanke@gmail.com

K.-U. Hahn
e-mail: k-u.hahn@dlr.de

P.G. Hamel (ed.), *In-Flight Simulators and Fly-by-Wire/Light Demonstrators*,
DOI 10.1007/978-3-319-53997-3_9

9.1 Modification and Equipment of Flight Test Vehicle

Dietrich Hanke
With contributions from Hartmut Griem and Hermann Hofer

9.1.1 Introduction and Project Definition Phase

As discussed in Chap. 7, the overall experience with the in-flight simulator HFB 320 operation was very positive. However, some limitations were encountered in its operation, mainly due to low payload, limited space for test equipment and crew, limited power supply, and high engine noise. Therefore, starting from 1977 the possibilities of higher-performance and efficient successor were explored. Technical requirements, covering a wide range of research applications and in-flight simulation, were then specified in the requirements catalog and formulated as a framework requirement for such a flight test vehicle [1].

The range of application covered the following research tasks: (1) Digital flight control, (2) Flying qualities and in-flight simulation, (3) Flight guidance and air traffic management, (4) Man-machine interfaces, (5) Modeling of aircraft and systems, and (6) Navigation and communication. In addition, the test vehicle was to serve as a test platform for system components such as sensors, antennas, actuators, avionics and navigation systems, data link systems, computer systems, operating elements, and display systems. Appropriate installation options and data interfaces were to be provided.

The aircraft was to have an electrical flight control system with high onboard computing power for experiments. In addition to a direct lift control, a direct lateral force control was also required, with the aim of enabling a complete six degrees-of-freedom simulation in flight.

Furthermore, based on the HFB 320 experience, a comprehensive system simulation in real-time of aircraft as well as its system components, partly as hardware-in-the-loop, was envisaged to enable testing of software functions under real-time conditions and release them for the flight test. A complete software development system was also to be provided for the real-time processes with configuration control of the modules.

Two aircraft were identified as suitable test vehicle, namely the short-range VFW 614 manufactured by VFW-Fokker in Bremen (see Fig. 9.1) and the business jet aircraft Grumman Gulfstream II, which was also used by NASA as an in-flight simulator for the training of the Space Shuttle astronauts (see Sect. 5.2.2.14). Some differences in these two types are found in the Table 9.1.

Fig. 9.1 VFW 614 (*Credit* www.vfw614.de)

The VFW in Bremen as manufacturer of the VFW 614 and Dornier in Friedrichshafen (as a client in cooperation with Grumman) for the Gulfstream II were requested to submit their offers in May 1979.

Due to discontinuation of the VFW 614 production after only 19 aircraft in December 1977, the choice of the VFW 614 as a test vehicle seemed uncertain. The VFW as a manufacturer, however, ensured support and maintenance of the VFW 614. Likewise, the Rolls Royce as an engine manufacturer also agreed to maintain the technical support for the engines M 45 H developed exclusively for the VFW 614. In addition, a concept was developed to ensure availability of spare parts and engines (eight engines with different life cycles were planned) for the expected period of the utilization of the test vehicle of at least 16 years.

With the German Federal Aviation Authority (LBA) as certification authority, it was agreed that the VFW 614 could be operated as a single item for the DFVLR, provided that DFVLR was recognized as a development organization. The last VFW 614 G17 assembled by VFW was mothballed after its acceptance flight testing in 1977 and was reserved for DFVLR as the test vehicle.

The technical proposals and cost estimates of both companies were received in 1979 and amounted to approximately 50 million DM. The offer of VFW also contained the realization of a lateral force device on the fuselage. This estimate was, however, far beyond the funds approved by the BMFT. Therefore, the DFVLR framework requirements and the scope of applications were revised [2, 3]. For this basic version with options for extensions, both companies were again asked to resubmit an offer, limiting the cost to 31 million DM. Both the companies submitted their offers in November 1980.

The final selection was made by DFVLR based on a comprehensive technical evaluation of both the offers, which ultimately led to the recommendation of the VFW 614. The important factors were the more spacious cabin, a large cargo door, more modern aircraft, good flying qualities, lower noise emission, and proximity of the manufacturer in

Table 9.1 Some details of candidate vehicles

	VFW 614	Gulfstream II
Production	Discontinued	Discontinued
Quantity	19	258
Inspection	Individual item	–
First flight	1972	1966
Manufacturer	German, VFW	American, represented by Dornier
Engines	Rolls-Royce MH 45 H	Rolls-Royce Spey
Bypass ratio	2.85:1	0.62:1
Related utilization	–	As space shuttle training aircraft (STA) at NASA

Bremen to Braunschweig. The takeoff and landing performance was yet another important criteria, because the runway in Braunschweig was then only 1300 m long. This was sufficient for the VFW 614 with full payload. Further technical information on the VFW 614 and its development history can be found in Refs. [4, 5].

Yet another important advantage was that the basic aircraft VFW 614 could be bought practically free of charge, since the development of the aircraft was largely financed by the Federal Ministry of Economics and Technology (BMWI). As a result, the investment funds could be made fully available for the conversion, without spending any on the procurement of the vehicle itself. In the case of a Gulfstream II, approximately 8 million DM would have been necessary just to procure a used machine.

After taking the decision for the VFW 614, in the course of 1981 the modifications on the airplane were specified in detail by VFW and coordinated with the DFVLR. It was decided to take over a number of development tasks by the DFVLR to realize the project within the cost limits. The breakdown of the development work and costs between VFW and DFVLR as well as the development period is shown in Fig. 9.2. The development and delivery contract of the new flight test vehicle, which was dubbed ATTAS (*Advanced Technologies Testing Aircraft System*), was signed in December 1981 by DFVLR and VFW.

Share of Development		Costs	
DLR	• System Specification • FBW-Computer-System	Project Definition	2.5
	• Avionic-Systems • Cockpit Modifications • Data Acquisition System • Systemintegration	Aircraft and Modifications	31.0
	• Ground Simulator	FBW - System	6.2
MBB	• Aircraft Modifications • Safety Systems • DLC-System • Wiring,	Avionic, Data Acquisition	2.5
	• Electrical Systems • Hydraulical Systems	Operation Facilities	1.6
LAT	Electro- Hydraulic Actuators	Ground Simulator	0.4
			45.0 Mio DM

Fig. 9.2 Work share and project costs

9.1.2 Project Organization DFVLR

A special project management was set up by DFVLR to carry out the major ATTAS project. *Heinz Krüger* was appointed as the general project development manager. The technical tasks within DFVLR were split between the institutes for Flight Mechanics and Flight Guidance and the Flight Test Department in Braunschweig as the main users.

The Institute of Flight Mechanics was assigned the task of the technical development of ATTAS as an in-flight simulator. *Dietrich Hanke*, who had already directed the development of the HFB 320 in-flight simulator, was appointed as the technical project director for the ATTAS.

The responsibilities for development, equipment maintenance, and operation in the DFVLR were defined as follows:

Institute of Flight Mechanics (FM): Cockpit instruments and panels, sidestick, fly-by-wire (FBW) computer systems, operating software and in-flight simulation software, ground-based development system, ground-based simulator, measurement system, sensors, and supervision of the actuation system electronics and of the experimental operating device.

Institute of Flight Guidance (FL): Noseboom, equipment and control cabinets, avionics systems, antennas, data recording/evaluation, telemetry, and autopilot operating unit and electronic displays.

Flight Test Department: Flight operation, maintenance, spare parts for the basic aircraft, supervision of actuation systems (LAT), and electrical and hydraulic power supplies. The new hangar in Braunschweig, already built in 1975/76, was sufficient to accommodate and work on ATTAS (see Fig. 9.3).

Development Operation: To meet the conditions laid down by the LBA, an airworthiness office (MPL), headed by

Fig. 9.3 New Hangar in Brauunschweig for ATTAS

Fig. 9.4 VFW 614 G13, D-BABM as a mockup at DFVLR (October 28, 1981)

Ludwig Tacke, was set up for the processing of approval of installations and modifications to the aircraft.

9.1.3 Project Organization MBB

After take over by the MBB Group at the end of 1981, VFW was continued under the name of MBB. Here, *Hartmut Griem* was appointed as the project manager, *Hajo Schubert* was responsible for the technology and *Gottfried Böttger* for the finances and administrative contracts. Particular emphasis was on equipment systems and installation, laboratory, ground and flight tests as well as all necessary cross-sectional tasks such as reliability, certification etc. The development of the electrohydraulic actuation systems was assigned by MBB to the company Liebherr-Aero-Technik (LAT) and of the test operating unit (Versuchsarten-Bediengerät—VBG) to the Bodenseewerk Gerätetechnik (BGT) [6].

9.1.4 Project Development

Immediately after the signing of the contract in December 1981, the development work began for the conversion and equipping of the new flight test vehicle ATTAS. The VFW 614 G13, which was previously flown by the Air Alsace, was handed over to the DFVLR as a mockup for the adaptation of equipment cabinets, modifications in the cockpit and attachment of the noseboom (see Fig. 9.4). It was transferred to Braunschweig on October 28, 1981. After completion of the installation and adaptation work, the G13 was scrapped in 1990, having removed the components as spare parts.

A working group was established to compare and select suitable onboard computer systems, particularly for the electronic flight control (Fly-by-Wire—FBW) that was an important part of the test vehicle. Whereby several key aspects had to be considered, such as the performance, reliability, airworthiness, and availability of interfaces such as ARINC 429 for the avionics systems and MIL BUS 1553B for connecting the electrohydraulic control systems.

The process computers of the companies AEG, EMM, DEC, and ROLM MIL-Spec Computers were investigated. Ultimately the "militarized" computer system was selected from ROLM from Mountain View, USA (Silicon Valley), which had developed a computer system in the standard form size ATR for harsh environmental conditions based on the operating system of the commercial computers of the company Data General. They were, therefore, airworthy, had all the necessary interfaces, and also enabled optical data communication between the various computers (FBW/L). At the same time, a software-compatible development system and a computer hardware for building the development simulator became available with the commercial computers of Data General.

A hardware-in-the-loop simulator was specified by the DFVLR, which enabled the developed software to be tested

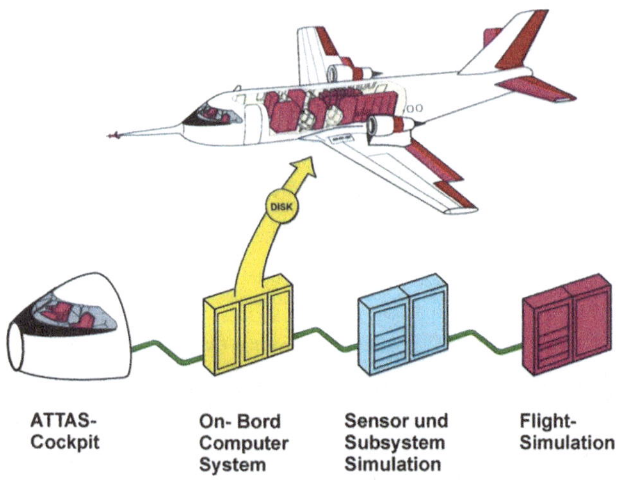

Fig. 9.5 Overall system ATTAS, aircraft, and ground-based simulator

and verified for Fly-by-Wire operation and the in-flight simulation under real-time conditions on the ground. Fortunately, the decision for the ROLM computers could be enforced in the DFVLR, which ultimately served reliably over the entire period of utilization of 27 years. The ROLM computers were delivered in August 1982 and integrated into the ground system.

The test aircraft together with the ground-based simulator was viewed as an ATTAS total system, which served the purposes of testing and preparing the flight tests (see Fig. 9.5). For the ground-based simulator, an original VFW 614 cockpit was procured and equipped with identical displays and controls, as provided in the ATTAS (see Fig. 9.6).

A very long noseboom with a length of 5.54 m was designed for the measurement of the angles of attack and sideslip and the airspeed in free stream. The noseboom was developed in 1986 by the DFVLR Technical Operations and manufactured using carbon fiber technology. Figure 9.7 shows the first installation of the noseboom on the ATTAS fuselage frame. During the first flight tests, undesirable oscillations of the noseboom were caused by an aeroelastic coupling between the flight log and noseboom. To overcome this problem, the noseboom frequency was increased by shortening the boom by 1 m, thereby the frequency closeness, a precondition for flutter, was eliminated.

The MBB removed all systems from VFW 614 G17 which were not needed (for example, the kitchen and the toilet), and carried out all arrangements that were necessary for the envisaged structural and technical modifications. They included (1) installation of the electrohydraulic actuators, including the safety-critical electronic units (*Actuator Electronic Units*—AEU) for activation, monitoring, and deactivation of the actuators, (2) direct lift control (DLC) with new, fast-moving trailing edge flaps, (3) column and pedal force simulation for the test pilots, (4) redesign of almost all cockpit panels as a result of the extensive

Fig. 9.7 Noseboom mounted on the fuselage frame

experimental and to-be certified basic instrumentation, (5) operating mode control unit and the safety devices for switching on and off the experimental controls, and (6) cabin preparation for the installation of all test facilities. For the proof of flutter safety, a complete static oscillation test had to be carried out.

The modifications to the G17 were carried out at the MBB plant in Lemwerder. The left fuselage side had to be opened in the cockpit area to install the control column force simulation system. The entire wiring was also carried out by MBB, which required considerable effort due to the high disturbance protection requirements for safety-critical systems (see Fig. 9.8).

The DLC flap system was investigated in the Dutch–German wind tunnel (DNW) in Amsterdam and the flap effectiveness was determined [7]. For this purpose, a still existing 1:5 scale wind tunnel model was modified at VFW and equipped with the DLC flaps.

After several months of ground testing, everything was ready for the first flight. Figure 9.9 shows the test team on their way to the first flight, which took place on February 13,

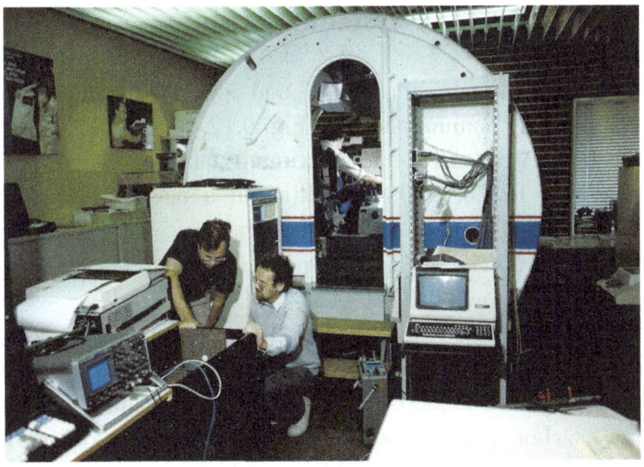

Fig. 9.6 ATTAS cockpit in ground-based simulator

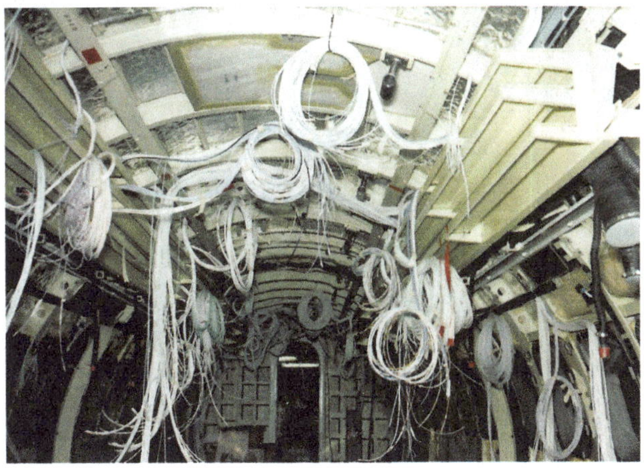

Fig. 9.8 Electrical cable wiring in the cabin

1985, 7 years after the first take-off of the G17. After a flight time of 3:20 h, the satisfied crew landed again in Lemwerder. The subsequent flight tests served the purpose of proving the functionality of the systems developed by MBB and their properties and performances specified by DFVLR. The verification tests were focused on (1) safety concept, (2) DLC-System, (3) symmetrical ailerons, (4) flight performance, (5) rudder booster, (6) landing flap position SP for the DLC-operation, and (7) acceptance. The aircraft was provisionally certified by the LBA and approved by DFVLR, subject to the still outstanding acceptance flight. At the end of the flight test phase (see Fig. 9.10), acceptance test by LBA was pending that was necessary for final certification.

The painting with the ATTAS signature on the vertical tail was made at the end of April 1985. Finally, on September 26, 1985 ATTAS was handed over to the DFVLR. The official transfer of ATTAS took place on October 1, 1985 and on October 24, 1985 the ATTAS development contract was officially declared as fulfilled [8–10].

Further Installations by DFVLR

After the aircraft was handed over to DFVLR, the installation of the FBW system and the testing of the interactions between all control components were carried out. Thereafter, it was possible to perform the first FBW flight on December 17, 1986 using the ROLM computer system. Based on an Executive Board decision, ATTAS was used initially in 1987 for aerodynamic tests on laminar flow. For this purpose, a laminar flow glove was applied over a limited area on the right wing. The transition from laminar to turbulent flow for a moderately swept wing under realistic flight conditions was determined in flight by means of pressure measurements and infrared measurement technology (see Fig. 9.11). As a result, ATTAS was not available to the main users, Institutes of Flight Mechanics and Guidance, until 1988.

Fig. 9.9 The flight test team on the way to first first flight [from *left*, *"HaLu" Meyer* (DFVLR), *Hajo Schubert*, *Dietmar Sengespeik*, *Bodo Knorr* (all MBB)]

Fig. 9.10 ATTAS flight test in Lemwerder (*Credit* MBB)

Fig. 9.11 Aerodynamic investigation with laminar flow glove

9.1.5 ATTAS System Description

The geometrical dimensions and performance data of ATTAS are summarized in Fig. 9.12. Compared to the standard VFW 614, the maximum take-off weight was increased to 20,965 kg, resulting in a new identification D-ADAM. Category A stands for aircraft weighing more than 20 tons. The flight envelope and the electrical operating modes are shown in Fig. 9.13.

There were no restrictions on the maximum speed (V_{MO}) or maximum Mach number (M_{MO}). The maximum altitude was limited to 25,000 ft because of the dismantled oxygen masks. For safety reasons, the minimum altitude above ground was set to 500 ft in FBW mode. It was demonstrated

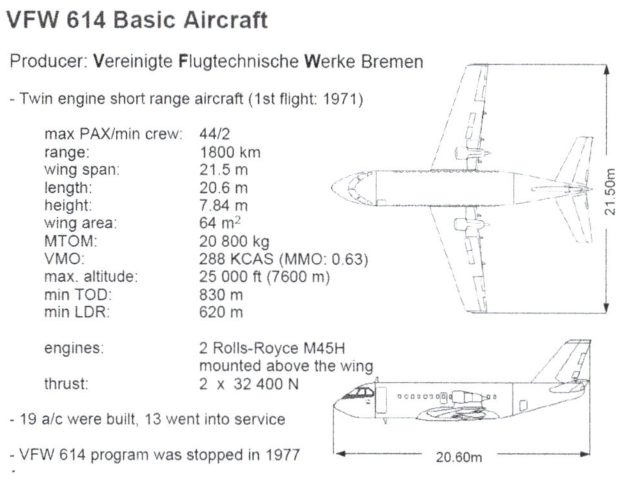

Fig. 9.12 ATTAS dimensions and performance data

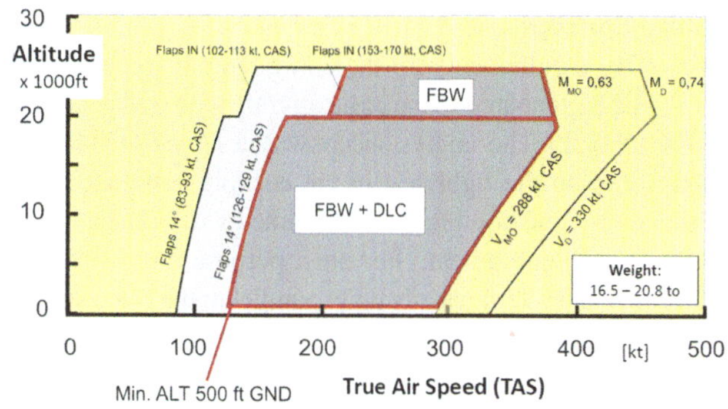

Fig. 9.13 ATTAS flight envelope

that the aircraft command could be safely taken over by the safety pilot at this altitude even in the event of a hardovers (maximum deflections) of all control surfaces.

The essential changes and conversions to the VFW 614 aircraft are illustrated in Fig. 9.14. They comprise of (1) safety pilot on the right with the basic control system, (2) experimental pilot on the left with electrical control (fly-by-wire) and experimental instrumentation, (3) digital computer interconnection system with optical bus communication, (4) experimental avionics system, (5) measurement system, data recording and telemetry, (6) electrohydraulic actuators for elevator, duplex; rudder, duplex; ailerons left and right, simplex; engines left and right, simplex; landing flaps, simplex; six direct lift flaps, simplex; electric autotrim system, (7) work stations for computer operator, measuring system operator, flight test engineer, and experimenter, (8) equipment cabinets for computer systems and actuator electronics, measurement system, avionics and data recording, and safety system, test electronics and hydraulics, (9) independent test electronics via in-flight APU and engines, (10) independent experimental hydraulics (2 circuits), and (11) noseboom for flight log sensor.

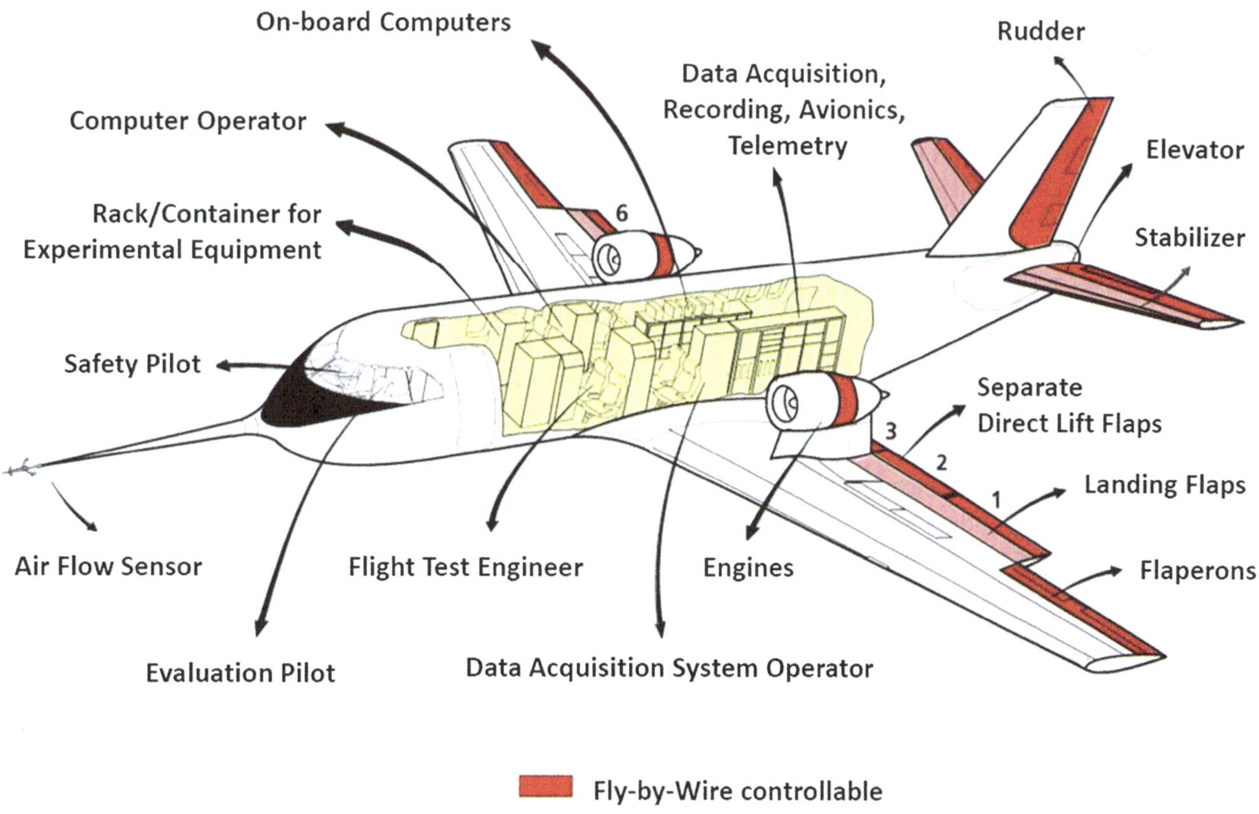

Fig. 9.14 Overview of system installations and modifications

Safety Concept

The safety requirements for the entire aircraft were changed for ATTAS compared to the basic aircraft approved according to FAR Part 25. The safety concept was designed such that the safety pilot on the right seat in the cockpit could access the basic control and equipment at any time by disconnecting the experimental equipment. For this purpose, a nonreactive experimental system was set up in parallel to the basic system. System faults were kept passive by fault detection mechanisms and redundancy, so that control surface hardovers could be avoided which could lead to dangerous flight conditions. As the control interceptors of the safety pilot moved always synchronously with the control surfaces, the monitoring of the system and of the experimental pilot was significantly improved and a takeover was possible at any time without large transients in the flight motion.

The safety installations included: (1) status error indication and acoustic warning, (2) mechanical 'back up' control system, (3) FBW-Off button on the safety pilot control column to disconnect the electrohydraulic actuators via hydraulically actuated clutches, (4) force sensors in the primary controls (pitch, roll, yaw axes). When a force threshold was exceeded by counter-reaction in the control elements by the safety pilot, the clutches were opened automatically, (5) force-limited electrohydraulic actuators, which prevented exceeding permissible load, (6) duplex redundant FBW system (fail passive), (7) automatic fault detection and fault display, (8) control surface redundancy in the DLC system and ailerons, and (9) emergency switch for switching off the experimental electronics.

The operating concept for switching on and off the actuation systems, as well as for the operating functions FBW and SIM, was adopted from the HFB 320, since it had proved highly successful. The mode operating device VBG was adapted to the specific characteristics of the system components. Likewise, the system status display in the glare shield for both pilots as well as the operation in the cockpit and the cabin were taken over.

Operating Modes

The ATTAS operation included three operating modes.

- **Basic mode**: Operation by means of safety pilot on the right with mechanical control of the basic aircraft.
- **FBW mode**: Operation by experimental pilot on the left with electrical control in selected axes.
- **SIM mode**: Operation by the experimental pilot on the left with electrical control and any controller functions in the experiment computer, for example simulation.

The FBW mode was created as an intermediate step for the transition from the basic to the SIM mode to simplify the switching operations between the modes and to ensure that all systems worked properly before the SIM mode was activated. Furthermore, it also avoided switching back directly from the SIM mode to the basic mode, because restarting the actuators was time-consuming (coincidence formation of the actuators with the respective surface position). In addition, autopilot functions were provided to the experimental pilots in a later phase in FBW mode, which made it considerably easier to set a stable flight condition required for the test. The functioning of the operating modes will be discussed in detail later in this section under the description of the experimental system.

Cockpit Modifications

Significant changes were made to the cockpit (Figs. 9.15 and 9.16):

Right Side, Safety Pilot

The safety pilot on the right side was provided with a FBW-Off button on the control column, status and fault indicators in the glare shield, additional indicators for the trim force of the automatic trimming system, display of the surface positions, locking of the DLC flaps and switches for passivating the hydraulics of the DLC actuators. In addition, an emergency switch was provided for switching off the entire experimental electronics and a warning light on exceeding the engine temperature.

Left Side, Experimental Pilot

On the left side for the experimental pilot, identical displays in the glare shield as those for the safety pilot were provided. In the center of the glare shield provision was made for an autopilot operating unit (*Experimental Flight Control Unit*—EFCU), which in the first phase contained a system made by BGT and was used in the HFB 320. It was later replaced by a system typical of Airbus A320. Only the control buttons and indicators were used, while the functions behind them were freely programmable and could be adapted to the respective research project.

The left instrument panel was completely redesigned. The basic instruments required for instrument flight operation were moved upwards to create space for two electronic displays, the *Primary Flight Display* (PFD) and the *Navigation Display* (ND). These electronic screens were similar to the standard displays from the Airbus A310 series (see Fig. 9.16). In addition, display for the engines, landing flaps and landing gear position of a simulated aircraft model were created, as well as rotary knobs for trimming the rudder and ailerons for the simulation model.

Center Console

The center console was also completely redesigned (see Fig. 9.17). Major changes involved the installation of a digital frequency selector (CDU) and an operating unit for

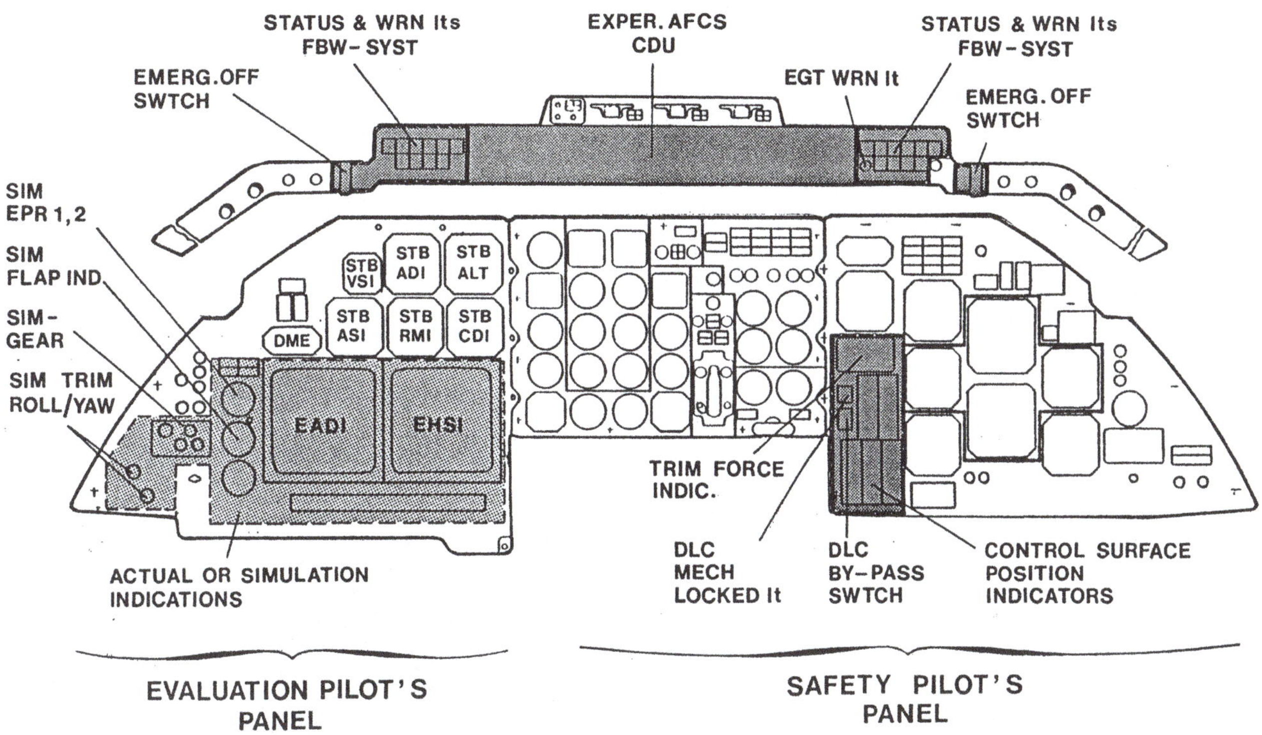

Fig. 9.15 Redesign of cockpit instrumentation

Fig. 9.16 Cockpit, left side

Fig. 9.17 Center console with VBG, experimental thrust and flap levers

the inertial reference system (IRS-CDU). Furthermore, the experimental operating unit (VBG) and additional thrust lever and a landing flap lever as input variables for the simulation model were provided.

Side Console Right

Rotary knob for manual activation of the DLC flaps with a reset switch was provided here.

Side Console Left

Heading and course selection knob was located on the side console on the left.

Control Devices

A conventional control column of the VFW 614 was available as a control device for the experimental pilot. The control column was equipped with two buttons for *Pilot Mark* and *SIM-OFF* (switching off the SIM-Mode). The control forces in all the three primary control axes were generated by an electromechanical control force simulation system KKS (Knüppelkraftsimulation) (see Fig. 9.18). The control force simulation system included spring packages, which were interchangeable with different stiffness. The control force gradients in the flight could be changed through articulated gearbox whose transmission ratio was controlled by an electric servomotor from the onboard computer.

For the basic operation of the VFW 614, for example for training purposes, it was possible to restore the original control configuration with dual control by connecting the control station to the right side using rods. In addition, a sidegrip operating unit (*sidestick*) was developed for the pitch and roll axes. (see Figs. 9.19 and 9.20). The control forces were generated by mechanical springs. In addition small hydraulic dampers were installed to improve the control feel. This *sidestick* was used as a standard equipment until the end of the ATTAS operations (see also Fig. 9.16). The FBW operation was possible with a control column or with sidestick. While using the sidestick the control column was removed.

Cabin Equipment

The cabin seats were completely removed, as well as the kitchen equipment and the toilet in the rear of the fuselage. The seat rails were reinforced to withstand the cabinet loads. In the front fuselage area opposite the entrance door, directly next to the freight door, space was provided for experiment-dependent fittings. Figures 9.21 and 9.22 show some equipment cabinets and other installations. The rear kitchen door on the right side was rebuilt as an emergency exit, so that in the case of emergency the entire crew could

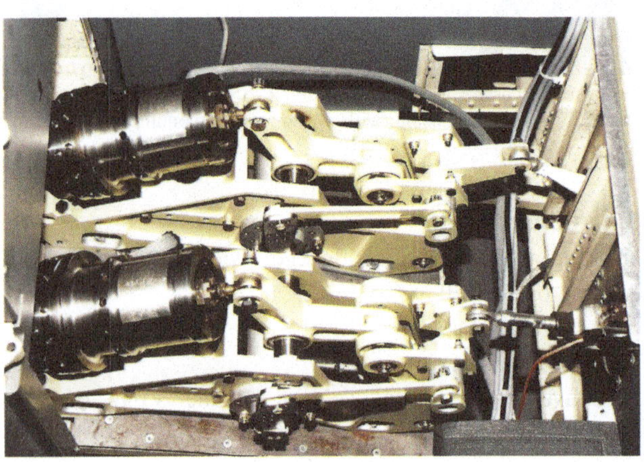

Fig. 9.18 Control force simulation system

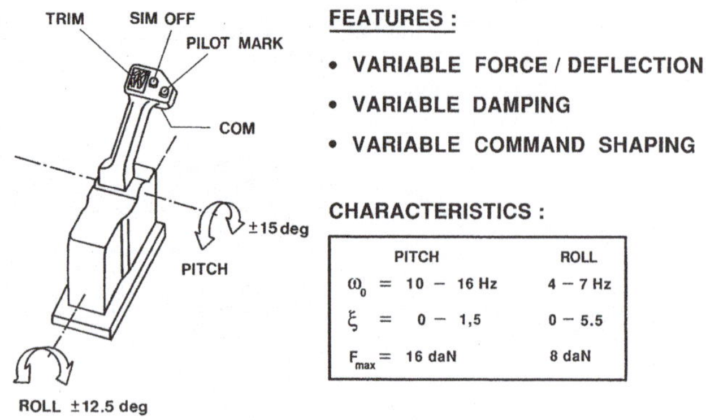

Fig. 9.19 Schematic of ATTAS sidestick with replaceable springs and adjustable dampers

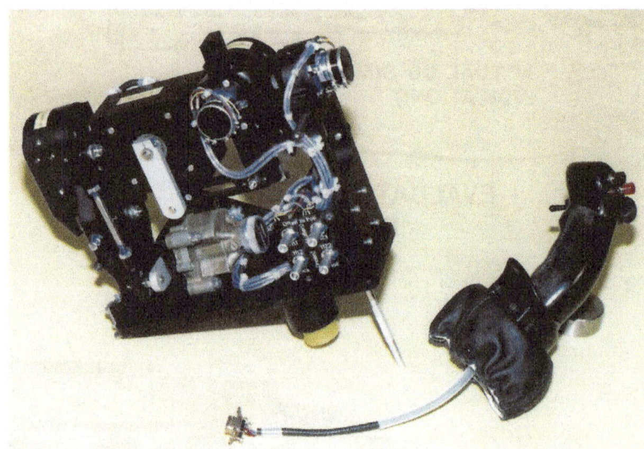

Fig. 9.20 Sidestick device

Fig. 9.21 Cabin view

Fig. 9.22 Measurement system cabinet being operated by *Adolf Zach*

leave the aircraft by parachute. Grab poles were provided on the ceiling to help getting there.

In the area of the hand luggage racks, three levels of cabling were provided: an area for current-carrying supply lines, an area for emitting lines and a region for interference-sensitive data lines. All areas were shielded by steel sheets to minimize interference due to electromagnetic interference. There were four operator stations, each with two seats, except for the operator of the measurement system which had only one seat, because the space for the emergency exit had to be left free. Thus, the maximum test team consisted of seven people besides the onboard mechanic on the jump-seat in the cockpit and two pilots.

Electrohydraulic Actuators

The electrically controllable control variables in FBW and SIM mode included the elevator, rudder, both ailerons, both engines, landing flaps, six direct lift flaps and elevator trim motor.

The elevator and rudder were controlled via duplex servo actuators in the input to the booster actuator of the basic aircraft. For safety reasons, the actuators were limited in their actuating force so that permissible loads were not exceeded. Figure 9.23 shows schematically the details of the duplex pitch control axis.

The right and left ailerons were controlled directly on the aileron tab via a single simplex servo actuator. It was possible to operate the ailerons symmetrically by disconnecting the connecting rope. Each engine was also controlled individually by a simplex servo actuator.

The landing flaps were actuated with a simplex servo actuator in the fuselage. To ensure synchronism they were

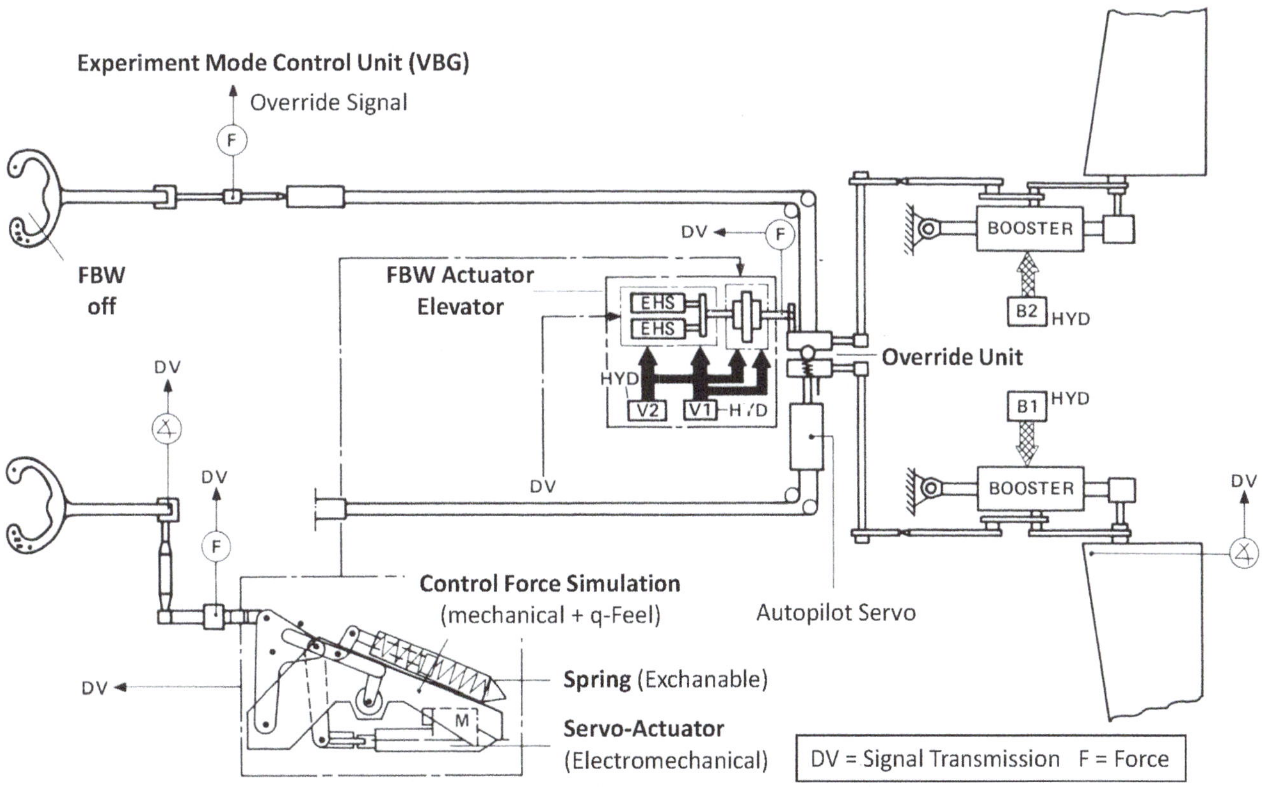

Fig. 9.23 Connection of electrohydraulic actuation system in elevator control

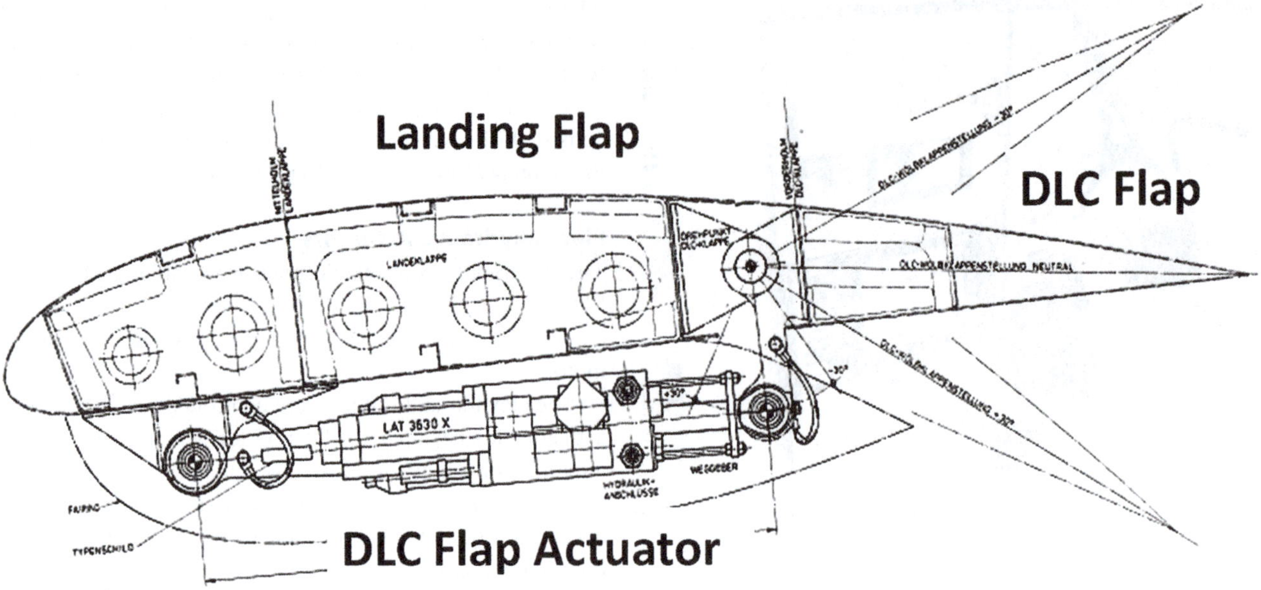

Fig. 9.24 Construction of DLC-flap with actuator on the landing flaps

connected with the mechanical linkage of the landing flap motor of the basic aircraft. The flap displacement in the FBW mode comprised of "fully retracted" up to "14° extended". While operating with the DLC system simultaneously, the landing flaps could be moved from the SP position to 14°.

Direct Lift Control

For direct lift control, additional flaps were installed at the trailing edge of the landing flaps. They served to decouple the heave and pitching motion. For this purpose, the landing flaps were shortened by 45% in its depth and the rear portions were replaced by additional flaps (see Fig. 9.24). The entire system consisted of three flap pairs on the left and right flaps. The DLC flaps could be deflected 35° up and down with a rate of 90°/s under aerodynamic loads. The introduction of three DLC actuation surfaces on each wing (actuation surface redundancy) resulted from safety considerations, namely to reduce the rolling moments in the event of a fault so that the rolling moment due to full asymmetrical deflection of a DLC pair could be compensated by the ailerons.

The electrical actuation was carried out pair wise (outside, center, inside). In the basic operation, the DLC flaps were mechanically locked in the neutral position so that the normal landing flaps were available. The synchronization of the flaps was monitored via a duplex DLC monitor module (asymmetry monitoring), which was part of the actuator concept of LAT. For deviations greater than three degrees or malfunctions, the flaps were automatically, hydraulically locked in the actual existing position. By reset function, the flaps could be brought back into the neutral position.

In the event of a failure of the reset function, the flaps could be switched force-free (hydraulic by-pass) so that they could float freely. While passing through the zero position, they were automatically mechanically locked. A landing with floating flaps was, however, also possible. The dual power supply in the electrical and hydraulic systems was distributed to the individual flap systems according to safety criteria (see Fig. 9.25). The DLC monitor module also monitored the passivation and locking status of all DLC actuators.

Autotrim System

In order to avoid stationary loads on the elevator in SIM mode and to avoid jerky changes of the elevator when switching back to basic operation, an autotrim system was developed and utilized as in the case of the HFB 320. The trimming function was based on the elevator deflection for force-free control, which the VFW had determined from previous flight tests.

Actuator Concept

Hermann Hofer

All control variables required for the experimental operation were operated by electrohydraulic actuators (EHS) of different redundancies developed by the LAT under the leadership of the development manager *Frieder Beyer*. The maximum force level was individually limited for each control axis by means of adjustable pressure relief valves, in

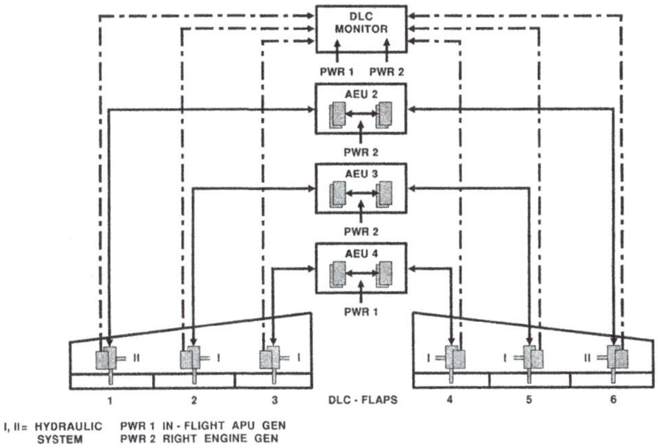

Fig. 9.25 DLC control surface redundancy

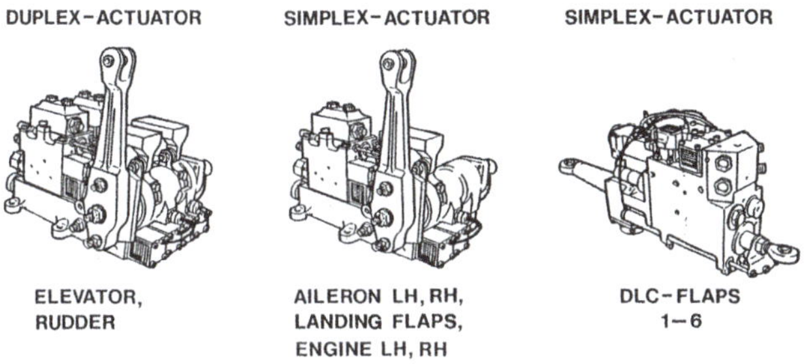

Fig. 9.26 Electrohydraulic actuators

Fig. 9.27 Duplex actuator

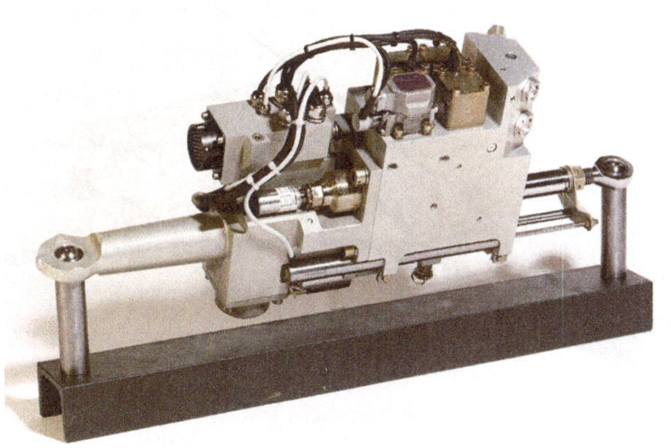

Fig. 9.28 DLC actuator (*Credit* LAT)

Fig. 9.29 Installation of elevator actuator in the tail cone (*Credit* MBB)

order not to exceed permitted loads. The FBW actuator system comprised of a simplex or duplex servo actuators with integrated hydraulic and mechanical interlocks, as well as signal and control circuit electronics (*Actuator Electronic Unit*—AEU). Figures 9.26, 9.27, 9.28, 9.29, 9.30 and 9.31 show the actuators and their installations. The actuator electronic boxes were located above the ROLM computer racks (see Fig. 9.21).

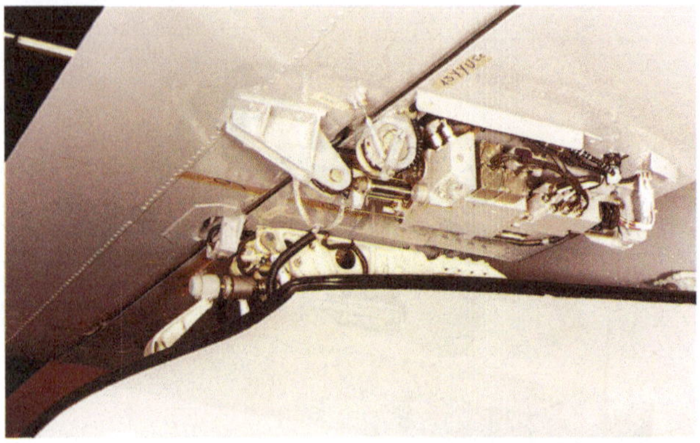

Fig. 9.30 Installation of DLC actuators (*Credit* MBB)

Fig. 9.31 Calibration of DLC flap deflection

The duplex servo actuator consisted of two identical actuator modules, whose piston rods were connected to a common output lever through a soft spring. Apart from this parallel connection at the output, the system was otherwise absolutely separated. Each actuator module was in itself a small electrohydraulic linear drive consisting of the cylinder, two-stage servovalve with position feedback, a bypass valve, and two inductive position transducers. The two actuator modules were mounted on a common base plate which contained the output lever as well as a hydraulically actuated clutch, which coupled the actuator to the mechanical control linkages.

The evaluation of the position feedback, the closing of the control loop, as well as monitoring were carried out in the control electronics. This was duplicated and monitored the correct processing of the input commands by comparing the feedback signals with an electronic model. In the case of deviations exceeding the defined limits, the affected actuator module was switched to by-pass. The advantage of this monitoring concept was a fast fault detection and thereby smaller transients. However, piston jam was not detected, but this error was extremely unlikely because of the high force levels.

The solenoid valves were de-energized in the event of unacceptable disparities between the measured servovalve position and the servovalve model. As a result, the pressure supply to the servovalve was interrupted. This opened the bypass valves and connected the two piston chambers of the affected actuator module. The defective control circuit was thus passivated by these measures. The second actuator module, however, continued to carry out the control commands of the higher level computer (with half the force), which allowed the uninterrupted experimental operation. This was an essential prerequisite for safety-critical flight tests, especially close to the ground.

Following another error in the second actuator module, it was also switched to bypass, thereby the overall function of the duplex actuator was now lost completely. The failure was displayed in the cockpit, and the safety pilot could disconnect the systems by opening the clutches.

The simplex actuator was basically similar to the duplex actuator, in which one actuator module was installed on the base plate. The clutch mechanism could be adopted identically.

The DLC flaps actuation was also designed as a linear electrohydraulic actuator and dimensioned for the predefined strokes and actuation forces. However, it had a hydraulic and a mechanical blocking mechanism as an additional function. On command, the hydraulic locking blocks the actuator in the instantaneous position to prevent asymmetry within a pair of flaps. The mechanical locking system maintains the actuator in the neutral position and thus prevents unwanted flap deflections during landing or when the landing flap is retracted. The locking mode was monitored with micro switches and fed back to the actuator electronics.

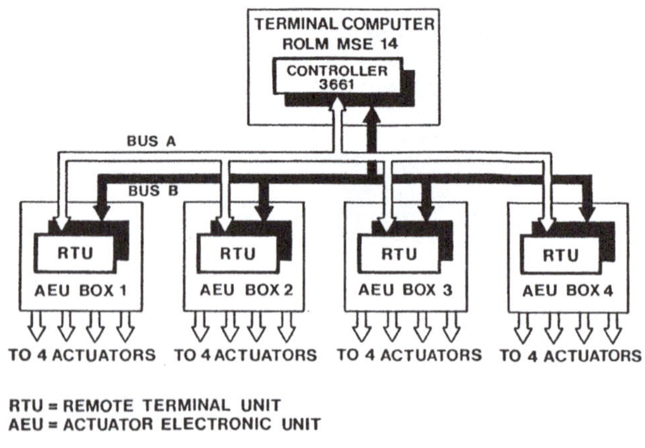

Fig. 9.32 Connection of actuators with the onboard computer system via duplex MIL-Bus 1553B

The entire electronics for the operation of the ATTAS actuators was combined in four identical *Actuator-Electronic-Units* (AEUs). The four AEUs communicated with the central data system via a duplicated serial and bidirectional data bus (MIL-STD-1553, see Fig. 9.32). Its main advantages were the deterministic transmission protocol, the galvanic separation of the bus subscribers among each other, and the availability of the necessary hardware.

Each *Actuator Control Unit* (ACU) was designed as a standard-4 MCU box and contained (1) A duplicated *Remote-Terminal-Unit* that selected the data from the MIL bus, decoded, and verified plausibility of the same before passing it on to the individual actuator electronics. In turn, the individual actual values as well as the system state were reported back to the higher-level computer system. The duplex MIL bus interface contained a current database, which received the commands of the FBW computer via the MIL bus and, in return, dumped system-internal data on the data bus upon request. The *Remote-Terminal-Unit* in each ACU thus constituted the only connection between the computer system and the actuator system. (2) Two microprocessor channels including sensor electronics, each calculated independently the control algorithm, monitored the servovalve by means of model calculations and performed the consolidation of the redundant sensor signals. Via intermediate registers, the processors exchanged the independently computed data and compared them with each other. If a channel detected a larger deviation, the affected actuator module was switched to bypass and the failure was displayed. Each actuator module, in conjunction with the two microprocessor channels, formed a self-error-detecting system.

Two power supply modules served to provide the secondary voltages for the analog and digital circuits. The primary supply of the ACUs was via a fail-safe 28 V DC bus system.

The distribution of the control channels to the four AEU boxes was selected in such a way that, in the event of a system component failure or a complete AEU being lost, only one control channel or, in the case of the DLCs, only one flap pair was affected. This resulted in a *Fail-Operational/Fail-Safe* behavior for the entire FBW system.

The entire AEU functionality was implemented in software and was developed in Assembler language. Like the hardware of redundant channels, the software was also similar, that is, designed the same way. The similar architecture of the FBW actuator system did not provide systematic protection against a generic error. This extremely unlikely event was, however, covered by the safety pilots, who could take over the aircraft at any time.

At the time of installation, the actuator system was an advanced concept using the latest available techniques, such as digital signal processing, data bus oriented system architectures, and high flexibility through software-based functionality. The concept of the ‘intelligent’ actuators, which close the control loop locally and carry out the redundancy management independently, had also proved to be successful. However, the actuator electronics design in the form of remote electronic units, which were installed centrally in the fuselage remote from the actuator, caused a considerable discrete cabling effort.

Electrical Power Supply

The electrical power supply for the experimental systems was duplex redundant, designed independently of the basic aircraft supply. The basic aircraft was supplied via the two generators of engines 1 and 2, each with 20 kVA. For ground operation, the power was supplied by the auxiliary gas turbine (*Auxiliary Power Unit*—APU), which also drove with 20 kVA the generator 3 or via a ground aggregate.

During the flight, generator 3 of the auxiliary gas turbine, which could also be operated in flight, and the generator 2 of the right engine, were available for the experimental electronics. In addition, a battery supply was provided for the test operation in case of failure of the DC power supply. Power supplies of 115/200 V—400 Hz, 26 V—400 Hz alternating current and 28 V DC were available.

Hydraulic Power Supply

The hydraulic power supply for the new electrohydraulic actuators was also duplex redundant. For this purpose, two separate hydraulic systems were additionally installed and the required pressure for each was generated by means of an electric pump and by the two basic hydraulic systems via *Power-Transfer-Units* (PTU). With the aid of the PTUs, the pressure was transferred without mixing the hydraulic circuits.

Onboard Computer Structure

Once again based on the experiences gathered with the previous in-flight simulator HFB 320, an onboard computer network system was designed to meet the following essential requirements: (1) adequate computing power for all real-time tasks expected in the future using a higher level programming language (targeted calculation cycle time of less than or equal to 20 ms), (2) sufficient redundancy for tests in safety-critical flight domains, such as landing, (3) airworthy hardware and interfaces for implemented systems (ARINC 429, MIL-BUS 1553B, etc.), (4) make available all operational functions to the user (experimenter), (5) provide all data (measured variables) required for the test, freely programmable computers to the user, and (6) a real-time capable ground-based simulator system, mostly compatible with the onboard system, for system development and software validation (operation and use).

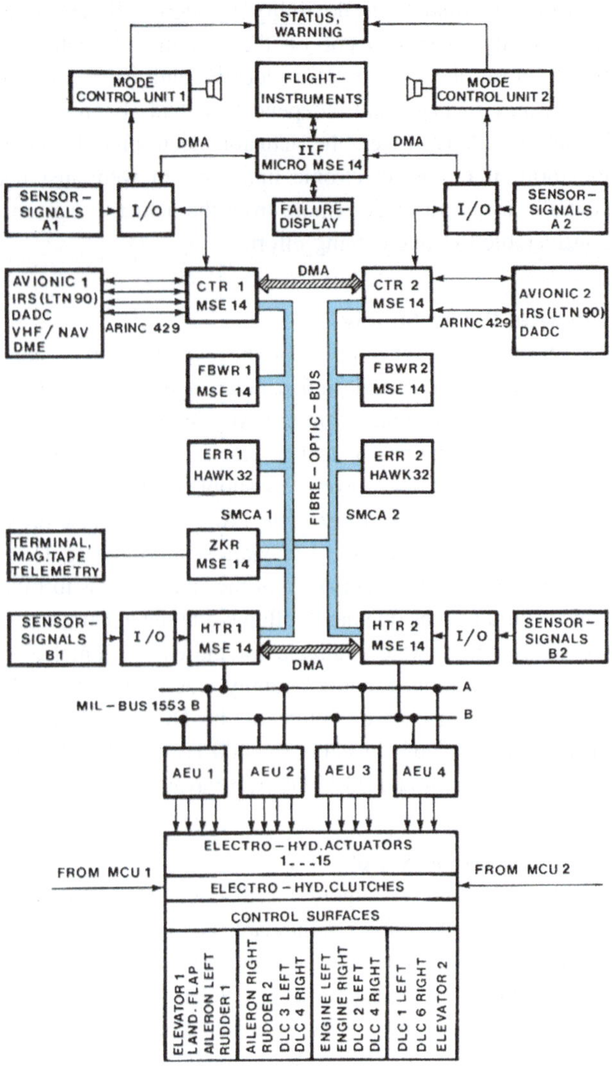

Fig. 9.33 Duplex onboard computer system

Fig. 9.34 Onboard computer components for one lane (simplex configuration)

The core of the Fly-by-Wire system was the onboard computer system with computer components from ROLM Mil-Spec Computers, on which all functionalities were implemented in computer programs (software). In order to meet the high demands of computing power with the processors available at that time, a multi-computer system with parallel processing was designed (see Figs. 9.33 and 9.34) [11].

The overall duplex system consisted of two parallel, identically constructed computes, in order to ensure a *Fail-Passive* behavior in the event of a fault. The two computers communicated over an optical serial bus system. Thereby it constituted a fly-by-wire/light system (FBW/Light). After delivery of ATTAS to DFVLR, only one computer branch was initially installed. The second one was subsequently retrofitted as part of the envelope expansion (see Sect. 9.1.8).

Functional Distribution in the Computer Network System

The four computers in each channel of the duplex system were assigned to the four function groups: (1) terminal functions, (2) control functions (fly-by-wire), (3) experimental functions, and (4) communication functions.

The terminal functions included all inputs, outputs, and interface to the connected autonomous systems. The cockpit terminal computer (CTR) established the connection to the measurement and display signals in the cockpit, with the tasks of the ARINC interface control and the feeding the flight instruments. In addition, the data communication was established in this computer with the experimental mode operating device, which represented the operator interface between mechanical control and experimental equipment. The rear terminal (HTR) processed the measurement and control signals from the rear of the aircraft and operated the MIL-BUS 1553B, the interface to the electrohydraulic actuation systems (EHS) for actuating all control surfaces and engines with a total of 15 actuation systems.

In the Fly-by-Wire computer (FBWR), the control functions were implemented between the pilot input and the positioning commands for the actuators. In addition, it catered to the switching operations for switching off the actuators (coincidence) and switching between SIM-mode and FBW-mode.

In the experiment and control computer (ERR), the experimental functions were freely programmable. The user was offered an interface to access the variable of the actuation system. The data relevant to the experimenter were made available to him in calibrated form (engineering units), which he could recalculate and feedback as control variables to the control system. Thus, the entire in-flight simulation functionality was processed in the ERR.

The central communication computer (ZKR) fulfilled the task of recording all signals appearing in the onboard computer interconnection system and transfer the data to the desired memories (magnetic tape) and connected telemetry receiver. The data recording was done continuously with a cycle time of 20 ms (1765 words), which corresponds to 88.25 kWords/s. A maximum of 160 kWords/s was possible.

Computer Communication

A ring bus system with optical fiber transmission was used for communication between the computers. The bus system, denoted SMCA (Serial Multiprocessor Communications Adapter), was developed by the ROLM, which allowed connecting up to 15 computers [12]. The maximum length of optical cables between the computers could be up to 2000 m.

A transducer box on each computer converted the computers signals into an optical serial signal. Each transducer was a transmitter and a receiver. The organization of sending and receiving was organized via a rotating "token". The transducers operated independently of the connected computers, so that the communication was maintained even in the event of a computer failure. The FBW and terminal computers were airworthy 16-bit computers of the type MSE 14 of ROLM MIL-Spec Computers in ATR format. The experiment computer (ERR) was a more powerful 32-bit computer of the type ROLM Hawk/32. The decisive factor for the choice of optical connection between the computers was the interference immunity against electromagnetic interferences.

The electrohydraulic actuation systems were connected via a MIL-bus 1553B each. The overall performance of the computer network system was 3.1 MFLOPS (*Mega Floating Point Operations per second*). For the fly-by-wire operations alone 1.8 MFLOPS were needed. The ERR had a computational capacity of 1.3 MFLOPS. The main memory in ERR was 2 MB.

A total of 417 measured variables and 262 output signals were processed, which were transferred via the interfaces such as A/D, D/A, ARINC 429, MIL-BUS 1553 B and digital I/O. In addition to the analog signals, 292 discrete input and 210 discrete output signals were also processed.

Functionality of Experimental System

The operation of the real-time program was divided into three modes: (1) Basic mode, denoted BASIS-MODE, (2) Fly-by-wire mode, denoted FBW-MODE, and (3) Simulation mode, denoted SIM-MODE. The corresponding states were displayed to both pilots in the glare shield.

In the BASIS-MODE, in which ATTAS could only be controlled by the right-seat safety pilot, the signal were connected with external devices and facilities, and the communication function was activated. In this mode, experiments could be carried out which did not have to access the controls (for example, device testing). ATTAS was used just as a test vehicle in such applications.

In the FBW-MODE, the aircraft was controlled electrically by the experimental pilot on the left. In this case all the control laws of the VFW 614 basic aircraft were reproduced 1:1. The main purpose of this mode was to facilitate easy handling of switching between mechanical and electrical controls. Thus, before the actuators were connected, the actuator position had to be matched to the current surface position (coincidence), so that a jerk-free connection could be ensured. In addition, monitoring and test functions were performed before switching to the actual experiment mode. Later, the FBW mode was supplemented by controller-assisted control laws (for example, a rate-command-attitude-hold function when using sidesticks) and autopilot functions to more easily and precisely establish a desired reference condition, such as altitude and airspeed, for experiments in the SIM mode.

The SIM-MODE provided the experimental condition. This enabled active interventions in flight controls in all axes with full authority, such as those required for in-flight simulation, autopilot functions or navigations functions.

Fault Detection and Handling in Duplex Computer Network System

The duplex computer system was implemented in the second expansion phase (see Sect. 9.1.7). Fault detection was performed by comparing the data in both computers, which communicated with each other via a Direct Memory Access (DMA). Each computer compared the data of the other with its input data. An error flag was set if a difference was detected over several computational cycles. In the case of error-free signals, an average value was formed from both signals and made available to both the computers. The data was processed bit by bit, which allowed the results to be checked for a parity bit.

Program Structures

Three basic program (software) structures were selected for the individual computers. The programs were divided into event-oriented (interrupt) and sequential flows. The interrupts were triggered from the cockpit terminal computer. Time noncritical functions were executed in the background for which the remaining processor time was used.

Procedures for Program Development

The program development was carried out based on a step-by-step plan, that was already applied in the of in-flight simulation development with HFB 320. The five development stages were characterized as: (1) analysis and theoretical design, (2) verification of the functions/simulation computations, (3) program integration/coding, implementation,

(4) validation in the system/real-time simulation in the ground-based simulator, and (5) investigations in flight test.

ATTAS software was developed based on a top-down approach. The software refinements passed through one after the other, and were completed with a review before starting the next development. Six development phases were selected, which were completed with the following documents and actions: (1) load sheet, (2) coarse design, (3) fine design, (4) encoding and module test, (5) integration and testing, and (6) validation.

Facilities for Software Development

The ground system and the ground-based simulator were mainly used for (1) development and testing of operational FBW software, (2) development and testing of user software, (3) maintenance of hardware and software, and (4) pilot training and preparatory flight test optimization.

The ground system (fixed-base simulator) was identical to the onboard computer system, the cockpit and all interfaces in the ATTAS (see Figs. 9.35, 9.36 and 9.37) [13]. The computer network system provided a powerful program development environment, in which a 32-bit computer of the type MV/6000 from Data General was the central component, with the operating system AOS/VS (*Advanced Operating System/Virtual Storage*). Since the onboard computer system was available for software development together with the ground system during the development phase, the components could be used optimally. Thus, the onboard computer system could be connected to the ground system via the optical bus. This allowed the developer to write the software program from his workstation and transfer it directly to the target computer in order to test it under real conditions on the target computer.

Fig. 9.36 ATTAS simulation cockpit (*external view*)

Programming Languages in Real-Time Operation

The onboard computer programs were developed in the higher programming language FORTRAN 77 (from Data General), using the real-time operating system ARTS (*Advanced Realtime Operating System*) of the company

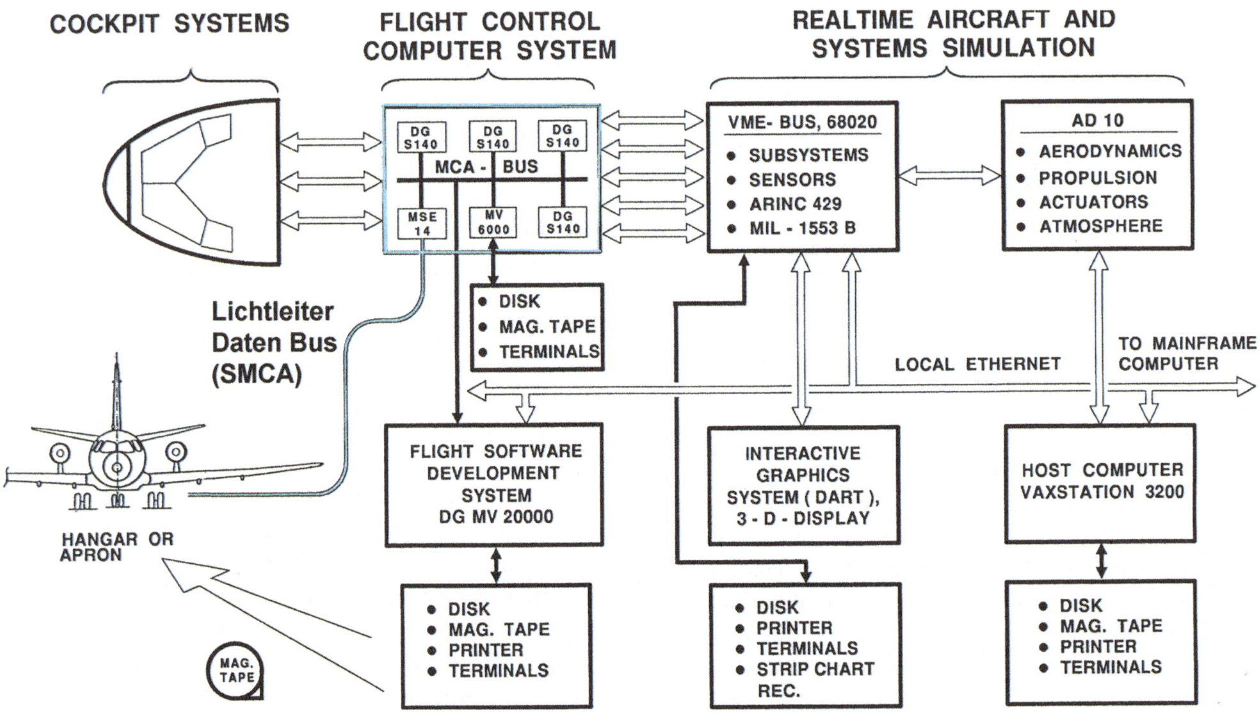

Fig. 9.35 ATTAS ground-based simulation system

Fig. 9.37 ATTAS simulation cockpit (*inside view*)

ROLM. Some interface drivers required special assembler programming.

Software for Program Development

The extensive software task required strict sequence planning and verification (*configuration management*). The software product TCS (*Text Control System*) from Data General was used to organize software development and communication between the developers [14]. The entire source code was documented in TCS files, including additional information such as date, comments on changes (who changed when, what, why). This allowed an unambiguous description of the modifications and past developments. Thereby all other affected program parts were automatically regenerated when a module was modified. On executing the *Build*, the system organized itself to create final versions resulting from the changes [15].

Software Validation

The ground-based simulator truly replicated the aircraft dynamics with all its sub-systems, so that the entire software could be tested under real-time conditions. All control and display elements of the cockpit and its hardware were mostly identical to those in the actual aircraft. The onboard computer network system was reproduced on the ground through same structure by software-compatible computers of the type Data General S140 and MV6000.

The VFW 614 aerodynamics, flight dynamics, electro-hydraulic control systems, including all data interfaces were simulated in real time on the hybrid computer EAI Pacer 600 and the multi-processor system AD 100 from Applied Dynamic Inc. (ADI). All interfaces were plug-compatible with the aircraft. Through the use of the fiber optics, it was possible to connect the ground system with ATTAS aircraft in the hangar, which was 500 m away. This enabled to include the onboard systems in the tests. This feature was invaluable during the extensive test phase.

Data Gathering

The data acquisition and onboard recording system (see Fig. 9.38) included all the data in the overall system, which included sensors for measurement of (1) flow variables, flight log, (2) air data (*Digital Air Data Computer*, DADC), (3) control surface and actuator positions, (4) engine data, (5) body-fixed rotational rates and linear accelerations, (6) aircraft attitude angles, (7) inertial data (*Inertial Reference System*, IRS), (8) navigation data, and (9) switch positions.

In addition, all the data from the actuators and selected data from the fly-by-wire system, for example, variables calculated in the FBW system from the experiment, were also gathered. The data was recorded onboard on a digital magnetic tape recorder via the central communication computer (ZKR). The data collection system was also connected to the PCM-telemetry equipment, through which a selected number of channels could be transmitted to the ground station (downlink). The data transmission to the aircraft (uplink) from the ground allowed the access to the experimental functions in FBW system, such as the navigation of the aircraft from the ground. The measuring system had a test and calibration computer connected via an IEEE 488 bus. Furthermore, a 8-channel recorder was available onboard to record selected data in analog format.

Avionic Systems

The onboard avionics included (1) digital air data computer (DADC), (2) inertial reference system (IRS, laser gyro), (3) instrument land system ILS, and (4) radio navigation systems, VHF omnidirectional range (VOR) and distance measuring equipment (DME).

A Dornier flight log was used to measure the angles of attack and sideslip, as well as True Airspeed (TAS) (see Fig. 9.39). The flight log was a high-precision instrument with a small propeller, which was mounted on gimbals, so that the propeller tip always pointed in the direction of the inflow. Through its orientation the direction of flow could be determined accurately. The true airspeed was measured by the rotational speed of the propeller. The flight log construction was, however, highly sensitive and as such it was not used under adverse weather conditions, such as rain and icing conditions.

Data Analysis and Data Handling

The onboard recorded data on the digital magnetic tape recorder were processed and analyzed offline using the data

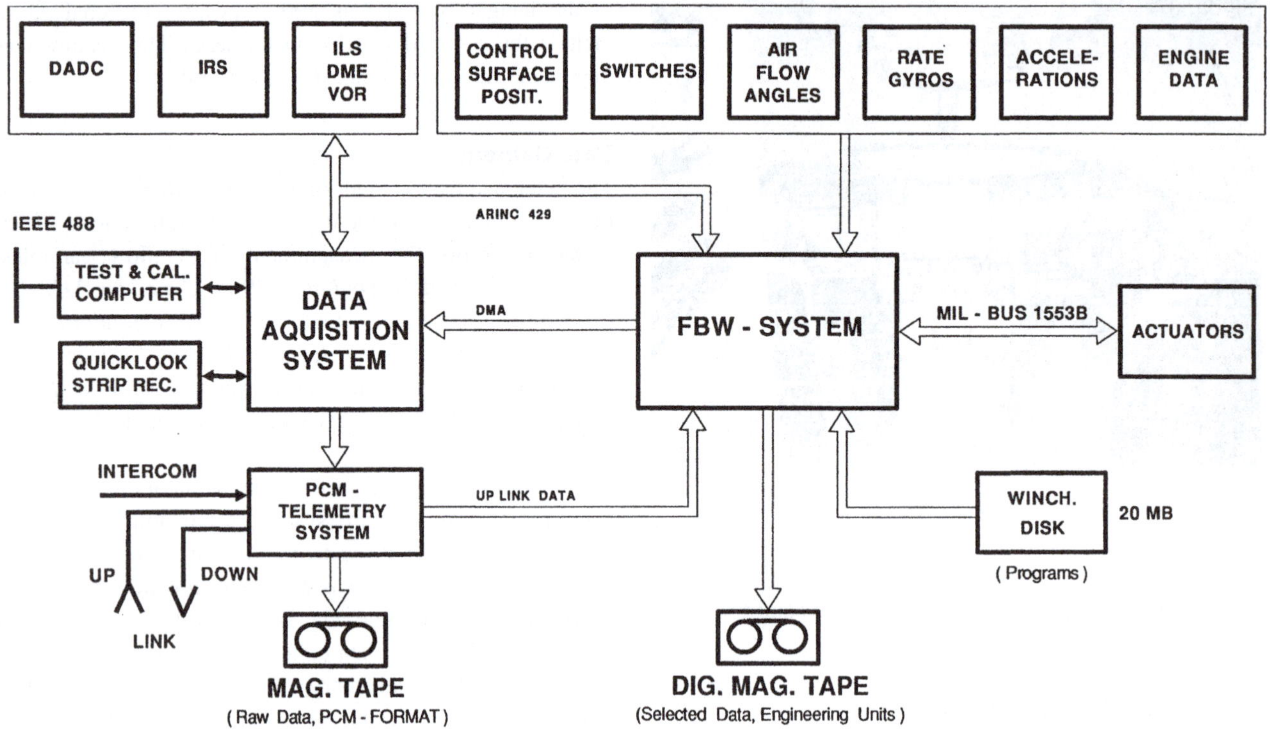

Fig. 9.38 ATTAS onboard measurement system and data recording

Fig. 9.39 Noseboom with Dornier flight log (*Credit* G. Fischer)

analysis program DIVA (*dialog-oriented experimental data analysis*), see Sect. 7.2.21.

List of Signal

The ATTAS signal list was a very important document describing all available signals. This list includes all the signals appearing in the overall system, which were unambiguously identified by sensor location, type of signal source, data interface, sampling rate, accuracy, measuring range, etc. It comprised of 3764 signals with 26 columns for data descriptions for each and 7 additional tables.

Telemetry Facility

The telemetry ground station had a large transmitting and receiving dish. The reception range was up to 250 km depending on the flight altitude. Selected data were transmitted in PCM format, which were recorded on magnetic tape in the telemetry ground station. In addition, two 8-channel recorders and several monitors were available in the ground station, so that the signals could be monitored in real-time. The telemetry link had its own radio connection to the aircraft so that the flight experiment could be overseen from the ground. The radio communication from the cockpit with the air traffic control system could also be listened to on the ground.

Antennas

ATTAS had all the antennas required for navigation (see Fig. 9.40), whose signals were available as measured

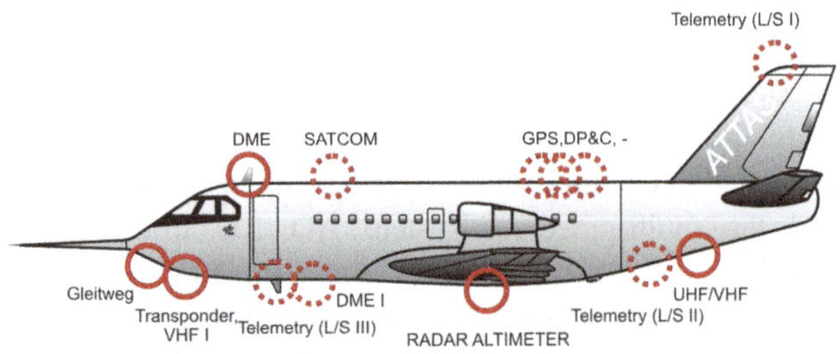

Fig. 9.40 ATTAS Antenna system

variables, such as (1) DME, (2) Transponders, (3) UHF, VHF, (4) Telemetry (L/S band), and (5) VOR and DME. A provision was made on the top of the fuselage to enable installation of user-specific antennas.

9.1.6 In-Flight Simulation/Model Following Control

As in the case of HFB 320, in the present case, too, the model following control with explicit model and the inverted model of the host aircraft dynamics in the feedforward loop were used in the ATTAS in-flight simulation (see Figs. 9.41 and 9.42). As already elaborated in Sect. 3.3, in this approach the motion equations of to-be-simulated aircraft (model) are calculated in real-time in the onboard computer, and the output variables of the model are fed to an inverse model of the host aircraft (here ATTAS). The inverse model then yields the control commands for all the control surfaces of the host aircraft so that the motion variables of the host aircraft match with those of the model. The deviations are automatically minimized by the feedback loop.

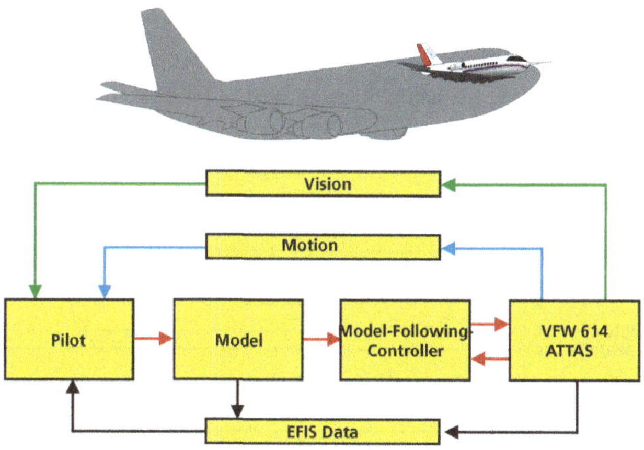

Fig. 9.41 Principle of in-flight simulation

The simulation quality in the present case was improved by the accounting for the nonlinear effects in the inverse model of the VFW 614. Figure 9.43 shows clearly the influence when nonlinearities were accounted for in the model description of ATTAS [16–18]. The results from the parameter estimation of the VFW 614 were incorporated in the feedforward loop to improve the model following quality.

A criterion in the frequency domain, based on unnoticeable dynamics, was formulated for the assessment of the model following quality. As long as the amplitude and phase of the frequency response remained within a defined range, the behavior was practically identical for the pilot. Thus, the ratio of the transfer function to the pilot input of a control axis of the VFW 614 in the SIM MODE to that of the model to the same input should remain within specified limits. In this way the model following quality could be verified in flight test. As a typical example, the result of an in-flight simulation of N 250 aircraft of the then Indonesian company IPTN (today: Indonesian aerospace) with ATTAS for the pitch axis is shown in Fig. 9.44. It shows that the error dynamics lie within the permissible band.

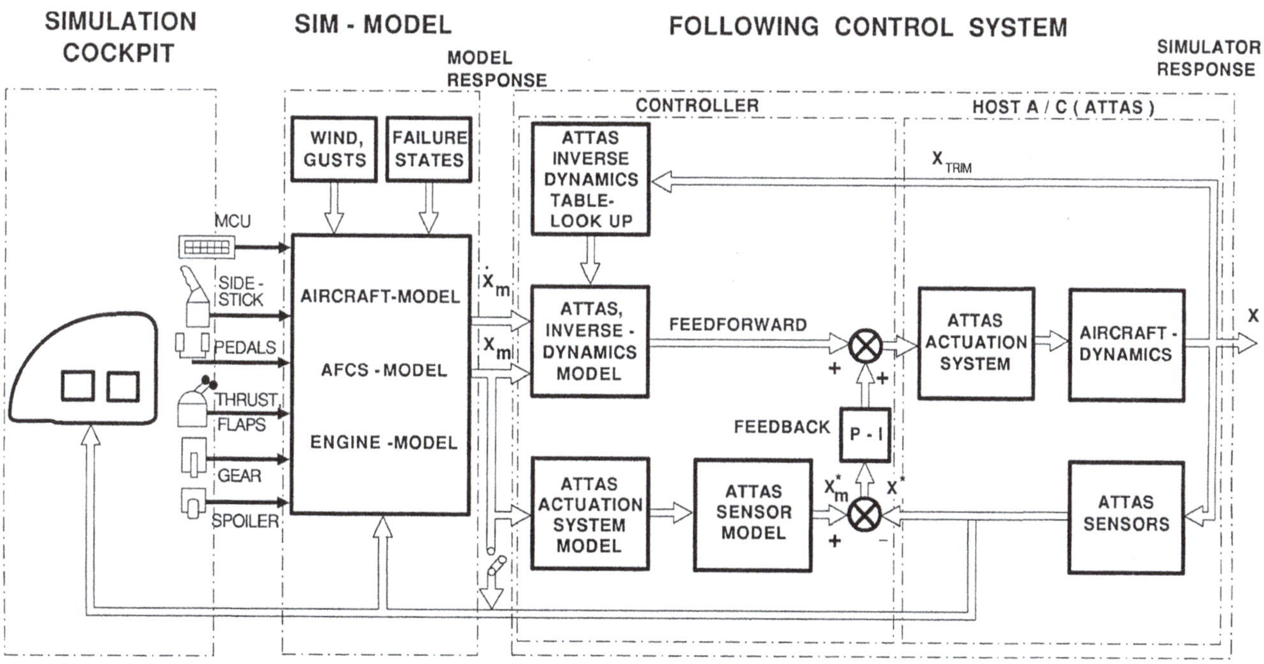

Fig. 9.42 ATTAS model following control

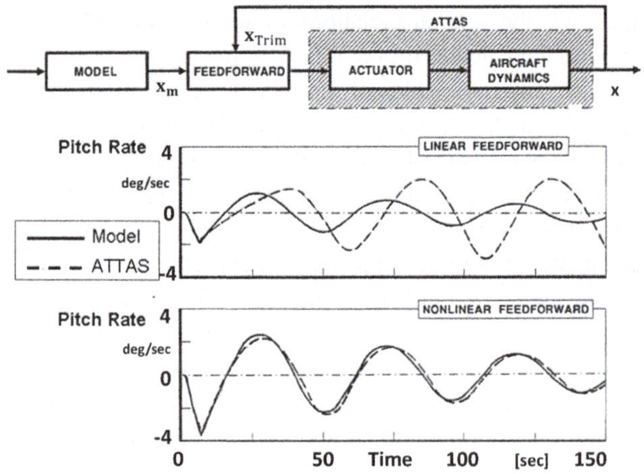

Fig. 9.43 Inprovement of simulation quality by account for nonlinearities in the feedforward loop

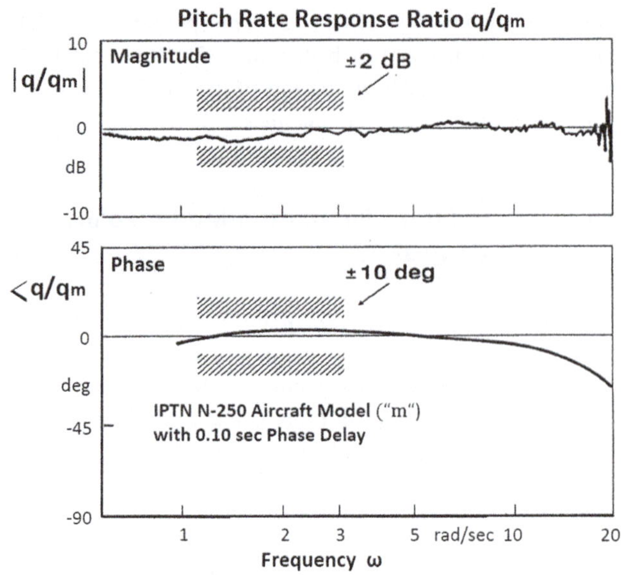

Fig. 9.45 Achieved model flowing quality in flight (model IPTN N-250)

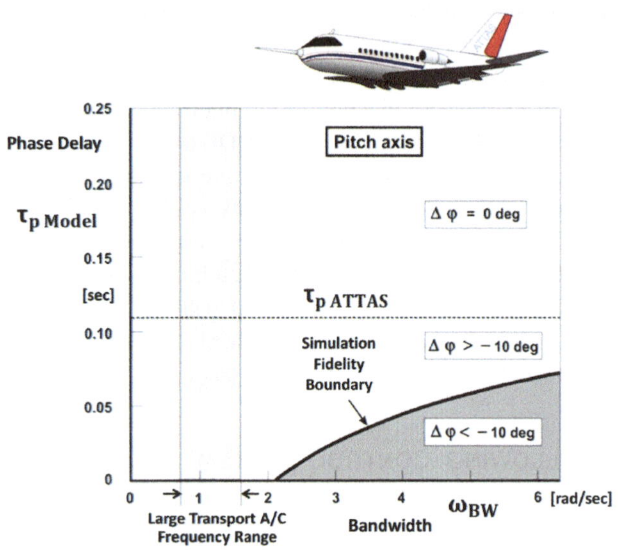

Fig. 9.44 Quality criterion in frequency domain for model following control

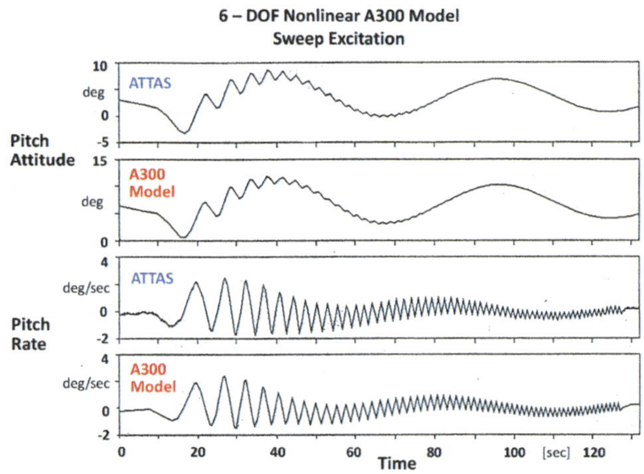

Fig. 9.46 Comparison of ATTAS and a 300 model to frequency seep input

In addition, a criterion was developed that defined the maximum permissible phase shift between model and simulator aircraft (ATTAS) as a function of frequency. The equivalent time delay in the system was used as a parameter for the phase shift, which resulted from the computation times and the actuator delays (see Fig. 9.45). Further tests with a complete Airbus A 300 model, as a typical example of a large transport aircraft, were carried out in order to demonstrate the model following quality (see Fig. 9.46). Details of some selected flight test programs for in-flight simulations are provided in Sect. 9.2.

9.1.7 ATTAS Upgrade

The ATTAS upgrade performed during 2000–2003 pertains to the improvement or replacement of components of experimental system to ensure operation till 2013. The procurement of new components was necessary because: (1) devices were no longer maintainable, (2) replacements for equipment were no longer obtainable, (3) device technologically became outdated and performance data were superseded, (4) it was essential to eliminate the technical deficiencies, and (5) new regulatory requirements had to be met. The investment volume for these upgrade measures amounted to 3.6 million DM (as of 2000).

The upgrade measures included:

(1) *New powerful experiment computer* (*ExEC, Power PC, VME bus*): The hitherto experiment computer (ERR) was replaced by a new experiment equipment computer (ExEC), which catered to the user applications [19]. It had a powerful Power PC 'PPC4' CPU and a VME bus interface for connection to the ROLM computers.

(2) *New digital air data computer* (*DADC*): In the basic system of the ATTAS, a new digital air data computer was required, which supplied the interfaces and signals required for the operation of the TCAS (*Traffic Collision Avoidance System*).

(3) *New flat screens in the cockpit* (Fig. 9.47): The hitherto graphic displays from the Airbus A310 program used in the ATTAS were now obsolete and had some limitations in the programming of the graphic symbols. They were, therefore, replaced by two larger flat screens from Siemens. The graphics processor allowed complete freedom in programming. The screens were 10.4 in = 26.4 cm diagonal in size.

(4) *New 5-hole probe* (Fig. 9.48): The previously used flight log had restrictions in flight operation during adverse weather conditions of rain and icing. Therefore, it was replaced by a heatable 5-hole probe. This required the development of new electronics for processing the data from the measured air pressures of the 5 holes (see Fig. 9.49). The operation with the flight log was still possible.

(5) *New TCAS* (*Traffic Collision Avoidance System*): The TCAS system was essential for the operation of ATTAS to meet the governmental regulations. When two aircraft come dangerously close to each other, the TCAS gives the pilots evasive instructions to avoid a collision.

Fig. 9.47 ATTAS upgrade—flat screens in cockpit

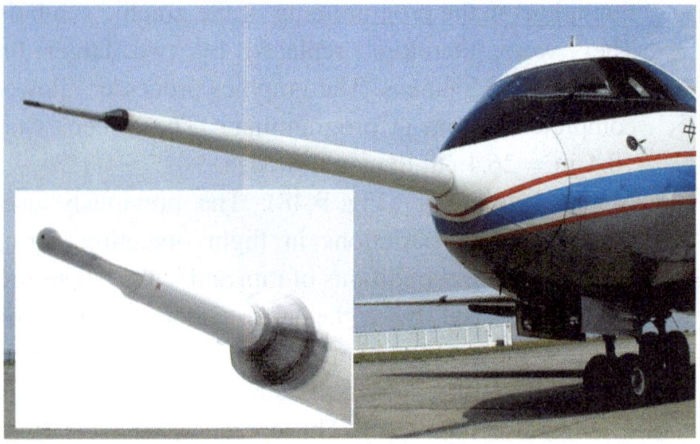

Fig. 9.48 ATTAS upgrade—5-hole probe for air data

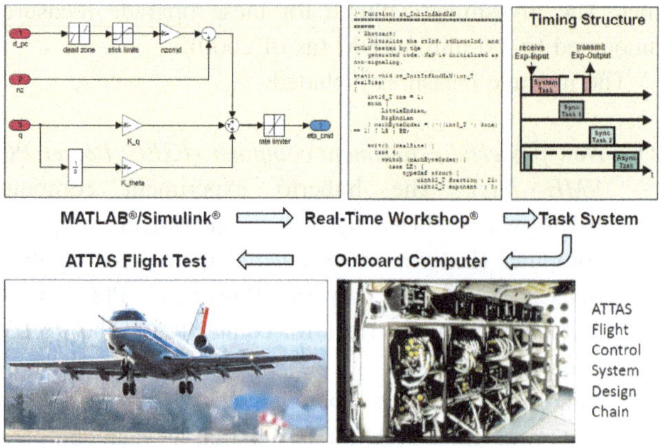

Fig. 9.50 ATTAS flight control system design chain

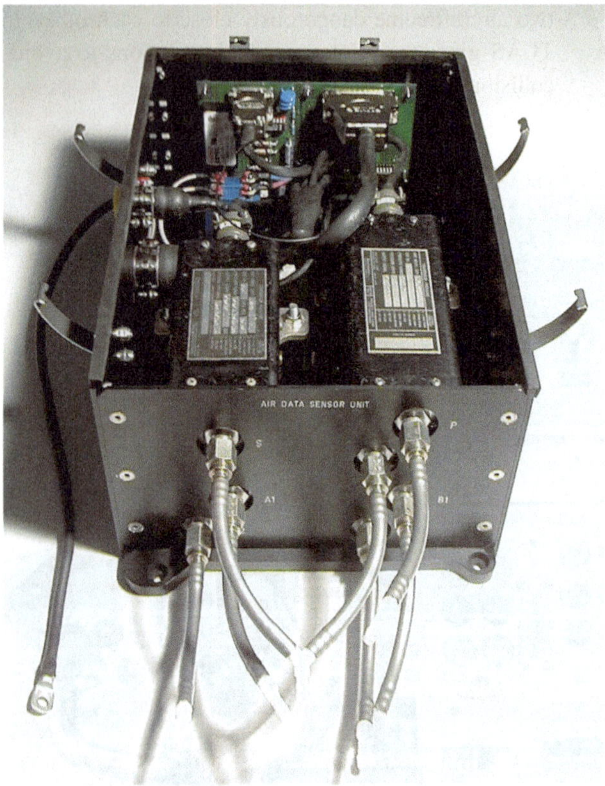

Fig. 9.49 ATTAS upgrade—electronics box for 5-hole probe

(6) *VHF/COM, Mode S transponder, based on new operating regulations.*

(7) *New visual system in ground-based simulator*: In order to improve the test preparation and optimize the test sequence with the pilot in the control loop, a visual simulation (LCD projectors) with a view angle of 120° horizontally and 45° vertically was developed in the ground-based simulator.

(8) *New user interface with MATLAB/SIMULINK*: Together with the new ExEC (Power PC), the interface to the user was improved considerably by the use of a model-based programming software MATLAB/SIMULINK for the modeling of technical systems due to its flexibility and simplicity [20]. SIMULINK enables the programming of dynamic processes by simple graphical linking of function blocks. For this purpose, a software was developed that allowed SIMULINK programs to run directly in the real-time application in the ATTAS experiment computer, see Fig. 9.50.

(9) *New Boot/Recording Server (BRS) and Quicklook* (see Figs. 9.51 and 9.52): On the one hand, the BRS served to provide the software for the various computers of the DV system (ROLM) and the ExEC, and on the other hand, to time stamp and record data streams from up to five different sources (DV system, measuring system, ExEC, twice video/audio) via the so-called 'front end LAN'. The data was recorded on a removable hard disk, replacing the obsolete magnetic tape technology.

These data were provided to arbitrary computers of the experimenter over the so-called 'experiment LAN' (see Fig. 9.53) and via the so-called 'client LAN' to several Quicklook PCs at the various operator stations. This gave the flight test engineer and the experimenters onboard a comfortable display of measurement and test data. With the BRS it was also possible to load the programs into the ROLM computers.

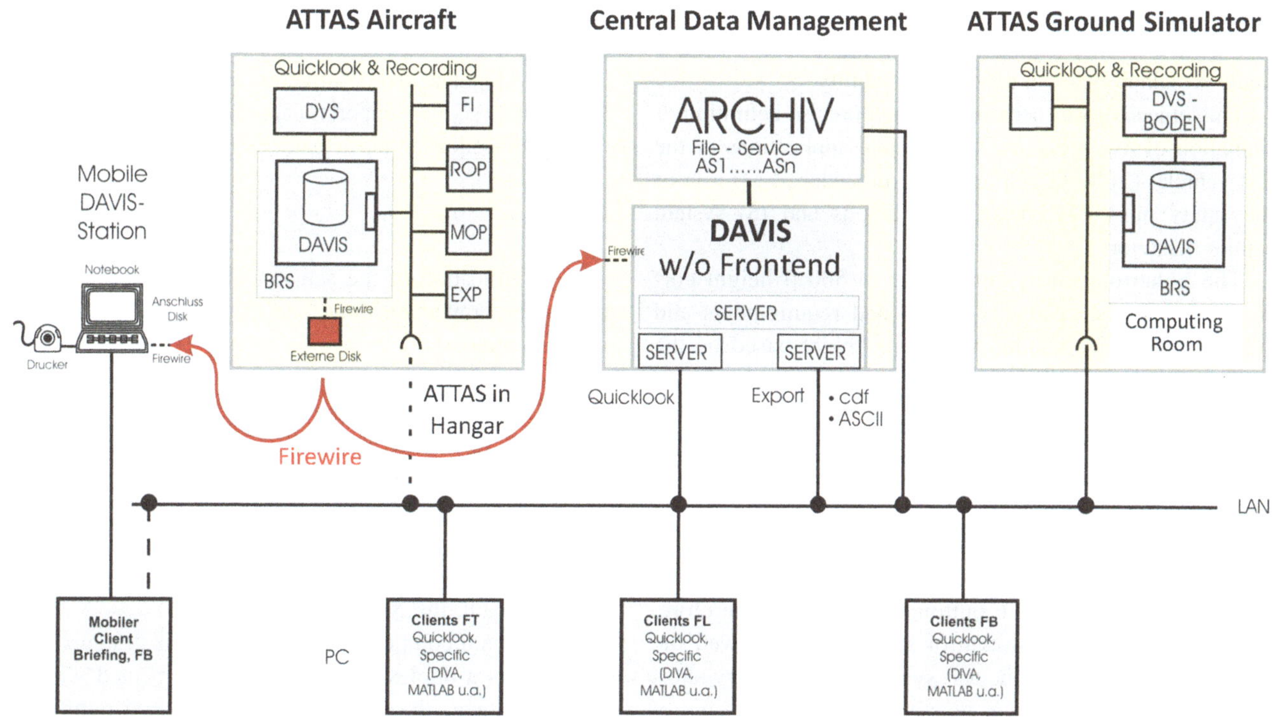

Fig. 9.51 ATTAS upgrade—onboard data gathering and display

Fig. 9.52 ATTAS upgrade—flight test engineer *Michael Preß* on FI cabinet with new operating console and display

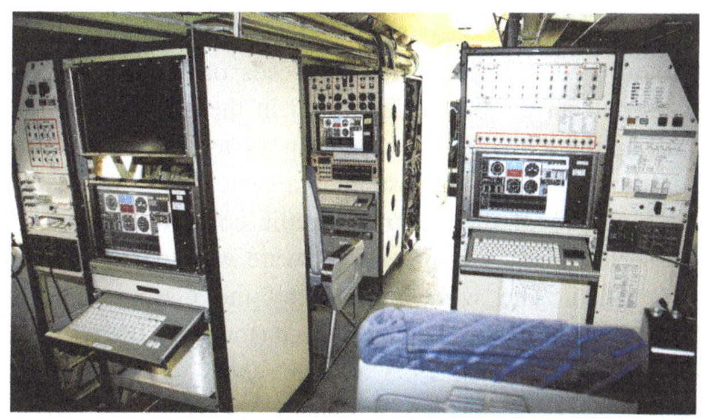

Fig. 9.53 ATTAS upgrade—flight data display on all work stations

9.1.8 ATTAS Envelope Expansion

To take into account the possible failures of the technical equipment, the flight domain of ATTAS aircraft in the electrical mode of operation was limited to a minimum 500 ft above ground. As a result, flight test were not possible below 500 ft up to landing in FBW or SIM mode. This flight domain was particularly important, in particular, for the assessment of the handling qualities, as in this phase the pilot experiences the highest workload and control activity and the system characteristics play a dominant role.

Therefore it was planned to carry out the technical modifications on ATTAS in order to open up the flight domain to landing in the SIM-mode. For this purpose, it was necessary to supplement and modify the existing experimental system and to certify the same according to the federal regulations. In addition, it was necessary to ensure that the risk to the equipment and crew in the case of system faults remained below a residual probability.

This objective was achieved through precise definition or limitation of five measures: (1) possible applications of the test vehicle, (2) flight and operating procedures, (3) role of the safety pilot, (4) operating conditions and (5) system failure behavior.

The hitherto security concept up to 500 ft height corresponded to the originally formulated requirements and was certified for operation up to this height based on the hardware solution FBW-OFF button on the control column. However, it turned out that for an operation below 500 ft with the worst case of simultaneous control surface hardovers the aircraft reactions were so large that they could not be controlled by the safety pilot without the risk of undesired ground contact. The reason for this was that it took too long a time to switch off the experimental system via the FBW-OFF button. ATTAS had, therefore, two force sensors in all control axes, which allowed the safety pilot to separate the test system within a very short time by instinctive counter-reaction to its moving control devices.

In flight tests with control surface hardovers and for the determination of the shutdown times and fault transients, it was found that it was not possible to handle such faults without limiting the operating speeds of the control surfaces and an amplitude limitation in the aileron. For this reason, the rates of the actuation systems were adapted based on the flight test results by a switchable limitation in the software of the electrohydraulic control system from LAT. The already approved software was therefore to be re-certified. The work, funded by the German Federal Ministry of Education and Research (formerly BMFT), was carried out between 1995 and the end of 1996 and was shared by DLR, DASA Bremen (formerly VFW, MBB) and LAT as follows:

DLR:

(1) Overall responsibility, (2) proof of the ability to take over in the event of a hardover, (3) specification of actuator rate limits, (4) integration of components developed by DASA and LAT, (5) installation of the second lane of the duplex on-board computer system, including software, (6) Sensor technology, (7) system tests, and (8) certification by LBA and MPL of DLR.

DASA:

(1) Development and construction of a reliable switch box, (2) Error analysis, (3) Support for approval, and (4) Supervision of the modifications to the actuator electronics at LAT.

LAT:

(1) Modification of the actuator electronics, (2) Supplementing the limiter functions, and (3) Software qualification for certification.

The components developed and integrated into the overall ATTAS flight control system for the expansion of the flight domain are shown in Fig. 9.54. The new switch-box was activated by the flight engineer before crossing 500 ft altitude through a LIMITER ON switch. This enabled the operating rate limitation in the AEUs. The activation was displayed to both pilots in the cockpit (GO lamp). In the event of a fault, the NOGO lamp flashed.

The onboard computer system, which was designed from the outset as a duplex system, was supplemented with the second computer. By comparing the input and output signals, a passive system behavior was established in the event of a fault. All sensors, air data computer, inertial reference system and signal processing boxes were duplicated. The duplex onboard computer system was not subject to approval. The duplex solution was used to improve the operation, the acceptance by the safety pilots and the correct functioning of the test system before reaching the altitudes below 500 ft [21].

Figure 9.55 shows the limiting values as well as the risk areas in the roll axis for the individual actuation systems determined from flight tests with simultaneous hardovers in all axes and from the transients occurring during the takeover. Figure 9.56 shows time histories of the aircraft reaction to hardover in all actuation systems. The first fly-by-wire landing in the electrical mode of operation took place on April 30, 1999 at the Berlin-Schönefeld airport. Thus, ATTAS was available for applications with experiments in the FBW or SIM mode up to the touch-down (see Fig. 9.57). An example of the different control activities during a manual landing in FBW mode with sidestick (direct link), that is, without controller support, and with a control column is shown in Fig. 9.58. The high sidestick activities during the landing are clearly visible, which were caused by the much higher control sensitivity. The measures for the envelope expansion were funded by the BMBF.

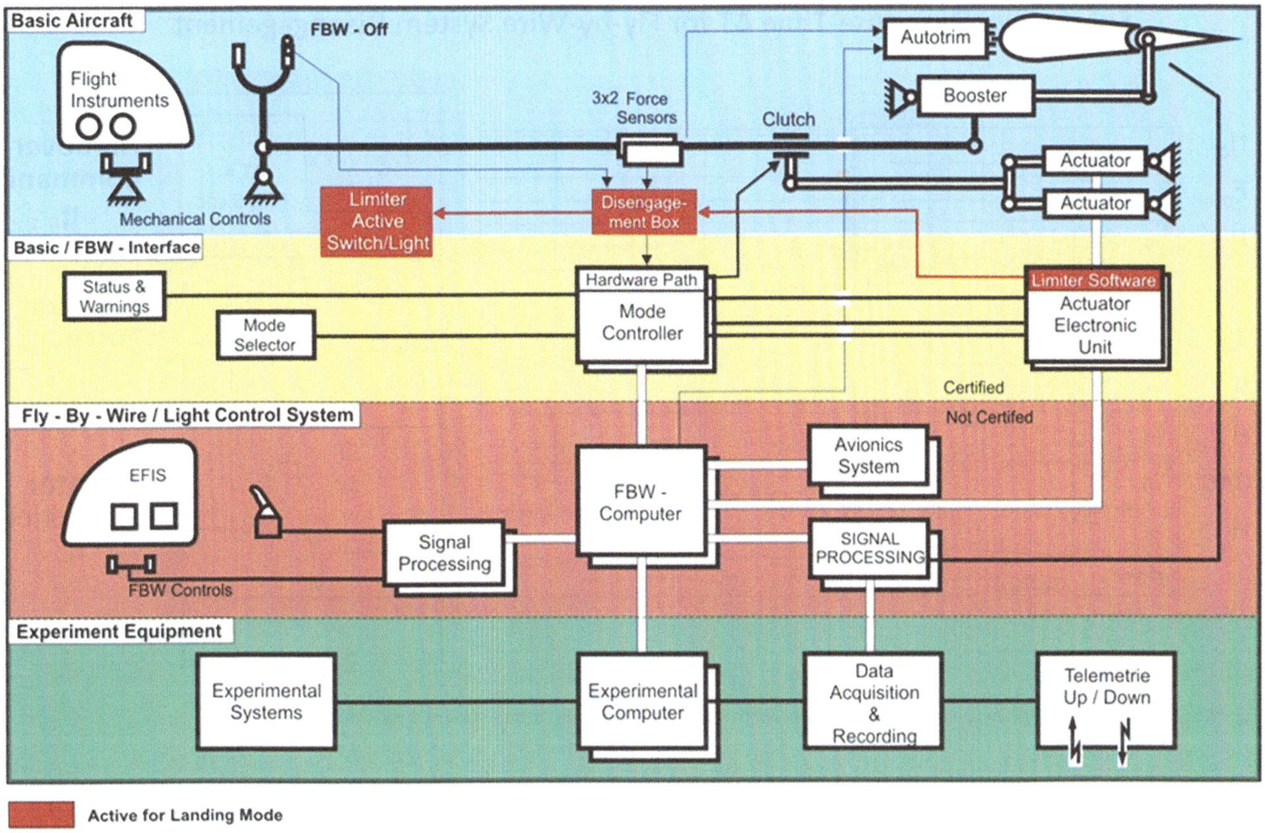

Fig. 9.54 ATTAS envelope expansion—flight control system modifications for safe landing in FBW-mode

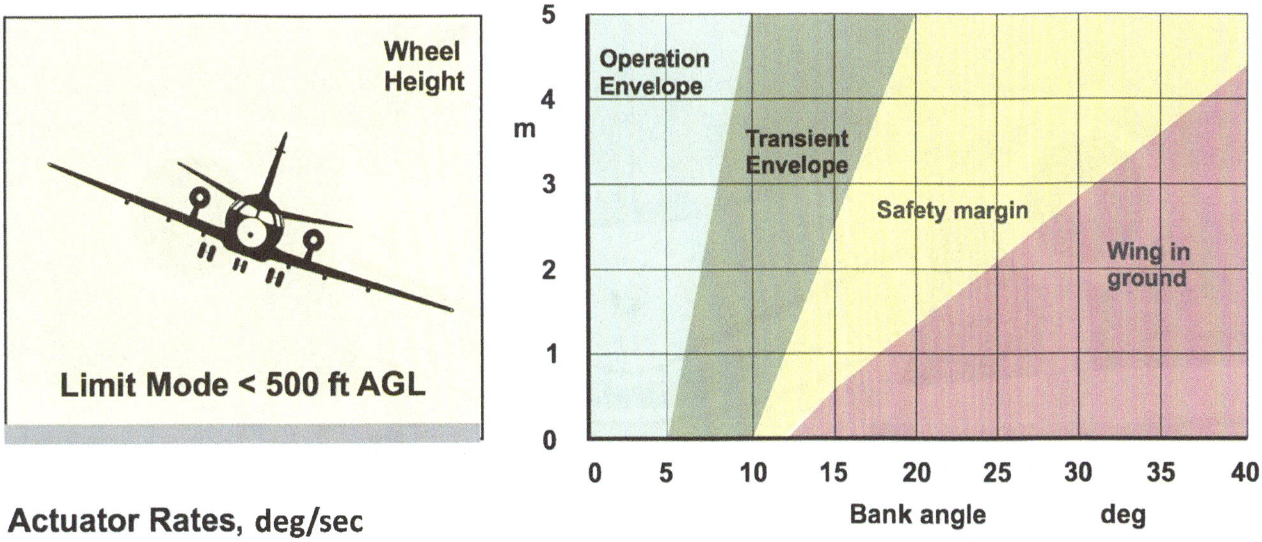

Actuator Rates, deg/sec

	Elevator	Aileron	Rudder	Ldg. Flap	Engine	DLC Flaps
Normal Mode	42	85	59	9	88	95
Limit Mode	25	55,45% Ampl.	30	0	31	15

Fig. 9.55 ATTAS envelope expansion—risk areas and actuation system rate limits for the landing

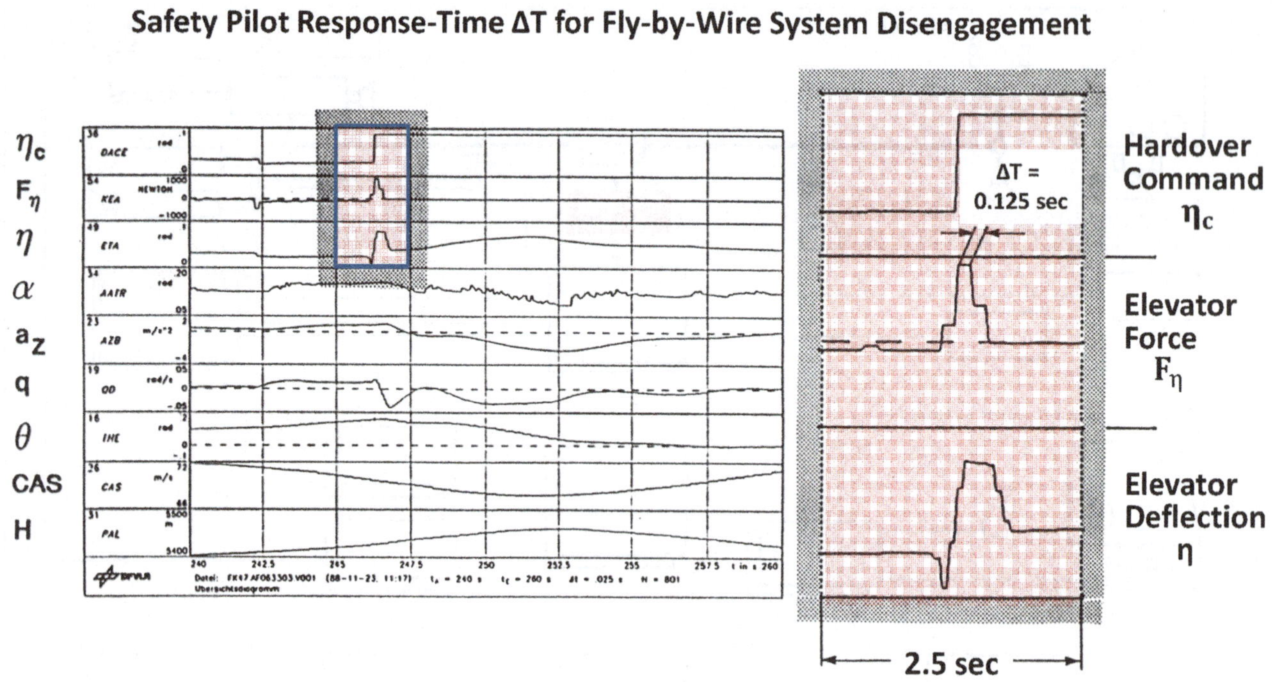

Fig. 9.56 ATTAS envelope expansion—transients in control surface hardovers during takeover by safety pilot

Fig. 9.57 ATTAS after landing in FBW-mode on April 30, 1999

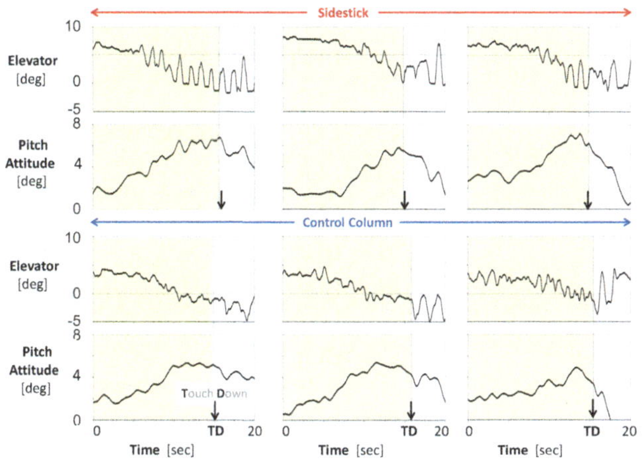

Fig. 9.58 Comparison of control activities during landing task with control column and sidestick

9.2 VFW 614 ATTAS Applications and Results

Klaus-Uwe Hahn
With contributions from 12 coauthors

9.2.1 Overview

As already pointed out in Sect. 9.1.4, after the modifications and developments over a period of four years the VFW 614 G17 became available to DFVLR starting October 1985. Initially it was, however, required to establish certain working conditions for the experimenters that enabled universal utilization of FBW/L control in the experimental mode up to in-flight simulation. Accordingly, the first intensive utilization over first few years focused on the comprehensive proof of reliable functioning of the FBW system, installation and certification of the nose boom with a high-precision flow sensor, and accompanying flight tests for system identification.

By means of system identification methodologies, a highly accurate flight-validated mathematical model of the ATTAS aerodynamics was determined, which forms the basis for the in-flight simulation (see Chap. 3). In 1989, the first tests were carried out for in-flight simulation. After its last flight as a part of an experiment on November 11, 2011, ATTAS had completed 2912.06 flight hours for research purposes with 3328 take-offs and landings over a period of 27 years. Subsequently, it was flown on December 7, 2012 to the German Museum in Oberschleißheim. The most important utilization programs which were carried out from 1985 to 2011 are chronologically summarized in Table 9.2. A few selected projects are briefly elaborated hereafter to illustrate the spectrum of ATTAS capabilities and utilization.

9.2.2 Hermes Spaceplane

Dietrich Hanke

Hermes was a spaceplane designed by the Centre Nationale d'Etudes Spatiales (CNES), the French Space Agency. It was to be launched into the orbit from the Ariane 5 launcher of the European Space Agency (ESA). The Hermes Spaceplane was to be used as a recovery transport system with a crew of three, similar to the Space Shuttle of the USA. In 1987, the Institute of Flight Mechanics of DFVLR was commissioned by the CNES to develop a concept and specification for an in-flight simulator for the Hermes Spaceplane. The in-flight simulator was to be used as a training device for the astronauts, particularly for familiarization during the difficult phases of manual approach and landing [22–24]. The in-flight simulator, called *Hermes Training Aircraft* (HTA), was to simulate the glide path from 37,000 ft altitude until touch down (see Fig. 9.59). A new technique based on target lighting array system called GRATE was to be used for flying qualities assessment and pilot training (see Sect. 12.3.2).

To arrive at a suitable host aircraft configuration for the planned HTA simulator aircraft, three candidates were investigated: (1) three engine business jet Dassault Falcon 50, (2) two engine business jet Bombardier Challenger and (3) two engine business jet Grumman Gulfstream IV. Two variations for the position of the evaluation cockpit were proposed (see Fig. 9.60): (1) one training and one safety pilot located in the basic aircraft cockpit and (2) an additional cockpit identical to the Hermes cockpit layout on top of the fuselage for a complete Hermes crew with the safety pilots in the basic aircraft cockpit. The performance calculation showed that the landing trajectory of the Hermes Spaceplane was only possible by using full in-flight thrust reversal, deployed air-brakes and extended landing gear of the host aircraft. Figure 9.61 shows the different host aircraft in the position to adapt the eye height of Hermes at touchdown.

Anticipating that DFVLR would develop a model following control for HTA, it was designed based on the Hermes design data and implemented in the ATTAS in-flight simulator. The objective was to evaluate the flying qualities of Hermes in level flight and to demonstrate the required in-flight simulation quality. The flight test measured

Table 9.2 Summary of VFW 614 ATTAS applications

Applications	Period	Participants
Hermes spaceplane	1987	DFVLR, CNES
Gust alleviation	1990–2011	DLR
Integrated ATM concepts	1991–1997	DLR, DFS, EUROCONTROL
Experimental flight management system	1991–1997	DLR
ATM demonstration	1992	DLR
Experimental cockpit	1992–1993	DLR
Aircraft pilot coupling experiments	1992–2010	DLR, WTD 61, University of Leicester
Flight control algorithms for a small transport aircraft	1993–1996	DLR, DASA
High-performance graphics generator	1995–1997	DLR, TU-Darmstadt, VDO-L
Artificial vision for all weather operation	1996–1999	DLR, EU FP4-BRITE/EURAM 3
Harmonized ATM	1997	DLR
Quick and robust autopilot design for automatic landing	1998–2000	DLR, EU FP4-BRITE/EURAM 3
Unmanned flight vehicle technologies	2000–2008	DLR, BWB
Pilot and flight test engineer training	2000–2008	DLR, ETPS, EPNER
Transport aircraft Dornier 728 Jet	2001–2002	DLR, Dornier
Wake vortex investigations	2000–2011	DLR, EU FP5-GROWTH
More autonomous aircraft in future air traffic systems	2003	DLR
Low-noise approach procedures	2005–2006	DLR
Optimized approach and landing procedures	2007	DLR
Parabolic flight	2008	DLR
Flight control with engines	2009	DLR
Steep approaches at Braunschweig-Wolfsburg airport	2010	DLR
Flying wing NACRE	2010	DLR
Parallel approaches at Braunschweig-Wolfsburg airport	2011	DLR

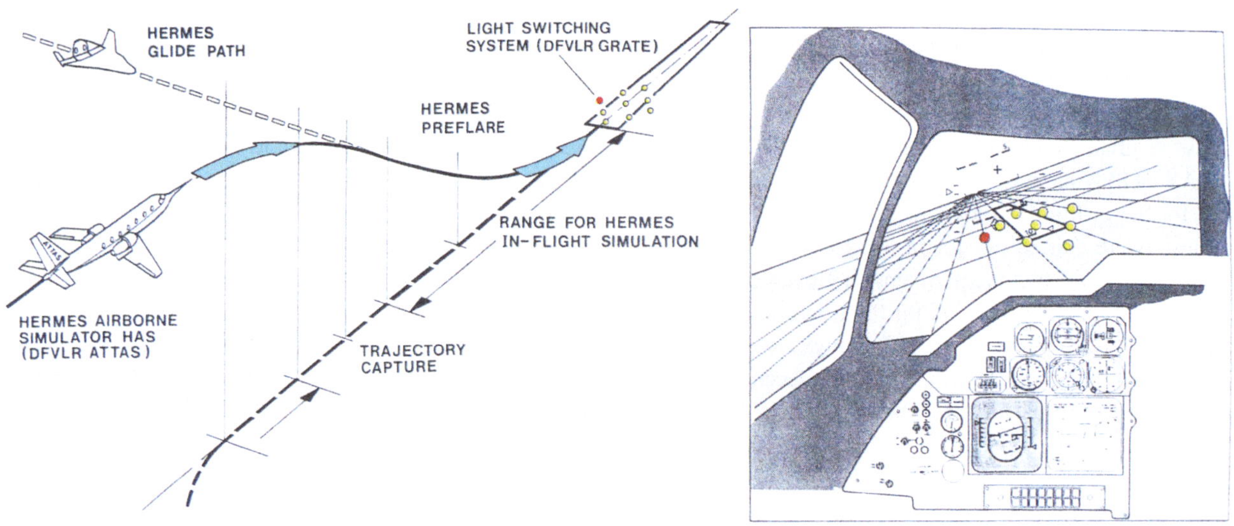

Fig. 9.59 Hermes precision approach and landing trajectory

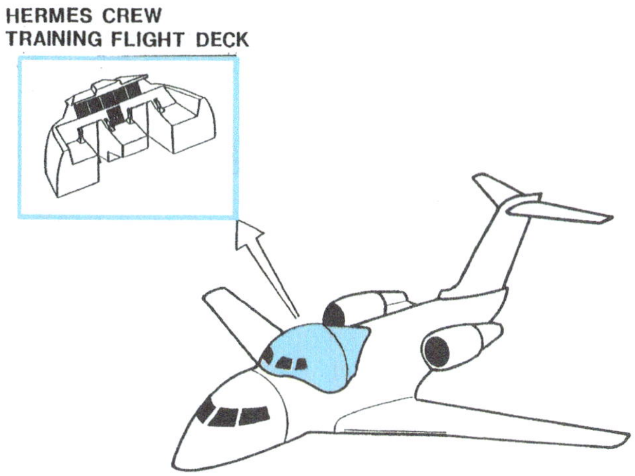

Fig. 9.60 HTA with added cockpit for crew training

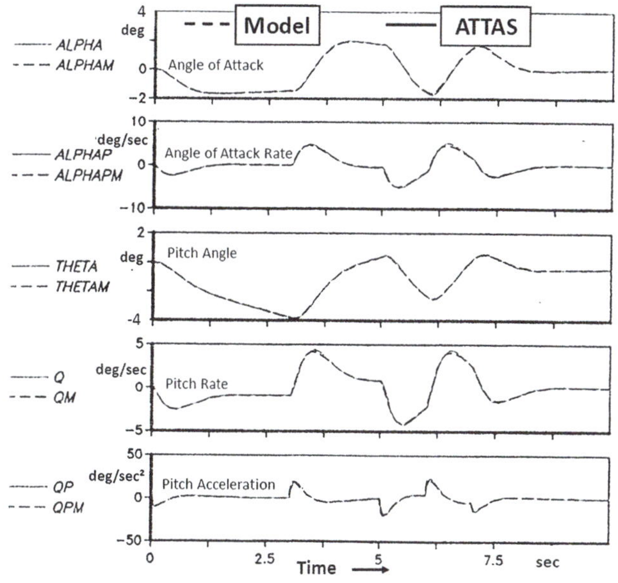

Fig. 9.62 Comparison of Hermes and ATTAS in-flight time histories

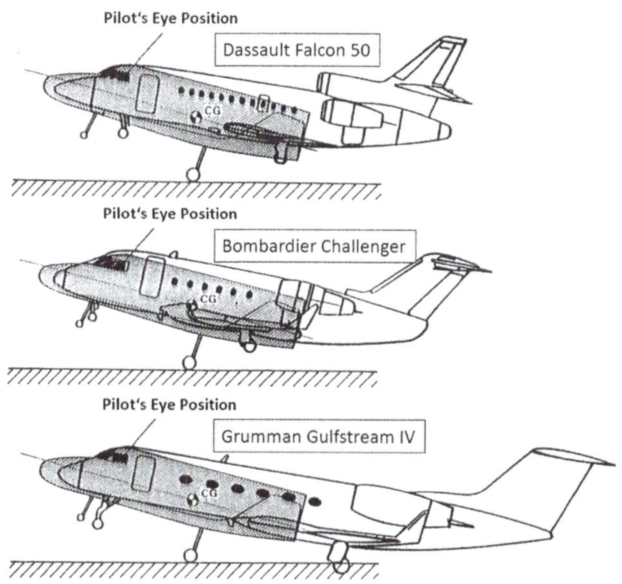

Fig. 9.61 Landing situation for the different host aircraft with respect to identical eye height of Hermes

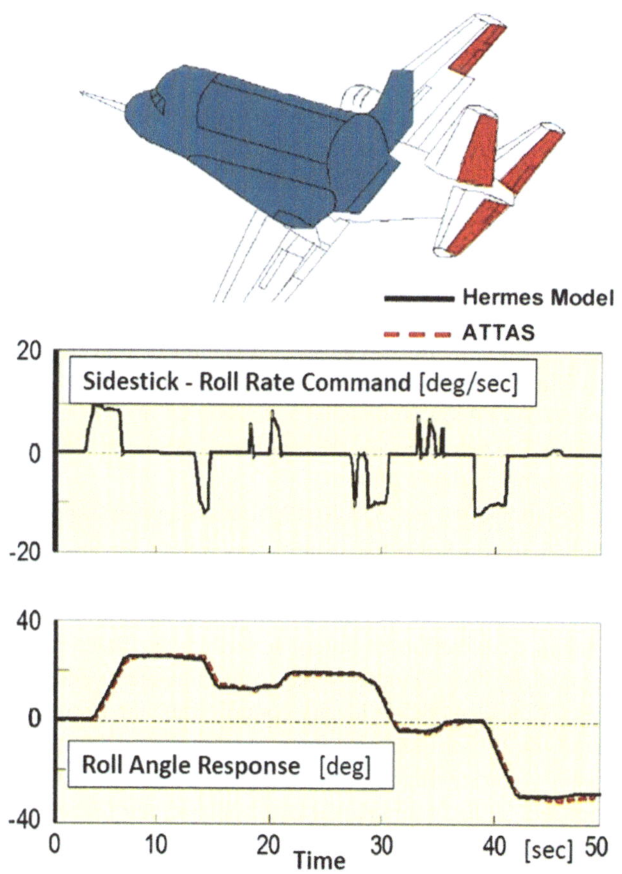

Fig. 9.63 Hermes in-flight simulation (comparison of time responses in the roll axis)

data showed an excellent model following quality as shown in Figs. 9.62 and 9.63, where the time responses in pitch and roll axes of the Hermes model and ATTAS are compared respectively. Both flight measured and the in-flight simulated responses are nearly identical.

Eventually, the Hermes project was cancelled because the weight of Hermes was not within the Ariane 5 rocket lift capabilities. Furthermore, the financial and political scenarios had changed in Europe (see also Sect. 11.5).

9.2.3 Fairchild-Dornier 728 Jet

Klaus-Uwe Hahn

During 2001 and 2002 ATTAS was deployed for a particularly interesting application, utilizing its capability of in-flight simulation to support the preparation of the first flight of the Fairchild-Dornier 728 Jet (Do 728 Jet) and its certification process. The 728 Jet (see Fig. 9.64) was conceptualized as a twin-jet passenger transport aircraft. In its basic version, it was designed to carry 75 passengers and to have a glass cockpit from Honeywell (EPIC System). It had a span of 27.12 m and a length of 27.04 m, and was developed almost to prototype maturity, despite recurring financial difficulties. Based on the anticipated excellent flight performance the German Lufthansa had secured in advance several procurement options. However, the development of the 728 Jet was discontinued just before its maiden flight in 2004 due to bankruptcy of the company. The fuselage of the first prototype was later employed for research on cabin air ventilation at DLR in Göttingen.

Fig. 9.64 Fairchild-Dornier prototype Do 728 Jet TAC 01

In 2001, the Institute of Flight Systems of DLR was contracted by Fairchild-Dornier to analyze the flying qualities of the 728 Jet and evaluate the flight control system concept. These included analytical studies applying flying-qualities and PIO criteria, using linear and nonlinear aircraft mathematical models [25]. In addition to these analytical studies, investigations were also carried out in a ground-based simulator and by executing numerous in-flight simulations with ATTAS. To perform the flight tests safely, accurate design and operation of flight-control laws of the 728 Jet was verified in advance including risks analysis of uncertainties in the aerodynamic parameters [26, 27]. To reproduce the control behavior accurately, the control column of ATTAS was modified to match that of 728 Jet. The control forces acting on the to-be-simulated aircraft were artificially emulated [28].

By simulating the 728 Jet flight characteristics applying explicit model-following control (see Chap. 3), the flying qualities were investigated by test pilots under real flight conditions [29]. The in-flight simulation of *Hermes Training Aircraft* described in the forgoing section was based on the completely linear model of the to-be-simulated vehicle as well as linear model-following control. Thus the excursions about the chosen reference flight condition were limited. In contrast, during flight experiments with 728 Jet a totally nonlinear in-flight simulation was implemented for the first time worldwide. For this purpose, the models of the to-be-simulated aircraft (the 728 Jet) and the model-following control were completely nonlinear. For example, the aerodynamics data tables from wind tunnel measurements were incorporated including all existing nonlinearities. The restrictions encountered in the linear simulation, namely small excursions around the reference condition, could thus be eliminated. Only through this innovative approach it was possible to meet the accuracy requirements.

A total of eight flights were carried out with a flight time of more than 17 h, in which 23 predefined initial flight conditions with different configurations, masses and centers of gravity were simulated. Thereby 25 switchable malfunctions were investigated and evaluated, including degraded flight control modes and mode transitions of the 728 Jet. Typical errors were for example: (1) engine failure, (2) flight controller in direct link mode (*Direct Law*, fly with "electrical control rod"), (3) yaw and pitch damper failure, (4) oscillating elevator, aileron and rudder, and (5) hydraulic failure.

When the flight controller is operated in the direct link mode (*Direct Law*), it implies the complete failure of the electronic flight control laws. In such a case the aircraft flies without any computer-aid, that is, with a direct proportional transmission of the pilot inputs to the control surfaces (flying with "electric rod"). The in-flight simulation quality with ATTAS had Level D/E quality, meeting the highest quality requirements for certification of today's training simulators [30]. As a typical example, Fig. 9.65 shows the time histories during lateral Dutch roll responses. The longitudinal motion variables are plotted on the left (from top to bottom: pitch command, thrust command, pitch acceleration, pitch attitude change, and airspeed variation), and the lateral-directional motion variables are shown on the right (from top to bottom: roll command, yaw command, roll acceleration, change of bank angle, and change of angle of sideslip). The resulting very high model-following quality is discernible from the figure. The same quality was also achieved for other types of maneuvers not shown here. This was the case for both the normal flight conditions as well as for the investigated system failures. This high simulation quality was the prerequisite for the utilization of the ATTAS 728 Jet in-flight simulation for the training of the test pilots prior to performing the planned first flight [31].

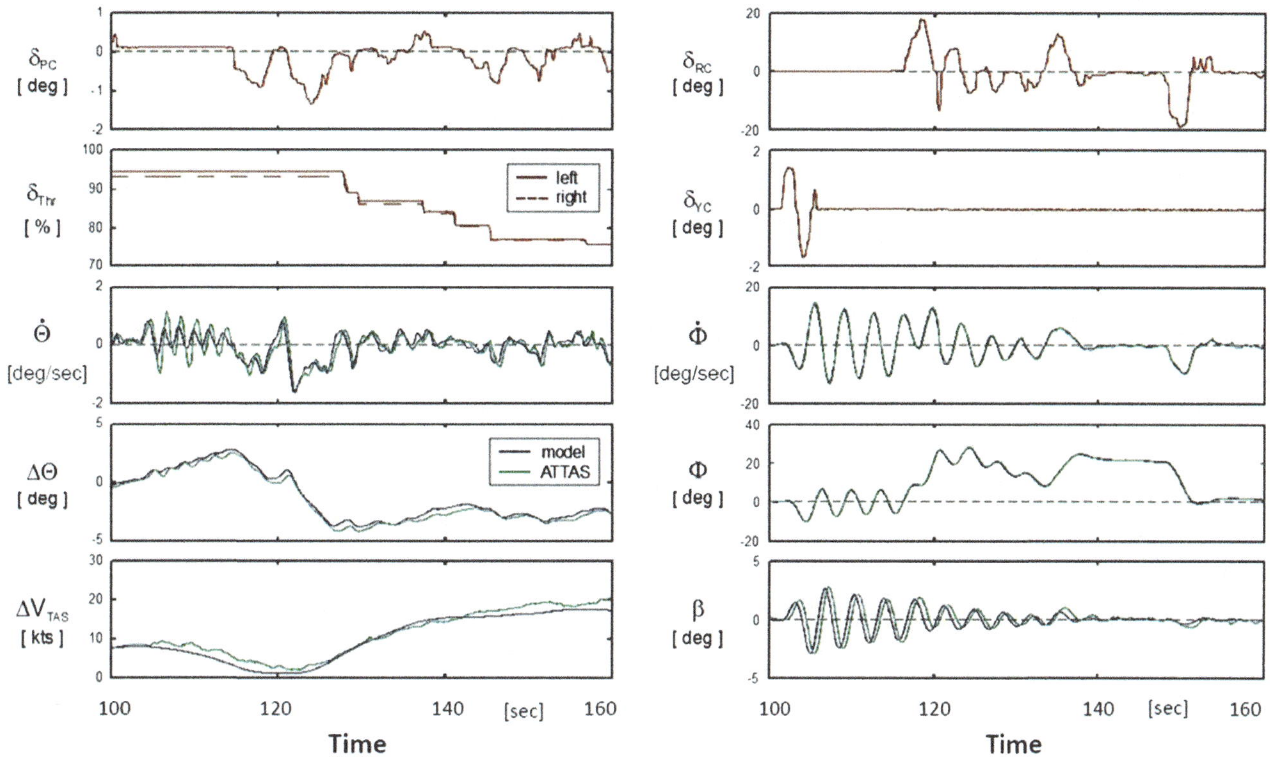

Fig. 9.65 ATTAS In-flight simulation of Do 728 Jet Dutch roll dynamics

9.2.4 Flying Wing NACRE

Jana Schwithal

The usual analytical methods and classical criteria for the flight mechanical assessment are often validated only for conventional aircraft configurations, consisting of fuselage, wings, and horizontal and vertical tail planes behind the main wing [32]. Their validity for different configurations is uncertain. Therefore, validity of analyses pertaining to flying qualities and safety, carried out during aircraft development phases, need to be verified in flight. Particularly in the case of unusual configurations, such as flying wings, flight test is an essential part of the development. However, without in-flight simulation (see Chap. 3) the flight tests would be feasible only after the construction of a prototype. But such a prototype is available at a very late stage in the development process. In addition, it must also be ascertained prior to the first fight that the aircraft dynamics does not exhibit a safety critical behavior. With the in-flight simulator ATTAS unconventional configurations can, however, be efficiently tested under real flight conditions to detect and rectify possible deficiencies during a very early stage of their development. The innovative approach of nonlinear in-flight simulation described in the foregoing section guarantees highest possible accuracy of the experimental evaluation. Accordingly, it was applied to the flying wing configuration that was developed within the EU-project NACRE (*New Aircraft Concepts Research*), (see Fig. 9.66) [33].

Fig. 9.66 European flying wing project NACRE

Such a flying wing aircraft, often called BWB (*Blended Wing Body*), is characterized by the lack of the conventional tubular fuselage. The passengers are accommodated in a specially designed center segment of the wing. In contrast to the conventional fuselage, this central part of the blended wing-body generates primarily the required lift. Contrary to a conventional vertical tail, this configuration had two relatively small vertical fins at the end of the middle wing segment. It did not have an elevator. The trailing edge flaps on the middle wing were used for longitudinal control. The flying wing configuration was designed as a large long-range aircraft with a maximum take-off weight of approximately 700 tons and a seating capacity of 750 passengers.

The handling qualities of flying wing configurations are often critically affected by the absence of the stabilizing tail and due to short lever arms of the control surfaces. On the other hand, an aircraft can be successfully introduced in the regular service, only when it is easily and safely controllable. For this reason, the elaborate evaluation of the flying qualities was carried out of the NACRE flying wing configuration. In addition to theoretical analyses with the help of flying qualities criteria, an in-flight simulation with ATTAS was utilized to ascertain through pilot evaluations the critical aspects pertaining to applicability of such criteria to unconventional configuration.

A total of three flight tests were carried out, in which the flight dynamics of the flying wing configuration was simulated with ATTAS and evaluated by two pilots. The first flight served the purpose of familiarization with flying the unusual configuration and with its dynamic behavior. For this purpose, the pilots carried out simple maneuvers and excited the aircraft's eigenmotion through control inputs. Hereby, the capability of ATTAS to replicate a significantly different configuration was continuously examined. As shown in Fig. 9.67, ATTAS and the flying wing configuration being investigated have a completely different geometry and differ significantly in size.

The flight experiments showed, however, that despite these differences the motion of the ATTAS host aircraft and that of the to-be-simulated aircraft model matched fairly well for all important parameters. With the exception of airspeed errors, resulting from the low ATTAS engine dynamics which was not directly sensed by the pilot, the in-flight simulation achieved the so-called Level-D quality, see Fig. 9.68. It was clearly demonstrated that such unconventional aircraft configurations could be realistically investigated with ATTAS.

In the two subsequent flight tests, the test pilots evaluated the flying qualities of the flying wing configuration during different maneuvers using the *Cooper-Harper* rating scale [34]. The flight tasks comprised of different turning flights and ILS approaches. The turning maneuvers were used to analyze the lateral dynamics, which were identified as critical in the preliminary analytical investigations. The ILS approaches served to evaluate the performance during precise flight path control. All maneuvers were carried out with the basic configuration of the flying wing.

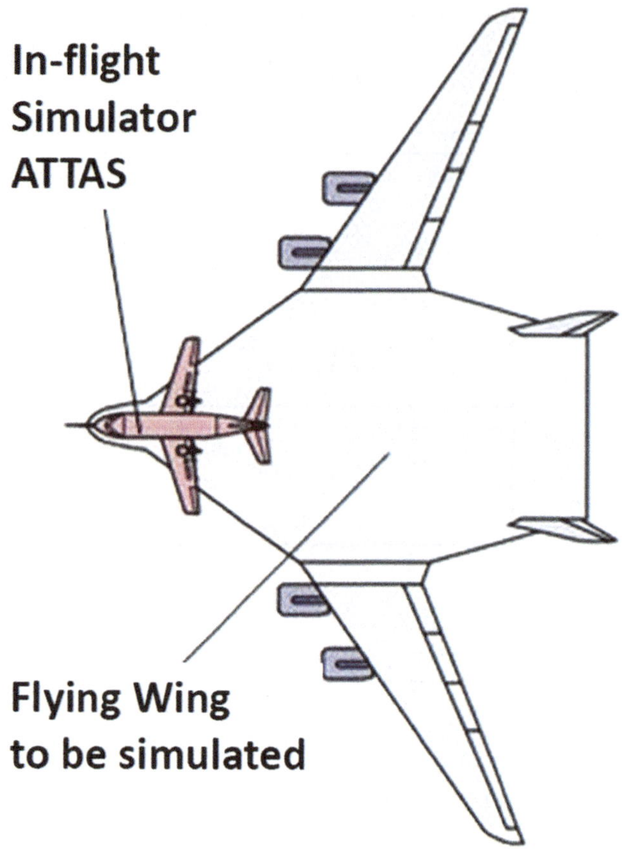

Fig. 9.67 Scale comparison between ATTAS and flying wing NACRE

Subsequently, the same flight maneuvers were repeated with a control-augmented variant of the flying wing. For this purpose, the aircraft model was equipped with a flight controller in order to support the pilots to control the vehicle and improve the flight characteristics. The pilot evaluation indicated that the unaugmented flying wing configuration was not controllable, resulting in strong deviations from the desired flight condition. For example, an undesired buildup of angle of sideslip was observed during curved flights. For the control augmented configuration both pilots noticed significantly improved flying qualities with reduced workload.

The improvements in the flying qualities were limited by inadequate yaw control due to the relatively small vertical tail and short lever arms, and due to the sluggish behavior of the aircraft about the longitudinal axis resulting from the large roll inertia due to the extreme wingspan. Satisfactory flying qualities were not achieved in all the cases [35]. The in-flight simulations with ATTAS demonstrated that further development work is necessary before the investigated flying wing configuration could be implemented into a real aircraft.

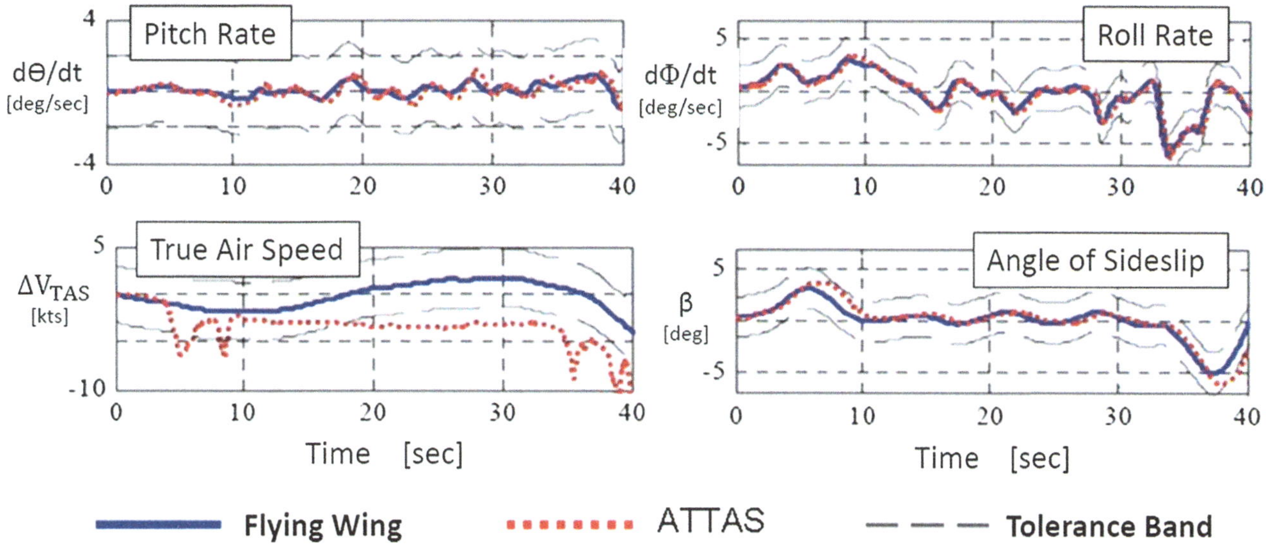

Fig. 9.68 Time history comparison of lateral dynamics between ATTAS and simulated NACRE model (with tolerance bands for acceptable Level-D simulator quality)

9.2.5 Gust Load Alleviation (1990–2011)

Klaus Uwe Hahn

LARS (*Load Alleviation and Ride Smoothing System*) was one of the first investigations that were carried out with the ATTAS at the beginning of nineteen-nineties. The objective hereby was to actively reduce the effects of vertical gust (vertical movements of air mass) on the aircraft motion; in other words, on the accelerations at the center of gravity, with the primary aim of improving passenger comfort. On encountering vertical updraft gusts the aircraft angle of attack increases and thereby the lift. This increased lift force leads to an additional load on the aircraft structure and to an upward acceleration of the aircraft and passengers. These effects reverse on encountering a downwind, resulting in downward acceleration, which is perceived by the passengers as unpleasant. It is similar to a fast downhill ride in a roller coaster. However, if the lift could be maintained constant in the presence of gust, the additional accelerations would disappear and thereby the unwanted aircraft reaction, too (see also Sect. 6.3.3). This can be accomplished by the use of special, fast responding trailing edge flaps on the wings, such as those that are available on the ATTAS, called the DLC (Direct Lift Control) flaps. The deflections of these DLC flaps change the wing profile and thus the lift.

The LARS concept is based on the principle of disturbance compensation. Figure 9.69 shows the computation of DLC flap deflection δ and elevator command η from the gust induced additional angle of attack α_W [36]. Thereby, the disturbance flow ahead of the aircraft is determined at first. In the case of ATTAS, the flow sensor on a noseboom provided this information. Assuming that the aircraft

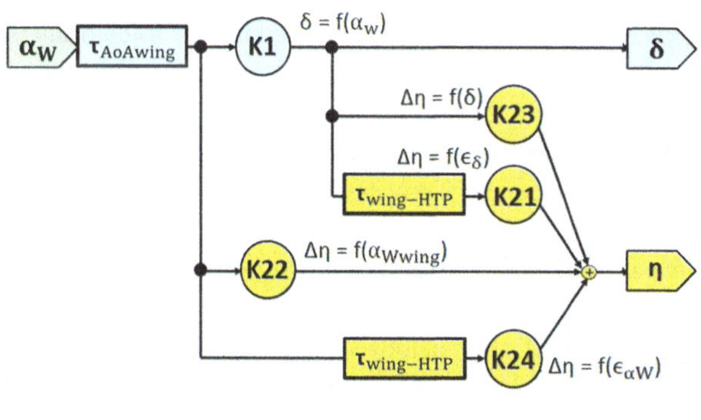

Fig. 9.69 LARS principle

aerodynamics is known, and therewith the effect of the disturbance, the deflection of the trailing edge flaps can be computed that is necessary to compensate the disturbance. The deflection of wing flaps leads to changes in the pitching moment as well as in the downwash behind the wings, and thereby the flow angle of attack at the horizontal tail. In a similar way the flow due to gust disturbances and its aerodynamic effects change, when they reach the horizontal tail. If the pitching moment balance of the aircraft is to be maintained in the presence of gusts, corresponding deflections of the elevator are necessary to compensate the additional pitching due to gust disturbances. The advantage of the gust disturbance suppression concept is that control interventions are necessary only when flow disturbances are measured. This so-called open-loop control system does not affect the aircraft flying qualities in case of a pilot intervention.

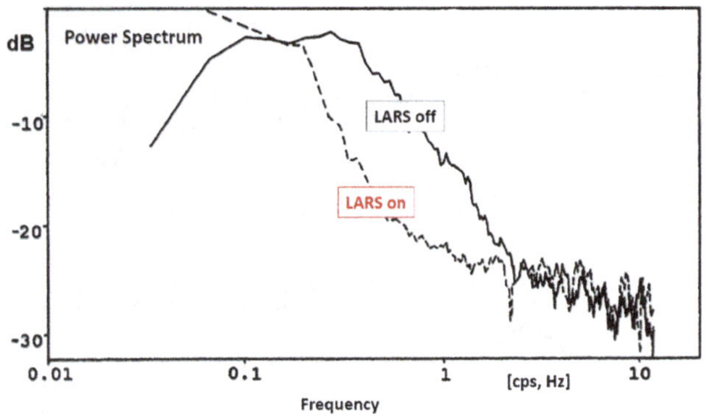

Fig. 9.70 Power spectral density of vertical accelerations

During 1990–1993, a large number of flight test were performed with ATTAS to successfully demonstrate the operational principle of LARS [37]. The vertical accelerations due to gusts could be reduced by 12 dB (that is, about 70%) in a frequency range of 0.2–2 Hz, that is particularly sensitive to passenger comfort (see Fig. 9.70). This highly promising result pertaining to vertical accelerations was, however, at the cost of more noticeable horizontal accelerations. This fact could also be verified in the simulations (see Fig. 9.71) [38]. The use of the thrust force to compensate for this side effect in the longitudinal axis was ruled out, because the thrust modulation was much too sluggish for these relatively high-frequency processes. However, even in this case ATTAS provided an alternative. The two outermost DLC flap pairs 1/2 and 5/6 on each wing were deflected oppositely as split flaps for drag control. Rapid longitudinal force changes could be achieved by activating these fast responding flaps. The inner flap pair (flap 3 and 4) was, as before, available for the generation of vertical forces. With this concept of a *Gust Management Systems* (GMS), the vertical and horizontal accelerations due to gust disturbances could be successfully reduced [39, 40], (see also Sect. 6.3.3.3).

Yet another application of ATTAS during 2008 and 2009 was the feasibility study of structural vibration damping with the aid of adaptive control in the EU project ACFA 2020 (*Active Control of Flexible 2020 Aircraft*) [41]. Thereby, using the flow sensor signal at the nose boom the first bending mode of engine nacelles at 3.5 Hz could be successfully damped.

Furthermore, in addition to the reduction of accelerations at the center of gravity based on the LARS approach, the possibility of damping the symmetrical wing bending was investigated within the EU project AWIATOR (*Aircraft Wing with Advanced Technology OpeRation*). This refinement of LARS to the GLAS concept (*Gust Load Alleviation System*) was carried out for a large flexible aircraft, whereby not only the passenger comfort could be improved but also the damping of the wing bending mode resulting in the

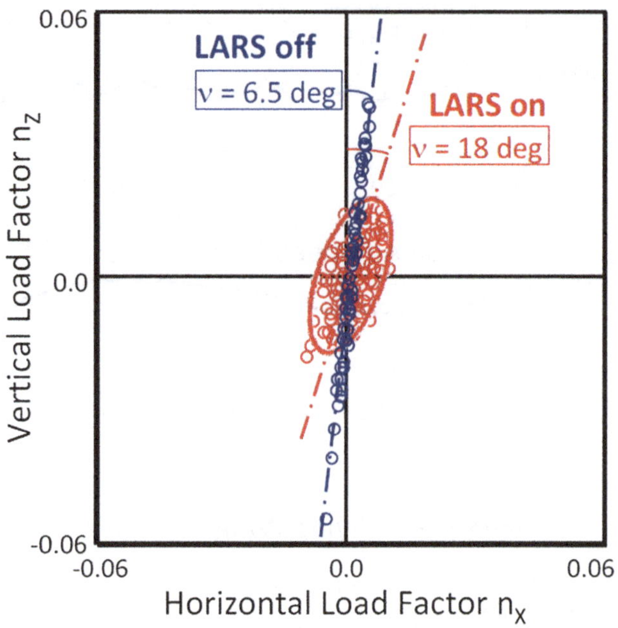

Fig. 9.71 Increase in the horizontal acceleration due to reduction of vertical acceleration by DLC flaps

reduction of the wing root bending moment [42]. The operation of the GLAS concept could be successfully demonstrated in computer simulations. Figure 9.72 shows the time histories of a vertical wind disturbance (on the left) and the resulting wing root bending moment (on the right). The red curve illustrates *without gust load reduction system*, the green curve *with the original (static) LARS concept*, and the blue curve *with the GLAS concept*. The reduction of the wing root bending moment is clearly discernible.

The flight demonstration of the GLAS concept was, however, not completed within the AWIATOR project. It was pursued in the succeeding project FTEG-InnoLA (*Flight Physics Technologies for Green Aircraft*—Innovations for the efficient simulation and testing process chain for loads and aeroelastics) as part of the 4th National German Aviation Research Program. Having adapted the GLAS design for an ATTAS in-flight demonstration, initial functional tests were performed in flight, followed by further flight tests to evaluate the system. The intended improvements could initially be achieved only in a very narrow frequency range. As such the control system had to be redesigned [43]. During the flight test on November 11, 2011, with an improved control system the atmospheric conditions were very stable and as such the system performance could not be evaluated satisfactorily. Another flight test with the GLAS concept was planned for the beginning of 2012. However, during the yearly inspection the ATTAS was grounded due to intolerable and irreparable engine defects. Thus, the flight on November 11, 2011 was the last mission of the VFW 614 ATTAS. Although ATTAS served as an experimental

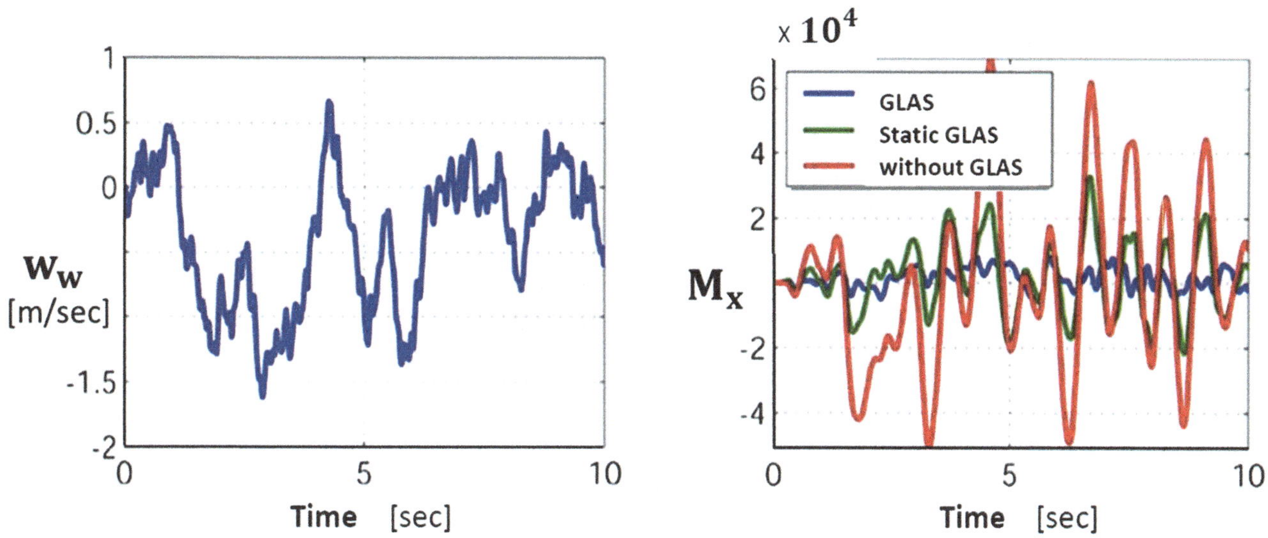

Fig. 9.72 Simulation results with GLAS (*gust load alleviation system*)

platform for numerous scientific applications, coincidently, the first and the last research mission of this extraordinary technology demonstrator aircraft were dedicated to gust load alleviation research.

9.2.6 Flight Control Algorithms for a Small Transport Aircraft (1993–1996)

Klaus-Uwe Hahn

An electronic flight control system was developed for a small 100-seater passenger transport aircraft within the technology program of the DASA (Daimler-Benz Aerospace Airbus). It was finally implemented on the VFW 614 ATD (Advanced Technology Demonstrator, see Sect. 6.3.7). For this purpose the flight control laws (FCL) were developed and flight tested on the VFW 614 ATTAS, jointly by DASA and DLR. The project dubbed SAFIR (*Small Airliner Flight Control Laws Investigation and Refinement*) was launched in April 1993 to address this task, with the objective of optimizing, validating, demonstrating, and evaluating the control laws in flight. Just six months later the first flight tests were carried out [44, 45]. The to-be-examined control algorithms were developed by DASA (*Robert Luckner*) using the development tool HOSTESS (*High Order Structuring Tool for Embedded System Software*) with automatic code generation [46].

In the first phase of the project, the standard flight control laws in the so-called *Normal Laws* mode were investigated, including automatic *Envelope Protections* (see Fig. 9.73). The FCLs were optimized iteratively as shown in Fig. 9.74. They were checked in the ATTAS ground-based system simulator prior to the flight trials. If necessary the adaptation

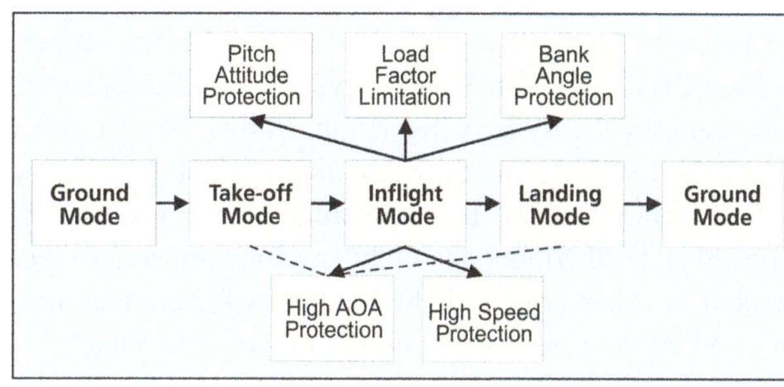

Fig. 9.73 Different envelope protection modes within the *normal laws* mode

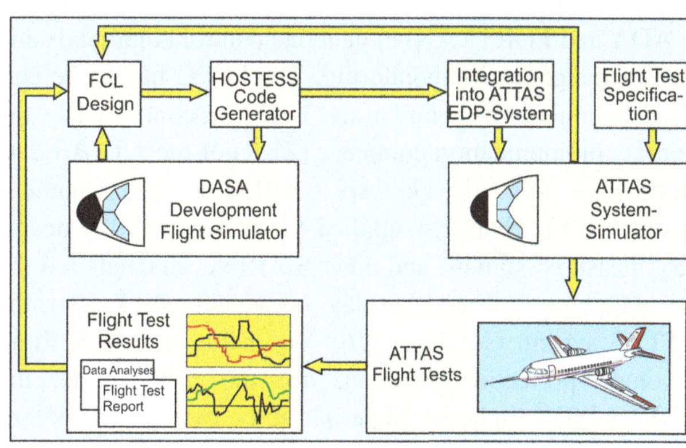

Fig. 9.74 Iterative optimization of flight control laws

could be carried out with HOTESS. Subsequently, after comprehensive ground testing the experimental software was approved for flight operations. This experimental software was ported as an independent program package into the ATTAS experimental and control computer (ERR) and thus integrated into the ATTAS-DV data processing system (see

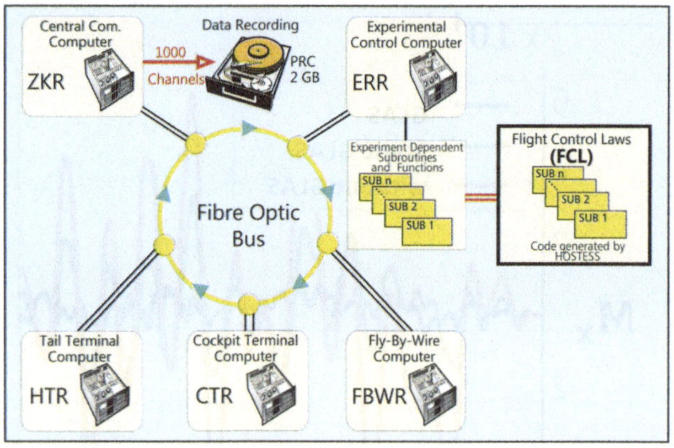

Fig. 9.75 Experiment configuration of SAFIR project (first phase)

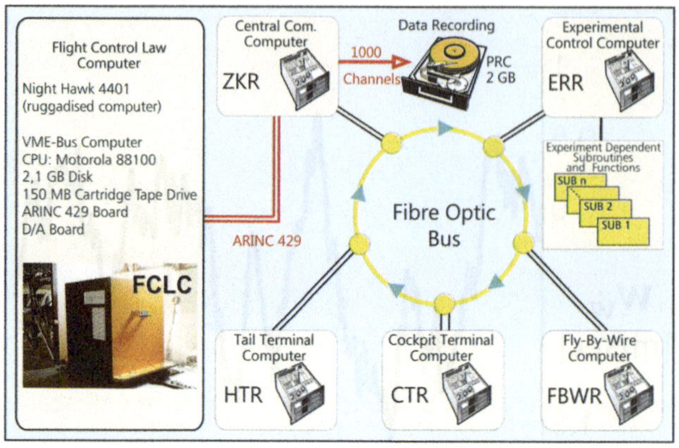

Fig. 9.76 Experiment configuration of SAFIR project (second phase)

Fig. 9.75). An experiment-specific interface supplied all the necessary information, coming from different sources including artificially generated sensor signals, required by the flight control law algorithms. The control commands calculated by the flight control algorithms were then returned to the ATTAS-DV system and converted into control surface deflections by the fly-by-wire system. During the first project phase, the focus was on the evaluation of the functionalities and assessment of the flying qualities in the normal laws mode [44]. Furthermore, the flight envelope protections were verified to avoid critical flight conditions for the maximum airspeed and the maximum permissible angle of attack.

In the second project phase, system-specific characteristics of the flight control system were tested and demonstrated [45]. For this purpose, the FCLs were programmed and implemented on the *Flight Control Law Computer* (FCLC) in two different programming languages (dissimilar software in ADA and FORTRAN) to generate control commands and reference signals for monitoring. The FCLC had to be certified for onboard operation and could be connected to the central communication computer (ZKR) of the ATTAS data processing system via six ARINC 429 channels. Three ARINC channels supplied the FCLC with the necessary measured signals, and other 3 ARINC channels fed the calculated control and display commands back into the ATTAS system (see Fig. 9.76). To enable testing of flight envelope protection functions for the bank angle, the ATTAS FBW flight envelope had to be extended to 45° of bank angle. installation of additional displays for angles of attack and sideslip to the pilots was mandatory to test the automatic angle of attack protection, which was close to the ATTAS stall-boundary.

To verify the control laws functional efficiency in the entire flight envelope, computer-generated synthetic signals as well as manual control inputs by the pilot were used. For the flying qualities assessment, the evaluation pilot had to fly a sequence of predefined flight maneuvers with course and altitude changes. The assessment of handling qualities of the control-augmented aircraft was carried out by the pilots using *Cooper-Harper* Ratings (see Fig. 2.6) [34]. During the period from 1993 to 1996, more than 300 test sequences were successfully tested in 12 test flights with a total flight time of 32 h. A total of nine different software configurations with 110 modifications were evaluated. The flight control laws investigated in the SAFIR project were later implemented in the industrial VFW 614 ATD project (see Sect. 6.3.7).

9.2.7 High-Performance Synthetic Vision System (1995–1997)

Dietrich Hanke

A Synthetic Vision System (SVS) with synthetic outside view was developed jointly by the Technical University of Darmstadt and VDO-Luftfahrtgeräte (VDO-L). This development was supported by the BMFT. The basic concept of the new vision system consisted of a 3D representation of the flight path and that of the surrounding terrain. Both of these were shown on the *Primary Flight Display* (PFD) as well as on the *Navigation Display* (ND), together with 3D representation of the terrain and the position in bird's eye view. In August 1997 this system was implemented and tested on the ATTAS [47].

The motivation for the development of SVS was the fact that there were various accidents in poor weather conditions, where the airplanes were flown directly into the mountains (*Controlled Flight Into Terrain*—CFIT), even when it was guided by air traffic control. The above approach was aimed at improving the situational awareness of the pilot and the flight safety during the entire Gate-to-Gate operation by combining the flight information with the synthetic terrain vision.

Depending upon the flight phase from taxiing, takeoff, cruise, landing, and docking at the gate, all necessary information was displayed by the SVS throughout the entire flight. Thus, suitable synthetic vision of the outside view was available to the pilot at any time. Furthermore, a predictor display, indicating the flight path eight seconds in advance, was provided in the PFD to assist the pilot. Color changes in the predictor were used to indicate flight operational limits. During the landing the flight path (ILS) was represented by a rectangle corridor (Channel Display), which the pilot had to follow until touchdown. Figures 9.77, 9.78 and 9.79 show the display formats for different flight phases as an example.

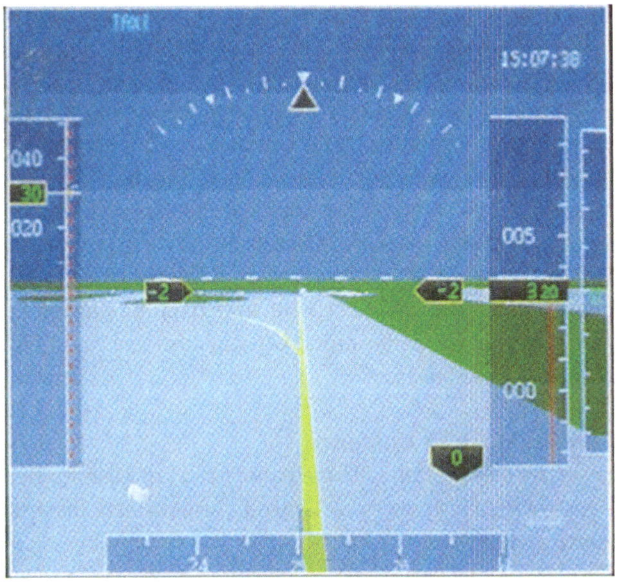

Fig. 9.77 Display in PFD: during taxiing

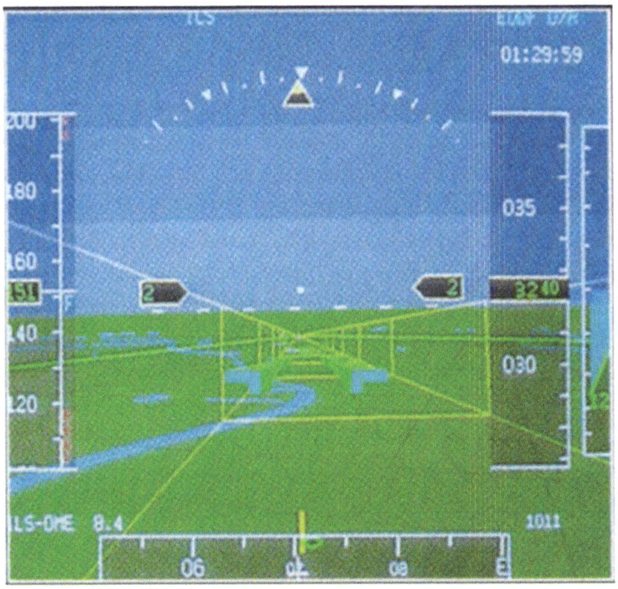

Fig. 9.78 Display in PFD: during landing approach with indicated approach corridor

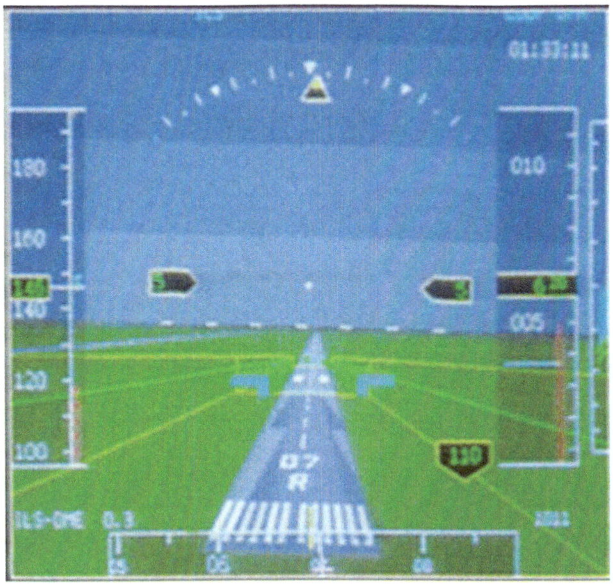

Fig. 9.79 Display in PFD: during final landing

Experimental System

The main components of the experimental systems were: (1) Primary Flight Display (PFD), (2) Navigation Display (ND), (3) Inertial reference system Honeywell H 746, (4) Navigation computer, (5) Data base computer Harris Nighthawk, and (6) High-performance graphic computer, Silicon Graphics ONYX.

A large commercial liquid crystal display (LCD) was mounted in front of the basic instruments as a cockpit display (see Fig. 9.80). The display could be folded down in order to provide the basic ATTAS instruments for take-off and landing. All experimental components were installed in two closed cabinets in the cabin near the loading door. The computer cabinet is illustrated in Fig. 9.81. Several interfaces provided the connection to the electrical power

Fig. 9.80 LCD display in ATTAS cockpit for synthetic vision

Fig. 9.81 Computer cabinet from external user in ATTAS

system and the communication with the on-board measuring system.

The experimental system was based on three advanced technologies: (1) Precision navigation with differential GPS, (2) World-wide 3D Geo data base, and (3) Commercial high-performance graphic workstation with 4D display software functions.

The aircraft position, airspeed, accelerations, and attitudes were calculated by the precision navigation system. The data from different sensors were compared and checked for plausibility. A filtering algorithm was developed to compute the precise aircraft position with an accuracy of less than a meter. Based on the precise position, appropriate datasets were selected for the terrain information from the Geo Database and transferred to the display computer.

In addition, the eyepoint of the pilot was calculated, which is needed for the exact synthetic 3D vision to be presented to the pilot. All vision data processed in the high-performance graphic workstation were displayed in real time to the evaluation pilot and the system operator in the cabin. The PFD and ND information were displayed on the large screen in the cockpit.

Flight Tests

The flight test campaign with 22 flights and 35 pilots was carried out at the international airport in Frankfurt in August 1997. The flight tasks comprised of the approach of beacons, virtual fixes, terrain collision situations, straight and curved landing approaches, and taxiing on the ground. All participating pilots were experienced active airline pilots. They were made familiar with the ATTAS aircraft, the flight test equipment, the 4D navigation guidance functions, the display symbology, and the flying tasks before the flight test campaign.

To briefly summarize the evaluation, the basic concept of the synthetic vision was well accepted by the pilots and it provided a significant improvement in the situational awareness compared to conventional display systems. The concept allowed manual or automatic landing approaches with high precision and surveillance. Furthermore, the synthetic vision turned out to be quite important for efficient taxi guidance to enhance the airport capacity also in a case of bad weather conditions.

Low-Level Flight Guidance

In yet another project, with the participation of Daimler-Benz Aerospace AG, Honeywell Control Systems and the Lufttransportgeschwader (LTG 61—Air Transport Wing of the German Air Force), the synthetic vision system was employed for flight testing of an autonomous manual low-level flight guidance system in 1996 [48]. The Harz mountain terrain in the northern part of Germany was digitized for an area of 217 × 68 miles and the data stored in the mission database onboard the ATTAS. The resolution of the terrain data was 1212 × 1181 ft with a sight cone of 60°. Taking into account the ATTAS performance data, 3D low-level missions at 150 ft above the ground were planned through the Harz Mountains. Although the actual test flights were carried out in flight altitudes of 10,000 ft, all information were displayed to the pilot as if he would be virtually flying at a height of 150 ft above the ground. The outside view from the cockpit was completely curtained-off. The synthetic vision provided to the pilot included the terrain topography, such as roads, lakes, cities and railroads. In a case of a risk of collision with the terrain the terrain color was changed to red.

The main guidance assistance was provided by the channel display for the flight path and the predictor. Without these functions, a manual low-level flight would not have

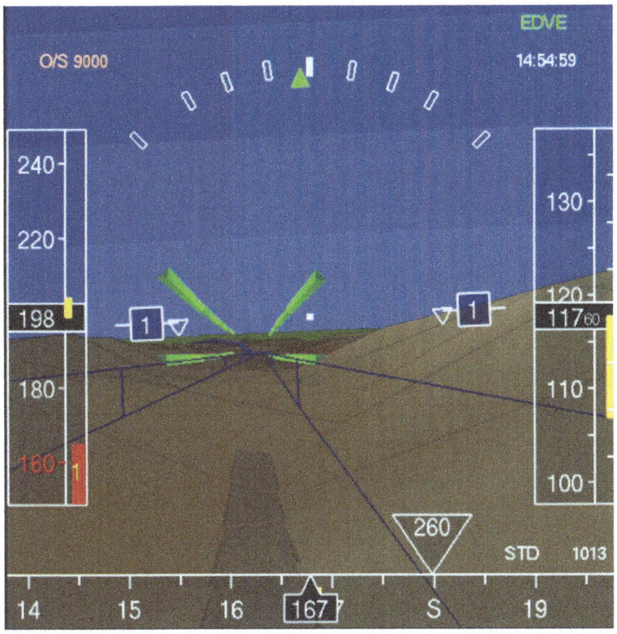

Fig. 9.82 Low-level flight information in PFD

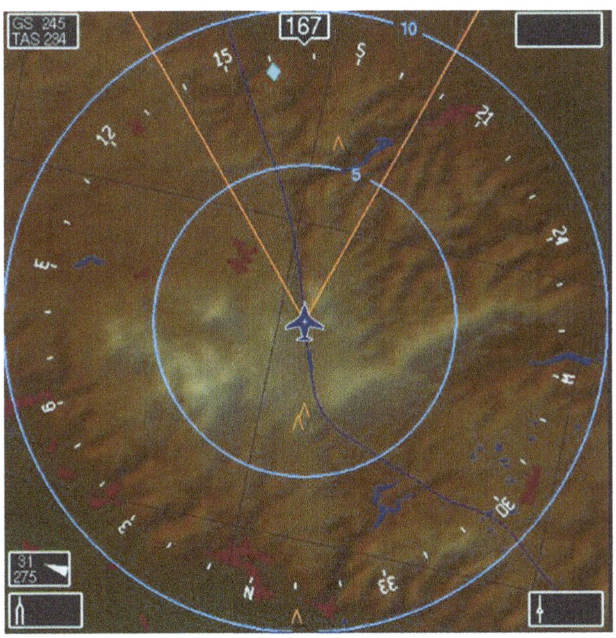

Fig. 9.83 Low-level flight information in ND

been possible. Figures 9.82, 9.83, and 9.84 depict typical low-level flight situations on the PFD and ND. Figure 9.85 shows a comparison of the commanded and actual flight path for the entire low-level flight route. It is apparent that the pilot flew the nominal flight path quite accurately.

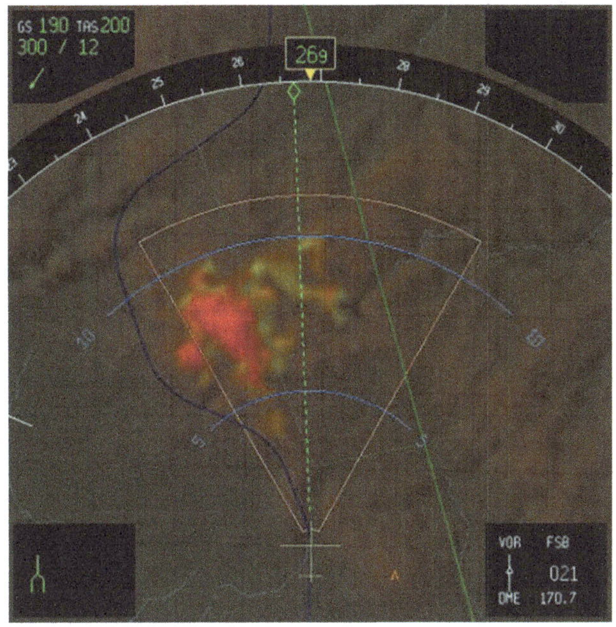

Fig. 9.84 Low-level flight with mountain obstacle in ND

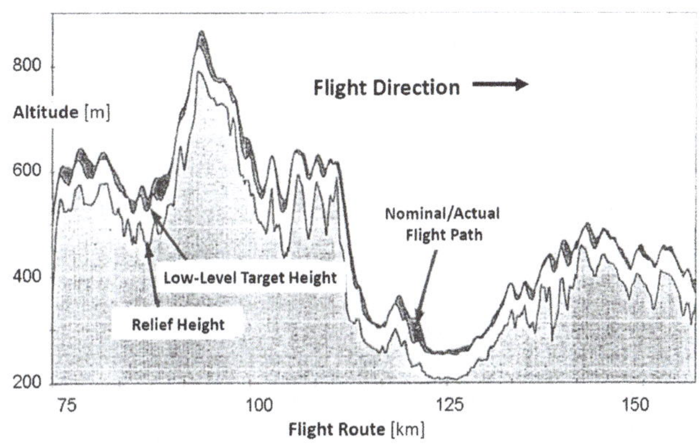

Fig. 9.85 Low-level nominal and actual flight path over the Harz Mountains

9.2.8 Enhanced Vision for All-Weather Flight Operation (1999)

Klaus-Uwe Hahn

As a part of the EU project AWARD (*All Weather ARrival and Departure*), pilot assistance systems were tested, which would allow a pilot to operate aircraft safely in the absence of the exterior view, regardless of the quality and precision of the flight guidance installations at the airport [49]. The objective was to carry out the flight operation with the equipment available onboard according to the "category" CAT III criteria, with a decision height (DH) of less than 50 ft and a visibility (RVR, *Runway Visual Range*) of 75 m only. For this purpose, two different concepts of pilot-assistance were pursued and evaluated, both based on the use of a differential GPS (*DGPS*), which is cost-effective compared to the conventional ILS (*Instrument Landing System*) or MLS (*Microwave Landing System*). One of these two concepts was a so-called SVS (*Synthetic Vision System*), which provides the pilot with an artificial synthetic exterior view (see Sect. 9.2.7). This system was tested in a ground-based motion simulator [50].

For the second pilot-assistance concept, a sensor-based artificial outer visibility was provided so that flight under quasi-visual conditions was possible. This system called EVS (*Enhanced Vision System*) was tested in flight using ATTAS. The basis for EVS was the availability of suitable sensors, which operate in a frequency range that expands the field visibility of the human eye so as to use additional, quasi-optical information. For this purpose, a *millimeter wave radar* (MMWR) and a *forward-looking infrared* (FLIR) camera were used, with which the necessary additional visual information was gathered [51]. The external view generated thereby was displayed on a HUD (*Head-Up Display*). Particularly, on a special transparent disk (*Combiner Unit*—COU) in front of the pilot's eye, on which also other useful information were provided. The projection on the transparent disk is such that a sharp image is formed when eyes focus far ahead of the aircraft. In this way, without having to focus differently, the pilot captures all the information, including sensor visual information, necessary for guiding the aircraft.

The overall EVS system included a large number of components, all of which were integrated at suitable places in the aircraft. They all had to be certified for flight operation. Figure 9.86 shows these components and their location of installation in ATTAS, where COU denotes the combiner unit, FLIR the forward looking infrared, MMWR the millimeter wave radar, HCP the head-up display control panel, HFDC the head-up display computer, MTR the mounting tray, OPU the optical projection unit, PU the processing unit, RCU the radar control unit, RPU the radar processing unit, RRCU the radar recorder control unit, and SMU the servo-mechanism unit. Figure 9.87 shows the two sensors mounted on the front bulkhead. The infrared camera (the small black cylindrical disk) on the top and the millimeter wave radar (large ocher-colored cylinder) can be seen in the figure. The square-sized pressure-tight plug plate, clearly visible in the figure, covered the hole cut in the bulkhead. It enabled the electrical connections between the sensors and the devices housed in the fuselage. To maintain the aircraft center of gravity within the permissible limits, the heavy special recorder for recording MMWR data had to be mounted far behind in the tail. The so-called EVS container was installed in the passenger cabin. Many smaller devices and system components were accommodated in this central part of the overall system, including the central processing unit (PU).

The sensor signals for the raster display on the HUD were processed in the PU. In this display, the information, which is usually presented in the primary flight display (PFD), was superimposed (in the *Stroke Mode*). Thereby, the pilot receives all the important information with a single glance forward in the direction of the flight, looking simultaneously

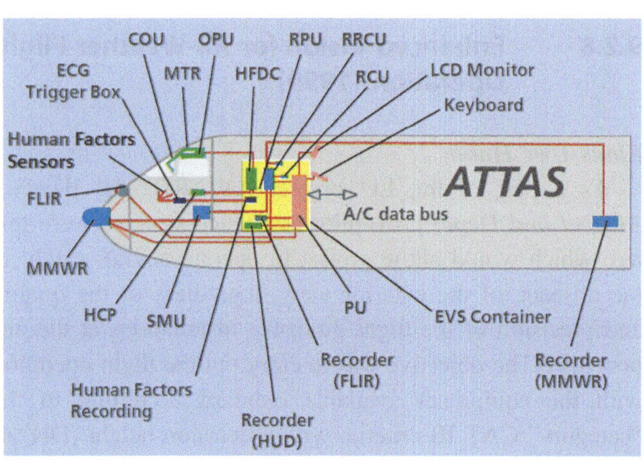

Fig. 9.86 EVS components and their installation in ATTAS

Fig. 9.87 Millimeter wave radar and infrared sensor on ATTAS

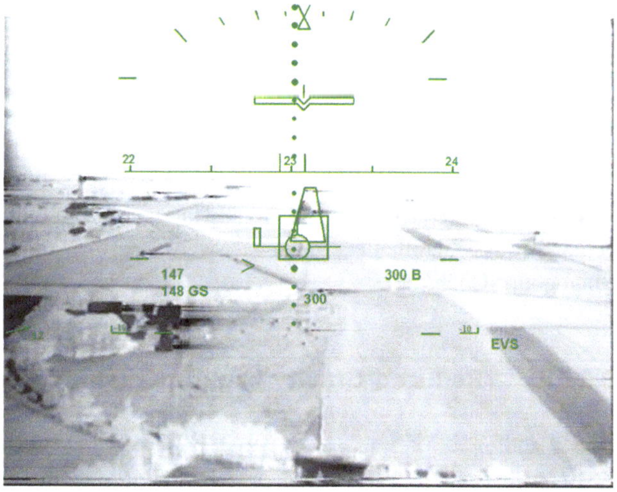

Fig. 9.88 Sensor-view information and flight guidance display in the HUD

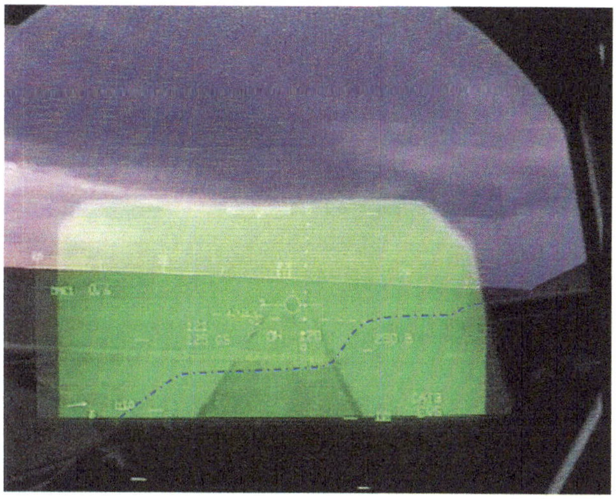

Fig. 9.90 View through the HUD

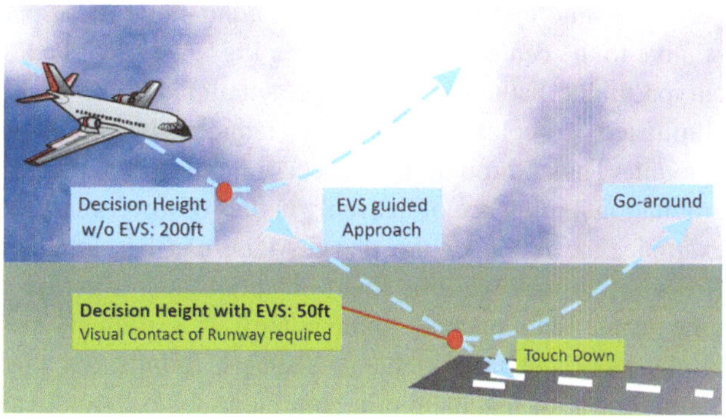

Fig. 9.89 Approach procedure with EVS meeting CAT III requirements using CAT I landing system

through the transparent pane (*Combiner Unit*) of the HUD and through the cockpit window. This data fusion is shown in Fig. 9.88.

The efficacy of an EVS is demonstrated in Fig. 9.89. Exemplary is the approach under poor visibility conditions to an airport, which is equipped only with an ILS of CAT I quality. According to CAT I conditions, this implies that when the pilot reaches the decision height (DH) of 200 ft above the runway, he must initiate the go-around if he cannot recognize the runway. On the other hand, with EVS, it is possible to provide the pilot with a supplemental vision of the outside, which is beyond the visual frequency range of human eye, prior to reaching the decision height. With this artificial visual information, the pilot could continue the approach to a height of 50 ft above the runway, even without a real outside view. This corresponds to a decision height according to the more precise CAT III requirements. Only at this new EVS decision height, the pilot should be able to recognize the runway in order to continue the approach. If the runway is not recognizable at this altitude, a go-around would have to be initiated despite EVS. Nevertheless, it is worth to note that high-quality CAT III approaches and landings are possible with EVS under considerably poorer visibility conditions, even at an airport that is equipped with just CAT I flight guidance aids.

Figure 9.90 shows the picture captured by a *Handycam* taken from the perspective of the pilot through the HUD. The information which is usually displayed in a PFD can be easily traced. The dashed-dotted blue line marks the contours of the instrument panel. Below this line, normally, the pilot would not have a view of the runway in front of him. However, since the EVS sensors are installed in the aircraft nose in front of the instrument panel, the view of the runway is not obscured. This information is presented to the pilot on the HUD, so that he can even virtually view through the instrument panel.

9.2.9 Fast and Robust Design of Autopilot Control Laws for Automatic Landing

Gertjan Looye

As a part of the European project REAL (Robust and Efficient Autopilot control Laws design, EU-FP5), an efficient design process for robust flight control laws was developed and applied to CAT-IIIb-capable automatic landing systems. The REAL project consortium consisted of Airbus, NLR, TU-Delft, and the DLR institutes of Flight Systems and Robotics and Mechatronics (now: System Dynamics and Control) [52, 53].

During the first phase of the project, a process based on NDI (*Nonlinear Dynamic Inversion*) was developed for controlling the aircraft attitude and TECS (Total Energy

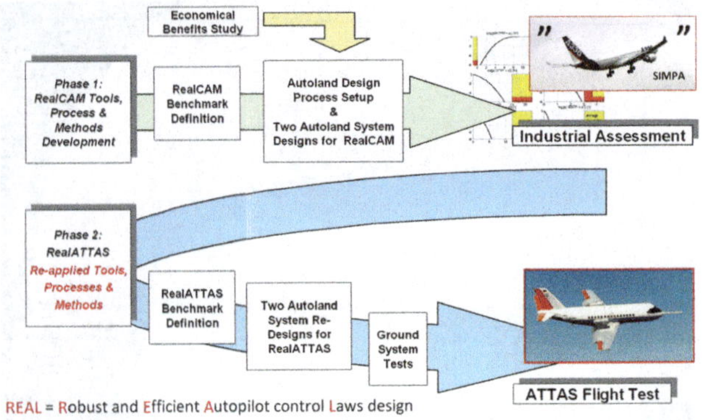

Fig. 9.91 Structure of the REAL project in sequential order

Control System) for decoupled tracking of flight path and airspeed references. Attitude control laws usually need to be individually tuned for each type of aircraft. NDI uses inverted model equations to adapt to the individual aircraft type, allowing for easy automation of the design when a sufficiently good model is available.

The software tool MOPS (Multi Objective Parameter Synthesis) was then used to automate the tuning of the control law parameters to meet the performance and robustness requirements. As part of the CAT-IIIb certification process, extensive Monte-Carlo analyses were performed. These results were directly incorporated in the parameter optimization. To demonstrate robustness, the developed process was first applied to the RealCAM (REAL Civil Aircraft Model), a simplified Airbus model, and then extensively evaluated in a qualified Airbus simulation tool (see Fig. 9.91, upper part).

The efficacy of the process was then successfully demonstrated by a re-design for ATTAS within a short time frame (see Fig. 9.91, lower part). The developed design process was adequately tested, as ATTAS has quite different flight characteristics and the resulting control laws had to be flight ready.

To minimize risks of flight testing newly developed flight control laws all the way to landing touch down, dedicated flight test procedures were developed. After successful landing tests on virtual elevated runways (starting at 500 ft, then stepwise reduced to 100 ft) at Hanover airport, six actual fully automatic landings were performed at Magdeburg-Cochstedt.

9.2.10 Reduced Gravity Experiments (2008)

Dirk Leißling

Investigations under reduced or zero gravity are important not only for space missions but also in other fields of natural sciences. A special flight test technique is needed in which the aircraft's acceleration cancels the earth's gravitational force. Such a condition can be created over a limited period of time in an aircraft through a parabolic flight path relative to the center of the Earth (see Fig. 9.92). With a less curved flight path, the relative gravitational force is not eliminated completely, but only reduced by a certain amount. Thus, through this special flight technique, the gravitational influence of heavenly bodies of lower mass, for example, the moon or the planet Mars, can be simulated. However, to maintain this flight condition over a period of time using only manual flight control is a difficult task. Furthermore, the unavoidable atmospheric turbulences affect the precision of maintaining a target load factor. Automatic control augmentation alleviates this problem significantly.

The beginnings of parabolic flights can be traced to the nineteen-fifties, when the effects of zero-gravity on human organism were investigated on future astronauts and

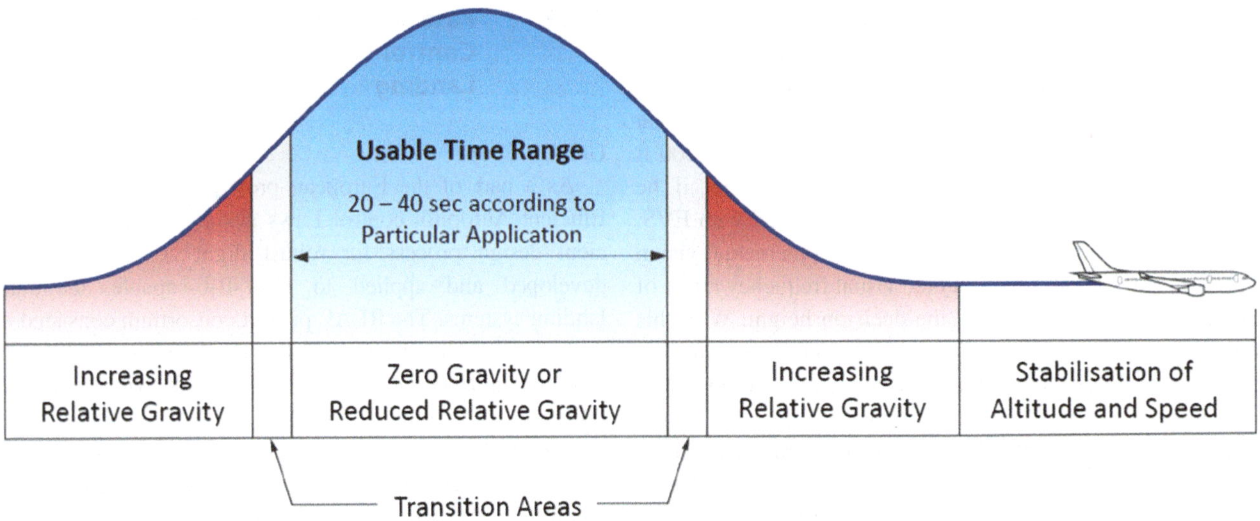

Fig. 9.92 Principle of parabolic flight

cosmonauts. Also, technical devices and systems could be tested for their operational capability in Space. After the nineteen-eighties, the scope of application was expanded to fields other than astronautics. Special aircraft used for parabolic flights were, for example, the Ilyushin 76 MDK in Russia or the Boeing KC-135A (1995–2004), McDonnell Douglas DC-9-C9B and Boeing 727 operated by NASA. In Europe, parabolic flights were performed from 1988 to 1995 with a Sud-Aviation SE 210 Caravelle. Since 1997, their role was taken over by the Airbus A300 ZERO-G of the company Novespace, which is operated—like its predecessors—with direct participation of the French space agency CNES (Centre National d'Études Spatiales) on behalf of the European Space Agency (ESA). With this aircraft, simulated zero gravity phase is maintained for approximately 20–25 sec [54–56].

Motivation for Parabolic Flights with ATTAS

Since 1999 DLR was one of the A300 ZERO-G users, with one or two campaigns every year. In May 2008, the project manager and program leader of DLR parabolic flights, *U. Friedrichs*, became aware of the alternative offered by the ATTAS, particularly of the technical capabilities based on the freely programmable flight control system. The short-term availability of this aircraft seemed to be more interesting than the commercially marketed A300 ZERO-G of Novespace. There was also a tangible interest in the utilization of ATTAS by the Institute for Geophysics and Extra-Terrestrial Physics of the TU-BS as well as by the DLR Institute of Space Systems in Bremen. For their experiments, a simulation of the gravitational acceleration on Mars planet was needed, which is 3.71 m/sec^2 (0.378 g), about one-third of the earth's gravity of 1 g.

The programmable flight control system provided by ATTAS was advantageous for "Martian parabolic flights" in terms of meeting the defined tolerances through better reproducibility compared to the manual control of other aircraft for such experiments. Furthermore, atmospheric disturbances could be compensated much faster and more exactly by an automatic controller. The experiments pertained to the greenhouse effect caused by the emission of dust particles (*solid-state greenhouse effect*) and the investigation of ground distortion properties under Martian conditions [57]. Having verified the technical feasibility based on geometry, mass, and electric power consumption, the first campaign was planned, comprising of 90 parabolic maneuvers in six flights with ATTAS [58].

Development of Control System

Initially, it was proposed to use an automatic control system only for the parabolic arc of the flight path, with manual initiation and termination. However, due to complexity, better reproducibility was possible only through automation of the complete maneuver. Accordingly, an automatic controller for the entire flight path was designed and parameterized based on the non-linear flight-mechanical model of the ATTAS. It was tested and evaluated under real-time conditions on the ground-based system simulator. A simple system was developed for lateral attitude control which adjusts the bank angle to zero degrees at the beginning of each maneuver and maintains it thereafter. The longitudinal controller was more complex and consisted of five different modules for the five main segments of the maneuver (see Fig. 9.93).

Phase 1: After manual activation of the SIM mode by the experimental pilot, the engine exhaust gas temperature (EGT) is checked. It is a measure of the available thrust power and must have a minimum value of 465 °C. This ensures adequate position at the beginning to perform the parabolic flight. If this condition is fulfilled, the pitch attitude is reduced continuously up to a value of −10° to increase the airspeed.

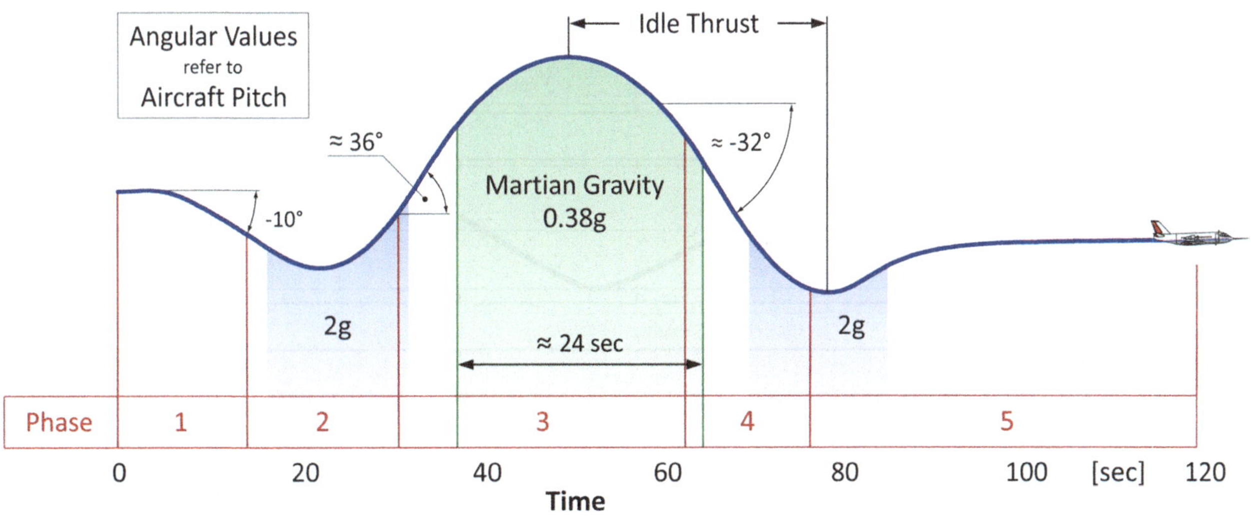

Fig. 9.93 Entire profile of Mars parabolic maneuver with controller for five segments

Phase 2: At a calibrated airspeed of 273 knots, the first flare maneuver is initiated reaching a load factor of about 2 g. Under these conditions the elevator actuators reach their force limitation; any further increase in the load factor is hardly achievable.
Phase 3: This controller module is activated on reaching a certain value of pitch angle which depends on the EGT, aircraft mass, configuration, and minimum permissible airspeed. It ensures that the airspeed converges to its minimum at the vertex of the parabola but never falls below it, thereby utilizing the velocity potential optimally. During the transition to the actual parabolic segment, the maximum pitch angle of the entire flight path profile is reached, which is roughly 36°. The controller then regulates the load factor to the desired value of 0.378 g, using the vertical acceleration measured by the inertial measurement unit. The controller parameters for both the proportional and integral parts vary as the nominal value is approached. On reaching the vertex of the parabola, the engine thrust is slowly set to idle to avoid an increase of the airspeed during the following descent.
Phase 4: On reaching a particular pitch angle corresponding to the actual aircraft mass, the parabolic segment is terminated and followed by a second flare maneuver. This ensures that the allowed maximum airspeed of 288 knots is reached during the fly-by-wire operation, but not exceeded. With a typical aircraft mass of 40,050 lbs (18,166 kg), the activation of this controller mode is initiated at a pitch angle of −30.85°. As a result of the system sluggishness and in order to get smooth transitions, the pitch angle continues to fall further for a short time before it starts increasing again in the course of the flare, reaching the pitch attitude of about −32°.
Phase 5: At a flight path angle of −5° the last section of the control system is activated. The aircraft is maneuvered to level flight again, whereby the engine thrust is set back to its original level at the lowest point of flare maneuver. At this moment a flight path angle of 0° is reached.

The autopilot status was displayed to the test pilot on a special display unit throughout the maneuver. In normal operation, the automatic trim function provided on the ATTAS by the horizontal stabilizer helps to avoid stationary loads on the elevator. This is, however, deactivated during the entire parabolic flight test to avoid generating any additional system disturbances.

Flight Test Results

On September 25, 2008, a flight test was carried out with a 7-man crew to test the implemented autopilot functions under real operating conditions. Starting at an altitude of 21,000 ft each time, three parabolic maneuvers were carried out successively. In each maneuver the Martian gravity of 0.378 g was held exactly for a duration of roughly 24 sec without exceeding any operational limits; the maximum acceleration error during this time was mostly less than 0.02 g [59]. Figure 9.94 exemplarily illustrates the performance during the third parabola. It is obvious that during the maneuver the possible speed range was optimally utilized. The green

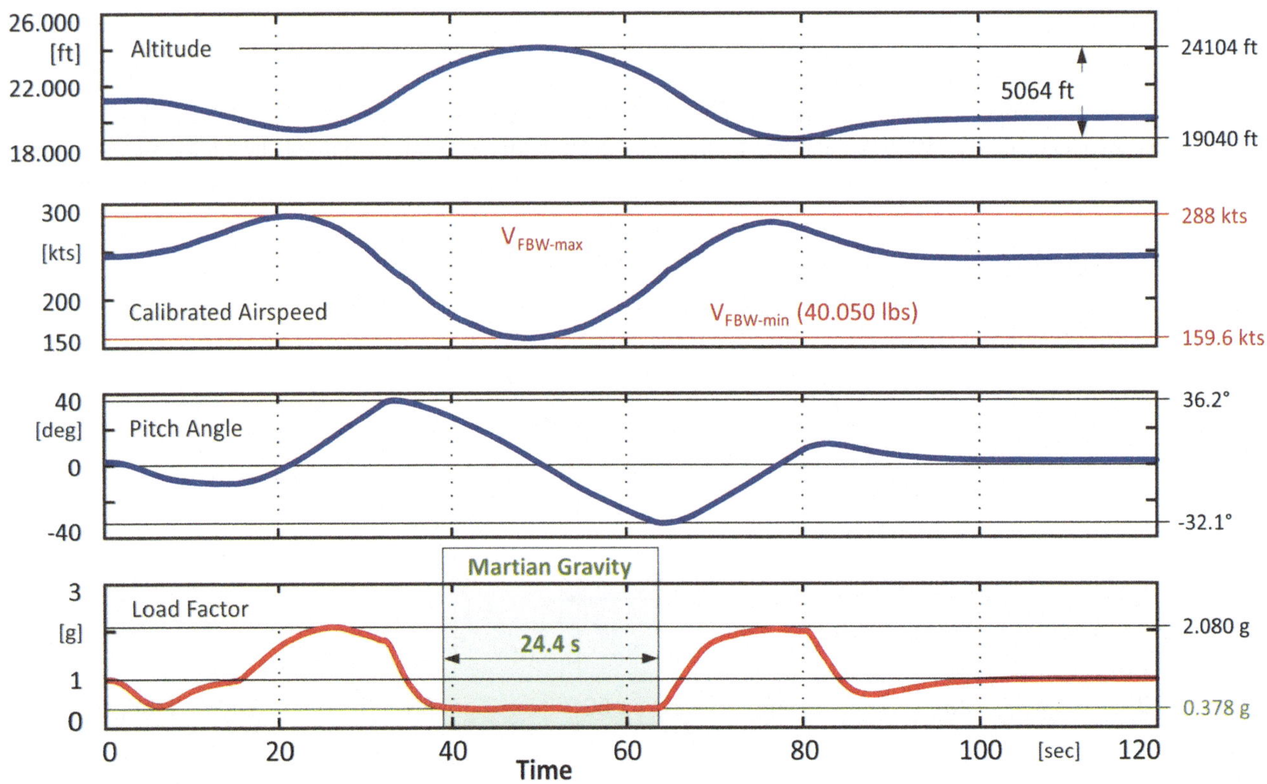

Fig. 9.94 ATTAS flight test F835, third parabola, aircraft mass 40,050 lbs

shadowed area shows the duration of 24.4 sec in which the "Martian gravity" of about 0.378 g was maintained.

During the three parabolas, the deviation of the measured acceleration from the nominal value was mostly less than the desired limit of 0.02 g. When this limit was exceeded temporarily, the maximum error of 0.03 g was considered to be within the adequate range. The reason for this exceedance can be found in the engine thrust reduction at the vertex of the parabola initiating a disturbing pitching moment. To eliminate this minor effect, a feedforward controller for the elevator channel was developed and used to compensate the disturbances caused by the engine thrust reduction. The functionality of this approach was demonstrated in the ATTAS system simulator subsequently. The reproducibility of parabolic flight with ATTAS was confirmed qualitatively as well as quantitatively by the excellent agreement of the three parabolic maneuvers.

9.2.11 Demonstration of Technologies for Unmanned Aerial Vehicles (2000–2008)

Dietrich Altenkirch

The capabilities of UAVs (Unmanned Aerial Vehicles) had been significantly increased by technological advances in the field of sensors, data links, and aircraft systems as a whole in the years up to 2000. The German Federal Armed Forces naturally pursued these developments and came to the logical conclusion that such UAS (Unmanned Aerial System) would be indispensable for reconnaissance in the future to obtain their own, independent information about a threat situation in foreign operations.

These operational scenarios, for example, a high-flying UAV, require participation in civil airspace to reach the target area. In Germany, the existing unmanned systems were allowed to operate only in military restricted areas. As a prerequisite for the development and later use of unmanned reconnaissance vehicles, the verification of technology and procedures for the participation of an UAV in civil aviation is necessary. Therefore, *Peter Hamel* (DLR) and active support from *Gerhard Morsch* (BWB) proposed to combine DLR and German industry skills in an UAV demonstrator program named WASLA-HALE (*Long Distance Airborne Reconnaissance—High Altitude Long Endurance*) using the ATTAS research aircraft of DLR. The EADS (European Aeronautic Defense and Space Company) and ESG (Elektroniksystem-und Logistik GmbH) were involved from the German industry. In addition, the DFS (Deutsche Flugsicherung) was involved as the national ATC organization.

The following tasks were anticipated to be carried out in the demonstrator program: (1) Development of criteria, guidelines and procedures for approval with national and international authorities and (2) Demonstration and proof of the suitability of procedures and techniques for the safe and disturbance-resistant operation of a UAV from the ground.

The aim was to minimize the risk of developing and operate an UAV in controlled airspace. The demonstration program using ATTAS as an experimental surrogate UAV was launched in two phases:

Phase 1: Definition of the implementation phase of the demonstrator program including first simulation studies.
Phase 2: Execution of the demonstration program. The selection of essential UAV-specific techniques and procedures, which have been tested and demonstrated with ATTAS, was carried out taking into account the technical conditions and the operational limitations of ATTAS. Apart from some areas of typical UAV missions, which were tested in 2 mission simulations, important standard and emergency procedures of flying in controlled airspace could be tested. The extensive basic equipment of the ATTAS was supplemented by additional experimental systems, which were already largely available with EADS and ESG. This included the FMS (Flight Management System, Missions Management System), the first one provided by the ATTAS basic equipment that was supplemented by EADS precision navigation system RAPIN+. Data link were installed for communication between a Ground Control Station (GCS), Air Traffic Control (ATC), and ATTAS.

The GCS at Braunschweig and Manching each consisted of the operator station and the data transmission system. The operator station as a workplace for the remote operator provided various operating and display facilities, which made it possible to investigate different concepts of an UAV ground control station. Ground-level mission management system was used to prepare the mission as well as support the mission leader during the mission. The data transmission systems were realized by extending the existing telemetry facilities.

Due to technical and operational restrictions, ATTAS could not demonstrate all phases and maneuvers of an UAV mission. Therefore, important preliminary work on the development and testing of standard and emergency procedures had been defined and evaluated in two simulation events. The flight tests on ATTAS were carried out in several flight test campaigns starting either from Braunschweig or Manching. The project objectives were achieved in three milestones:

Milestone 1: Operation in Temporary Restricted Areas (TRA) only:

- Detection of standard procedures and functions
- Flight schedules with FMS

Milestone 2: Operation in TRA's:
- Change of flight plan during the mission
- Demonstration of emergency procedures

Milestone 3: Proof of UAV operation in common airspace:
- Proof of a complete WASLA-HALE mission, including emergency procedures
- Transfer of the UAV control to a second Ground Control Station.

Figure 9.95 shows the components involved in the experimental UAV system for a ground test at Braunschweig airport with the subsystems: (1) UAV-Board System (ATTAS), (2) GCS-Braunschweig, (3) GCS-Manching, (4) Stationary Data Link Station Braunschweig, (5) Data Link, and (6) Mobile Data Link Station Manching.

In this test scenario, the functions of the UAV board system were checked with the two ground control stations and the data link stations as well as the transfer procedures between the two stations. The ground test included the entire functional chain and enabled a high test depth. With this complex experimental UAV system with ATTAS as a real flying aircraft in controlled airspace several test flights in the North German airspace were performed from Braunschweig airport. Shortly after the conventional start of ATTAS, the pilots passed over the control to the remote operator in the ground control station Braunschweig. There was a trained pilot, who took over the radio communication with ATC as well as the guidance of ATTAS through the airspace acting on instructions of ATC. The pilots in the ATTAS cockpit were only responsible for monitoring the ATTAS basic systems, in order to be able to intervene in the event of a fault.

Many flight routes led to the TRA 202 near Bremen, where airspace was cleared by the ATC to carry out test maneuvers. Following the successful test flights from Braunschweig, two typical UAV flights were carried out at the site of the WTD 61 Flight Test Centre at Manching on June 3, 2004. The flights had both standard procedures as well as emergency procedures. The first flight (test No. 428) from Braunschweig to Manching had the following test targets:

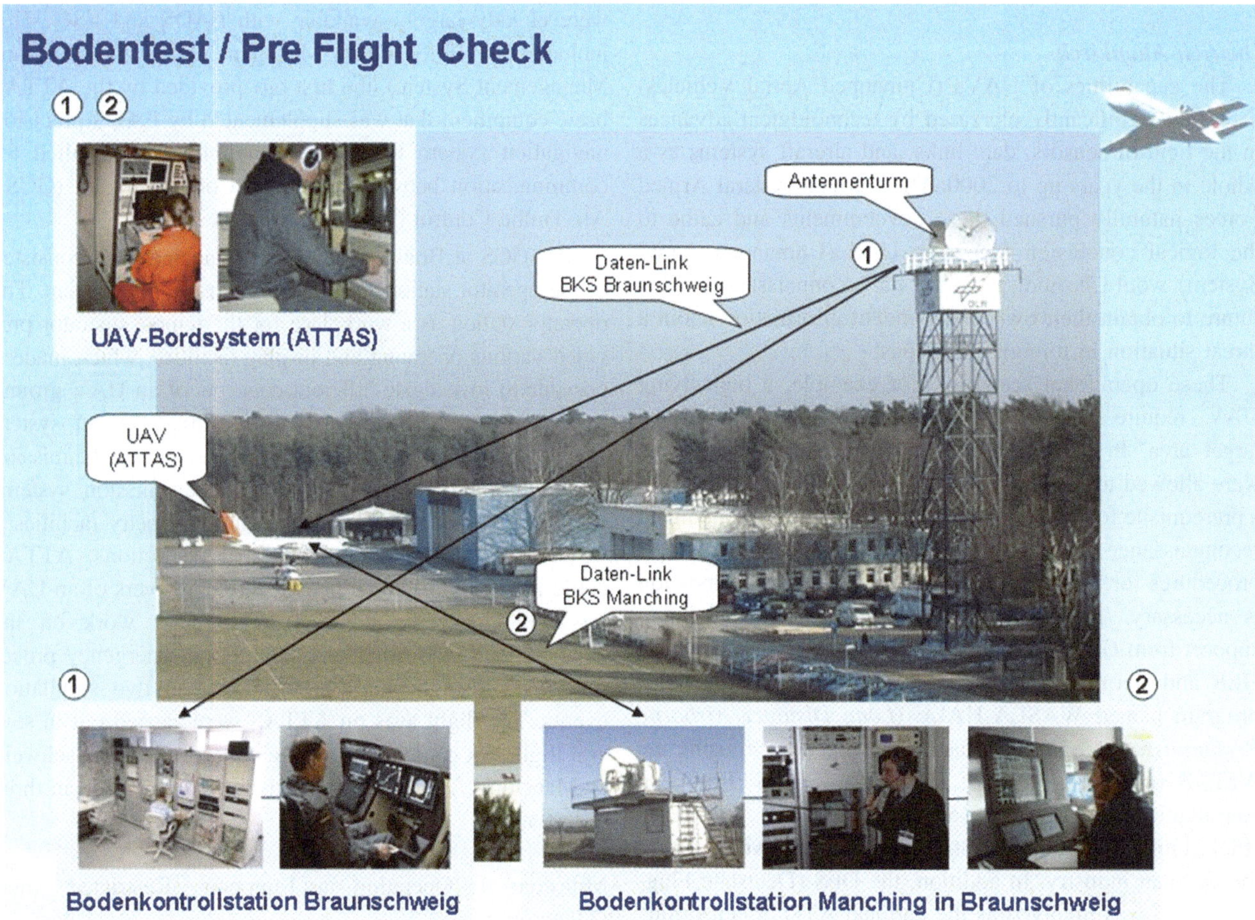

Fig. 9.95 ATTAS-UAV components on the ground

Flight route: Braunschweig 26—GALMA—KULOK—RUDNO—Manching 25L.

Objectives: UAV tests with transfer of UAV control to a second GCS:

- UAV Control acting on ATC Commands,
- Radio communication with different ATC-centers via the UAV.

After takeoff from Braunschweig, the control was transferred to the GCS-Braunschweig. This generated the route given in the IFR flight plan and activated it in the UAV experimental system. This was followed by a change of control between the ATC center Bremen and ATC center Berlin. At the position of the waypoint TABAT (see Fig. 9.96), the remote control was given by an active disconnect from the GCS-Braunschweig, and the GCS-Manching was able to take the remote control with a Reconnect. The information about the flight route deposited in the onboard systems was handed over to the GCS-Manching and could be modified as needed for the further flight. After the transfer to the ATC-Munich, a remote flight with a fully automatic approach to the simulated ILS ETSI 25L was carried out.

In a second flight demonstration (test No. 429) at Manching, standard and emergency procedures were demonstrated under control of the GCS Manching. Figure 9.97 shows the route from Manching with the following test objectives:

Flight route: Manching—TRA 210—Manching 25L

Objectives: UAV test with GCS–Manching

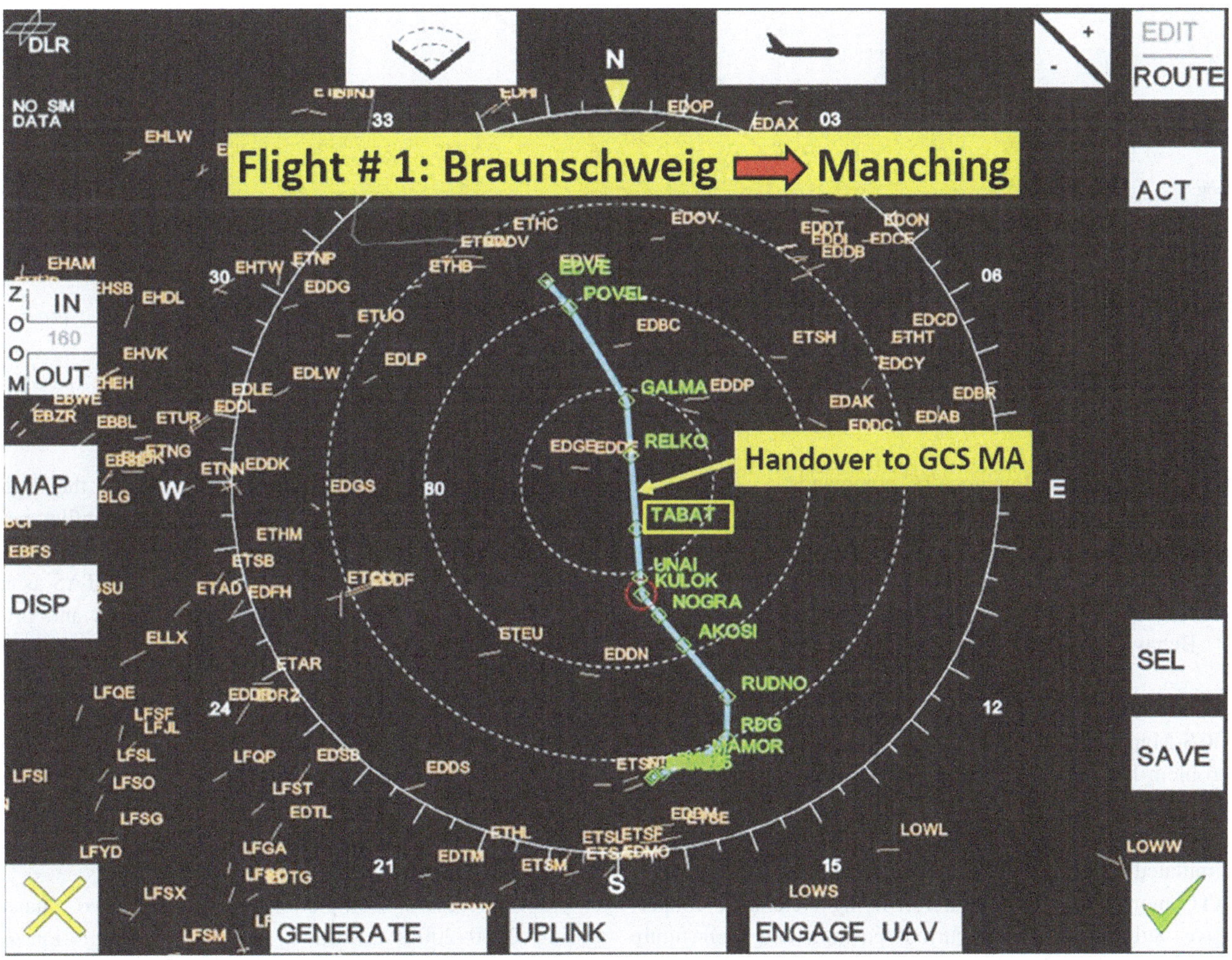

Fig. 9.96 WASLA-HALE transfer flight Braunschweig-Manching

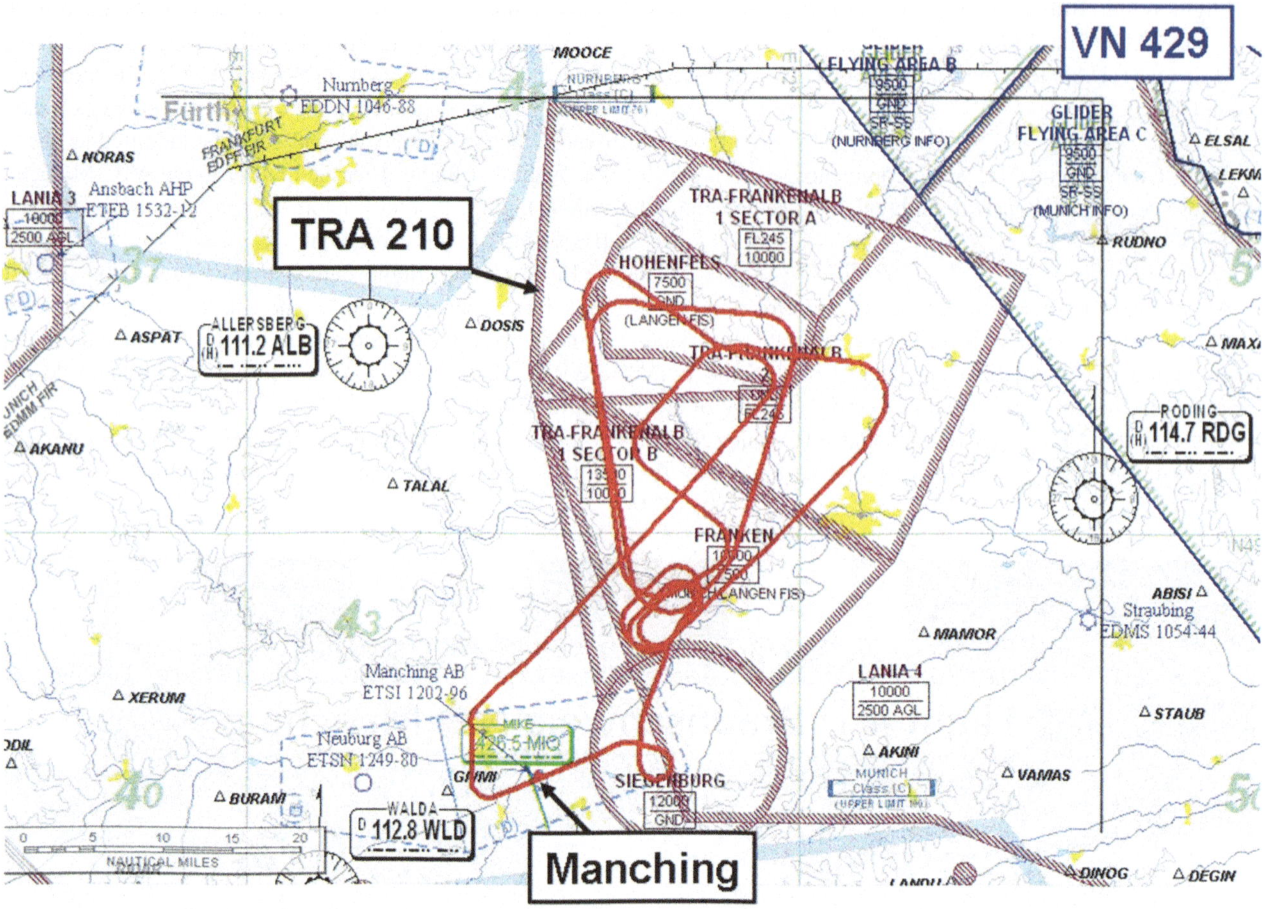

Fig. 9.97 WASLA-HALE flight route at Manching

- UAV control following ATC commands
- Solving a simulated traffic conflict with ASAS (Airborne Separation Assurance System)
- Resolution of a bad weather conflict
- Autonomous evasion of an aircraft on a collision course
- Behavior at data link loss.

After transferring the UAV control to the GCS-Manching, the TRA 210 was entered with the problem-free guidance of the 'TRA monitor', according to the radar vector specifications. Afterwards, the FMS route in the TRA was re-activated by the GCS and a scenario was simulated in the case where the UAV is in airspace without ATC monitoring, but the aircraft flying there were cooperative and therefore their current position and their future Flight path was known. The ASAS in the experimental system onboard ATTAS analyzed the flight paths of the simulated air traffic in this case with regard to possible conflicts and produced 3 solutions, taking into account different targets. A solution variant was chosen, the flight path was planned accordingly and the traffic conflict was resolved.

With the completion of Phase I and II with ATTAS as an experimental UAV, the demonstrated techniques and procedures were evaluated with regard to the operational use of UAVs in the controlled airspace. Recommendations for regulatory requirements were issued. Based the results obtained, the BWB finally provided an outlook on future work on the complete integration of unmanned aircraft, also under VFR conditions (Visual Flight Rules), which essentially require the Sense and Avoid functionality (detection and avoidance of a collision).

After successful completion of the first two phases during 2000–2004, a further step in the integration of UAVs into the general airspace was undertaken in Phase III. The airspace was expanded and the necessary UAV abilities were examined in the case of participation in the general air traffic, that is, in uncontrolled airspace.

For this purpose, special procedures for the management of a UAV were investigated in cooperation with ATC and also the problem of the sense and avoidance of unmanned aircraft.

To be able to operate UAVs in controlled airspace for civil aviation, certain safety requirements had to be met. For some time, the elaboration of regulatory provisions had been under consideration, as they were discussed in various organizations such as EASA, EUROCONTROL and FAA. Initial proposals have already been presented. At the beginning of Phase III, however, a higher-level requirement could be identified, for example, during the UAV operation in civil airspace, a 'level of safety' comparable to manned traffic (equivalent level of safety—ELOS) must be guaranteed. This directly resulted in the requirement that an UAV must be able to see or recognize, and be able to evade, similar to a conventional human-controlled aircraft.

During a collision scenario of two aircraft, several sensors onboard generate an image of the surrounding airspace according to their field of view and recognize, track and, if possible, identify and classify the detected objects. If it is calculated that one or more of the objects are on collision course or fall below a minimum distance, a corresponding evasive maneuver is initiated to resolve the conflict.

Furthermore, the Sense and Avoid capability of a HALE/MALE UAV system was specified in Phase III and was demonstrated as an example in simulation and flight experiments in the ATTAS UAV experimental system. For this purpose, sensor capabilities have been specified, selected, and a radar and EO (electro-optic) sensor has been integrated into the ATTAS as an experimental UAV. The integration of radar and EO sensors is shown in Fig. 9.98.

Fig. 9.98 Sense and avoid sensor integration in ATTAS

The Do 228 and the DR 400 of the DLR were used as non-cooperative aircraft, which were intended to represent both larger and faster, as well as smaller, slow aircraft, which also consisted of different materials, namely aluminum and lumber.

The UAV experimental system on board ATTAS of the WASLA-HALE study phase II was extended by components which recognize the conflict by an approaching non-cooperative air transport operator and lead to a rule-based evasive maneuver. The avoidance command was directly connected to the ATTAS flight control system. In several flight tests in 2007, the performance of the sensors and the entire Sense and Avoid system was determined and a large number of automatic evasive maneuvers were successfully performed with the two non-cooperative Do 228 and DR 400 aircraft.

Finally, for the phase III flight tests, a scenario was chosen in which the collision angle was continually reduced to 0°, that is, the Do 228 as intruder and ATTAS flew directly towards each other. For security reasons, the flights took place exclusively under FL 100 under VFR conditions. The ATC monitored the flight and gave information on the surrounding traffic. Also for security reasons, an altitude difference of +500 ft relative to ATTAS was chosen for the Do 228. The position and height of the intruder could be tracked on the navigation display as TCAS information at any time by the ATTAS pilots. On the radio, instructions were given to the crew of the intruder aircraft in order to achieve a lateral, timely overflight at the VOR Hehlingen accounting for the current wind conditions. Figure 9.99 illustrates the flight paths of ATTAS and Do 228 with varying collision angles.

An object was detected by the onboard Radar for the first time at a distance of approximately 8 nm. Several successful automatic and/or autonomous evasive maneuvers were carried out by the ATTAS-UAV onboard system and the Sense and Avoid functionality of the system under the selected conditions. A final evaluation clearly showed that the experimental system was able to successfully avoid an object on collision course while maintaining minimum distances.

Based on the results achieved in Phase III, European regulations were created under which an operational Sense and Avoid system could be developed and allowed to be used in the future. They enabled full integration of UAV in the general European and possibly international airspace. However, the solution to this problem will still take a few years, as evident from the failure of the Euro-Hawk project of the German Federal Armed Forces in 2013.

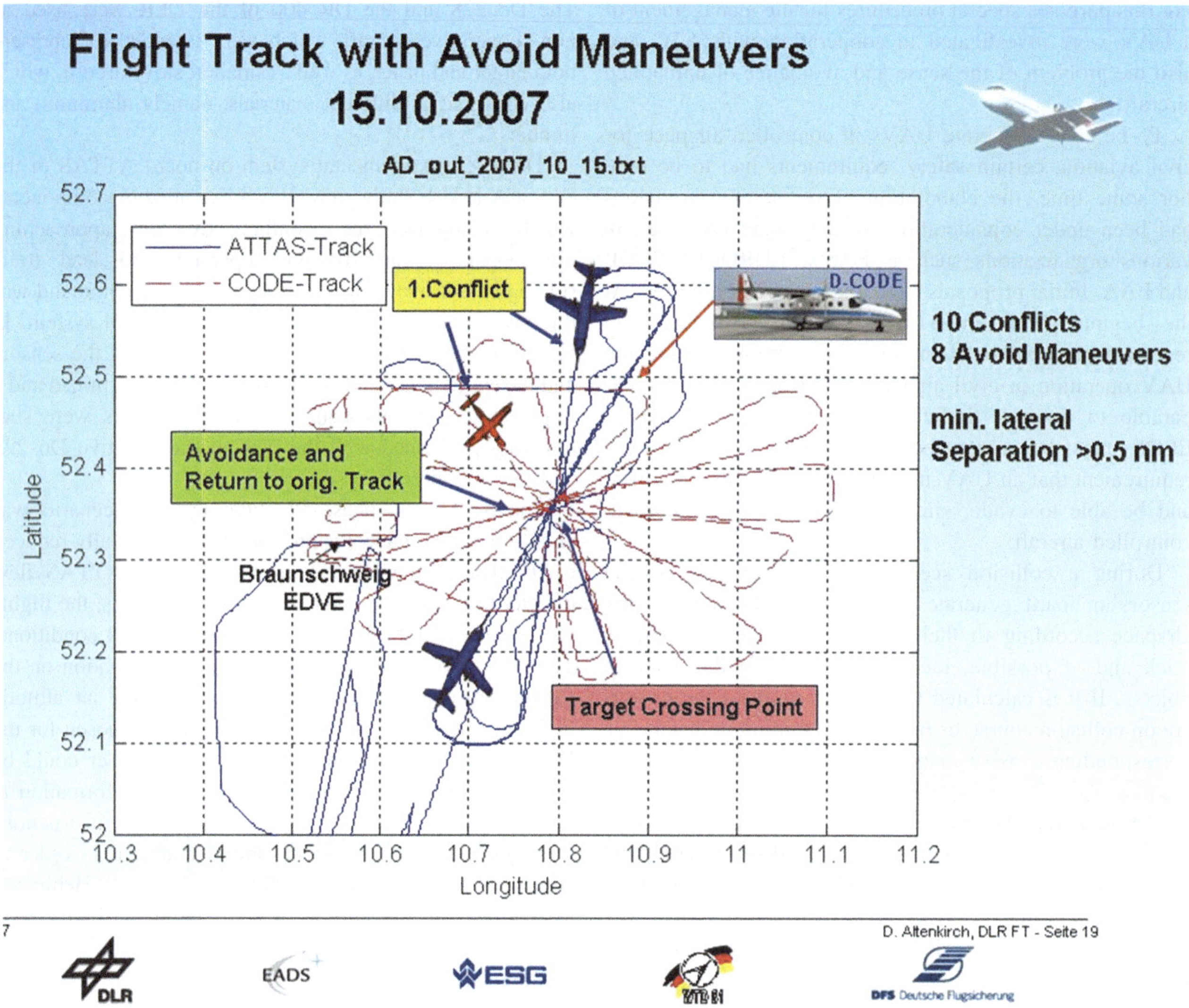

Fig. 9.99 ATTAS with sense and avoid system changed course to avoid collision with Do 228

9.2.12 Aircraft-Pilot Coupling Research (1992–2010)

Dietrich Hanke and Oliver Brieger

9.2.12.1 Introduction

Early nineteen-nineties, during the development of new fighters equipped with digital fly-by-wire flight control systems several accidents occurred which were caused by *Pilot Induced Oscillations* (PIO) or also called *Aircraft Pilot Coupling* (APC). The two most sensational accidents were the total crashes of the Swedish JAS 39 Gripen [60] and the Lockheed Martin YF-22 Raptor prototype [61], where the tail touched the ground during landing and the aircraft was severely damaged. Nearly all new aircraft equipped with digital flight control, military or civil, have been more or less prone to this problem and posed challenges in the development.

The cause of all these accidents was traced to an insufficient rate of operating the aerodynamic control surfaces (*Rate Limiting*). Control surface rate limiting may become extremely dangerous, because it is triggered only under specific conditions. In such a case, the aircraft control behavior is changed suddenly and unpredictably so that the pilot can lose control. Because all actuators are rate-limited the flight control design must guarantee that the commanded rates do not exceed the maximum possible actuator rate.

In general, aircraft with reduced natural stability for performance enhancements need increased control augmentation in order to provide acceptable artificial stability (see Sect. 6.1.2). These performance enhancements are at the cost of increased control activity. Further to maneuvering commands, the control deflections and rates may become so large that actuators could reach their maximum allowable deflection rates. In the US and also at DLR the research activities addressing this problem were focused on theoretical investigations, followed by flight tests in order to understand, describe and develop solution possibilities.

9.2.12.2 Rate Limiting Element Onset Parameter

The describing function of a rate limiter in the frequency domain was developed which could be used in the common stability analysis procedures [62–65]. The concept of a Rate Limiting Element (RLE) was introduced by *Dietrich Hanke*, which was valid for all rate limiting elements such as actuators and equivalent software functions. Furthermore, the flying qualities parameter called the *Onset Frequency* was defined in terms of amplitude and frequency at which the rate limiting element became active (onsets). The *Onset* describes the beginning of the pilot-aircraft instability. The *RLE-Onset* parameter and the describing function have been accepted internationally [66, 67].

Figure 9.100 shows the input/output behavior of a rate limiting element. At the *Onset-Frequency* the behavior changes abruptly. The commanded input signal (sinusoidal) is transferred to a triangle output signal (*Sinusoidal Input/Triangle Output*). The output signal shows large phase lag and reduced amplitude (see Fig. 9.101). The sudden change of the control behavior drives the pilot to increase his control commands because the aircraft did not follow his command anymore. The aircraft control becomes unstable (*Out of Phase*) which could finally result in an uncontrollable state of flight.

Figure 9.102 shows a block diagram of the control chain from the pilot to the actuator. When the input rate command R is larger or equal to the maximum rate R* of the rate limited actuator, the rate limiting element becomes active (RLE-Onset). This unsteady behavior is symbolized by a switch. After the onset, the signal follows the upper path shown in the same figure.

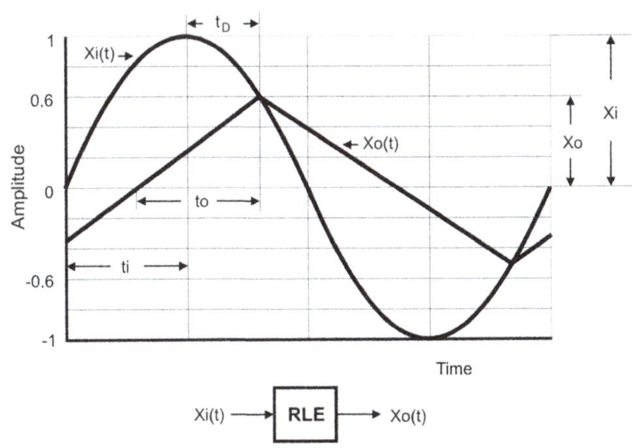

Fig. 9.100 Time history of the non-linear behavior of a rate limiter (RLE-Rate Limiting Element)

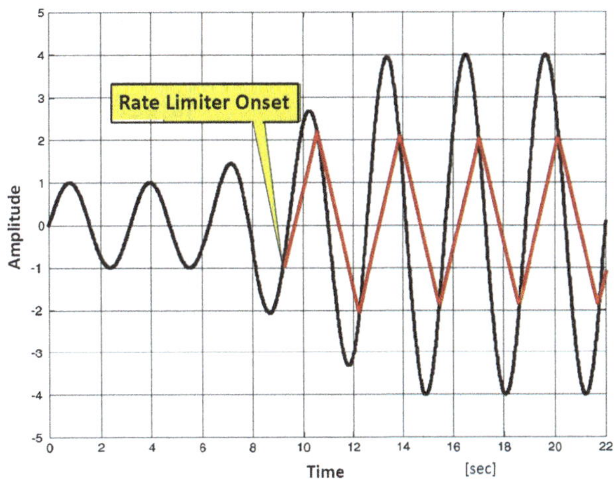

Fig. 9.101 Rate limiter onset in a case of increased amplitude (frequency constant, input signal *black*, output signal *red*)

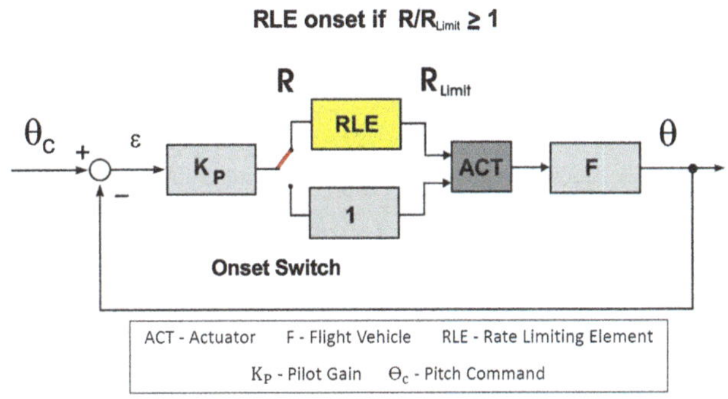

Fig. 9.102 Block diagram of a pitch control loop with rate limiting

In order to identify the minimal acceptable rate limit in the control path, flight trials were carried out on the ATTAS aircraft. The rates of the actuators in all control axes were limited by programming the onboard computers. The total pilot/aircraft system was evaluated under different flight conditions and tasks. In order to avoid the PIO problem induced by the phase delay of the rate limiter, a phase compensation function was developed which limits the input amplitude such that the rate limitation of the actuator is not exceeded [68–71]. Thereby the actuator response is always in phase with the commanded signal but with lower amplitude.

The piloting task consisted of rapidly aligning the ATTAS with high precision from a vertical and lateral offset position to a leading target aircraft (Do 228 of DLR). This task required high pitch and roll control activities and drove the actuators into rate limitations (see Fig. 9.103). Figure 9.104 shows that the phase compensator worked as predicted and the output signal remained in phase with the input signal, hence avoiding a PIO tendency.

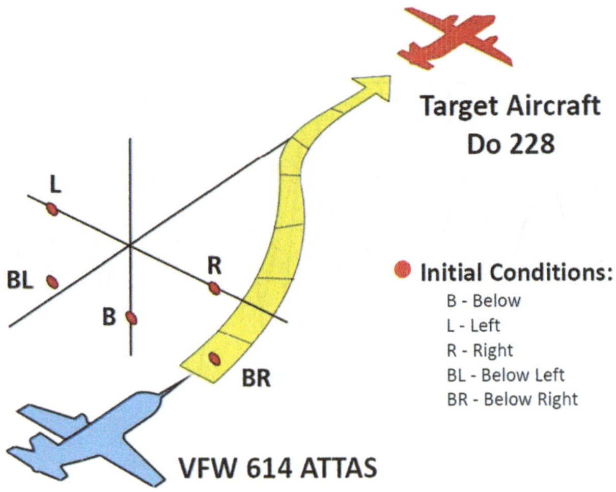

Fig. 9.103 Flight task of aligning to a flying target from vertical and lateral offset positions

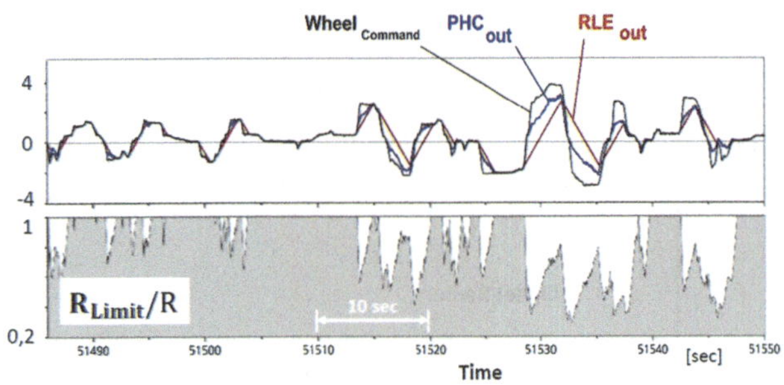

Fig. 9.104 Flight test results with phase compensation PHC (*white areas* represent the auto-activation of the compensator in case of rate limiting exceedance)

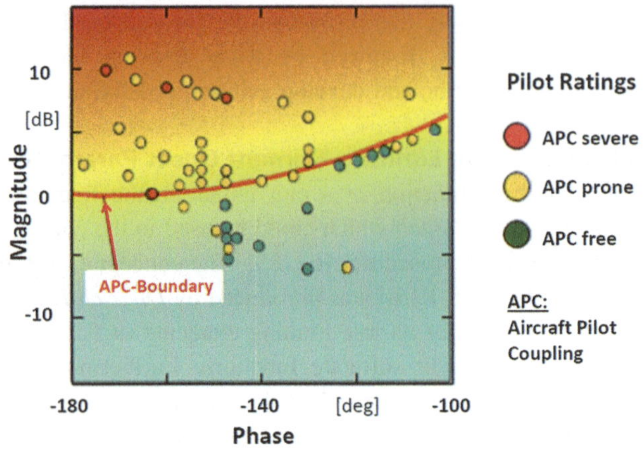

Fig. 9.105 Comparison of pilot rating with the OLOP-Criterion

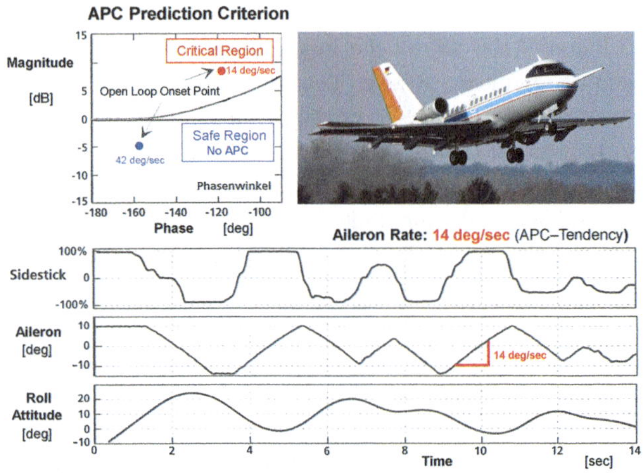

Fig. 9.106 Lateral APC tests during ATTAS flight envelope expansion

9.2.12.3 OLOP Stability Criterion

In the nineteen-nineties, a stability criterion was developed by *Holger Duda* from DFVLR, which is known as the OLOP-Criterion [72–82]. This criterion could be used to evaluate the stability of control loops with nonlinear behavior of rate limiting. Based on an approach of calculating the describing function of nonlinear control loops the possibility of so-called *Jump-Resonance* phenomena after the onset of a rate limiting actuator was verified. It was shown that the position of the onset point of the rate-limiting element in the *Nichols-Diagram* provided a good basis for the development of a nonlinear stability criterion, referred to as the OLOP-Criterion (*Open Loop Onset Point*). It was verified by using published flight data of PIO occurrences, such as YF-16, X-15, Space Shuttle landing approaches, and ground-based simulation experiments of Saab Military Aircraft.

In cooperation with the Swedish Aeronautical Research Organization FFA (today: FOI) specific rate-limiting trials were carried out in a ground-based motion simulator. Various aircraft models were used, flown and evaluated by experienced test pilots. The comparison of the pilot evaluations with the OLOP-Criterion is shown in Fig. 9.105. Each point in the diagram represents the application of the criterion, the pilot rating, and the experiment. The green area below the boundary indicates absence of APC tendencies, whereas points above the boundary (red area) illustrate severe APC-problems.

Finally, flight tests were carried out with the in-flight simulator ATTAS in order to define the minimum acceptable rate where no APC occurs. During the flight tests, APC occurred for an aileron deflection rate of 14°/s. With an increased rate of 32°/s the pilot/aircraft system was free of APC, as it was also predicted by the OLOP-Criterion (see Fig. 9.106). The ATTAS maximum basic aileron actuator rate of deflection is around 85°/s. For the ATTAS landing in fly-by-wire mode the limiting aileron rate was set to 55°/s, with a deflection limit of 45% of the maximum amplitude (see also Fig. 9.55). Also the OLOP criterion was internationally accepted as a valid tool in stability analysis of flight control systems with rate limitations [67, 83].

9.2.12.4 Saturation Alleviation Flight Tests

Yet another approach, an Anti-Windup Phase Compensator (AWPC) was developed and tested in flight with ATTAS. This work was accomplished in cooperation with the University of Leicester and the German Air Force Flight Test Center (WTD 61) under the project named SAIFE (Saturation Alleviation in Flight Experiment) in 2006 and 2010.

In the flight trials, the maximal rate of the aileron actuator of ATTAS was reduced to half of the original value to simulate a hydraulic failure. The reduced actuation rate showed a strong PIO tendency in the roll axis during aggressive maneuvering in the landing approach. In extensive flight tests [84–92] with pursuit tasks, offset-approaches were performed, where the aircraft flies from a lateral offset to the centerline of the landing strip. Through the application of AWPC, the PIO tendency could significantly be reduced. Figure 9.107 illustrates the influence of the compensator on characteristic flight data of the roll motion without and with the compensator. It is evident that with the compensator, the pilot could carry out the landing approach with no control problems.

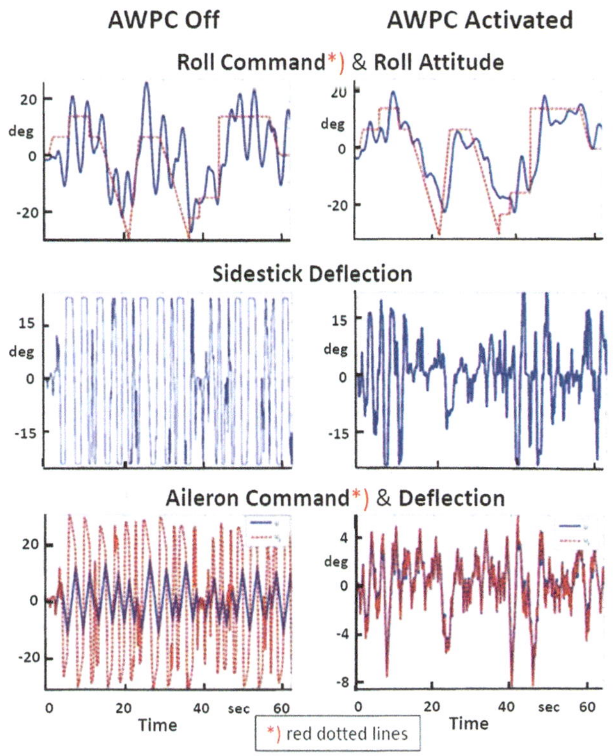

Fig. 9.107 Characteristic roll response data from SAIFE flight tests (AWPC—anti-windup phase compensator)

9.2.13 Wake Vortex Experiments (2001–2011)

Klaus-Uwe Hahn

The wake vortex phenomenon has been known since the beginning of aviation, as it is inseparably coupled to the dynamic lift force generated by the wings. It is well known that the vertical (lift) force, necessary for sustained flight, is generated due to the difference in pressures on the upper and lower surfaces, as a result of airflow over the wing profile in the longitudinal direction. Thereby cross flows accrue in the span wise direction, too, resulting in movement of air-layers towards the fuselage above the wing and towards wing-tips below the wing. This cross flow develops into a strong tip vortex behind the aircraft on the left and right wings. Their direction of rotation is pointed upward around the wing (see Fig. 9.108) [93]. The distance between the two

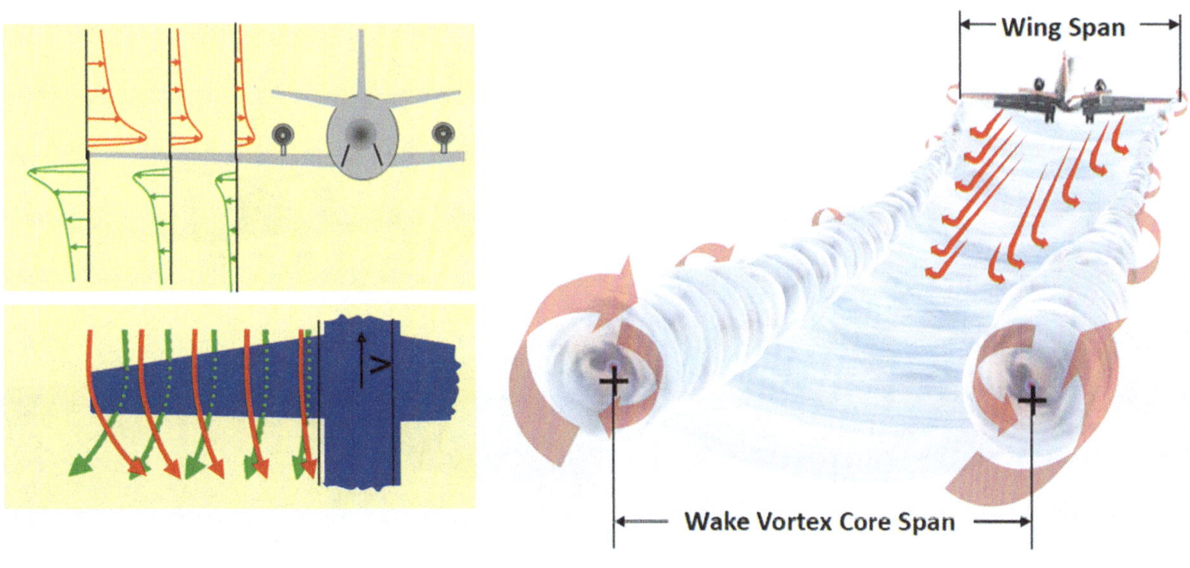

Fig. 9.108 Cross flow on the left wing and evolution of tip vortices (view from behind the aircraft)

counter-rotating vortices is slightly less than the wing span. This vortex pair produces a characteristic flow field, behind the aircraft, which is referred to as a wake vortex (see Fig. 9.109). As seen in Fig. 9.110, the visualization of the wake vortex generated by ATTAS over the runway, clearly confirms its occurrence in practice. The vortex wake becomes weaker when it ages, but it can exist for up to several minutes, depending on the atmospheric conditions.

Although this phenomenon was known for a long time, it became relevant only in the nineteen-sixties due to the increasing air traffic and the introduction of larger transport aircraft (Boeing B-747). The aircraft flying behind a B-747 experienced increasingly strong turbulences. This led to intensive theoretical and experimental investigations of the wake vortex phenomenon. To avoid entering into undecayed wake vortices, wake turbulence separation minima were introduced in air traffic, which solved initially the problem of encountering wake vortices for many years [94]. The introduction of even larger and heavier airplanes (Antonov AN 225, Airbus A-380), the increasing air traffic density, the limited capacity at large airports, and parallel runways, however, called for more comprehensive research.

ATTAS was used to investigate the wake vortex phenomenon for the first time during 2000–2002 as part of the European S-Wake project (*Assessment of Wake Vortex Safety*) [95]. To reliably and realistically model and simulate flying into wake vortex, flight tests were carried to gather flight data during such wake vortex encounters. The ATTAS aircraft was used as a lead aircraft, generating the vortex wake (ICAO separation class: medium). The fully instrumented Do 128 of TU-BS and Citation II of NLR flew behind (separation class: light), entering intentionally into the ATTAS wake that was made visible using a smoke generator mounted on the left wing (see Figs. 9.111 and 9.112). The wake encounters were flown at different distances of 0.5 to 1.5 nautical miles behind the ATTAS (see Fig. 9.113). A large number variables were measured and

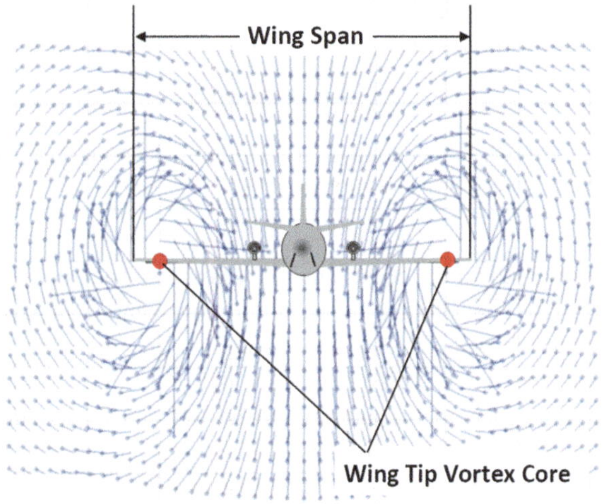

Fig. 9.109 Flow field of a wake vortex behind the aircraft (view from behind on the aircraft)

Fig. 9.110 Visualization of ATTAS wake flow field

Fig. 9.111 Smoke generator on ATTAS left wing

Fig. 9.112 Visualization of left wake vortex of ATTAS

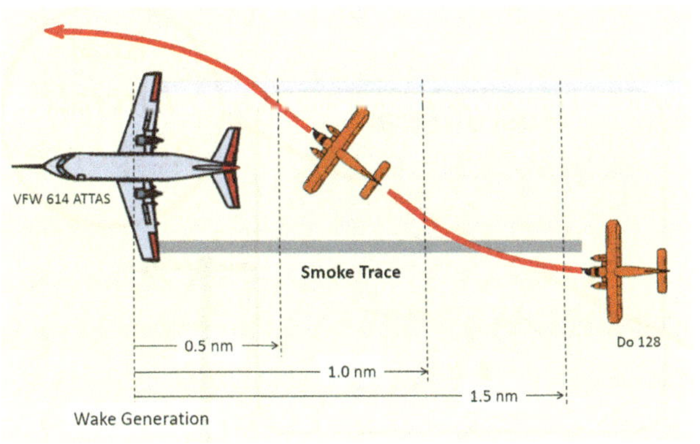

Fig. 9.113 Schematic of encounters flown with Do 128 in the ATTAS wake vortex

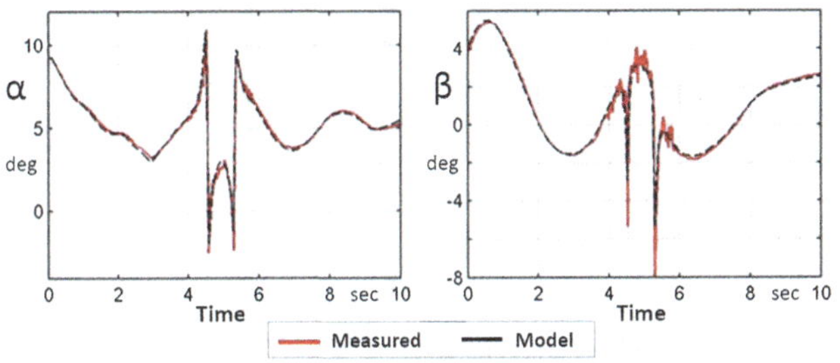

Fig. 9.114 Comparison of flight measured flow angles with the identified model for the flow field of a wake vortex

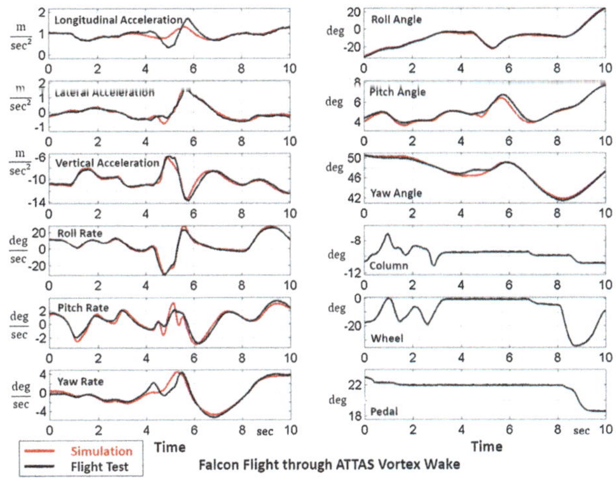

Fig. 9.115 Comparison of motion variables during wake vortex encounter

recorded for offline data analysis applying parameter estimation methods [96–98]. Thus, the flow field of the ATTAS wake and the effect of the wake vortex flow field on the follower aircraft (aerodynamic interaction model, AIM) were precisely modeled, determined and simulated.

Similar flight tests were later carried out with the DLR Falcon 20E as a follower aircraft, encountering ATTAS wakes. These tests were specifically designed to validate the AIM for aircraft with swept back wings. As a typical example, Fig. 9.114 shows a comparison of the angle of attack α and angle of sideslip β disturbances measured during a wake encounter with those from the identified model. Figure 9.115 shows typical variables of the Falcon response dynamics measured in the flight test, comparing them with the simulation results using the AIM identified from flight data. Considerable acceleration, rate and attitude responses can be seen in all axes resulting from the encounter as observed for the time segment from 4.5 to 6 s, whereby no pilot control inputs were applied. The aforementioned approach of modeling and estimation of the parameters of generated wakes and its effect on the follower aircraft was subsequently applied to the conventional transport aircraft [99].

The in-flight simulation capability of ATTAS was also used in the wake vortex investigations to determine the risk boundaries in which such encounters were acceptable to the pilot to continue the flight operation safely. The flight tests were carried out for approach configurations, whereby the direction of entry into a wake is almost parallel to the vortex axes. For such an encounter scenario, a vortex generates primarily a rolling motion of the follower aircraft. To assess the encounter severity, an assessment measure, called *Roll Control Ratio* (RCR), was applied which accounts for the disturbance and controllability of the aircraft about the roll axis. It is the ratio of the required roll-control input for compensating the roll moment induced by the wake to the maximum available roll-control moment of the aircraft. For idealized, quasi-stationary conditions, the values of RCR > 1 imply that the rolling moment induced by the vortex wake cannot be completely compensated by control inputs. Under operational conditions, however, other influences play an important role, particularly the delayed pilot reaction. Nevertheless, the investigations yielded that the RCR is a suitable parameter for establishing risk boundaries. For better reliability of RCR limits, experiments were initially carried out in a ground-based motion simulator and finally in the airborne simulation with ATTAS. For this purpose, with the help of in-flight simulation, ATTAS encounters into wake vortices were simulated during real landing approaches in flight, whose effect on the flight behavior was to be assessed by the pilot after each approach. The simulated encounters were started at different heights above the runway threshold.

The principle of in-flight simulation of wake vortex encounters is depicted in Fig. 9.116. The wake vortex induced disturbance could be accounted for in two different ways, namely as time or space dependent. The first option

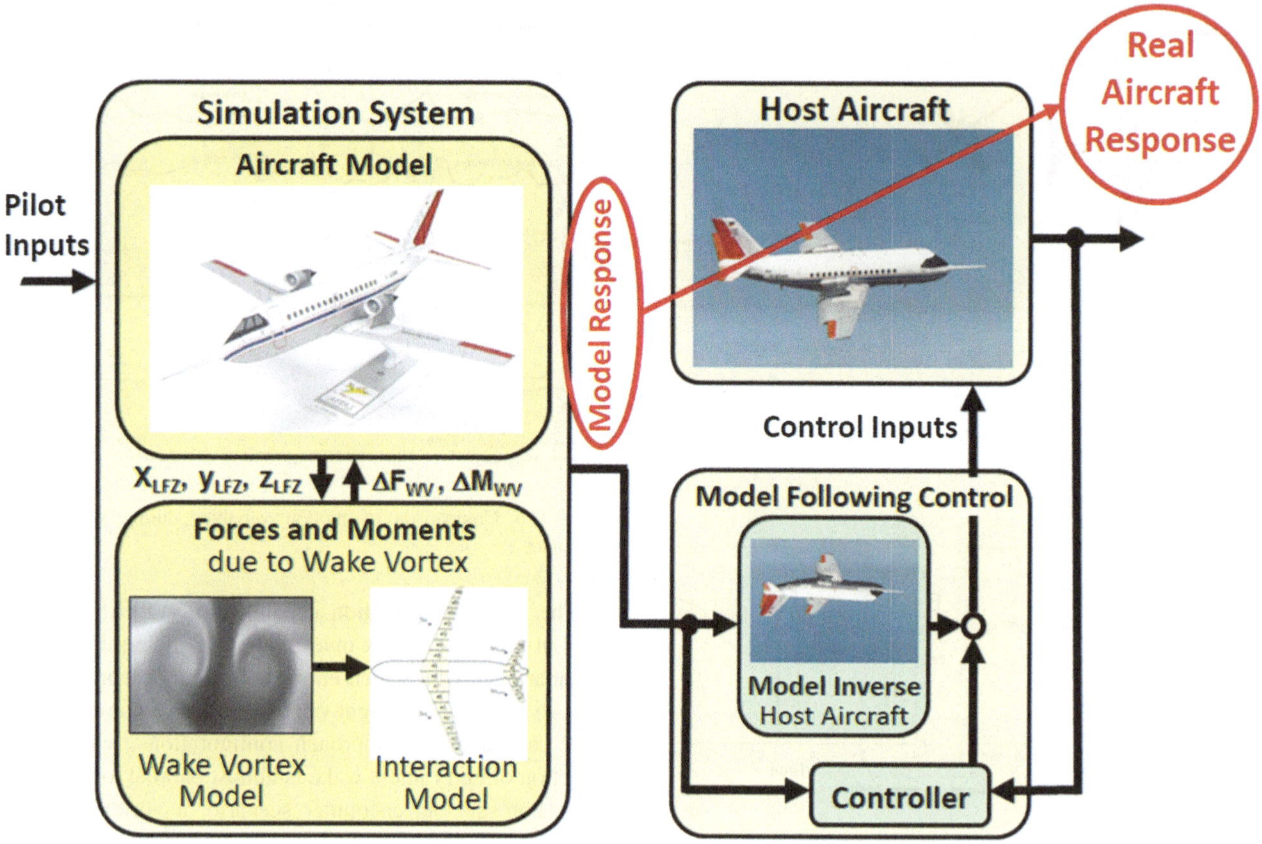

Fig. 9.116 Principle of in-flight simulation of wake vortex

offered the advantage that the evolution and strength of the disturbance could exactly be predetermined and reproduced. The second approach, on the other hand, was more realistic because the progression and strength of the disturbance depended on the actual intrusion of ATTAS into the wake vortex and thus influenced by the pilot reaction and the resulting control inputs. However, precise pre-determination of definite vortex strengths and the reproducibility of the simulated encounters were possible only to a limited extent. Based on the ground-based simulation and flight tests results, it could be concluded that for RCR of less than 0.2, the pilots had no difficulties while flying through the wake vortex [100–103]. The aforementioned tests were carried out for a straight line progression of wake vortices. Without going into any further details, it would suffice to mention that encounters based on curved wake vortices were also investigated in airborne simulations with ATTAS [104]. These results indicated potentials of PIO-tendencies during certain wake vortex encounter scenarios.

The wake vortex encounter situation can be further improved by control augmentation, such as active control systems. This approach is depicted in Fig. 9.117. For this purpose, an algorithm was developed, which reconstructs the flow field from air data measurements by a forward-looking sensor (LIDAR, RADAR). If the flow field is known, the AIM can be used to compute the resulting forces and moments and thereby determine the effect of the wake vortex flow field on the incoming aircraft even before the aircraft flies into this flow field. The necessary control inputs can be calculated and commanded automatically to alleviate the aircraft reaction due to wake vortex disturbances. The effectiveness of such an automatic control system was demonstrated in three in-flight simulation campaigns. During the first two campaigns (2006 and 2009), only the three primary control surfaces (ailerons, rudder, and elevators) were used for active wake control.

Although just 20 and 16 encounters, flown during these two flight phases respectively, were insufficient to arrive at a statistically significant conclusion, the encounters with active wake control were assessed by the pilot far better than normal manual control. A further improvement of the encounter situation and thus the pilot evaluation was achieved by the adding direct lift control (DLC) (see Fig. 9.118) [105].

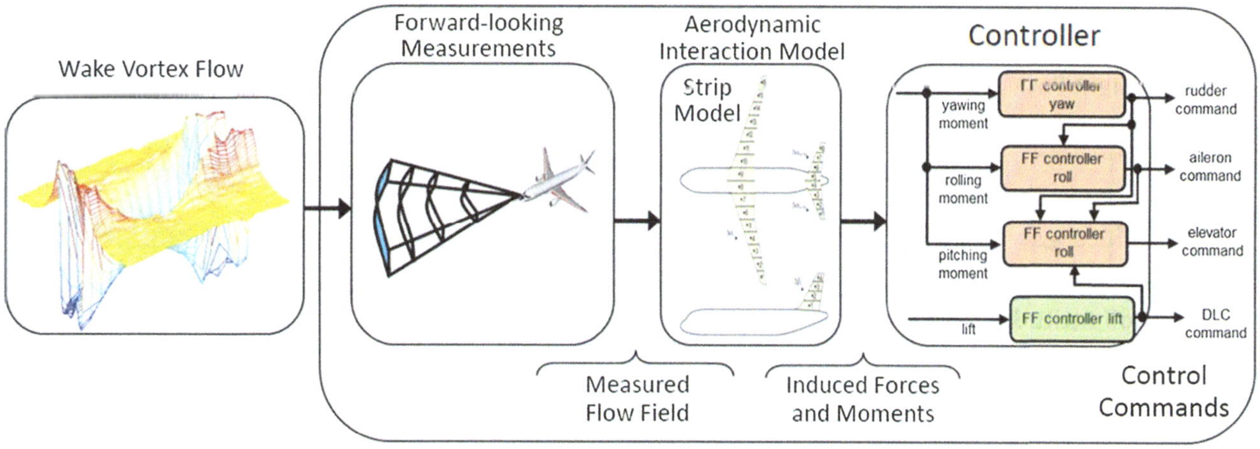

Fig. 9.117 Active wake control concept for alleviation of wake vortex encounters

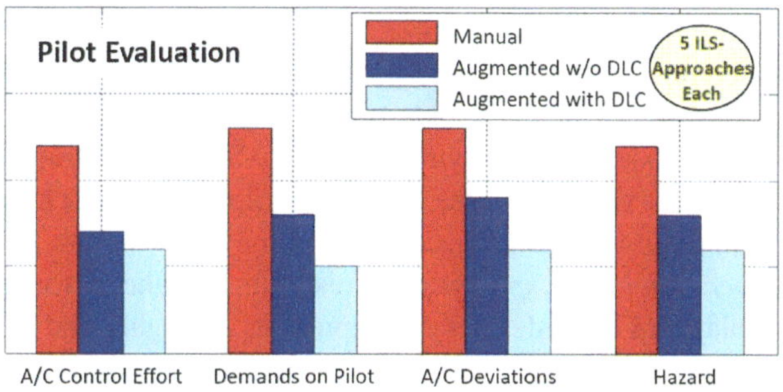

Fig. 9.118 Pilot assessments of wake vortex encounters without and with active wake control (control variables: ailerons, rudder, elevator, and DLC flaps)

9.2.14 Aircraft Emergency Thrust Control (2006–2009)

Nicolas Fezans

The total failure of aircraft primary controls can lead to catastrophic consequences. Hence, it is necessary to make such failures extremely improbable through high redundancy. However, during the last four decades, there have been a few incidences which were attributed to such an exceptional primary control failure or to other external influences. Of these accidents, at least two were with the civilian aircraft in regular operations: Japan Airlines Flight 123 (Boeing 747 SR-100, JA8119) on August 12, 1985 near Tokyo and United Airlines Flight 232 (McDonnell Douglas DC-10-10, Sioux Gateway Airport) on July 19, 1989, in Sioux City, Iowa, USA. Some accidents with military and civilian aircraft were due to failures of primary control caused by external damage [106]. It was, therefore, only logical that in the early 1990s, the large research organizations like NASA addressed this issue and carried out appropriate research programs on emergency thrust-only flight controls (TOC) [107]. This research was also pursued further at DLR during 2006–2009 to develop a fault-tolerant flight control system based on engine thrust. The control strategies were based on the model predictive control theory [108] and on a structured antiwindup controller [109]. This research program led during 2009 to an in-flight demonstration with ATTAS.

The emergency flight control concept enabled a safe landing only by regulating the engine thrust (throttles-only control) in the event of a complete failure of all primary control surfaces (elevators, ailerons, and rudder).

Although it has been possible in the past for pilots to partially retain aircraft control through manual thrust regulation over a short period of time, such a task is extremely difficult and error-prone. The demonstrated flight-controller allowed a significant reduction of the pilots' workload. The variation of the total thrust force resulted in the ascending or sinking flight, whereas the asymmetrical thrust between the left and right engines allowed heading changes resulting from aerodynamic couplings of yawing and rolling motions. The response of the aircraft to such an asymmetrical thrust command remained very sluggish and posed special demands on the piloting skills. For this reason, the controller was designed in such a way that the pilot had to use the sidestick as a "*high level*" command in terms of desired changes in the flight path angle (stick forward/back) and required bank angles (stick left/right). A simplified block diagram illustrating this concept is shown in Fig. 9.119 [109]. The commanded values were displayed beside the actual flight path and roll angles on the Primary Flight Display (PFD). This extended display is important for maintaining the situation awareness because of the slow aircraft reaction to a thrust variation.

Starting November 2009, the functioning of this flight control concept was flight tested with ATTAS under

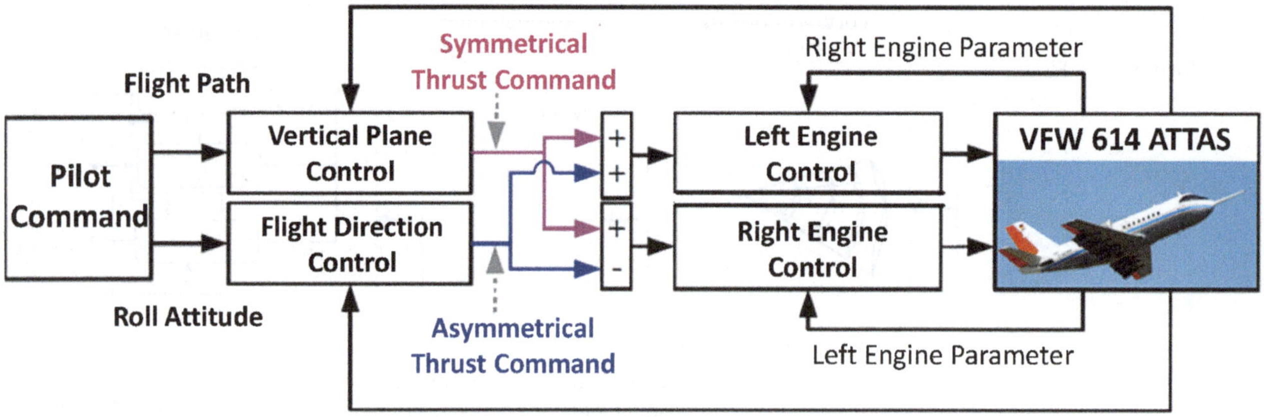

Fig. 9.119 Simplified throttles-only-control system architecture

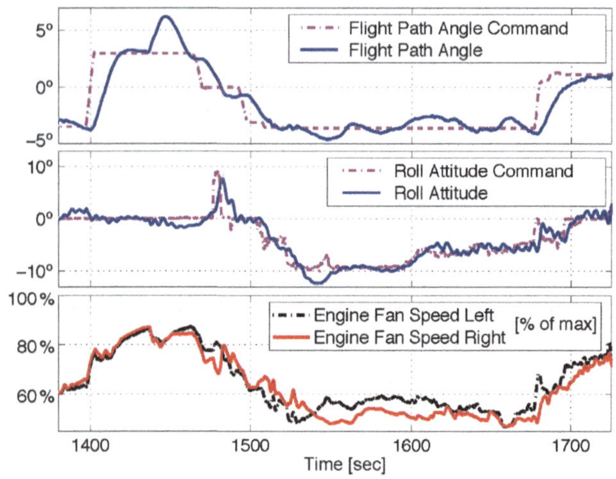

Fig. 9.120 Time history comparisons of flight path and roll attitude input commands and actual ATTAS responses

realistic conditions. Initially, the system response to different pilot inputs was tested; a few minutes from these tests are shown in Fig. 9.120. Despite slight turbulence and the poor aircraft maneuverability due to simulated failures of the primary control surfaces, the pilot commands were followed quite well with the help of engine thrust control. The commanded altitude and heading changes were flown slowly, but with sufficient precision. Subsequently, several landing approaches with go-around were carried out with different controller settings. With this "emergency controller", it was possible to perform satisfactorily landing-approaches at Braunschweig Airport with an acceptable workload for the pilots. The investigations in the ground-based simulator and the flight tests with ATTAS proofed that the pilots were almost always able to approach the runway and land the TOC aircraft. Without such a controller, it was possible to perform the same task only in exceptional cases.

9.2.15 Experimental Cockpit (1992–1993)

Volkmar Adam and Uwe Teegen

As already pointed out in Sect. 9.1.5, the simulation (SIM) mode allowed flying the ATTAS under instrument flight rules (IFR) from the *Experimental Cockpit* (ECOCK), installed in the cabin directly behind the front cockpit. The *Experimental Cockpit* was for research into the advanced display and input devices and represents the right-hand side of a modern transport aircraft cockpit. It comprises of an interface to the experimental onboard systems, providing easy access to all relevant flight data that is used for experimental purposes. Just behind the pilot seat, a supervisor workstation is provided, which allows controlling the running of experiment and taking physiological measurements. All modifications to ECOCK, that is, installation of new input or output devices, were performed in the laboratory prior to the flight trials. For purpose of system checks and preparation of test schedules, ECOCK can be connected to a flight simulator representing ATTAS and relevant onboard systems.

After takeoff and after engaging the Fly-by-Wire-System, the controls can be switched over to ECOCK giving the pilot (nearly) full authority of the aircraft. The pilot can then fly the aircraft either manually using sidestick and throttle levers or fully automatic via AFCS and FMS.

ECOCK made its first flight on November 25, 1992. The pilot used command control modes for pitch and roll axes and some autopilot modes. *Primary Flight Display* (PFD), *Navigation Display* (ND) and *Engine/System Display* (EngD) were running on 5″-*Cathode Ray Tubes* (CRT) during that flight campaign. A *Flight Control Unit* (FCU), similar to the hardware used in the Airbus A320, was installed in the year 1993. Four 13″-flat panel displays, driven by two Silicon Graphics workstations, replaced the four 5″-CRT and their symbol generator hardware (see Fig. 9.121).

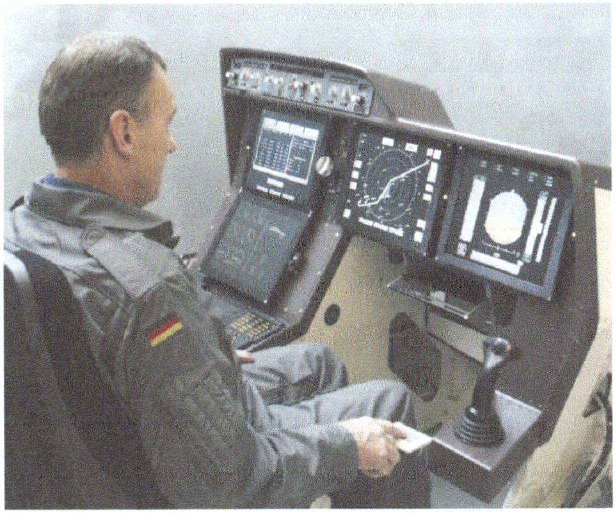

Fig. 9.121 Experimental cockpit with touch pad installed at the *right* armrest

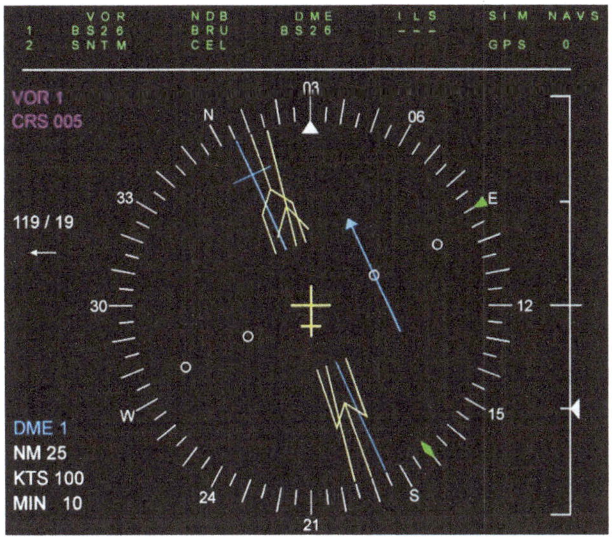

Fig. 9.123 Navigation display (Rose mode configured for VOR-navigation)

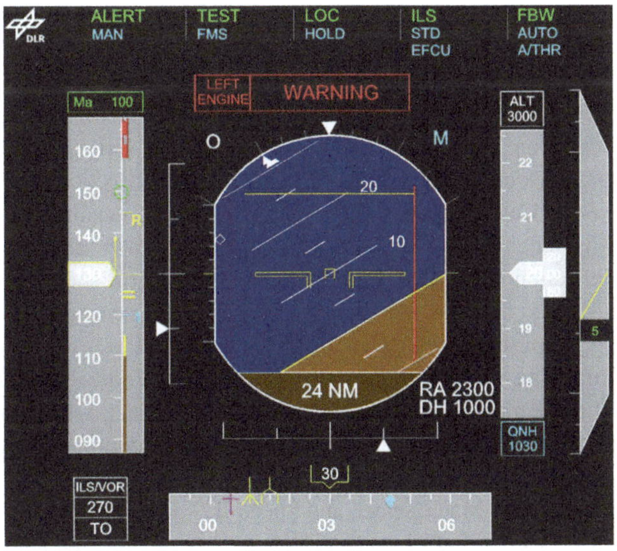

Fig. 9.122 Primary flight display

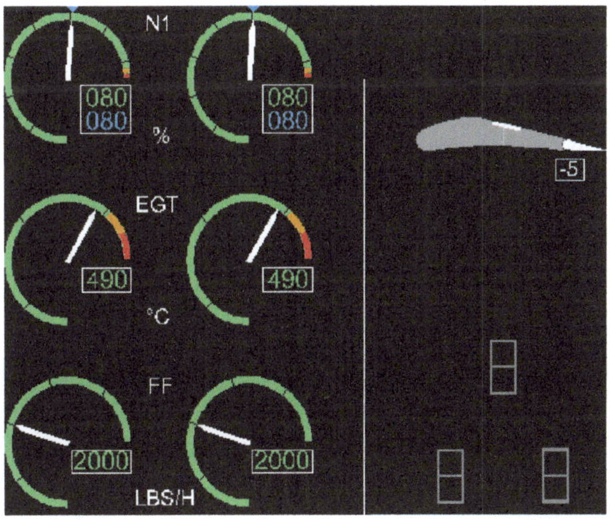

Fig. 9.124 Display of engine parameters, flap position and landing gear (*rectangular symbols*)

Primary Flight Display

The display format of PFD was similar to that commonly used in modern transport aircraft. It shows aircraft state, warnings, limit values, autopilot mode etc. (see Fig. 9.122).

Navigation Display

The *Navigation Display* (ND) could either be used in *Rose-Mode*, or *Horizontal Display Mode* or *Vertical Display Mode*. In *Rose-Mode* (Fig. 9.123), HSI (*Horizontal Situation Indicator*) symbology is shown. *Rose-Mode* for VOR or ILS Navigation was selected via display control unit on the left-hand side of the FCU.

Engine and Systems Display

The essential parameters of both engines, namely turbine speed N1, *Exhaust Gas Temperature* (EGT) and *Fuel Flow* (FF) are displayed on the *Engine Display*. The positions of landing flaps, spoilers, and landing gear are shown on the right-hand side (see Fig. 9.124).

Flight Control Unit

The *Flight Control Unit* (FCU) allowed to engage autopilot and autothrottle modes and to enter target values, for example, *HDG Select/Hold, ALT Select/Hold, CAS Select/Hold, FMS Guidance*, etc. (see Fig. 9.125).

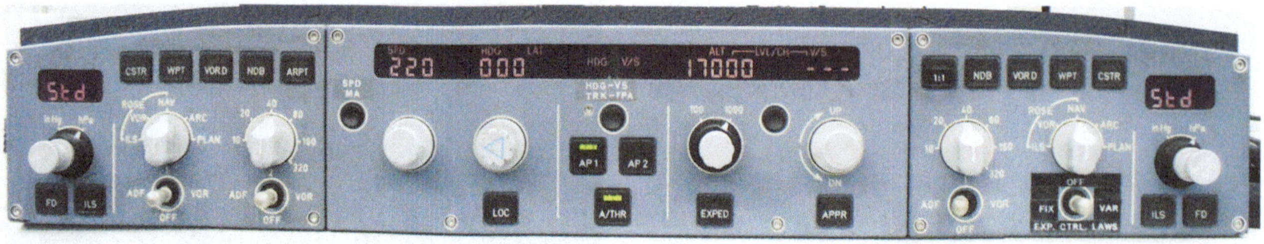

Fig. 9.125 Flight control unit

Radio Management Unit

The *Radio Management Unit* (RMU) allows entering frequencies of onboard COM/NAV equipment (VHF-COM, VOR, DME, ILS, ADF) and entering the transponder code.

9.2.16 Automatic Flight Guidance System for ATTAS (1991–1993)

Volkmar Adam and Hayung Becker

The development of the experimental AFGS (*Automatic Flight Guidance System*) for ATTAS was based on the know-how gained during the development and flight testing of the Integrated Flight Guidance System (IFGS) for HFB 320 (see Sect. 7.3.9). The first version of AFGS ATTAS did even use the former mode controller ADB (*Autonomes Digitales Bediengerät*) (see Fig. 9.126). Flight Testing of AFGS for ATTAS began in 1991. The initial flight tests were mainly dedicated to validating and refining aerodynamic and engine model data used in the ground-based simulator.

Fig. 9.126 ATTAS cockpit with ADB on the glare shield

In 1993, several *Flight Control Units* similar to the A 320 hardware were acquired for installation in the ATTAS front cockpit, ECOCK and ATTAS ground-based simulator. It was regarded essential to provide an autopilot mode controller. The experimental AFGS provided command control modes for pitch and roll axes as basic modes and a great number of autopilot and autothrottle modes to automate many phases of a flight [110].

Fly-by-Wire (FBW) pitch mode and FBW roll mode were activated as soon as the SIM-Mode of the ATTAS FBW-System was engaged. In this case, deflection of the sidestick corresponds to a commanded pitch or roll rate. If deflection was nearly zero, then the present vertical speed and bank angle were stabilized. If sidewards deflection of the sidestick was zero and bank was nearly zero, heading hold was engaged automatically. Likewise, when forwards deflection of the sidestick was zero and vertical speed was also zero then altitude hold was automatically engaged.

By pushing button AP1 ON, the pilot had access to the following autopilot modes: Altitude Select, Vertical Speed, Flight Path Angle, Heading Select, Track Select, Direct To and LOC Land. By pushing button ATHR additionally the following autothrottle modes were enabled: Go Around, CAS Select, Mach Select, and Speed ILS. If all three buttons AP1 ON, AP2 ON and ATHR ON are pressed then the guidance commands of the Experimental Flight Management System (EFMS) are used as input to the control system.

Due to the design of the control system as a coupled multi-variable system with dynamic feed-forward control of control variables, it was possible to adhere very precisely to commanded airspeed and altitude, as well as to the commanded horizontal route with low control activity.

9.2.17 Experimental Flight Management System (1991–1997)

Volkmar Adam and Uwe Teegen

Flight Management Systems (FMS) are commonly used to plan an optimized flight path from departure to the destination airport and to guide the aircraft automatically along the predicted flight path. However, very often this functionality cannot be used to the full extent when traffic density is high and restrictions are imposed by ATC on an individual flight for safe separation. It is anticipated that in the future more advanced *Air Traffic Management* System (ATM), ground-based planning computers will be connected to onboard 4D-FMS via a data link for exchange of planning information. Such integrated ATM system can negotiate restrictions generated in a ground-based computer with an advanced onboard 4D-FMS [111, 112].

The experimental FMS (EFMS) was developed within the *Program for Harmonized ATM Research in Eurocontrol* (PHARE). It was configured as a flexible research tool which can readily be adapted to specific requirements [113]. It was not intended to achieve a complete simulation of an operational FMS but only to develop and implement functions relevant to the execution of planned experimental investigations and demonstrations. As such many standard functions of operational FMS were missing [114–116]. However, several innovative functions were supported, such as:

- Planning of flight trajectories taking into account a wide range of constraints affecting the vertical profile as well as the arrival time at certain way-points.
- Precise guidance along a 4D-trajectory or within a 4D-tube which provides the aircraft a specific maneuver margin along the trajectory. Accordingly, the 4D-tube represents the clearance given by air traffic control.
- Negotiation of trajectories and constraints with a ground-based air traffic control planning computer, via an automatic data link.
- Transmission of meteorological data measured on board the aircraft to a ground-based dynamic meteorological database.
- Sampling of route-related meteorological data from this meteorological database for purposes of airborne flight planning.

Trajectory Prediction

A major function of the EFMS was to predict a flyable trajectory which meets ATC imposed constraints. To do this, the EFMS generated the trajectory conforming to the route and altitude constraints by means of a simulation of aircraft motion using aerodynamic, engine, wind, and air temperature data as well as performance parameters and relevant aircraft operational procedures.

The lateral route was made up of great-circle sections between way-points and arcs with a fixed radius at way-points. The vertical profile consisted of a sequence of quasi-optimized flight phases. The climb was predicted at high power setting and a quasi-optimized *Calibrated Airspeed* (CAS) schedule whilst the descent was planned at near idle power setting. Airspeed and altitude profiles were planned and modified iteratively such that all altitude and time constraints were fulfilled wherever possible. In order to adhere to required arrival times at specific way-points, the CAS profile was modified accordingly. In the final phase of the flight from Metering Fix to Approach Gate within the TMA, the flight path length was also suitably modified, that is trombone or fan type path stretching was applied.

Guidance

EFMS guidance was a continuous control process which provided updated guidance commands every 150 ms to the AFCS. Lateral guidance steered the aircraft along the route by bank angle command which was a function of the present cross-track deviation and present track angle deviation. A prediction of bank angle required during the turn was included as a feed-forward term.

Vertical guidance comprised of several selectable guidance options for the climb, cruise, and descent. With regard to an economical climb at high power setting with minimal thermal cycling, it was decided to fly an open climb at constant thrust and CAS schedule as is common practice in transport aircraft operation, that is, altitude and time are not controlled in a climb.

The aircraft was operated at the same thrust setting and CAS schedule applied for the prediction of the climb profile. This lead normally to deviations from the predicted altitude and time profile which depended on the accuracy of the meteorological forecast (the wind and air temperature) and aircraft performance data (aerodynamic drag, engine thrust, aircraft weight) used for trajectory prediction. However, if there are no strict ATC constraints which require a more precise tracking of the altitude and time profile during a climb, there is no reason to apply a higher control effort.

Full 4D-control commenced at the Top of Climb (TOC) and was employed throughout cruise and descent. A simple algorithm calculated an incremental CAS command according to the prevailing time deviation. During the cruise, altitude was controlled by elevator and CAS by throttle. During the descent, CAS control was achieved through the elevator whilst total energy was controlled on thrust by an algorithm, which was part of the experimental AFCS. In order to provide some margin for reducing thrust, the descent was normally planned at a low value of thrust, rather than being at flight idle setting.

Airborne Human Machine Interface (AHMI)

4D-trajectory generation and negotiation capabilities call for efficient I/O-devices that ease pilot interaction regarding flight management. Support of immediate pilot action within new cockpit procedures while maintaining the safety standards for aircraft operation was a major goal of the PHARE AHMI project. To approach this goal graphical representation of the constraint list, 4D-trajectory, and 4D-tube on the screen, combined with object-related input capabilities, have been chosen and implemented. The *Navigation Display* (ND) was enhanced by a touch-pad input device and graphical display objects which the pilot may act on directly.

The advanced *Navigation Display* incorporated the *Horizontal Display* (HD) and the *Vertical Display* (VD) with input capabilities. The general display layout is illustrated by Figs. 9.127 and 9.128. Regarding the essential information, the HD conformed to the conventional EFIS ND layout, whereas the VD was a new development. The ND was operated through two separate display modes, namely the PLAN and MONITOR modes.

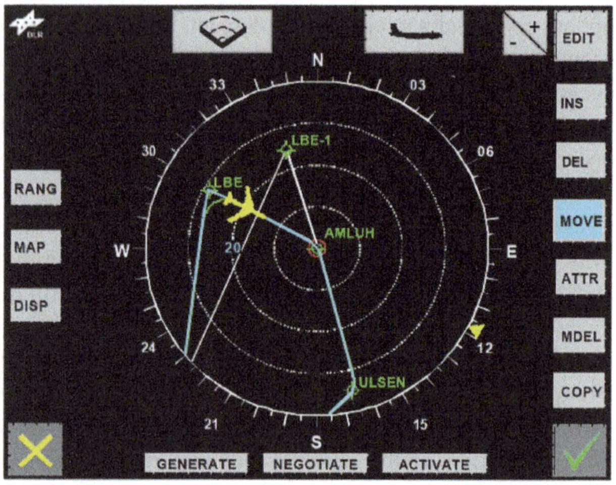

Fig. 9.127 Horizontal display in PLAN mode

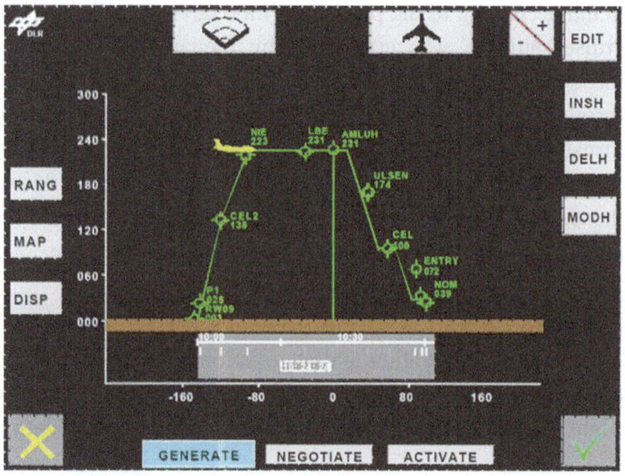

Fig. 9.128 Vertical display in monitor mode

PLAN Mode

The PLAN mode supported constraint list modification, that is, flight planning and enabled the pilot to initialize and edit the constraint list representing the basis for any 4D-trajectory prediction. In this mode, the HD was oriented north up like a geographical map (see Fig. 9.127). The EDIT menu provided for editing the constraint list included insertion, deletion, and modification of all types of constraints.

VD showed the vertical flight profile versus distance with the reference way-point centered on the screen. The distances between constraint list way-points conformed to those of the HD due to identical display ranges. Altitude was scaled automatically depending on altitude range within the selected range distance. However, a mere constraint list did not include a vertical profile unless a trajectory had been generated.

MONITOR Mode

The MONITOR mode supported the pilot in monitoring flight progress with respect to the active 4D-trajectory and the 4D-tube representing the ATC clearance and contract between aircraft and ATC. In this mode, the aircraft symbol was fixed near the bottom of the screen. The display represented the area in front of the aircraft in an angular range of approximately 150° which corresponds to the standard EFIS ND representation. On the VD, the aircraft position was fixed near the vertical scale. The display showed the predicted vertical profile (see Fig. 9.128). Altitude was scaled automatically as in the case previous mode.

Representative Trial Flight Result

As a typical example of EFMS, Fig. 9.129 shows the performance during a descent. Before reaching the *Top of Descent* (TOD), the 4D-trajectory was updated to compensate for any prevailing time error. The update was an essential prerequisite for the accurate tracking of the descent profile. The descent was initiated by reducing thrust to near idle setting, while elevator maintained the airspeed. After establishing the descent, thrust controlled the total energy. Actual CAS was following the demanded value, which was calculated from the predicted CAS and an incremental CAS command provided for the compensation of any prevailing time error. Due to an increasing tailwind component in en-route descent, the ground speed increased by about 10 kts, which led to an increase in time deviation up to −4 s (early). This deviation led to a CAS reduction which in turn yielded an altitude deviation since thrust had reached idle power limit and air brakes were not applied. The altitude deviation was about 400 ft at the end of the en-route descent.

At 9000 ft, a level flight sub-phase was entered due to an altitude constraint at the Metering Fix. A trajectory update was performed to compensate the time deviation of −4 s by path stretching, that is, the length of the remaining flight path to the Approach Gate was appropriately modified.

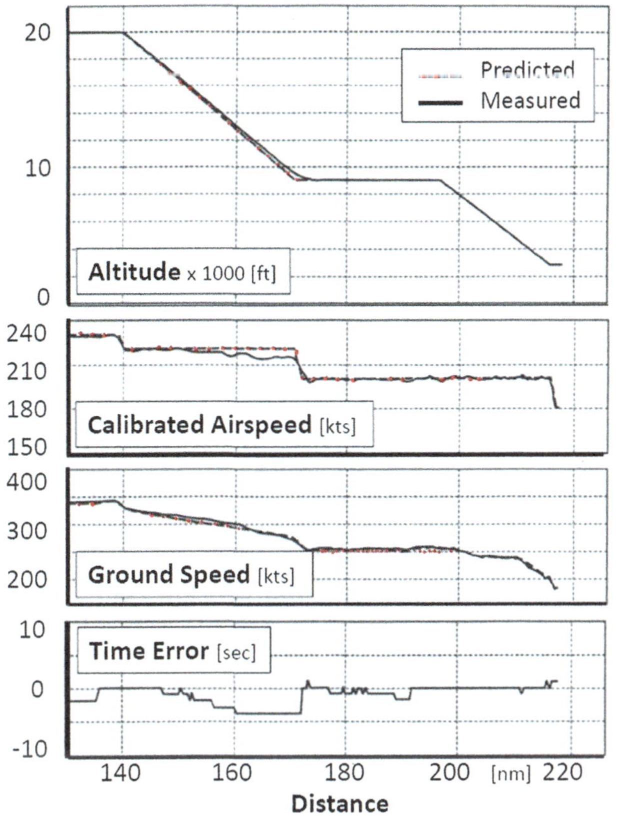

Fig. 9.129 Stepwise descent from cruise altitude to approach gate altitude

Before starting the next descent phase, the trajectory was updated again and the prevailing time deviation of −2 s was nullified. The descent from 9000 to 3000 ft followed the altitude profile with only minor deviations (less than 70 ft). A final trajectory update was performed at the entry way-point of the path stretching area. Time deviation in TMA descent was less than 1 s, leading to an arrival time deviation of 1 s (late) at the Approach Gate. The guidance accuracy shown in this example was typical for the cases in which the meteorological forecast agreed with actual wind and air temperature. All flight trials with ATTAS performed in May and June 1997 showed similar results.

The EFMS was used by several PHARE partner organizations for research into more advanced ATM systems. It was installed onboard the BAC 1-11 aircraft of Qinetiq (former DERA *Defense Research Agency*), onboard the Cessna Citation of NLR and also in a B 747 cockpit simulator of the *Eurocontrol Experimental Center* (EEC). After the end of the PHARE program, the EFMS was further developed and it was renamed as AFMS (*Advanced Flight Management System*). The next flight experiment with AFMS onboard ATTAS and BAC 1-11 was dedicated to research into *Airborne Separation Assurance* (ASAS) (see also Sect. 9.2.18). It was also used for UAV-guidance in controlled airspace (see also Sect. 9.2.11) and for advanced approach procedures (see also Sect. 9.2.19).

9.2.18 Trajectory-Oriented Airborne Separation Assistance (2003)

Bernhard Czerlitzki

Trajectory-oriented, time-based air traffic operations, data link communication, and *Airborne Separation Assistance Systems* (ASAS) play an important role in the optimum utilization of the airspace. New air and ground-based procedures are needed for the implementation of the so-called ASAS delegated maneuver. It offers the opportunity to delegate the surveillance of the surrounding air traffic and the trajectory planning for an evasive maneuver to the cockpit crew. ASAS provides the pilot with early warning of any conflicting aircraft, allowing the conflict to be resolved on board the aircraft in a strategic, rather than a tactical way.

The two important ASAS procedures are (1) Longitudinal Spacing 'Merge Behind' and (2) Lateral Spacing 'Pass Behind'. In the first case, a new allocation of tasks between controller and flight crew is envisaged as one possible option to improve, in particular, the sequencing of arrival flows. It relies on a set of new spacing instructions. The flight crew can be tasked by the controller to maintain a given spacing with respect to a designated aircraft. In the second case, the objective is to provide the controller with a new set of instructions to solve a conflict when two aircraft are at the same altitude on intersecting tracks. The controller directs one of the aircraft crews to pass behind the other while maintaining a given minimum spacing.

Cockpit Display of Traffic Information (CDTI) is driven by *Automatic Dependent Surveillance-Broadcast* (ADS-B) information coming from surrounding aircraft, which broadcast their identification and state vector. CDTI gives a graphical representation of the traffic on the Navigation Display in the form of position, range and airspeed of each aircraft, and thus enhances the crew's situational awareness. CDTI is a component part of all ASAS functionalities. When instructed by the controller, the pilot identifies the target on the CDTI and initiates the delegated task. The *Flight Management System* in the aircraft automatically generates an optimal trajectory, in accordance with the instruction, and then guides the aircraft through the maneuver.

In 1999, a consortium of equipment suppliers, research organizations, and service providers started a joint project MA-AFAS (*More Autonomous Aircraft in the Future ATM System*), partly funded by the European Commission under the Fifth Framework Program. The main focus was on the investigation of an integrated air/ground system combining trajectory-orientation, data link communication, and airborne separation assistance as complementary components of a modernized ATM system.

The activities covered the development and integration of a 4D-FMS avionic package composed of a VDL-4 data link, a CMU (*Communication Management Unit*) and a FMU (*Flight Management Unit*). The functionality included, for example, (1) ASAS applications using CDTI and ADS-B, (2) Generation and negotiation of 4D-trajectories and subsequent automatic 4D guidance, (3) Taxi Management (taxi route, CPDLC messages, clearances, CDTI, runway alert and runway incursion), (4) Communication with AOC (*Airline Operations Centre*), (5) Precision approach procedures (SBAS, GBAS), and (6) Communication (VDL-4 data link)

The DLR task was to validate the MA-AFAS functionality both in a ground-based flight simulator and in the ATTAS. Figure 9.130 shows the hardware components of the MA-AFAS experimental system incorporated into the ATTAS cabin. An IHTP (*In-House Test Platform*) laptop PC was also connected via Ethernet to the PC cards in the FMS cabinet, allowing additional monitoring of the system behaviour, including the emulation of the MCDU (*Multi-purpose Control and Display Unit*) and to detect any possible problems being encountered. The ground platform A-SMGCS (*Advanced Surface Movement Guidance and Control System*) was used to run the taxi management test procedures at the Braunschweig-Wolfsburg airport [117].

In March 2003, a joint five-day campaign was carried out in the Italian airspace, over the Mediterranean Sea between Rome and Sardinia. Successful trials were conducted by using the two research aircraft ATTAS and BAC 1-11 (QinetiQ, formerly DERA) for the demonstration of ASAS delegated maneuvers in actual practice. Rome Ciampino was the departure and arrival airport for all test flights [118]. In a realistic scenario, the two aircraft were positioned at the same height on intersecting tracks. The future conflict was observed by an air traffic controller at Roma Air Traffic Control Centre who instructed the BAC 1-11 to pass behind the ATTAS at a distance of 6 nm. The FMS automatically generated an optimal conflict resolution, in accordance with the instruction, and then guided the aircraft through the lateral evasive maneuver at a constant flight level when activated by the pilot. Subsequently, a longitudinal spacing maneuver 'Merge Behind' was carried out along the test route. ATTAS served as the target, while the BAC 1-11 aircraft had to maintain a given spacing from the ATTAS. These maneuvers were successfully repeated under various wind and weather conditions.

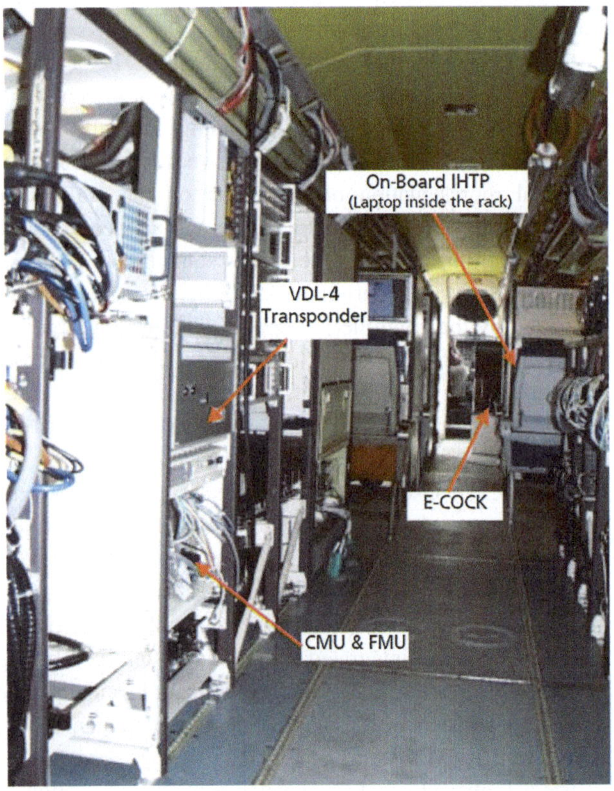

Fig. 9.130 Installation of the MA-AFAS experimental system into ATTAS

9.2.19 Noise Abatement Procedures (2005–2006)

Alexander Kuenz

One limiting factor for the volume of air traffic is aircraft noise. Principally, noise can be reduced by revising the noise sources, typically the engines and the airframe. Alternatively, noise can be reduced by applying low-noise departure and arrival procedures. Investigations on aircraft noise abatement flight procedures have already been performed in the early 1970s with the HFB 320 FLISI. At that time, a potential for noise reduction was proved for steeper approaches (“*keep-ém-high-policy*”) and high speed low drag overflights (see Sect. 7.3.2). A significant reduction of noise exposure can be achieved using a *Continuous Descent Approach* (CDA) procedure. It requires an aircraft to descend continuously from *Top of Descent* (TOD) until touchdown (see Fig. 9.131) [119].

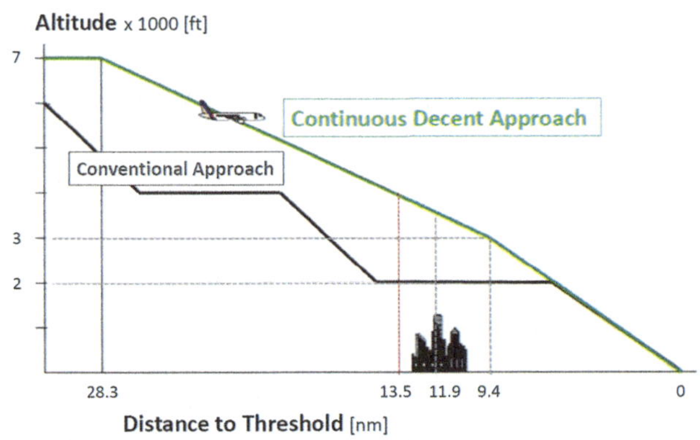

Fig. 9.131 Continuous descent approach compared to a conventional approach

When applying a CDA, noise reduction results from two characteristics. First, the continuous descent leads to a higher altitude profile compared to a standard *Low Drag Low Power* (LDLP) procedure by eliminating the intermediate level flights. The higher distance between the noise source and noise recipient on the ground results in a better absorption. Second, descents are flown with reduced thrust (ideally at idle thrust) leading to decreased noise emissions at the source.

In today's operational practice, CDA procedures are only performed if air traffic control gives the clearance and the pilot is willing to perform a CDA. Today, application of CDA procedures imposes deviations on arrival times, necessitating an extra two minutes margin for touchdown time scheduling. Therefore, CDAs are usually performed only under low traffic conditions. Due to the complex prediction of the precise TOD position CDAs are usually not flown with idle thrust.

Both advanced arrival management systems on the ground and 4D-capable *Flight Management Systems* (FMS) on board are necessary to integrate the CDA procedure under high traffic conditions. A precise 4D-FMS-guidance fulfilling a ground-predicted required time of arrival helps to eliminate negative effects of the CDA procedure on airport's throughput.

Three different types of *Noise Abatement Procedures* (NAP) have been compared involving the LDLP, *Advanced Continuous Descent* (ACDA), and *Steep Continuous Descent* (SCDA) procedure. DLR's Advanced FMS (AFMS) proved a highly accurate predictability of 4D trajectories in flight trials with the A330-300 Full Flight Simulator of *Zentrum für Flugsimulation Berlin* (ZFB) and DLR's test aircraft ATTAS for all three implemented NAPs. Investigations concentrated on both automatic and manual flight characteristics, rating the performance based on time and altitude deviations. Since there is no unique definition of the procedures in use, the applied NAPs are described below:

LDLP (Low Drag Low Power)

The aircraft starts idle descent at FL80 with constant speed using cruise flight configuration (flaps, slats, and gear in). When passing 3000 ft, the speed is decreased in a cruise flight segment, and flaps and slats are extended in subsequent steps. After interception of the ILS glideslope, speed is decreased furthermore. Landing gear is extended at 1800 ft above ground level.

CDA (Continuous Descent Approach)

The aircraft starts its shallow descent at FL80, thrust is close to idle. During descent, the descent angle stays constant. Speed is decreased subsequently while adapting speed and slat configuration according to aircraft specification. The ILS glideslope is intercepted in descent mode without any cruise flight segments. As for the LDLP, the ILS is intercepted at 3000 ft from below. After interception of the ILS glideslope, speed is decreased furthermore. Landing gear is extended at 1800 ft above ground level.

ACDA (Advanced Continuous Descent Approach)

The ACDA is similar to the CDA. However, the complete descent is flown in idle thrust. Therefore, both descent rate and descent angle are not necessarily constant during descent.

SCDA (Steep Continuous Descent Approach)

In this case, the aircraft starts its descent much later than the ACDA. The initially constant speed is decreased when passing 7000 ft, allowing extraction of flaps, slats and landing gear at rather high altitude. This high drag configuration results in a steep descent profile. Keeping the speed constant contributes to a high descent rate. The aircraft intercepts the ILS glideslope in about 2000 ft from above. Once intercepted, speed is decreased until reaching touchdown speed.

All approaches were performed at the airport in Braunschweig. Just before each flight trial, the latest wind forecast had been uploaded into the AFMS. For the automatic flight trials, the AFMS guidance commands were directly executed by the AFCS. Also flaps and slats were operated directly by the AFMS. Due to safety regulations, the landing gear was extended by the pilot, according to a countdown provided on the ND. For manual trials, guidance assistance was provided on the PFD.

Flight trials with ATTAS resulted in very high prediction and guidance accuracy, comparable to the results from simulator trials with the A330. Figures 9.132, 9.133 and 9.134 show altitude and speed deviations of ACDA, LDLP and SCDA flight trials with ATTAS. For more than

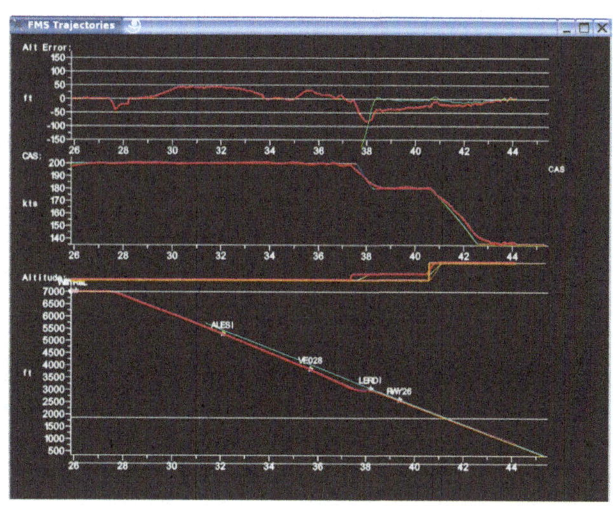

Fig. 9.132 ACDA ATTAS

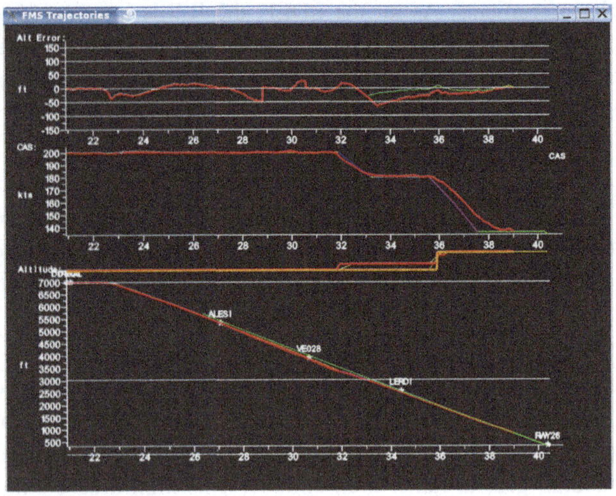

Fig. 9.133 LDLP ATTAS

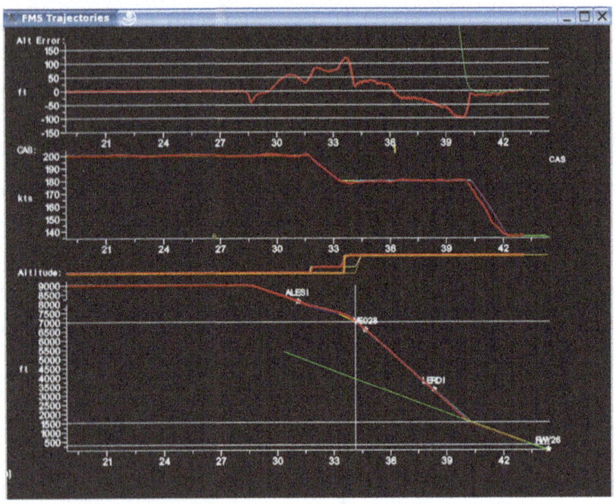

Fig. 9.134 SCDA ATTAS

30 approaches, the precision was, typically, less than ±150 ft altitude error and ±3 sec time deviation at the touchdown point. Even under the worst case conditions with mini jet-streams not covered by the wind forecast and when strong winds from south-west created a constant downdraft in the lee of Harz mountains, touchdown times have been fulfilled with a time deviation of 10 sec [119]. The manual trials proved that highly accurate approaches are possible with aircraft not providing an appropriate connection between AFMS and AFCS.

Without the connection between autopilot and AFMS, the pilot nevertheless relies on an accurate prediction of the TOD. Furthermore, the pilot needs assistance tools to follow the predicted 4D trajectory precisely. Figure 9.135 depicts the "birdy"-symbol on the PFD, showing the pilot how to guide the aircraft. The "birdy" gives commands for both pitch and roll angle.

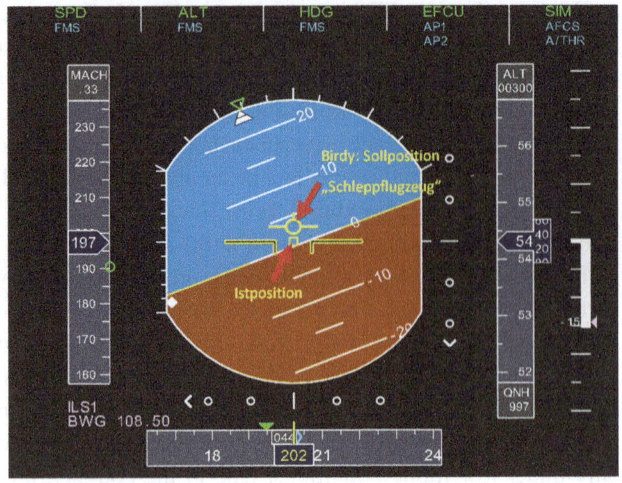

Fig. 9.135 Primary flight display with guidance symbol "birdy"

The pitch command is calculated from a synthetic descent rate guiding the aircraft back to the correct altitude. If the actual descent rate is bigger than the synthetic one, the "birdy" moves upward, telling the pilot to pull the sidestick and vice versa. If the actual roll angle is smaller than the synthetic one, the "birdy" moves to the right, asking the pilot to push the sidestick to the right and vice versa.

As long as thrust is not at idle limit, the pilot has to adapt thrust according to the speed requirement (green dot on speed tape). Once in idle, speed adaptation is performed by means of the pitch angle.

Flight trials showed that the achievable accuracy when following manually a given 4D-trajectory is comparable to the automatic mode using the autopilot. However, there was a substantial difference in the pilot's workload. Using the automatic guidance functionality the pilot was able to focus his attention on monitoring flight progress. Manual guidance supported by the "birdy" was demanding and he could hardly fulfill other tasks at the same time; giving him additional inputs or tasks led directly to deviations from the trajectory.

9.3 ATTAS Retired

Dietrich Hanke

After 26 years of operation, the ATTAS was grounded on June 27, 2012 after irreparable microcracks were observed on the engine during yearly maintenance and functional replacement engine was not available. As a token of appreciation by the German research and industrial community, it was proposed that the ATTAS, together with its entire equipment, be donated to German Museum of Science and Technology in Schleissheim near Munich. The hurdle faced in this apparently simple proposal was how to transport the grounded aircraft over roughly 600 km from Braunschweig which was the home of ATTAS in the past. However, by making one of the eight replacement engines temporarily operational, it became possible to obtain a special permission from LBA to make just one flight to Schleißheim. This final flight was performed on December 7, 2012 by the two DLR skilled pilots Hans-Jürgen Berns and Stefan Seydel. As precautionary measures, the aircraft weight was reduced by de-installing some computers, since the runway in Schleißheim was very short, just 808 m. Figure 9.136 shows ATTAS during the perfect final landing at Schleißheim. The voluntary fire brigade did not need to intervene, instead, she preferred to pose with the museum team in front of the new exhibition object (Fig. 9.137).

On October 15, 2013, the ATTAS was officially handed over to the German Museum for display as an example of successful symbiosis of research and industrial organizations, reliable utilization of modern technology, and dedicated efforts over more than two and half decades.

As evident from Sect. 9.2, a variety of challenging research projects were successfully carried out with ATTAS. It was the focal point of research activities pursued at DLR

Fig. 9.136 Last perfect landing in Schleißheim

Fig. 9.137 ATTAS after the arrival at the German Museum in Schleißheim

in the field of flight mechanics, making significant contributions to international aeronautical research and culminating into international recognition. It was a prestigious symbol of aeronautical research in Germany and testified the successful development and use of a unique flight test demonstrator in Europe.

References

1. Onken, R.: Spezifikation zum DFVLR Flugversuchsträger (Stand September 1979), DLR-IB 153-79/34 (1979)
2. Hanke, D.: Rahmenforderungen für den Flugversuchsträ-ger ATTAS (Advanced technologies testing aircraft), Stand Mai 1980, DLR-IB 111-/80 (1980)
3. Thomas, F., Onken, R., Hanke, D.: Versuchsflugzeug der DFVLR für In-Flight Simulation und Flugführungstechnolo-gie. 2. BMFT-Statusseminar, Garmisch-Partenkirchen, 8./9. Oktober (1980)
4. Morgenstern, K., Simberger, G.: VFW 614. Wilhelm Gold-mann, München (1977). ISBN 3-442-03473-6
5. Wenz, F.-H., Zistler, F.-P.: VFW 614—Deutschlands erstes Kurzstrecken-Düsenverkehrs-flugzeug. Stedinger, Lemwerder (2003). ISBN 3-927697-33-8
6. Griem, H.: Flugversuchsträger ATTAS für die DFVLR. DGLR-Jahrestagung, Hamburg 1984, MBB-UT-17-84 (1984)
7. Mercker, E., Krahl, H.: ATTAS Windkanal-Messungen mit einem VFW 614-Modell, DNW-TR-82.07 R, Dezember (1982)
8. Anon.: ATTAS, Flugversuchsträger für die DFVLR, Ab-schlussbericht TE 13, Apr 1986
9. Hamel, P., Krüger, H.: ATTAS – Der neue Erprobungsträ-ger, 4. BMFT-Statusseminar, München, 28–30 Apr 1986
10. Hanke, D.: Der Fliegende Simulator und Technologieträ-ger ATTAS der DFVLR, Luft- und Raumfahrt 7 Heft 1/86 (1986)
11. Hanke, D.: Data processing concept for the DFVLR research aircraft ATTAS. In: Proceedings of 3rd European Rolm Users Group Conference, Braunschweig, 7 May 1985; Bad Soden, 8–9 May 1985
12. Lange, H.-H., Hanke, D., Nietschke, W.: Einsatz eines Rech-nerverbundsystems mit optischer Buskommunikation im DFVLR-Flugversuchsträger ATTAS, DGLR Jahrbuch 1986, S.86-151 (1986)
13. Saager, P., Heutger, H.: Ground-based system simulation for an inflight simulator aircraft. In: Proceedings of the 15th ADIUS Annual Conference, Ann Arbor, Mich, USA, pp. 203–219, 12–15 June 1994
14. Hanke, D., Lange, H.-H.: The role of system simulation for the development and operation of ATTAS. In: Proceedings of the Flight Simulation Conference, Moscow, Russia, 11–16 Aug 1992
15. Lange, H-H., Bauschat, J.-M.: Der ATTAS Software Pool, ASP, ein Mittel zum Software Management und zur Validierung von Flugsteuerungssoftware—Ein Erfahrungsbericht, DGLR-Bericht 94-02, pp. 197–208 (1994)
16. Henschel, F.: Über Regelungskonzepte zur In-flight simulation unter Berücksichtigung von Nichtlinearitäten und Totzeiten in den Stellsystemen, DFVLR-FB 85-24 (1985)
17. Chetty, S., Henschel, F.: Model following flight control system design for ATTAS, DFVLR's new flight simulator. In: Proceedings of AIAA Guidance and Control Conference, Monterey, Cal/USA (1987)
18. Bauschat, J.-M.: On the application of a nonlinear in-flight simulation technique. In: Proceedings of the European Control Conference ECC 91, Grenoble, July 1991
19. Bauschat, J.-M.: Der Experiment-und Regelrechner des Flugversuchsträgers ATTAS. Die Einbindung von Benutzer-Software, DLR-IB 111–93/42 (1993)
20. Gestwa, M., Leißling, D., Bauschat, J.-M.: Development of experiment function for the flying test-bed ATTAS using MATLAB/SIMULINK with a modified real-time workshop. In: Model Based Design Conference, München, 8. Juni 2005
21. Hanke, D.: ATTAS-Flugbereichs-erweiterung, Schlussbe-richt, DLR-IB 111/97–18 (1997)
22. Hanke, D., Rosenau, G.: Hermes training aircraft, technical specifications, DLR-IB 111-91/03
23. Hanke, D.: Simulation fidelity requirements for HTA, DLR-IB 111-13/90-11 (1990)
24. Hanke, D., Gagir, G.: Hermes training aircraft. In: Proceedings of the 2nd European Aerospace Conference on Progress in Space Transportation, Bonn-Bad Godesberg, FRG, 22–24 May 1989, ESA SP-293, Aug 1989
25. Hering, R., Mönnich, W.: Nonlinear Matlab/Simulink Simulation of the Fairchild/Dornier 728JET, DLR-IB 111-2001/39 (2001)
26. Hanke, D., Heinecke, T., Lange, H.-H.: 728JET Flight Control System Evaluation. DLR-IB 111-2001/25 (2001)
27. Leißling, D., Gestwa, M., Bauschat, M: In-flight simulation in support of an aircraft certification process. In: AIAA Guidance, Navigation, and Control Conference and Exhibit, Austin, TEX (US), 11–14 Aug 2003
28. Otte, T.: 728JET Stability and Control—In-flight Simulation (IFS) with ATTAS. Persönliche Notizen (2013)
29. Wilmes, V.T.: Numerical Handling Qualities Evaluation of 728JET In-flight-simulation Testing. DLR-IB 111-2001/33 (2001)
30. N.N.: AC 120-40B—Airplane Simulator Qualification, U.S. Department of Transportation, Federal Aviation Administration (1991)
31. Leißling, D., Mönnich, W.: Quality of the ATTAS/Fairchild Dornier 728JET In-flight Simulation. DLR-IB 111-2002/10 (2002)
32. Anon.: Flying Qualities of Piloted Aircraft, MIL-HDBK-1797A, Wright-Patterson AFB (1997)
33. http://ec.europa.eu/research/transport/projects/items/nacre_en.htm
34. Cooper, G.E., Harper, R. P.: The Use of Pilot Rating in the Evaluation of Aircraft Handling Qualities, NASA TN-D5153, Apr 1969
35. Ehlers, J., Niedermeier, D., Leißling, D., Verification of a flying wing handling qualities analysis by means of in-flight simulation. In: AIAA Atmospheric Flight Mechanics Conference and Exhibit, Portland, USA, Aug 2011
36. König, R.; Hahn, K.-U.: Load alleviation and ride smoothing investigations using ATTAS. In: Proceedings of the 17th Congress of the International Council of the Aeronautical Sciences, Stockholm, Sweden (1990)
37. Hahn, K.-U.; König, R.: LARS—Auslegung und Erprobung eines fortschrittlichen Böenabminderungssystems mit ATTAS, Deutscher Luft-und Raumfahrtkongress, DGLR Jah-restagung, Jahrbuch 1991 I, Berlin, Germany (1991)
38. Hahn, K.-U.: Beiträge zur Realisierung eines Gust management systems, Deutschen Luft-und Raumfahrt-Kongress 1992, DGLR-Jahrestagung, Jahrbuch II 1992, Bremen, Germany (1992)
39. Hahn, K.-U.; König, R.: ATTAS flight test and simulation results of the advanced gust management system LARS. In: AIAA Atmospheric Flight Mechanics Conference, Hilton Head, South Carolina, USA (1992)
40. König, R., Hahn, K.-U., Winter, J.: Advanced Gust Management Systems-lessons Learned and Perspectives. AGARD CP-560, Paper 13 (1994)
41. Wildschek, A., Maier, R., Hahn, K.-U., Leißling, D., Preß, M., Zach, A.: Flight test with an adaptive feed-forward controller for alleviation of turbulence exciting structural vibrations. In: AIAA Guidance, Navigation, and Control, Chicago (2009)
42. Hecker, S., Hahn, K.-U.: Advanced gust load alleviation system for large flexible aircraft, Deutscher Luft-und Raum-fahrtkongress 2007. In: First CEAS European Air and Space Conference, CEAS 2007-110, Berlin, Germany (2007)

43. Krüger, W.: Innovation für eine effiziente Simulations-und Versuchsprozesskette für Lasten und Aeroelastik (FTEG/InnoLA)-Abschluss-bericht, Institut für Aeroelastik, DLR-IB 232-2013J 02, März 2013
44. Buchholz, J. J., Bauschat, J.-M., Hahn, K.-U., Pausder, H.-J.: ATTAS and ATTHeS in-flight simulators, AGARD CP-577, Paper 31 (1996)
45. Heintsch, T., Luckner, R., Hahn, K.-U.: Flight Testing of Manual Flight Control Functions for a Small Transport Aircraft (Project ATTAS-SAFIR), AGARD CP-593, Paper 26 (1997)
46. Kröger, A.: Data Flow oriented control law design with graphical language HOSTESS. In: Conference Proceedings, International Conference and Exhibition, 8th international event on Civil and Military Avionics, Heathrow (1994)
47. Lenhart, P.M., Purpus, M., Viebahn, H.: Flugerprobung von Cockpitdisplays mit synthetischer Außensichtdarstel-lung. DGLR-Jahrestagung, Bremen (1998)
48. Sigl, W., Purpus, M., Viebahn, H.: Erprobung und Bewer-tung von neuartigen Flugführungsdarstellungen vom Boden-versuch bis zur integrierten Flugerprobung, Deutscher Luft-und Raumfahrtkongress, Dresden, 24–27 Sept 1996
49. Leppert, F.: The AWARD program. SPIE Congress, Orlando (1998)
50. Mayer, U., Kaiser, J., Gross, M.: AWARD's synthetic vision flight guidance display as basic display for free flight environment. IFAC World Congress, Beijing (1999)
51. Lo Presti, Riparbelli, G., Morizet, B., Mayer, U., Hahn, K.-U., De Reus, A., Crawford, S., Poussin, G., Graillat, B., von Viebahn, H.: AWARD—All Weather Arrival and Departure, Synthesis Report, SA-0-SR 001, Brite-EuRam III, BRPR-CT96-0197, BE-1655, Bordeaux, France, June 2000
52. Looye, G., Joos, H.-D., Willemsen, D.: Application of an Optimization-based Design Process for Robust Autoland Control Design, AIAA-2001-4206 (2001)
53. Bauschat, M., Mönnich, W., Willemsen, D., Looye, G.: In: Flight Testing Robust Autoland Control Laws, Guidance, Navigation and Control Conference, Montreal, Canada (2001)
54. Reinke, N., Friedrich, U., Kuhl, R. Müller, M. Ruyters, G.: Die DLR-Parabelflüge—Forschen in Schwerelosigkeit. DLR-Projektbroschüre, Aug 2009
55. N. N.: Parabolic Flight Campaign with A300 ZERO-G, User's Manual, 5.2 edn. Novespace, Juli 1999
56. N. N.: JSC (*Lyndon B. Johnson Space Center*) Reduced Gravity Program User's Guide. NASA, JSC, Aircraft Operations Division, März 2013
57. Blum, J., von Borstel, I., Brucks, A.: Experimental simulation of martian regolith. Experimentbeschreibung, Institut für Geophysik und extraterrestrische Physik, TU Braun-schweig, DLR-Institut für Raumfahrtsysteme, Juli 2008
58. Leißling, D.: Sitzungsprotokoll des Betriebsausschusses ATTAS 13/2008. Deutsches Zentrum für Luft und Raumfahrt, Aug (2008)
59. Preß, M.: Mars-Parabeln und DLC-Test Flugversuchsprotokoll, Deutsches Zentrum für Luft und Raumfahrt, Sept 2008
60. Rundqwist, L., R. Hillgren: Phase Compensation of Rate Limiters in JAS 39 Gripen, AIAA Paper 96-3368 (1996)
61. Dornheim, M.A.: Report pinpoints factors leading to YF-22 crash. Aviation Week Space Technol **137**(19), 53–54 (1992)
62. Hanke, D.: The Influence of Rate Limiting Elements in Flight Control Systems on Handling Qualities. DLR-IB 111-93/61 (1993)
63. Hanke, D.: Handling Qualities Analysis on Rate Limiting Elements in Flight Control Systems. AGARD-AR-335, Paper 11 (1995)
64. Hanke, D.: Contribution to Rate Limiting Caused Control Problems. Workshop on Pilot-in-the-loop Oscillations, Braunschweig, 12–13 June 1997
65. Hanke, D.: Criterion for Predicting Aircraft-pilot Coupling Caused by Rate Limitation. DLR-IB 111-98/25 (1998)
66. Klyde, D.H., Mitchell, D.G.: Investigating the role of rate limiting in pilot-induced oscillations. In: Proceedings of AIAA Atmospheric Flight Mechanics Conference and Exhibit, Austin, Texas, pp. 1–12, Aug 2003
67. McRuer, D.T. (Chair) et al.: Aviation safety and pilot control. In: Understanding and Preventing Unfavorable Pilot-Vehicle Interactions, US National Research Council. National Academy Press, Washington D.C. (1997)
68. Hanke, D.: Phase compensation: a means of preventing aircraft-pilot coupling caused by rate limitation, DLR-FB 98-15 (1998)
69. Hanke, D.: Flight Test Data Evaluation of Rate Limiting Caused PIO's, DLR-IB 111-2000/34 (2000)
70. Hanke, D.: Flight test evaluation and data analysis of rate limiting induced PIO's. In: AIAA Guidance, Navigation, and Control Conference and Exhibit, pp. 1–21. Austin, TX, (US), 11–14 Aug 2003
71. Hanke, D.: Preventing rate limiting induced aircraft-pilot coupling by phase compensation. In: (1) ASTEC'03, 8th International Symposium, Zhukovsky (Moscow Region), Russia, 26–28 Nov 2003, (2): DLR-IB 111-2004/50 (2004)
72. Duda, H.: Systembewertung von reglergestützten Flugzeugen unter Berücksichtigung von Stellratenbegrenzung im Frequenzbereich. Zeitschrift für Flugwissenschaft und Weltraumforschung (ZfW) **17**, 343–350 (1995)
73. Duda, H.: Prediction of pilot-in-the-loop oscillations due to rate saturation. J. Guid. Navig. Control, **20**(3) (1997)
74. Duda, H.: Flight control system design considering rate saturation. Aerosp. Sci. Technol. **4**, 265–275 (1998)
75. Duda, H., Duus, G., Hovmark, G., Forssell, L.: New flight simulator experiments on pilot-involved oscillations due to rate saturation. J. Guid. Control Dyn. **23**(2), 312–318 (2000)
76. Duda, H.: Effects of Rate Limiting Elements in Flight Control Systems—A New PIO-criterion, AIAA-Paper 95-3204 (1995)
77. Duda, H., Hovmark, G., Forssell, L.: Prediction of Category II Aircraft-pilot Couplings—New Experimental Results, AIAA-Paper 97-3499 (1997)
78. Duda, H.: Fliegbarkeitskriterien bei begrenzter Stellgeschwindigkeit, DLR-FB 97/15 (1997)
79. Duda, H., Duus, G., Hovmark, G., Forssell, L.: New Flight Simulator Experiments on PIO due to Rate Saturation, AIAA-Paper 98-4336 (1998)
80. Duda, H., Duus, G.: Recent Results of APC Testing with ATTAS, SAE Aerospace Control and Guidance Systems Committee Meeting (1999)
81. Duus, G., Duda, H.: Analysis of the HAVE LIMIT Data base using the OLOP Criterion, AIAA-Paper 99-4007 (1999)
82. Duus, G.: SCARLET 3—A Flight Experiment Considering Rate Saturation, AIAA 2000-398714-17, Denver (2000)
83. Gilbreath, G. P.: Prediction of pilot-induced oscillations (PIO) due to actuator rate limiting using the open-loop onset point (OLOP) criterion, Thesis, USAF Air Force Institute of Technology, WPAFB, Ohio (2001)
84. Brieger, O., Leißling, D., Kerr, M., Postlethwaite, I., Sofrony, J., Turner, M.: Flight testing of a rate saturation compensation scheme on the ATTAS Aircraft, DGLR-2006-111. Braunschweig, Germany (2006)
85. Sofrony, J., Turner, M.C., Postlethwaite, I., Brieger, O.M., Leißling, D.: Anti-windup synthesis for PIO avoidance in an experimental aircraft. In: 45th IEEE Conference on Decision and Control, San Diego, CA, USA (2006)
86. Brieger, O., Kerr, M., Leissling, D., Postlethwaite, I., Sofrony, J., Turner, M.C.: Anti-windup compensation of rate saturation in an experimental aircraft. In: American control conference, New York, NY, USA (2007)
87. Kerr, M., Postlethwaite, I., Sofrony, J., Turner, M.C., Brieger, O.: Flight Testing of low-order anti-windup compensators for improved handling and PIO suppression. In: American Control Conference, Seattle, WA, USA (2008)

88. Brieger, O., et al.: Flight testing of a rate saturation compensation scheme on the ATTAS aircraft. Aerosp. Sci. Technol. **13**, 92–104 (2009)
89. Ossmann, D., Heller, M., Brieger, O.: Enhancement of the nonlinear OLOP-PIO-criterion regarding phase-compensated rate limiters, AIAA-AFM-6207. Honolulu, HI, USA (2008)
90. Brieger, O., Turner, M.C.: Flight test for PIOs on advanced technologies testing aircraft system (ATTAS), SAE aerospace control and guidance systems committee meeting # 102, Grand Island, NY, USA (2008)
91. Kerr, M., Marcos, A., Penin, L, Brieger, O., Postlethwaite, I., Turner, M. C.: Piloted assessment of a fault diagnosis algorithm on the ATTAS aircraft, AIAA-GNC-5760, Chicago, IL, USA (2009)
92. Brieger, O., Kerr, M., Postlethwaite, I., Turner, M., Sofrony, J.: Pilot-involved-oscillation suppression using low-order anti-windup: flight-test evaluation. AIAA J. Guid. Control Dyn. **35** (2), 471–483 (2012)
93. Schlichting, H., Truckenbrodt, E.: Aerodynamik des Flugzeuges, Teil 2, 2, neubearbeitete edn. Springer, Berlin (1969)
94. International Civil Aviation Organization (ICAO), Doc 4444-RAC/501 Rules of the Air and Traffic Services, 13th edn (1996)
95. de Bruin, A.: S-Wake assessment of wake vortex safety—publishable summary report, NLR-TP-2003-243 (2003)
96. Krag, B., Flight test report, DLR IB 111-2001/40 (2001)
97. Fischenberg, D., Results of flight test data analysis, SWAKE-TN-222_1 (2002)
98. Fischenberg, D.: Bestimmung der Wirbelschleppen-charakteristik aus Flugmessdaten, DGLR-JT2002-170 (2002)
99. Fischenberg, D.: A method to validate wake vortex encounter models from flight test data. In: Proceedings of the 27th International Congress of Aeronautical Sciences, Nice (2012)
100. Hahn, K.-U.: Coping with Wake Vortex, 23rd Congress of the International Council of the Aeronautical Sciences, Toronto, Canada, Sept. 2002
101. Hahn, K.-U., Schwarz, C., Friehmelt, H., A Simplified Hazard Area Prediction (SHAPe) model for wake vortex encounter avoidance. In: 24th International Congress of Aeronautical Sciences (ICAS), Yokohama, Japan, Aug–Sept 2004
102. Hahn, K.-U., Schwarz, C., Safe limits for wake vortex penetration. In: AIAA Guidance, Navigation and Control Conference and Exhibit, Hilton Head, South Carolina, Aug 2007
103. Schwarz, C. W., Hahn, K.-U., Fischenberg D.: Wake Encounter Severity Assessment Based on Validated Aerodynamic Interaction Models, AFM/ASE/GNC/NST Conference, Toronto, Ontario, Aug 2010
104. Vechtel, D.: In-flight simulation of wake encounters using deformed vortices. Aeronaut. J. **117**(1196) (2013)
105. Hahn, K.-U., Fischenberg D., Niedermeier, D., Horn, C.: Wake encounter flight control assistance based on forward-looking measurement processing. In: AFM/ASE/GNC/NST Conference, Toronto, Ontario, Aug 2010
106. Terry Lutz, T., Greeves, B.: Throttles-only-control (TOC)—10 Steps to a Survivable Landing Following Loss of Normal Flight Control, Interpilot, pp. 32–34 (2004)
107. Burcham, F., Fullerton, C.G.: Development and Flight Test of an Emergency Flight Control System using only Engine Thrust on an MD-11 Transport Airplane, NASA/TP-97-206217 (1997)
108. Almeida, F.A., Leissling, D.: Fault-tolerant Model Predictive Control with Flight Test Results on ATTAS, AIAA-2009-5621, Aug 2009
109. Fezans, N.: simple control law structure for the control of airplanes by means of their engines. In: Advances in Aerospace Guidance, Navigation and Control, pp. 151–162. Springer, Berlin, (2011)
110. Becker, H.: Das "automatic flight control system" des ATTAS für ATM Experimente, DLR-IB 112-2004/55 (2004)
111. Adam, V., et al.: A conceptual model of a future integrated atm system and novel functional requirements for a future flight management system. GARTEUR Working Paper FM WP (89) 004 (1989)
112. Adam, V., et al.: Integration of flight management and atm systems. Final Report of GARTEUR Action Group FM (AG) 03 (1990)
113. Kirstetter, B.: PHARE—Concept and programme. In: DLR-Mitt. 89-23, Paper 14 (1989)
114. Adam, V., Kohrs, R.: On Board Planning of 4D-Trajecto-ries. AGARD CP-504, Paper 16 (1992)
115. Adam, V., Czerlitzki, B., Kohrs, R., Rataj, J.: Beschrei-bung von Planungsalgorithmen für das Experimentelle Flight Management System, DLR-IB 112-92/43 (1992)
116. Adam, V., Ingle, G., Rawlings, R.: Experimental Flight Management System. AGARD CP-538, Paper 24 (1993)
117. Ludwig, T.: MA-AFAS D39 Annex A—DLR Taxi and Flight Trials, DLR (2003)
118. QINETIQ (UK): MA-AFAS D39—Flight Test Validation Report, QINETIQ/S&E/AVC/ CR031041-D39 (2003)
119. Kuenz, A., Mollwitz, V., Edinger, C., Becker, H.: Lärmminderungspotential und Kapazitätsauswirkungen von „Continuous Descent Approach"-Verfahren, Forschungs-verbund Leiser Verkehr, Abschlussbericht zu LAnAb EA 1632, DLR-FL, TUB-ZFB (2007)

Author Biographies

Dietrich Hanke was a research scientist at the Institute of Flight Systems at DLR in Braunschweig (1970–2005). He received his Dipl.-Ing. degree in Mechanical/Aeronautical Engineering from the Technical University of Braunschweig (1970). His main research interests are flying qualities, flight and system dynamics, flight control, active sidestick, and in-flight simulation (fixed-wing aircraft). From 1972 to 1984 he was project leader for the development of the HFB 320 FLISI as an in-flight simulator. In 1981 he became Head of the new established Flight Simulation department. In this position, he was responsible for the technical development of the VFW 614 ATTAS in-flight simulator and demonstrator aircraft, as well as for the development, operation, and maintenance of the Fly-by-Wire/Light system. Also, he headed several research projects with the HFB 320 aircraft (Direct Lift Control, Command Control), the ATTAS aircraft (Hermes Space Plane, Rate Limiting) and several international cooperation programs with the AFFDL and TsAGI, Russia. After his retirement he founded the Maui Ultra Fins company dealing with the development and production of high performance windsurfing fins.

Klaus-Uwe Hahn is a research scientist at the Institute of Flight Systems (formerly Flight Mechanics) at DLR in Braunschweig since 1989. He is currently head of Flight Dynamics and Simulation. He received his Dipl.-Ing. degree in Mechanical/Production Engineering (1975) from the FH Braunschweig-Wolfenbüttel and in Mechanical/Aeronautical Engineering (1981), and Dr.-Ing. degree (1988) from the Technical University of Braunschweig. In 2008 he was appointed Professor at the Technical University of Hamburg-Harburg. His main research interests are flight performance, flight dynamics, flight control, flight test (fixed-wing aircraft). From 1989 to 2011 he was involved in a variety of flight tests with the VFW 614 ATTAS. He is a member of the DGLR, a member of the Planning and Advisory Board of the SAE Aerospace Control and Guidance Systems Committee (ACGSC), and a senior member of AIAA.

Helicopter In-Flight Simulator EC 135 FHS

10

Jürgen Kaletka

With contributions from Heinz-Jürgen Pausder and Wolfgang von Grüngahen

J. Kaletka (✉)
Braunschweig, Germany
e-mail: juergen.kaletka@dtv-hifi.de

P.G. Hamel (ed.), *In-Flight Simulators and Fly-by-Wire/Light Demonstrators*,
DOI 10.1007/978-3-319-53997-3_10

10.1 Introduction

The helicopter in-flight simulator Bo 105 ATTHeS, described in Chap. 8, was operated by DLR from 1982 to 1995. In 1993, it was decided to replace ATTHeS with a new airborne simulator. The definition, selection, and development of the new simulator, based on an EC 135 as the host vehicle, are described in detail in this chapter. Further on, some selected results from typical utilization programs are presented. As the helicopter was initially defined as an active control technologies (ACT) demonstrator as well as a helicopter in-flight simulator (*Flying Helicopter Simulator*—FHS) it was called EC 135 ACT/FHS.

10.2 FHS Definition and Planning Phase

10.2.1 How It Began

A broad application spectrum of the Bo 105 ATTHeS provided a deep experience on the benefits of helicopter in-flight simulators. They included the definition and evaluation of flying qualities criteria for modern helicopters, the training and education of test pilots and flight test engineers, and the design and evaluation of new cockpit and display technologies. In addition to ATTHeS, the German Air Force Flight Test Center (WTD 61) in Manching operated a BK 117 (AVT) for cockpit component tests. However, it was not possible to modify its flight characteristics.

The development of new and complex control and cockpit technologies requires early testing of the components in a realistic environment, ideally in flight. It allows a detailed evaluation and analysis pertaining to pilot workload, safety aspects, operational benefits, and technical and economical risks. It became apparent that appropriate test facilities were needed to reduce development costs and risks. To be prepared for the realization of new key technologies for future European rotorcraft there was the need for a test vehicle with a much wider application range. Initiated by the DLR Institute of Flight Systems, a Memorandum of Agreement (MoA) was signed in 1993 between Eurocopter France, Eurocopter Germany, and DLR. It was entitled "Development and Operation of an Active Control Technology Demonstrator and Flying Helicopter Simulator ACT/FHS". "*This MoA was motivated by the need for future helicopter test facilities in order to replace the DLR Bo 105 ATTHeS In-Flight Simulator and to support the Eurocopter ACT Demonstrator Policy*". The required application areas for industry, research organizations, and government organizations were:

- *Technology integration and demonstration,*
- *Flying qualities evaluation and flight control systems research, and*
- *Support for government agencies and flight test centers.*

In 1994, a national working group was set up with members from Eurocopter Germany, DLR, and WTD 61. After one year of extensive deliberations, this working group generated a definition study on the development of an "ACT Demonstrator-Flying Helicopter Simulator (ACT/FHS)" with the major sub-tasks:

- Definition of spectrum of utilization,
- Definition of an appropriate system architecture,
- Selection of a suitable test vehicle, and
- Planning of the development phase.

The final report was the main basis for the decision of the MoA partners to develop the ACT/FHS [1]. Main parts of the study are presented in the following sub-sections.

10.2.2 Definition of Applications

It was agreed that ACT/FHS had to be designed for a wide application spectrum to meet the requirements of industry, research originations, and test centers. Priorities were focused on in-flight simulation, system development and integration, and technology demonstration (see Fig. 10.1). The formulated ACT/FHS application requirements were also compared to the capabilities of actually existing test helicopters:

	Industry	Research	Test centres
In-flight simulation			
• helicopter simulation	○	●	●
• flying qualities	●	●	●
• pilot training	○	●	●
System development			
• system architecture	●	●	○
• control systems	●	●	○
• cockpit systems	●	●	○
• ACT components/functions	●	●	○
• mission packages	●	●	●
Technology demonstration			
• functional aspects	●	○	○
• operational aspects	●	○	●
• operational benefit	●	○	●

● First priority ○ Second priority

Fig. 10.1 ACT/FHS utilization spectra

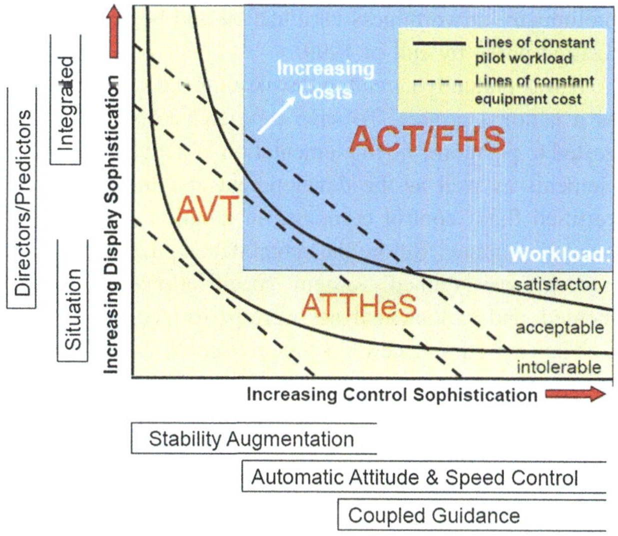

Fig. 10.2 Comparison of requirements and user areas

- Bo 105 ATTHeS, DLR, (see Chap. 8),
- BK 117 AVT, WTD-61,
- AS 365 Dauphin 6001, Eurocopter France (see Sect. 6.2.4.3),
- BK 117 FBW Experimental Helicopter, Kawasaki Heavy Industries (see Sect. 6.2.5.2), and
- JUH-60A RASCAL (*Rotorcraft Aircrew Systems Concepts Airborne Laboratory*), NASA (see Sect. 5.2.2.17).

The envisaged ACT/FHS utilization potential was compared to the national test vehicles ATTHeS and AVT. It became clear that the new helicopter can only fulfill the various requirements with a freely programmable active control system and modular installation equipment to allow fast changes and implementation of new elements for future cockpit and mission technologies (see Fig. 10.2).

10.2.3 ACT/FHS Concept

An essential and largest part of the study was concerned with a very detailed proposal for the technical concept of the helicopter. The objective was to specify the layout of an open architecture modular system with a high degree of variability. The specified system had to be designed to support, test, and evaluate new components from their experimental status in the design phase until their final version for a serial production (*criticality should include non-essential and essential to critical*). It was suggested that, for flight tests, the helicopter be always flown by two pilots,

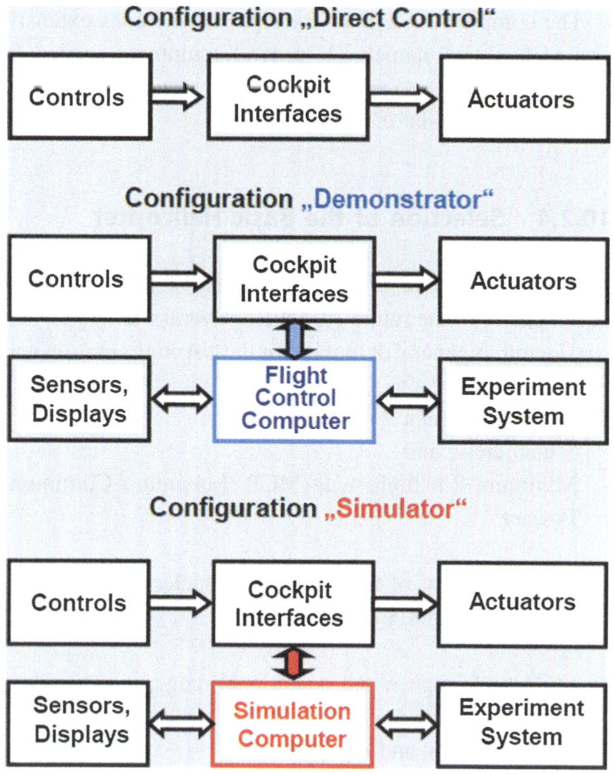

Fig. 10.3 Proposed system concept configurations

a safety pilot and an evaluation pilot who conducts the tests. It was mandatory that the safety pilot was always in a position to take over the control of the helicopter on his own decision, independent of the actual flight condition.

Modular and hierarchical system architecture with a standardized interface was proposed. As shown in Fig. 10.3, the hierarchy is composed of three configuration levels:

1. Direct Control using the standard mechanical control; safety pilot in command.
2. Demonstrator Configuration: Fly-by-Wire/Light direct control, no modification of control inputs; evaluation pilot in command.
3. Simulator Configuration: an experimental computer can modify the pilot control inputs; evaluation pilot in command.

A workstation for a flight test engineer was to be provided behind the two pilots. For a later expansion phase, this seat can optionally be modified for a second experimental pilot. Emphasis was placed on a system layout that allows fast and easy modifications and the installation of new components (both, hardware and software) in the future.

The complete ACT/FHS concept also includes extensive ground facilities, namely (1) ground equipment needed for the helicopter operation, (2) a mobile ground station, and (3) a system simulator.

10.2.4 Selection of the Basic Helicopter

For the selection of a suitable standard helicopter for the development of the future ACT/FHS several candidates were considered. A general definition for the test configuration was:

- Payload between 250 and 500 kg,
- 3 man crew, and
- Minimum 2 h flight with MCP (Maximum Continuous Power).

The assessment of the candidate vehicles was structured into the following six criteria:

- flight performance and operational range,
- flying qualities, agility,
- space for crew and equipment,
- suitability for use in the operational environment,
- development and operational risks, and
- economic efficiency and costs.

Considered helicopter candidates were:

- EC 135,
- BK 117 C+,
- Tiger PT1,
- Dauphin 365 N2,
- Super Puma, and
- NH 90.

The EC 135 was selected, as it demonstrated a well-balanced and homogenous evaluation result, in particular, pertaining to the technical and economic criteria. In addition, it incorporated the state of the art technology, especially with its bearingless main rotor system (high dynamic response capability) and digital engine control.

10.2.5 Schedule and Costs

During mid-1995 the working group finalized a detailed proposal for the ACT/FHS development. It included project structure, responsibilities and work-sharing, cost estimates, and time schedule. The proposal included: acquisition of the basic EC 135 by DLR by mid-1997, first flight with the direct (mechanical) control system by end of 1998, and preliminary airworthiness certification and begin of the utilization phase by end of 1999.

It was mutually agreed to develop the research helicopter as a national project. To save time and costs, it was suggested to postpone the implementation of active rotor control elements as well as the development and integration of a certified flight control computer in addition to the experimental computer. Before the actual development started, the concept was refined, system specifications were documented, and various analyses were performed.

The revised and new documents served as the basis for the contract to develop the new in-flight simulator which began in 1996 [2, 3].

10.3 From Serial to Research Helicopter

10.3.1 Introduction

The development of the FHS was started in 1996 in a close cooperation between Eurocopter Germany, LAT (Liebherr-Aero-Technik, today: Liebherr Aerospace), and DLR. As a host vehicle, the Eurocopter EC 135, S/N 28, was selected and acquired by DLR in 1997. The modifications of the basic EC 135 were planned and conducted by all three partners. As already pointed out in Sect. 10.1, the helicopter was called ACT/FHS. However, for better readability, the abbreviation FHS is used hereafter in this chapter.

The conversion of the original EC 135 helicopter into the FHS research platform required significant modifications (see Fig. 10.4). Therefore, the empty EC 135 hull was taken from the production line and transferred to the Eurocopter Germany prototype construction department. Here, the integration of the standard and FHS specific components was undertaken and all further modifications were implemented. From the beginning of the FHS development, the cooperation between all partners was essential to meet the objectives of very different future applications for both research and technology programs. It was anticipated to design and build a vehicle with a high application oriented flexibility and adaptability to cover a wide range of user requirements in order to support various national and international research and technology programs.

The first FHS flight using the mechanical control system took place in August 2000. Two years later, the helicopter had passed successfully extensive flight tests for all components and mode conditions. Ready for use, it was delivered to DLR in November 2002 and received the aircraft registration D-HFHS. The FHS operational system was complemented at DLR Braunschweig by two ground stations: a ground-based simulator for the preparation and support of individual flight test programs and a mobile

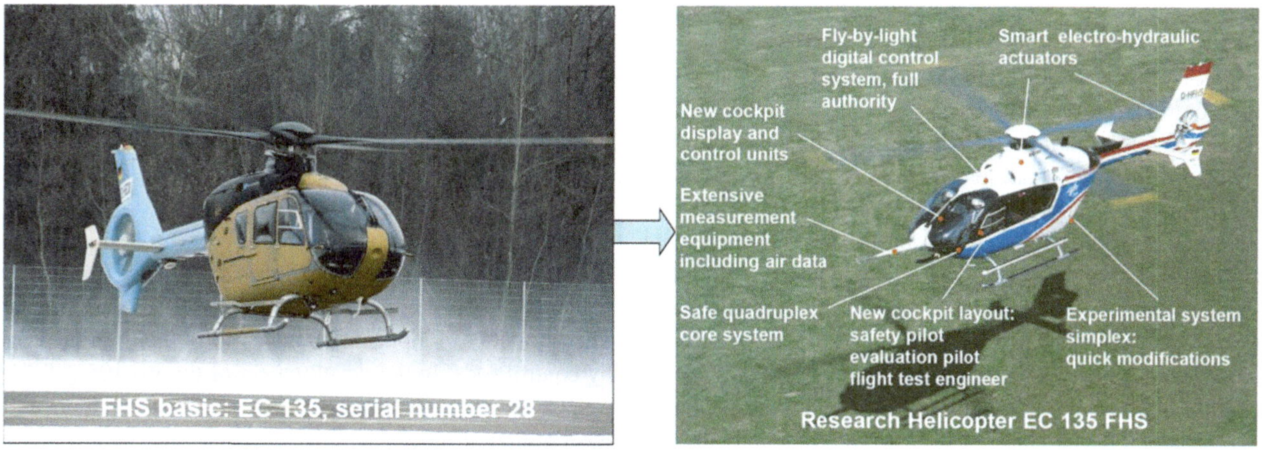

Basic EC 135: Light multi-role helicopter, twin-engine

Length	12.19 m	Maximum speed	259 km/h	Hover ceiling	3045 m
Height	3.51 m	Cruise speed	230 km/h	Maximum altitude	6095 m
Main rotor diameter	10.20 m	Main rotor RPM	395 U/min	Maximum range	635 km

Fig. 10.4 Standard EC 135 becomes FHS

Fig. 10.5 FHS presentation at the European Rotorcraft Forum 2003 at Lake Constance

telemetry station for communication, flight test control, data recording and evaluation. In 2003, FHS was presented to the international public at the Berlin Air Show (ILA) and during the European Rotorcraft Forum (ERF) in Friedrichshafen, Germany (see Fig. 10.5), [4].

10.3.2 Application Domains

The FHS was to be used to examine the feasibility of new technologies, to evaluate their pros and cons, and to demonstrate the benefits of new helicopter concepts [5, 6]. It was designed for a wide application spectrum and it could support all phases during the development of a new system from the first layout until the final testing. There are three major application areas covering the spectrum of user needs:

Airborne Simulation

In the case of in-flight simulation, the pilot control inputs are first fed to an onboard computer. According to the implemented program, the inputs are modified and transferred to the control actuators. Flying simulation gives the possibility to change the dynamic characteristics of the basic helicopter in a way, such that the pilot has the impression of flying a different vehicle. Such modifications could just be the variation of a single parameter, for example, an increase of a time delay between pilot input and actuator response. This allows the demonstration of pilot induced oscillations (PIO) tendencies or rotor-pilot coupling, as described in Sect. 8.4.1. Such effects can support the qualification and training of pilots. More demanding and more complex tasks could also be simulated, such as the dynamic behavior of a completely different type of helicopter. This new helicopter may not even exist in reality but can still be in a design phase. The pilot can fly and test it and give his evaluation comments. In comparison to ground-based simulators, the pilot flies the vehicle in a true airborne environment with real visual and motion cues. The in-flight simulation is not only an excellent tool for basic and applied research in handling qualities, controls, displays, and human factors, it will assist in the design, development, and evaluation of future helicopters before their first flight. This avoids the expensive modifications in the development process of a real helicopter at a later stage. For the fast changes required in a research environment, a high degree of flexibility must be provided for the airborne simulation role.

Development and Testing of New Systems

A further application area of the FHS is the development, implementation, and evaluation of new electronic flight-control systems. Many aircraft still use a mechanical control system consisting of a sequence of rods and/or cables to link the pilot controls to the hydraulic actuators that move the control surfaces. These systems are relatively heavy, require careful routing through the aircraft and cannot be adapted to flight conditions. Their major advantage is demonstrated reliability. However, techniques to transmit pilot inputs by electrical signals have been developed and are now in use also for non-experimental aircraft. The required reliability is obtained by multiple individual signals to provide redundancy but still at less weight than a mechanical control system. With such a digital Fly-by-Wire control system the pilot control inputs are immediately converted into electrical signals. Now, the conventional pilot controls can be replaced by more effective and intelligent devices such as control inceptors. As the FHS is equipped with a redundant Fly-by-Wire/Light system, it is a perfect tool for the development and evaluation of new control systems like active sticks. Active sticks are programmable and they offer a wide range of applications. Pilot control forces are adapted to the actual flight conditions. Tactile cues like vibrations, breakout forces, and soft-stops provide warnings, prevent unintended inputs, and inform on aircraft operating limits. In comparison to optical or acoustic signals, the haptic feedback is immediately sensed by the pilot and he can react faster and more intuitively. It is anticipated that active control systems will reduce pilot workload and will help to make flying less stressful and safer.

The programmable onboard computer allows the testing of new control law concepts. The FHS programmable multi-function display can help to define the most appropriate information to be displayed for the pilot with respect to the actual flight condition and task.

Technology Demonstration

The third key area for the FHS utilization is the integration and qualification of innovative technologies, like active control components, new flight control laws, and advanced cockpit systems. Technology demonstration encompasses evaluating and proving the functionality and operational benefits of new technologies up to the point of certification. Also, these applications need a high flexibility for component and system integration including both hardware and software modifications.

An important example of innovative technologies for helicopters is the FHS control system itself. It was for the first time that a full authority digital Fly-by-Light control system was implemented. It was the primary control for all flight conditions including the start and landing phases. For this purpose, a new system architecture was developed. Here, a major emphasis in the design was placed on two essential factors, namely high safety standards according to the stringent civil certification requirements and at the same time, maximum flexibility for configuration changes to meet user needs. The FHS control system will be described later in detail.

10.3.3 EC135 Becomes FHS

In the series production of the EC 135 and Bo 105 the helicopters were equipped with a mechanical control system. The pilot control inputs are transferred to the rotor actuators by control rods. In the development of the Bo 105 ATTHeS (see Chap. 8) the standard control system was maintained. In the simulation mode, the control inputs were calculated by the onboard simulation computer. They were fed to electro-hydraulic actuators that were connected by clutches to the standard control rods (see Fig. 8.5 in Chap. 8). In the definition phase of the FHS, it became clear that this concept will not meet the needs of future users. Therefore, the mechanical control system was completely replaced by a full-authority digital control system using Fly-by-Wire/Light technology. The system architecture was specified to meet two essential requirements:

Safety: The standard operation of the helicopter is the Fly-by-Light mode at all flight conditions, including low altitude, transition, start, and landing. This configuration had to comply with civil certification requirements. A modified mechanical control system was still installed but should only serve in the case of emergency.

Flexibility: For the conduct of user programs and in particular for the simulation task, it is absolutely necessary to easily change control laws or models or to implement new hardware components. Even during a flight test campaign, some modifications should be allowed. This flexibility is only possible without stringent safety and certification constraints.

Obviously, these two requirements are contradictory. The safety of electronic systems is based on multiple redundancies for all components of hardware and software. By continuously comparing the redundant signals it is possible to detect failures and malfunctions and to disconnect the faulty channel. Here, at least a triple redundancy is needed to compensate for errors. It is evident that the development of such a control system is quite complex and entails a large amount of work, cost, and time. Once it is designed, built, tested, and certified, it is practically "frozen". Modifications are no longer possible without starting a new documentation, testing, and certification process. On the other hand, flexibility implies fast and uncomplicated modifications according to user needs: at best no redundancy, no extensive testing, and no certification. In other words: a more vulnerable system.

One of the most demanding tasks during the FHS design phase was how to build a control system that fulfills all aircraft safety regulations and still allows the use of less reliable hardware and software components for experiments?

FHS System Architecture

The FHS control system uses a hierarchical architecture and was installed in two associated onboard units, as illustrated in Fig. 10.6. It consists of a “core system”, which provides the required safety, and an “experimental system”, which gives the flexibility for modifications [7]. The core system meets civil certification requirements with a probability of catastrophic failures less than 10^{-9} per flight hour. It was achieved by quadruplex redundancy in all components together with the dissimilarity of both hardware and software. The heart of the core system is the core system computer. It is the central interface that receives the control signals from both pilots and flight state signals from all onboard sensors. It communicates with the experimental system, which can modify the control commands from the evaluation pilot. The hierarchical architecture becomes obvious when the responsibilities of the core and the experimental system are compared. All signals are fed to the core system computer and based on this comprehensive information the core system checks all data and finally decides whether the resulting control inputs are acceptable. Only then they are applied to the hydraulic smart actuators. The experimental system offers a lot of freedom for the individual user programs. It calculates new control input signals and sends them to the core system computer without a detailed data check. In principle, the core system can be considered as the “boss” who gives the final OK. The experimental system is his “employee” who develops new ideas but is allowed to be wrong. Some additional functions of the core system are addressed below. As the core system is quadruplex with dissimilar DO-178B Level A certified software in the core system computer and the smart actuator electronics, it is obvious that any later changes of the core system will require a significant effort, in particular with respect to testing, documentation, and qualification. Consequently, the core system should not be modified unless it is absolutely necessary.

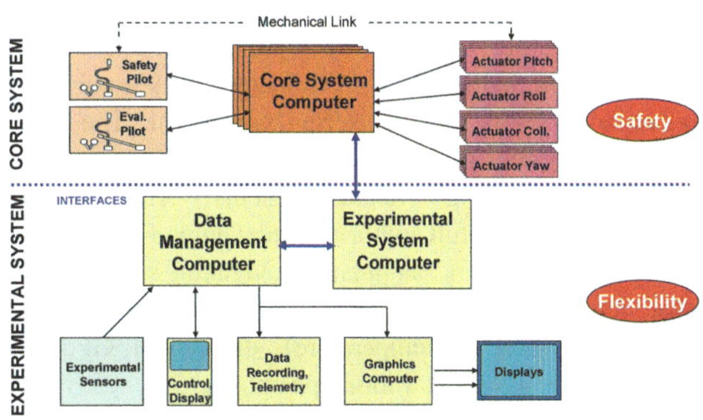

Fig. 10.6 FHS system architecture

The main elements of the experimental system are the experimental computer and the data management computer. The first one communicates with the core system computer. It receives the evaluation pilot command signals, modifies them according to the programmed control laws and transfers them back to the core system. The data management computer collects all data provided by basic sensors and by sensors in the experimental system and transfers them to the telemetry, the onboard data recording, and to the graphics computer that controls the displays. In contrast to the core system, the experimental system is only simplex to allow relatively easy and fast modifications. The criticality level is “minor”, which implies that the system may fail and produce errors. Therefore, several safety features are implemented in the core system computer to avoid critical helicopter flight responses due to unrealistic control inputs.

All components of the core system are quadruplex, beginning with the sensors for the pilot control motions up to the hydraulic smart actuator electronics. Redundancy is also provided for the other helicopter components. There are two independent hydraulic systems, two electrical generators, and four backup batteries. The FHS has also a battery operated auxiliary hydraulic system, which allows pre-flight checks and preparations without any external equipment.

Figure 10.7 illustrates the technical realization of the architecture and some of the major helicopter modifications. As also shown in Fig. 10.8, the EC 135 cabin accommodates a three-person crew with a safety pilot in the left pilot seat and an evaluation pilot in the right pilot seat (unlike most fixed wing aircraft). A flight test engineer station is located behind the two pilot stations. Both pilots have conventional

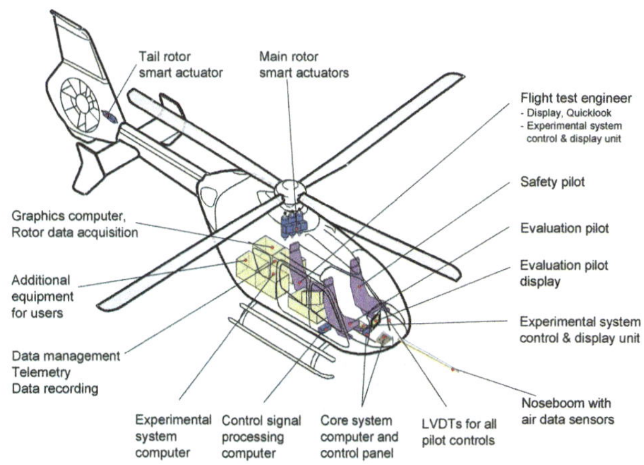

Fig. 10.7 Main FHS modifications

Fig. 10.8 Cockpit view

controls (stick, collective, pedals). The control positions are measured by linear variable displacement transducers (LVDT) with four sensors for each pilot control. LVDT has high resolution and measurement accuracy. The sensor works within an electrical field without contact or friction between the LVDT's core and coil assembly, providing a fast dynamic response and has a long mechanical life. The electrical outputs can directly be used without amplification and are sent to the four core system computers.

The FHS layout was based on the use of the Fly-by-Light control system for all flight conditions. Consequently, the original mechanical control system of the EC 135 was not implemented. However, for the safety pilot an additional mechanical link from his controls to the hydraulic actuators was installed as a backup in case of an emergency. Instead of conventional solid rods, flexball cables were chosen. In principle, they are similar to the better known Bowden wire cables. They have an inside wire and can only transmit pulling forces. The interior of a flexball cable is more complex. A flexible central blade can be moved between two lines of balls imbedded in bearing cages (see Fig. 10.9). A flexball cable reacts for tension and compression. It has a

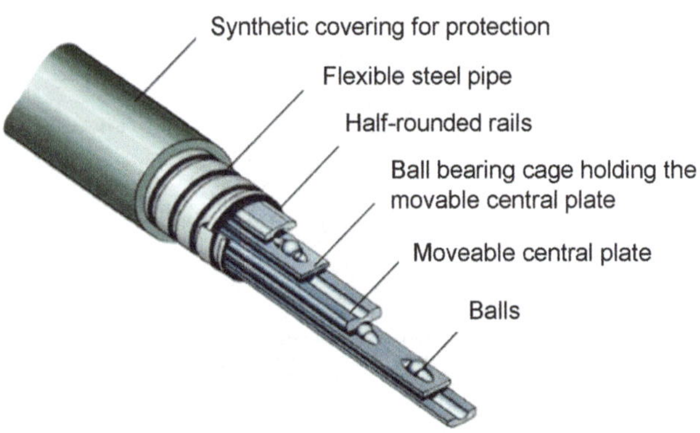

Fig. 10.9 Structure of flexball cable

high mechanical efficiency with very low backlash even with long routings. It can easily be installed and has a high flexibility. Maintenance or lubrication is not needed. However, there are constraints with respect to allowed minimum bending radii.

In the FHS the cables for the main rotor actuators were routed within the windscreen center frame. The cable for the tail rotor actuator is below the cabin floor.

The FHS cockpit is shown in Fig. 10.10. The safety pilot panel (left side) is equipped with an *Avionique Nouvelle* glass cockpit with standard instrumentation. In the center console between the two pilots is a control unit for the core system. Both the evaluation pilot (right side) and the flight test engineer (seated behind the two pilots) have a freely programmable multifunction 10-inch experimental display and a control panel for the display. The units are identical but independent from each other; hence the pilot may select navigation instruments on his display while the flight test engineer can choose a quick-look from recent flight measurements. The flight test engineer has also access to the experimental system, for example, for changing configurations and parameters. The flight test engineer seat is located in the center of the cabin so that he can also observe the cockpit instruments and has a free view to the outside. His workstation is on his right-hand side (see Fig. 10.11).

The four core system computers are located in two separate housings, each with its own cooling system, under the cabin floor. Each housing contains two computers, which are dissimilar in hardware and software. The original hydraulic actuators were replaced by FHS specific smart actuators underneath the main rotor and close to the tail rotor. The actuator electronics receive the control commands from the core system computers via optical fibers. Most of the components of the experimental system were installed in the cargo compartment behind the flight test engineer [8]. They were mounted on three aluminum pallets. The pallets were fixed on rails and could easily be removed from the helicopter or reinstalled. It allowed fast modifications and testing in the laboratory or on the fixed-base simulator. A fourth pallet is free for user specific equipment. As a research helicopter, the FHS is fully instrumented with a number of redundant sensors and measuring equipment. The instrumentation system mainly includes two air data units, two attitude and heading reference systems (AHRS), a radar altimeter, FADEC (full authority digital engine control) data, linear accelerometers, an inertia navigation system (INS), nose boom air data (static and dynamic pressure, angles of attack and sideslip, temperature), differential GPS, and control input signals at various positions.

FHS Operational Modes

Pertaining to the signal flow and the pilot in command, the FHS has three commonly used control modes: (1) safety

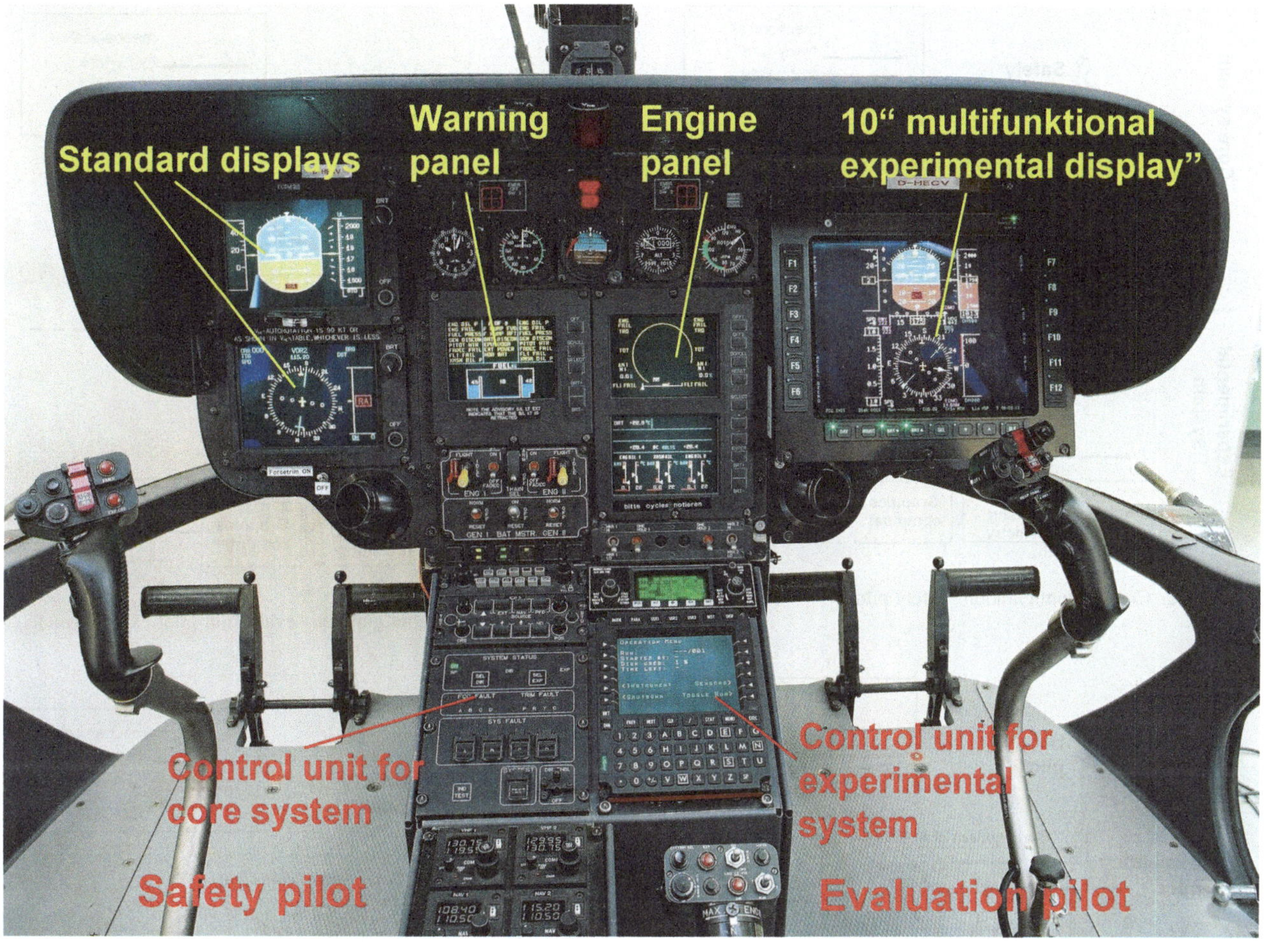

Fig. 10.10 Cockpit panel

Fig. 10.11 Workstation for flight test engineer

pilot mode, (2) evaluation pilot direct mode, and (3) evaluation pilot experimental mode. In a fourth mode, the mechanical backup can be used. The mechanical link is not intended to be a standard control mode, but it plays an important role in the evaluation pilot modes. The corresponding data flow for the individual modes is shown in Figs. 10.12, 10.13, 10.14 and 10.15. For simplification and better understanding, only one pilot control element (stick) and one data channel are presented in these figures. However, it has to be kept in mind that all four pilot controls have identical equipment and that all core system components are quadruplex redundant, from the sensors, measuring the pilot inputs, up to the actuator electronics.

Safety Pilot Mode

As shown in Fig. 10.12, the control positions, measured by LVDTs, are transmitted by electrical wires to the core system computer and demodulated. As the computer is located close to the sensors, only short wires were needed. The core system computers send the inputs via optical fibers to the actuator electronics, which control the hydraulic valves and consequently the actuator motion. The distance from the core system computers to the actuators is longer (in particular to the tail rotor actuator) so that full advantage is taken of the fiber optics technology. The mechanical flexball cables are attached to the pilot controls and they follow the

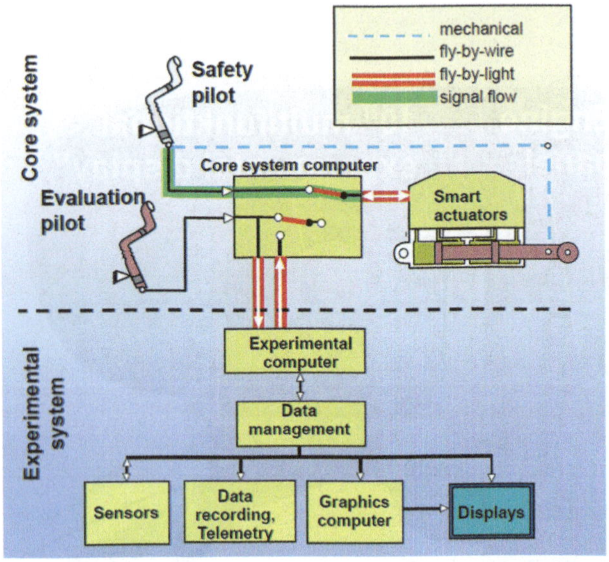

Fig. 10.12 Control configuration "safety pilot"

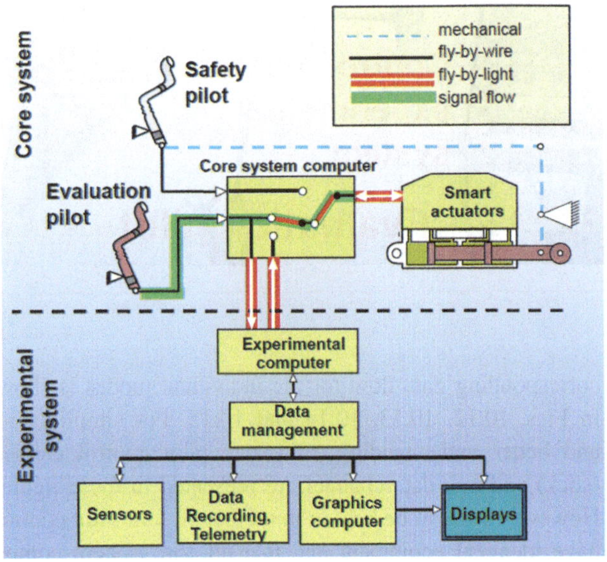

Fig. 10.13 Control configuration "evaluation pilot direct"

control motions. However, in the safety pilot configuration a hydraulic clutch in the smart actuators decouples the cables from the actuators and they have no effect. It is possible to use the experimental system, for example, for data recording and telemetry. But its output channels are switched off and the core system computers do not accept any signals from the experimental system.

Evaluation Pilot Direct Mode

In this operation mode, the pilot in command is the evaluation pilot, see Fig. 10.13. Similar to the safety pilot mode, the measured control positions are transmitted to the core

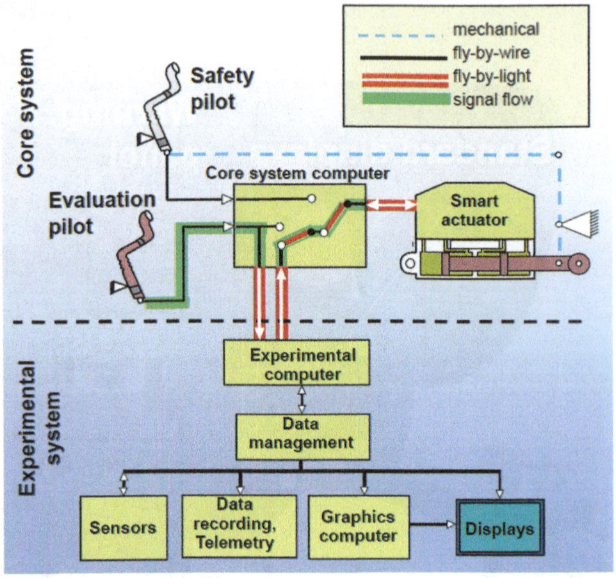

Fig. 10.14 Control configuration "evaluation pilot experimental"

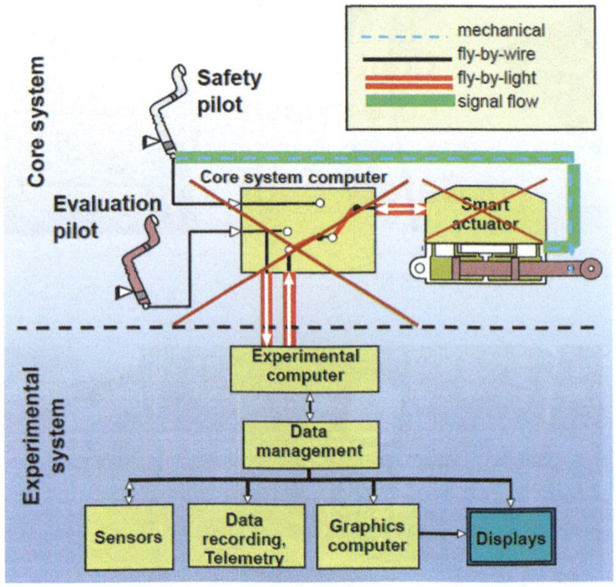

Fig. 10.15 Control configuration "safety pilot mechanical"

system computer and sent via the optical fibers to the actuators. As there is no mechanical link to the evaluation pilot controls, the pilot flies the helicopter in a pure Fly-by-Light mode. In contrast to the safety pilot mode, the hydraulic clutch in the actuator is closed. Now, the actual positions of the actuator piston rods are back driven by the flexball cables to the safety pilot controls. Apart from the emergency case, this is the second major role of the mechanical control system. It synchronizes the safety pilot control positions with the actuator positions. For the evaluation pilot direct mode, it means that the control positions for both pilots are the same. The experimental system has no influence.

Evaluation Pilot Experimental Mode

As in the previously described mode, in the present case too, the evaluation pilot is in command, see Fig. 10.14. But now the experimental system is fully engaged. The control inputs are received by the core system computer and transferred to the experimental computer. Here, the control inputs are modified according to the implemented user software and sent back to the core system computer. After a detailed data check they are transferred to the actuators. The actuator motion and consequently the FHS dynamic response is now due to the modified inputs and no longer directly related to the pilot inputs. The evaluation pilot flies the helicopter with modified flight characteristics in a pure fly-by-light mode. The hydraulic clutch is closed. The actual positions of the actuators are back driven by the mechanical link to the safety pilot controls. Therefore, his control positions always correspond to the motion of the helicopter control surfaces. They are in agreement with the hydraulic actuator outputs and different from the evaluation pilot controls. Whenever the safety pilot takes over the command and the FHS is switched into the safety pilot mode his controls are automatically in the correct position and he can continue the flight without any further synchronization.

Safety Pilot, Mechanical Control System

This configuration, shown in Fig. 10.15, does not belong to the "normal" operation of the helicopter as the FHS is developed and certified for the Fly-by-Light mode in all flight conditions. It will only be used in an emergency case if the optical system fails. The safety pilot controls are directly connected to the hydraulic smart actuators by the mechanical flexball cables. All other components of the core system are inactive. The pilot flies the standard EC 135 with a mechanical control system. This configuration can intentionally be selected by the safety pilot, for example, for testing or training purposes. In the worst case, when a severe error is encountered in the control system that cannot be compensated or corrected, it leads to a full breakdown. To avoid a fall back into the mechanical control mode, various procedures have been implemented in the core system computer and the smart actuator software to detect and eliminate wrong or unrealistic data channels. It is evident that such a malfunction in the Fly-by-Light system is a highly critical situation for the primary control system. It will require an intensive investigation and most probably a new effort to keep or renew the certification while the helicopter is grounded.

Role of the FHS Crew

Various procedures have been implemented in the core system to detect data errors and to eliminate or alleviate their influences. The efficacy of these techniques was successfully demonstrated during the FHS testing phase. Nevertheless, the human capabilities like awareness, judgement, and reaction should not and cannot be replaced. Therefore, the FHS crew and in particular the safety pilot are essential in the FHS safety concept.

Flight Test Engineer: The flight test engineer keeps track of the planned flight test program. Before starting a new test he informs the pilots about details of the test and the required flight condition. He has a working station with a multifunctional display and has direct access to the experimental computer. He can select any pre-programmed configuration, change parameters and configurations. During and after a test the flight test engineer documents comments and can make a first evaluation of the test data. He also communicates with the crew in the ground station.
Evaluation pilot: The evaluation pilot conducts the individual flight experiments as the pilot in command. He is in close contact with the flight test engineer and the ground crew and gives evaluation comments on the actual test. Like the flight test engineer he has a multifunctional display connected to the graphics computer of the experimental system. Various information can be displayed like flight instruments, camera signals, supporting graphics as help for the test conduction, and quick-looks of recorded measurements.
Safety pilot: Although each individual test scenario is tested and evaluated on the ground-based simulator, critical situations can arise in the experimental mode, for example, due to hardware failures or non-realistic software commands. Therefore, the safety pilot continuously observes the motion of his controls, the helicopter response, and the flight condition. During the experimental mode, he is flying "hands-on". He can immediately take over the command by pressing a button or by overriding the control forces. Then, the core system switches to the "safety pilot" mode which is still in the Fly-by-Light mode. Due to the mechanical control system feedback, the control positions of the safety pilot are always in the correct position. To evaluate and prove that the safety pilot is able to react fast enough to critical situations, a major part of the FHS flight test program was used to generate both single axis and multiple axis runaways in the experimental computer. It was demonstrated that (1) the limiters in the core system computer are able to decelerate the control inputs, (2) the safety pilot is able to immediately obtain control, and (3) the safety pilot is able to stabilize the helicopter without difficulty and without significantly losing altitude.

The safety pilot is responsible for the total flight, including the intervals where the evaluation pilot is in command. Due to this responsibility and the specific safety task for the FHS, the pilot must have a test pilot qualification and FHS flying experience. Therefore, the safety pilot will, in general, be provided by DLR, independent from the individual user of the helicopter.

Switching between Operational Modes

The appropriate modes are selected by using the core system control unit and switches on the safety pilot and evaluation pilot controls. The control unit for the core system (see Fig. 10.16) is located on the center console between the two pilots and can be observed by all crew members. It provides switches to select a mode, to start test routines, and to test and reactivate disconnected data channels. Lamps inform on the actual mode, switching status, errors, and warnings. A change of the control mode is announced and confirmed by an additional acoustic signal.

A new control mode is selected on the core system control unit or by switches on the pilot controls. According to the complexity of the three control modes they are ordered from "low" to "high", which means from "safety pilot" to "evaluation pilot direct" and to "evaluation pilot experimental". The respective conditions for switching are outlined in Fig. 10.17.

For the transition to higher modes (for example, from safety pilot mode to evaluation pilot mode) the new mode is first pre-selected and the evaluation pilot controls are synchronized with the current actuator position or the position obtained from the actual model in the experimental computer. The evaluation pilot controls are driven by the trim motors. During this process, lights flash on the core system control unit for pilot information. A continuous light confirms successful synchronization. Then, the actual mode change is activated by the pilot in command by pressing a button on his collective lever. Due to the synchronization and an additional fading function transition errors during mode change are avoided.

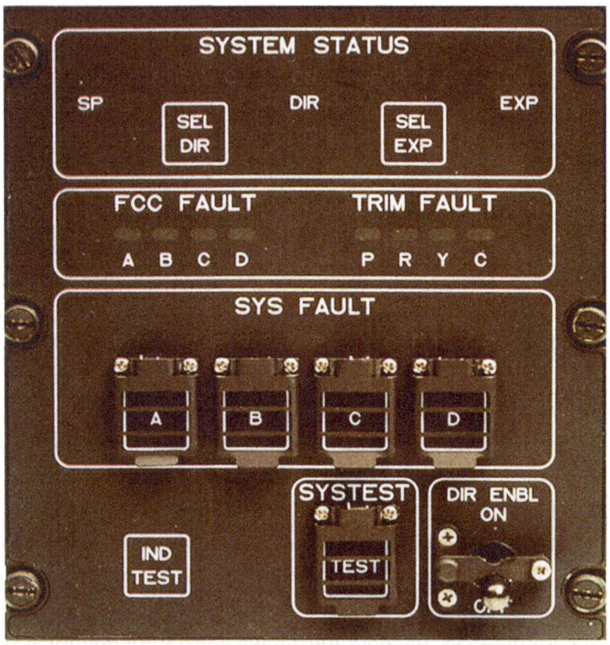

Fig. 10.16 Core system control unit

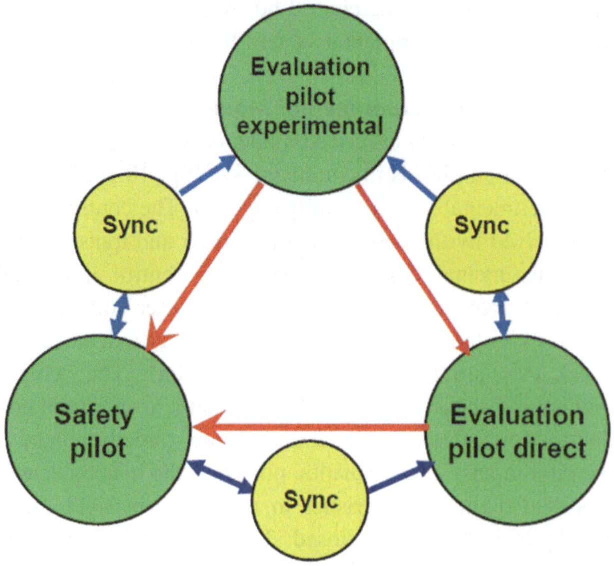

Fig. 10.17 Conditions for configuration changes

Switching to "lower" modes (for example, from experimental mode to safety pilot mode) does not require synchronization as the controls are already in the right position. The desired mode is immediately active. This fact is essential as it allows the safety pilot to take over control of the helicopter without delay by either pressing a button or by overriding the control forces.

10.3.4 Technical Details

More detailed information on the core system computer, the smart actuators, and optical data transfer are provided hereafter.

Core System Computer

The core system computer (see Fig. 10.18) is the heart of the FHS control system. It initiates control mode changes, generates the command signals for the smart actuators, and performs most safety functions. To provide the required safety, the computer layout is also based on the concept of redundancy and dissimilarity. The core system computer consists of four functionally identical lanes. The hardware is housed in two segregated boxes with its own cooling system. They are installed at different locations beneath the cockpit floor. To avoid system inherent failures, dissimilarly is applied for both software and hardware. Each box contains two dissimilar hardware lanes, with one lane based on a

Fig. 10.18 Core system computer

microcontroller and the other one on a signal processor, built by different manufacturers.

The core system software was developed following the rules for Level "A" functions, according to RTCA/DO-178B and ARP 4754. System requirements were translated into two dissimilar software requirements. Software design and verification were performed by two different teams, each team designing the software for both hardware variants against one of the two software design documents. This leads to four dissimilar sets of software. All software was written in the C language.

All lanes run asynchronously with a cycle time of 2 ms. To detect any abnormalities each lane has a number of continuous tests and watchdog timers installed. An optical cross-communication between the lanes exchanges mode switching information. Figure 10.19 shows a diagram of the signal flow within the core system computer. The signal path starts at the LVDT control position sensors for the two pilots. The signals are demodulated and A/D converted, fed to the

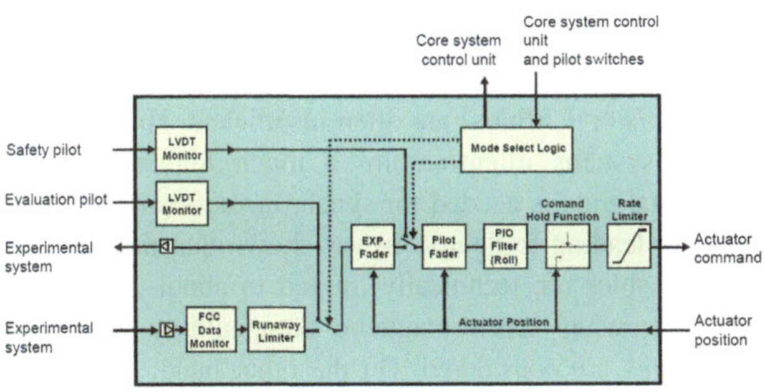

Fig. 10.19 Signal flow in core system computer

LVDT monitor, and checked. The evaluation pilot control positions are passed to the experimental flight control computer, where they may serve as an input to the control law. In the "evaluation pilot experimental" mode, control input signals from the experimental system are sent back to the core system. Because of the low reliability of the (simplex) experimental system, these signals can be wrong. Therefore they are checked by the data monitor for parity (parity bit), validity, and update. In addition, the signals are passed through a runaway limiter to prevent faulty input signals from structurally damaging the helicopter. The algorithm used by the runaway limiter restricts the actuator rate for large and fast signals, but not for slow or for fast short signals. As such, the runaway limiter provides the largest possible flight envelope protection without endangering the aircraft. The definition of the limiter values is based on simulation results, existing flight data, and data from specifically conducted flight tests with the FHS helicopter. The evaluation uses the relationship between maximum control actuator speeds and amplitude, and duration of the control input. Three sets of limiters with different restriction levels are currently defined. The most restrictive limiter permits experimental mode operations throughout the flight envelope. The other two limiters are less tight but have altitude and speed restrictions. Flights without a runaway limiter would be only allowed with a safer experimental system.

After the runaway limiters, the input signals pass through automatic fading functions, PIO filter, and rate limiters before they are sent to the hydraulic smart actuators. When the actuator speed is limited there is a slight risk for a PIO (pilot induced oscillation) tendency in the roll axis. Therefore, a PIO filter reduces the phase shift and eliminates this risk. At the end of the control path, a rate limiter restricts the maximum speed of the actuator output to avoid pressure drops in the hydraulic system. Finally, the core system computer also monitors and controls the evaluation pilot's trim system.

Hydraulic System—Smart Actuators

The FHS has four identical smart actuators: three main rotor actuators mounted on the cabin ceiling below the main rotor and one tail rotor actuator in the vertical fin to control the Fenestron®. Figure 10.20 shows the three actuators for the main rotor (longitudinal, lateral and collective control). The upper part (black housing) contains the electronic components and the actuator software. The lower part contains the electro-hydraulic components with electrical rotary torque motors, control valves, hydraulic cylinders, and the actuator shaft. The mechanical linkage seen in front of the figure is connected to the flexball cables of the mechanical control system. According to the actual control modes it switches to the corresponding function of the flexball cables: "safety pilot": the flexball cables have no function, "evaluation

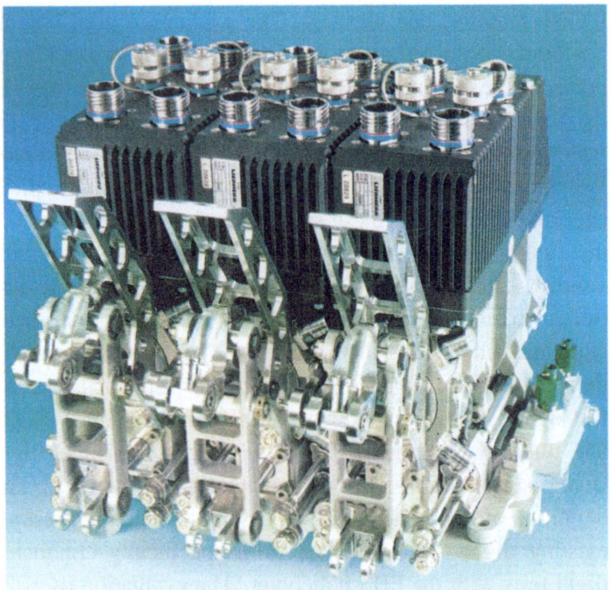

Fig. 10.20 Main rotor smart actuators

pilot": the flexball cable moves the controls of the safety pilot and "safety pilot mechanical controls": the flexball cables connect the safety pilot controls directly to the control rods of the hydraulic actuators.

A functional schematic of the smart actuator is given in Fig. 10.21. The requirements for the actuator control electronics and the software development is practically identical to the design of the core system computer as described above. The quadruplex hardware is dissimilar and the software was written by two different teams. But the components are installed in a single common housing. The actuator electronics receive the set point of the control inputs via the Fly-by-Light connection from the core system computer. First, voting and consolidation are performed. For each channel, the redundant signals are compared, eliminating data failures. Each actuator data lane drives one coil of the quadruplex electrical rotary torque motor. The torque motor is mounted on a single control valve shaft, which controls

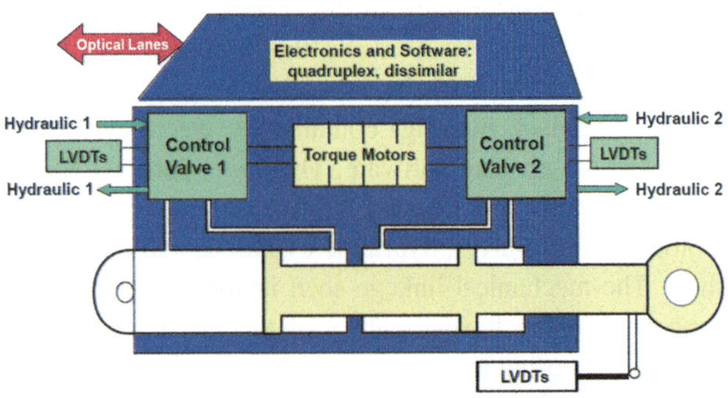

Fig. 10.21 Structure of the electro hydraulic smart actuator

both valves. The actuator is controlled digitally with a two level cascade loop controller: the outer loop is controlling the actuator position and the inner loop is controlling the direct drive valve position, which is proportional to the actuator rate. The control valve position commands, as well as the measured control valve position signals, are consolidated across all channels to avoid force fighting on the control valve shaft. By limiting the actuator position error before consolidation, undetected hardware or software failures in a single channel can be compensated by the remaining healthy channels. Control of the outer loop is performed with a cycle time of 2 ms; the inner loop control has a cycle time of 400 μsec.

The smart hydraulic actuator was specially designed for the FHS helicopter. It mainly consists of a tandem cylinder assembly driven by a quadruplex direct drive valve assembly. It is controlled by the quadruplex actuator control electronics. The FHS has two independent and segregated hydraulic systems. Each hydraulic system is connected to one of the two control valves and supplies one camber of the tandem cylinder. The motion of the piston rod is measured by four LVDTs and sent to the actuator electronic as feedback information. The signals are also sent to the core system computer and are available in the experimental computer.

The smart actuator assembly is one compact unit. The redundancy concept permits a malfunction of one hydraulic system and simultaneously the loss of two electrical lanes without a major performance deficit. The performance of the smart actuators is comparable to that of the standard EC 135 main rotor actuators.

Optical Data Transfer—Fly-by-Light

A triplex redundant optical data transfer was already installed for the tail rotor control in the Bo 105 ATTHeS (see Sect. 8.3.2). For safety reasons, the standard mechanical control was not removed. Results and experience from ATTHeS, and also those from similar programs in France and the US, revealed the high potential of electronic data transmission. But they also showed the deficits of certain Fly-by-Wire systems that still hinder an increasing industrial application, namely data transfer rate, weight, and immunity to electromagnetic interference.

Data transfer rate: Data transfer rates of actually used data bus standards like ARINC429 (100 kbit/s) or MIL-STD-1553 (100 kbit/s or 1 Mbit/s) are often insufficient. High dynamic control systems often require a much higher data rate. Additional time is needed for synchronization and bus management leading to delays. It was shown that data rates for copper cables are technically limited to about 2 Mbit/s, in particular when electromagnetic compatibility (EMC) requirements are considered. On the other hand, optical data transmission offers significantly higher rates.

Weight: Usually most flight computers are installed in the center of the aircraft in a common housing. From here thick, long, and heavy cable harnesses are distributed to sensors, instruments, actuators, etc. The use of individual decentralized computers, smart devices (like smart actuators), and optical cables can lead to a considerable reduction in weight and needed space.

Electromagnetic Compatibility (*EMC*): Helicopters operating at low altitudes, close to the ground, and close to ships are often within an electromagnetic field with an intensity of more than 200 V/m, with an increasing tendency in the future. To protect a full authority Fly-by-Wire control system with a failure probability of 10^{-9} per flight hour from electromagnetic interferences requires an enormous and expensive effort. Fiber optic cables are immune to electromagnetic interference which is another major benefit of Fly-by-Light solutions.

The present research helicopters with electronic control systems still keep their original mechanical control system for safety reasons. The experimental control system was switched on whenever it was required. Consequently, there was no need to develop the new technology to fulfill the high safety standards for a full authority control system. However, to obtain acceptance and confidence, the electronic control system must prove its reliability and usability as a stand-alone system. Here, the FHS helicopter played a successful role as a technology demonstrator with a primary full authority Fly-by-Light control system [9–11]. The experience gained is a sound basis for future helicopter flight control developments.

10.3.5 Ground Facilities

The FHS system also includes a ground-based system simulator and a mobile data/telemetry ground station to support the flight tests. The ground station consists of two modules, the telemetry station and the data evaluation station. They are installed in two containers, which can be transported to the actual flight testing site to allow FHS operation, including ground support at user sites or at air fields. The telemetry station has an automatic aircraft tracking antenna with a video camera and communication equipment. PCM data, sent by the FHS, are received, recorded, and transferred to the data evaluation station via Ethernet.

The data evaluation station offers work places for three engineers. Each place is equipped with a PC based data station to allow real-time data monitoring by quick-look or appropriate software tools during the flight tests. Communication with the helicopter flight test engineer and evaluation pilot is conducted by the responsible test engineer on the ground. Based on the preliminary data checks he decides whether a test was successful or has to be modified and repeated. Data provided by the telemetry link can be recorded in the ground station. But for a more detailed evaluation, the onboard recorded data will usually be preferred. After landing, the data from the disk is transferred to a computer in the container to allow the full range of project-oriented off-line evaluation. In addition, the PCs can be used to develop and modify the evaluation software. Thus, two major objectives in the FHS flight test data concept were fulfilled. Firstly, the required real-time information to control the tests is provided to the user during the flight tests. And secondly, at the end of the flight, he has access to both his own and DLR developed software tools to conduct a detailed data analysis and evaluation.

The ground-based system simulator is primarily designed as a hardware and software-in-the-loop test facility for the FHS. It replicates the flying environment of the FHS with a real cockpit. It is a fixed base simulator without motion and with a large field-of-view visual system (see Fig. 10.22) [12]. Pilots are provided with a cockpit that is very similar to the one in the FHS. It includes side-by-side seating for the safety pilot and the evaluation pilot and offers the same displays, control units, and pilot controls. All functions of the core system computer are represented including switching between the operational modes. The EC 135 helicopter dynamics, the core system computer, and some sensors are simulated. Here, an emphasis was placed on a precise mathematical model of EC 135 that realistically represents the helicopter dynamics. A hardware duplicate of the actual complete experimental system is installed in the simulator. It also serves as a spare unit for the helicopter, if needed. In addition, further hardware components can be connected, that is, from external users. Before any new hardware or software is installed in the helicopter, it is first tested in the ground-based simulator. The simulator is independent of the

Fig. 10.22 FHS ground-based simulator: approach to DLR research campus Braunschweig

FHS, so it can also be used when the helicopter is deployed in flight tests. It offers a perfect test environment through all phases of a flight test program: (1) development, test, and preparation environment for engineers, (2) tests and verification of new hardware and software components before implementation in the helicopter and before flight and (3) pre-flight training and briefing of the crew, in particular when new pilots are involved.

10.4 FHS Research Programs

10.4.1 Introduction

During the first decade (2003–2012) of operation, the FHS flew 960 h for different programs. In general, the research vehicles usually need larger periods on the ground to prepare new flight tests and to implement and check required modifications or additional components. Therefore, the flight time of almost 100 h per year demonstrates the highly effective FHS utilization. Some earlier DLR projects requiring in-flight simulation, which were discontinued after the Bo 105 ATTHeS accident in 1995, could now be pursued with the availability of FHS.

The global activities are coordinated by the FHS user committee that includes representatives from German ministries, Eurocopter Germany, and DLR. The individual projects are planned, scheduled, and coordinated by the FHS Board, consisting of the involved DLR institutes and the project leaders of the experiments. For the future FHS applications, some standards were defined to specify the interactions between different design approaches, test procedures, and implementation in the helicopter. Here, requirements and interests of both DLR and external users were taken into account. As an example, Fig. 10.23 outlines the steps from the conceptual design of a control system for the realization and evaluation in flight. Interfaces for software and hardware were reviewed and extended to allow an easy and flexible access for external users, who may also provide their own hardware and software.

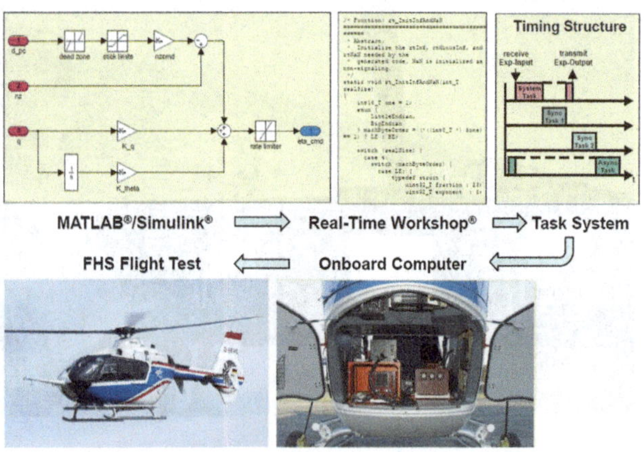

Fig. 10.23 FHS flight control system design chain

The development of nonlinear, generic mathematical helicopter models for simulator applications and flight test preparations was already started during the FHS design phase. The in-flight simulation concept is based on the "model following control system" approach. It requires high fidelity state space models, which were determined by system identification techniques. For this, a comprehensive flight test program from hover to maximum speed was conducted to gather the required data. Classical mathematical models describe the motion of a rigid body in equations for forces and moments for the three axes: longitudinal, lateral and vertical. Such six-degrees-of-freedom models can only represent the low-frequency range of helicopter flight dynamics up to about 10 rad/sec. They neglect the effect of the rotor dynamics. Consequently, the model calculates an immediate linear or angular acceleration response due to a control input. In reality, however, the first reaction is the main rotor tilt due to stick inputs. Then, body accelerations build up with a delay, similar to a second order system response. However, most control laws rely on a correct initial response. Adding an equivalent time delay for the model response is only a very rough approximation. This is why six-degrees-of-freedom helicopter models are often not appropriate for the intended purpose. Therefore, the FHS mathematical model was extended by including rotor degrees of freedom using an implicit formulation for blade flapping and a parametric formulation for the blade regressive lead-lag motion. Through such an extension the model response agreed with flight data for a frequency range of up to about 30 rad/sec, which is adequate for the control system design and application. For a better fit of the vertical response, an implicit formulation of the dynamic inflow (describing the inertia effects of the rotor induced airflow) completed the modeled states. Figure 10.24 demonstrates the quality of different model complexities compared to flight test data. The frequency responses for the vertical acceleration due to collective control inputs are presented for models without and with dynamic inflow effects [13].

Calm atmospheric conditions occur only infrequently. Usually, gusts and winds are encountered during flight testing. Therefore, an emphasis was placed on the development of empirical turbulence models, which can be used in both ground-based and in-flight simulations, with the aim of giving the pilot a realistic feeling of flying in real turbulence for hover and low speed. Additionally, these models are used to present deterministic disturbances for the control system design. The models were derived from flight test data collected under different turbulence conditions, which were recorded by anemometers at the test location. A predicted response of the helicopter due to the pilot stabilization inputs

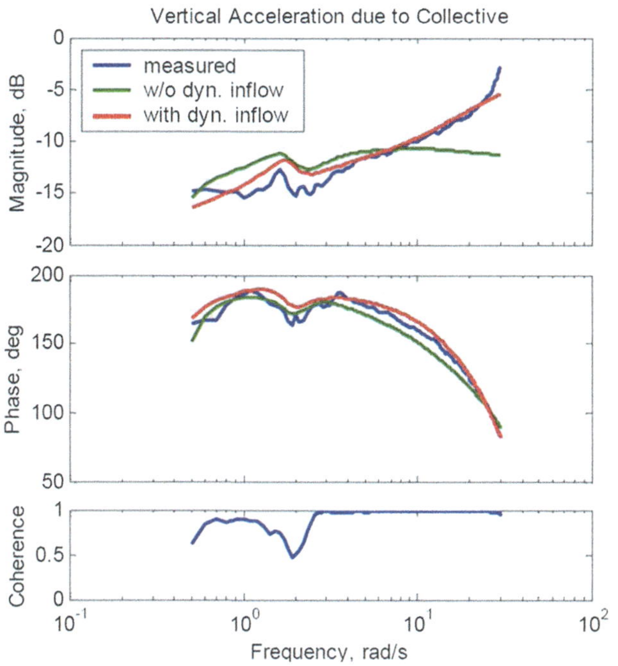

Fig. 10.24 Improved mathematical model accuracy through dynamic inflow modeling

Fig. 10.25 FHS and ACT-IME crew

was subtracted from the actually measured one. The remaining random response signals were converted to equivalent control input signals. During a flight in a calm air or piloting a ground-based simulator, these equivalent inputs can be added to the actual pilot control inputs, giving the pilot the impression of flying in turbulent air. Hence such models are denoted as "*control equivalent turbulence input models*" [14].

Helicopters are often used to transport larger payloads to remote locations. The loads are either attached to a load hook by a rope or a sling system beneath the aircraft, or they are carried by a hook or a winch on the side of the helicopter. However, the load can reveal an uncontrollable dynamic behavior and can reduce the helicopter's overall stability. The interaction between aircraft and load depends on various factors like weight, shape of the load, airspeed, and rope length. The pilot must react quickly to the motion of the sling load in order to keep the entire helicopter/sling load system stable. It can lead to dangerous situations, where sometimes the load has to be dropped to avoid an accident. To help the pilot maintain control over the helicopter, a flight director display was developed. It indicated the required control inputs to effectively damp the load pendulum motion and to allow maneuvering without exciting oscillatory load modes. The display was successfully tested in flights. Based on the experience with the flight director, the development of an automatic control system for load carrying and positioning was first started in ground simulations and then consequently prepared for flight tests. Two different alternatives for the implementation of a load stabilization algorithm were evaluated: (1) as an add-on to classical stability augmentation systems or autopilots with limited authority and (2) as a fully integrated component, interacting with the aircraft control system (see Sect. 10.4.5) [15, 16].

In 2004, the first demonstration of a successful FHS test program by an external user was the comprehensive flight test program ACT-IME (Active Control Technology to Improve Mission Effectiveness). It was conducted by Eurocopter France. Advanced mission adapted control strategies, developed by Eurocopter, were evaluated. The complete program including software development, implementation in the FHS ground-based simulator, and flight test and evaluation, was fully under the control and responsibility of the external user. Interface definitions and implementation support were provided by DLR so far as needed. Essentially, it was demonstrated that an external user can independently and under his own responsibility conduct tests with the FHS, without sharing any information, recorded data, evaluations or results with DLR. Figure 10.25 shows the joint flight test crew after the last flight.

DLR also continued some research programs with FHS that were started with the Bo 105 ATTHeS, for example, techniques for variable flying qualities, control system development, and the test pilot and flight test engineer training. Typical examples of such applications are presented in some details in the following, namely (1) control system, (2) active inceptors (sidesticks), and (3) pilot assistance.

10.4.2 Model Following Controller

Based on the experience gained from the Bo 105 ATTHeS testbed, the main emphasis was placed on the design and

Fig. 10.26 Principle of in-flight simulation

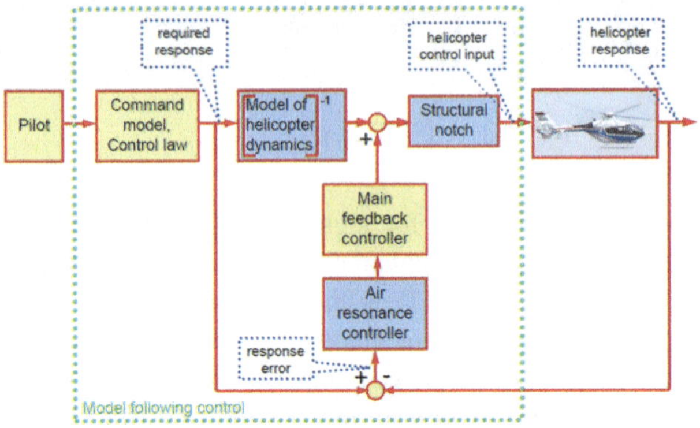

Fig. 10.27 Block diagram of FHS model following control system

optimization of the model following control system for the in-flight simulation (see Fig. 10.26). It gives the FHS the potential to change its inherent EC 135 dynamic flight characteristics. The principle approach for the control system is shown in Fig. 10.27 [17]. Here, the key element is the "model of helicopter dynamics" representing a mathematical model for the FHS dynamics. If the mathematical model exactly describes the helicopter behavior, the inverted model neutralizes the original EC 135 dynamics. The pilot flies the command model as defined in the forward loop. Deficiencies due to model inaccuracies and external disturbances are corrected by a feedback loop. A number of parameterized command models are available in the experimental system computer. They can easily be retrieved during flight by the evaluation pilot or the flight test engineer. In comparison to helicopters with articulated rotors, helicopters with a bearingless main rotor like the EC 135 are more sensitive to the so-called air resonance phenomenon. Principally, it is a coupling effect between the lead-lag motion of the main rotor blades and the body modes. For the EC 135, it can occur in flight, when the regressive lag mode couples with the fuselage roll motion mode. It is noticed by oscillations in the roll motion. To avoid resonance problems, in particular with higher feedback gains, an air resonance controller was added to the feedback loop (see Fig. 10.27) [18].

According to the development contract, the FHS had some constraints in the flight envelope when it was delivered

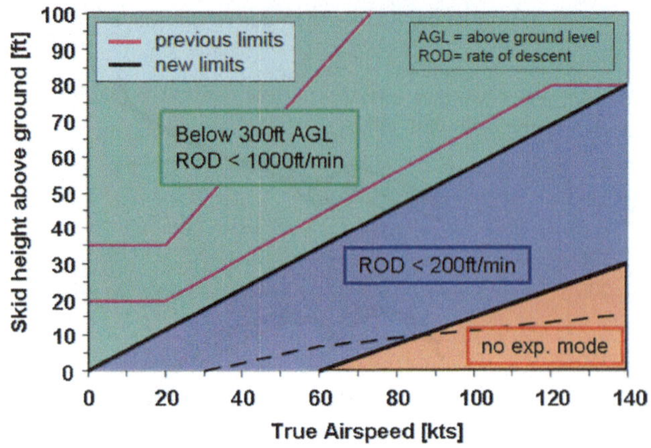

Fig. 10.28 FHS flight envelope for flight in experimental configuration

in 2002. Flights in the "evaluation pilot experimental" mode were not allowed at low heights (below 20 feet over ground) and at low speed. Extensive tests were conducted to document the time needed for the safety pilot to gain control after full control inputs (e.g. from the evaluation pilot or external perturbation). Based on the measurements from these so-called runaway tests, the certification for the flight envelope was extended (see Fig. 10.28). On May 23, 2008, the first landing of the FHS with an engaged experimental system was performed [19].

10.4.3 Active Inceptors (Sidesticks)

In modern aircraft, a wide selection of information about the vehicle and the flight conditions is available. For an optimal support of the pilot, it is essential to select the information he needs for the actual flight situation and present it to him in a most effective way. It can be considered as an interface between the aircraft, the environment, and the pilot. Two factors play an important role for the pilot in assimilating the information, namely (1) he has a very sensitive feeling of accelerations (this is particularly important as helicopters show strong linear and rotational acceleration responses) and (2) he can immediately react to changes in his field of view (horizon). Both of these actions are performed intuitively and subconsciously, in other words at no extra cost. On the other hand, additional information required by him for controlling the helicopter has to be generated and provided explicitly, which involves additional measures such as hardware and software. This is the field of a new generation of pilot controls. With the classical mechanical inceptors, the pilot is controlling the vertical motion using the collective lever in his left hand. He corrects the yawing motion with the pedals at his feet. Furthermore, the pitch and roll motions are controlled by the right hand with the cyclic stick between the

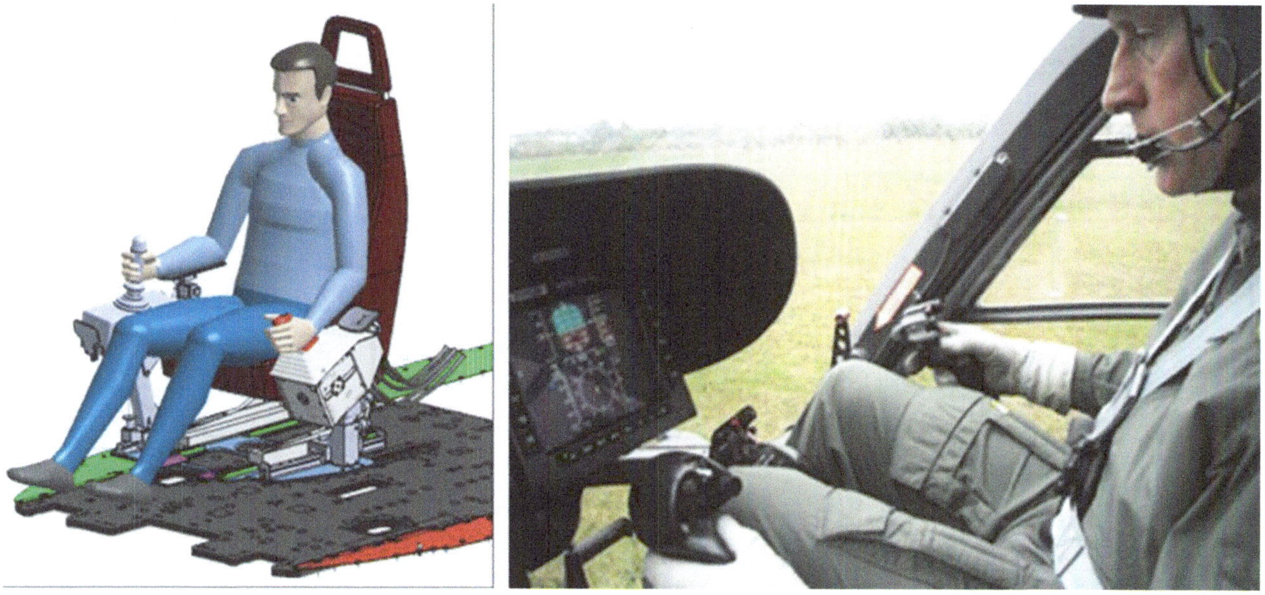

Fig. 10.29 FHS-cockpit with 2 sidesticks, side-by-side configuration

pilot's knees. Helicopter motions are highly coupled. An input in one control generates responses in all axes so that the pilot is always simultaneously working with all his controls, which means with both hands and his feet.

Advanced helicopter flight control systems will feature active inceptors (for example side-sticks), where the forces felt by the pilot are generated by electric motors. They expand the classical, vision-centered human-machine-interface using haptic information. By local variation of their force-feel characteristics, additional information can be transferred to the pilot in an intuitive and effective way. In its role as technology demonstrator, the FHS is ideally suited for the assessment of new inceptors because they can be installed as controls for the evaluation pilot. The upgrade of the FHS and the integration of a sidestick began in 2004 with a feasibility study together with Airbus Helicopter Germany. In 2007, the FHS cyclic control stick was replaced by a 'Goldstick' from Stirling Dynamics Ltd. [20, 21]. After an experimental acceptance study, a second sidestick for the heave axis was obtained from LAT. It was flown successfully in September 2009. The new cyclic stick is now on the right side of the pilot and the classic long pole stick was replaced by a short pole stick, capable of adapting and changing its force profile based on mission requirements. The standard collective lever was removed. The vertical motion (up-down) is now controlled by a short pole active stick at the left side of the pilot. The obvious ergonomic advantage is that the pilot can sit more upright. It also results in several further improvements. In addition, the left-hand sidestick allows the control of two degrees of freedom (forward-backward and left-right) and can optionally be used for yaw control. In flight tests, it was rated as an "intuitive control". Furthermore, this 'side-by-side' configuration improves the ergonomics, the comfort, and the crash safety (see Fig. 10.29).

Active inceptors offer many advantages, among them the ability to adapt the control forces to the actual flight condition and to the status of the flight control system. The sticks can be designed to always provide an optimal control force, leading to improved handling qualities and higher mission effectiveness. The haptic feedback, the so-called 'tactile cueing', is a significant feature of the active sticks. By carefully shaping the profile of the control forces, the pilot can be informed on flight envelope limits, helicopter load limits, or obstacles without having to monitor continuously the limit displays on the cockpit panel. This is essential for flights under visual conditions, where the aerial surveillance is an additional pilot task. So the sidestick helps to improve the situational awareness. When the pilot applies a force to the active inceptor it responds dynamically and the inceptor displacement controls the augmented helicopter. By closing the feedback loop in the inceptor control system, it is possible to indicate the limits mentioned above to the pilot by adding cues or varying force gradients (see Fig. 10.30). An overview of the features of active inceptors and their usage in the feedback-block is outlined in Fig. 10.31. It has to be pointed out that all characteristics and details are freely programmable. They can be adapted to the specific helicopter configuration and actual flight situation. This possibility opens a large variety of solutions and needs criteria for optimization.

For activities pertaining to tactile cueing, so-called demonstrator functions were developed and tested in flight in 2007. These demonstrator functions included load-factor

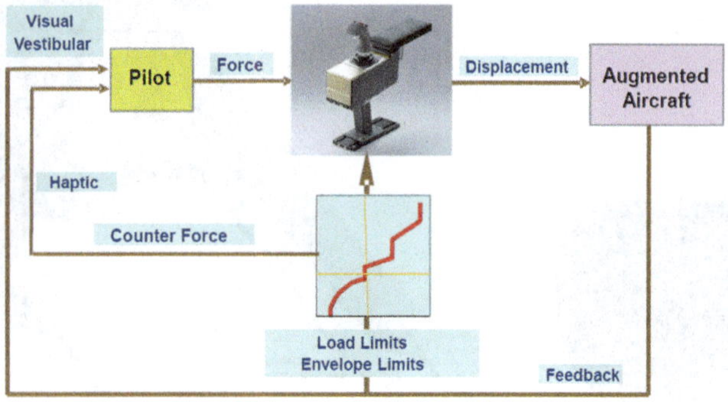

Fig. 10.30 Extended pilot-inceptor-aircraft loop

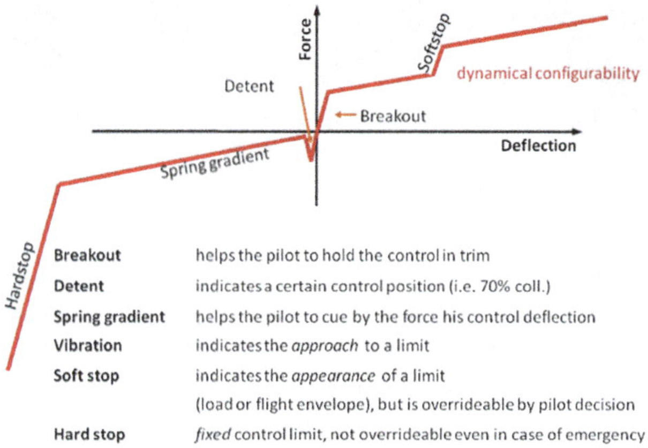

Fig. 10.31 Features of active inceptors (stops, forces diagram)

limitations, mast bending limitations, and tactical guidance to fly a standard 360 degree turn by using a soft stop. In critical situations, soft stops can be overridden by the pilot. A haptic vortex ring state protection (sink rate limitation) developed in cooperation with the ONERA was successfully demonstrated in flight in 2010. As a further application, a torque protection cue was developed in cooperation with Eurocopter and demonstrated in flight. Another activity is related to obstacle avoidance. It supports the pilot flying in obstacle scenery close to the ground. Considering the active inceptor technology as part of the overall active control technology, and using an integrated approach, these functions can be imbedded into more comprehensive pilot assistance systems.

Under the umbrella of the US-German Memorandum of Understanding for Cooperative Research on Helicopter Aeromechanics, a task considering 'Handling Qualities for Actively Controlled Rotorcraft' was formulated (see also Sect. 12.3.3). DLR and the U.S. Army Aeroflightdynamics Directorate performed common and complementary in-flight and ground-based simulator studies. The objective was gaining insight into the influence of the dynamic inceptor parameters (damping and natural frequency) on the handling

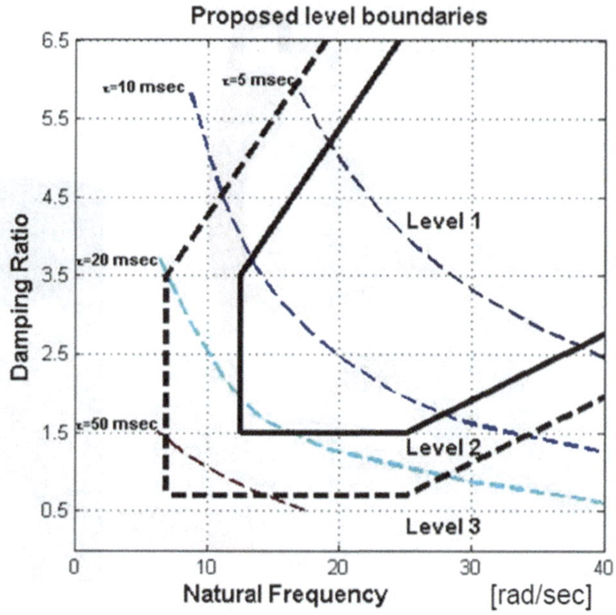

Fig. 10.32 Proposed level boundaries for active inceptors

qualities of helicopters. For the evaluation in flight, the considered mission task elements were "hover" and "forward flight slalom". Objectives were to provide design guidance for rotorcraft with active inceptors and to identify a methodology for integrating inceptor characteristics into the optimization process of the entire system to improve handling qualities. The used test vehicles were the FHS equipped with two active sidesticks and the JUH-60 RASCAL (see Sect. 5.2.2.17).

Several flight test campaigns were performed. The dynamic inceptor parameters (damping and natural frequency) were systematically changed and handling qualities were evaluated to define the requirements for active inceptors [22, 23]. The pilots stated that they preferred (1) short delay between their control inputs and the initial response of the aircraft and (2) high damping to allow a quick and precise control of the stick position without any danger of overshooting. With the general requirement for higher damping values, the first proposal for level boundaries was generated. They are given as bold lines in Fig. 10.32. To show the influence of time delays, selected contour lines are added to the diagram. Level 1 indicates the region of satisfactory and level 2 for acceptable handling qualities.

10.4.4 Pilot Assistance Systems

The objective of helicopter pilot assistance is to support the pilot with suitable technologies to reduce his workload and increase the probability of a successful mission. The challenge in the definition of an assistance system is that it has to

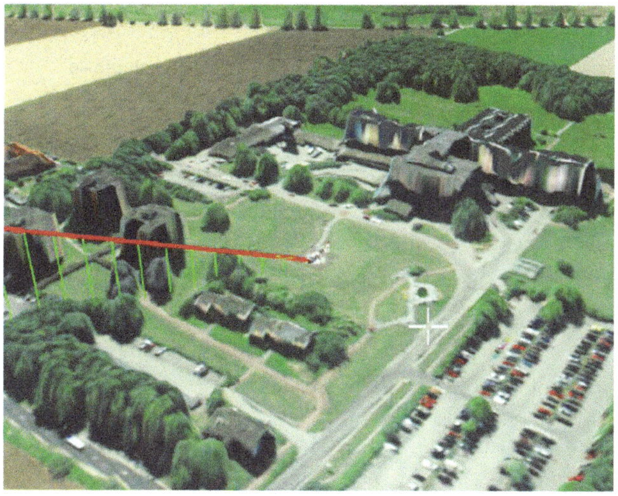

Fig. 10.33 Example of a high resolution stereo image with embedded departure route

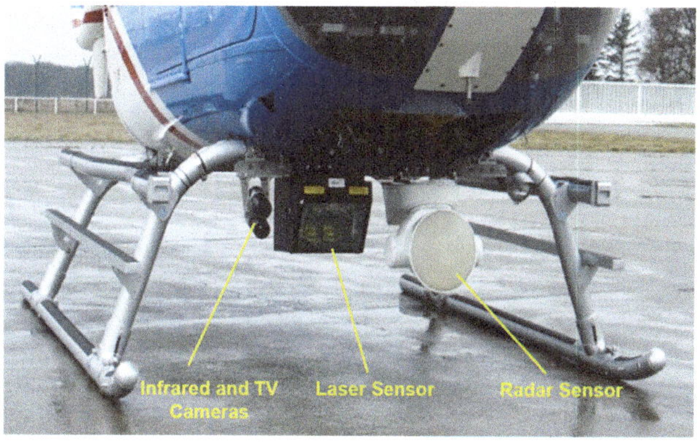

Fig. 10.34 Sensors for the All Flight program

be adapted to (1) the helicopter configuration, (2) the actual flight task, and (3) the pilot capability. In April 2003, a common DLR/ONERA project called PAVE (pilot assistant in the vicinity of helipads) was initiated. It concentrated on automatically and manually flown landing approaches and departures, emergency procedures, as well as noise abatement flight profiles. For increased situation awareness, high-resolution stereo images were integrated into a virtual landscape. An example for a high-resolution stereo image with embedded departure route is given in Fig. 10.33. Various supporting modules were developed including an intuitive planning module for easy flight plan changes and a guidance module for automatic flight modes. A flight director display showed the deviations from the pilot-defined trajectory. Flight testing started in 2006 and the PAVE project was finished in 2007 by a successful demonstration of an automatic flight for an emergency medical services mission [24–26].

A follow-on project was ALLFlight (Assisted Low Level Flight and Landing on Unprepared Landing Sites). It aimed at operating a helicopter under degraded visual environment conditions with optimal handling qualities for the entire flight from start to landing. The required hardware for the tests included a high landing skid equipped with sensors to detect ground contact, a beam for sensor installations, and four external sensors (ladar, radar, TV camera, and infrared camera). Figure 10.34 shows the additional sensors on the FHS. An additional computer was installed as part of the FHS experimental system. It was needed for the extensive navigation task and for the calculation and presentation of maps, terrain, obstacles, and possible landing trajectories. Flight tests began in November 2011. The measured data were used for an online obstacle free trajectory planning. Individual algorithms for the three flight segments start, en-route, and landing were derived. The algorithms consider all helicopter limitations and typical procedures (for example, CAT A start and landing) of piloted operations. To improve pilot acceptance of the automated trajectory planning, 68 pilots from various operators were interviewed for their trajectory planning preferences. As an example, Fig. 10.35 shows a map of obstacles and the terrain profile with suggested landing flight paths.

In addition to infrared and TV data, radar and ladar measurements were obtained (see Fig. 10.34). In principle, the two last sensors provide redundant information. Both are detection and ranging systems, where signals are sent out and their reflections from any objects are received and processed. The more familiar radar is based on electromagnetic waves. It is best suited for the detection of larger objects. Ladar is an optical system and uses laser technology. The main difference is that it operates in higher frequency bands. It has a higher resolution and is able to detect smaller objects like electrical wires. By the so-called "data fusing" procedure, the measurements of both sensors are combined to take advantage of the benefits of the two systems and to give the pilot the best possible image (see Fig. 10.36) [27–29]. In 2012 the displayed fused data in combination with the selectable flight control parameters were tested in flight [29–31].

Another approach to support the pilot when flying in low visibility conditions like fog, brown out and white out, or even in dawn or during the night is based on a helmet-mounted display (HMD). Therefore, the Elbit's JedEye™ helmet mounted display system (see Fig. 10.37) was installed in both the FHS and a ground-based simulator, the DLR Generic Cockpit Simulator GECO with a collimated vision system. The integration of such a helmet in the research helicopter offers the possibility to increase the situation awareness especially under degraded visual condi-

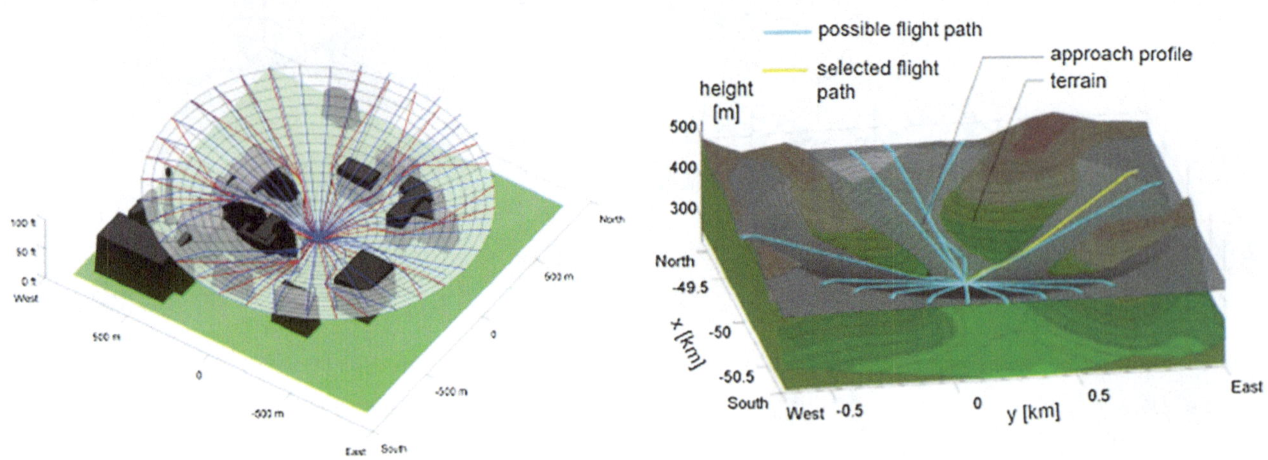

Fig. 10.35 Map of obstacles (*left* buildings, *right* terrain) and possible landing trajectories

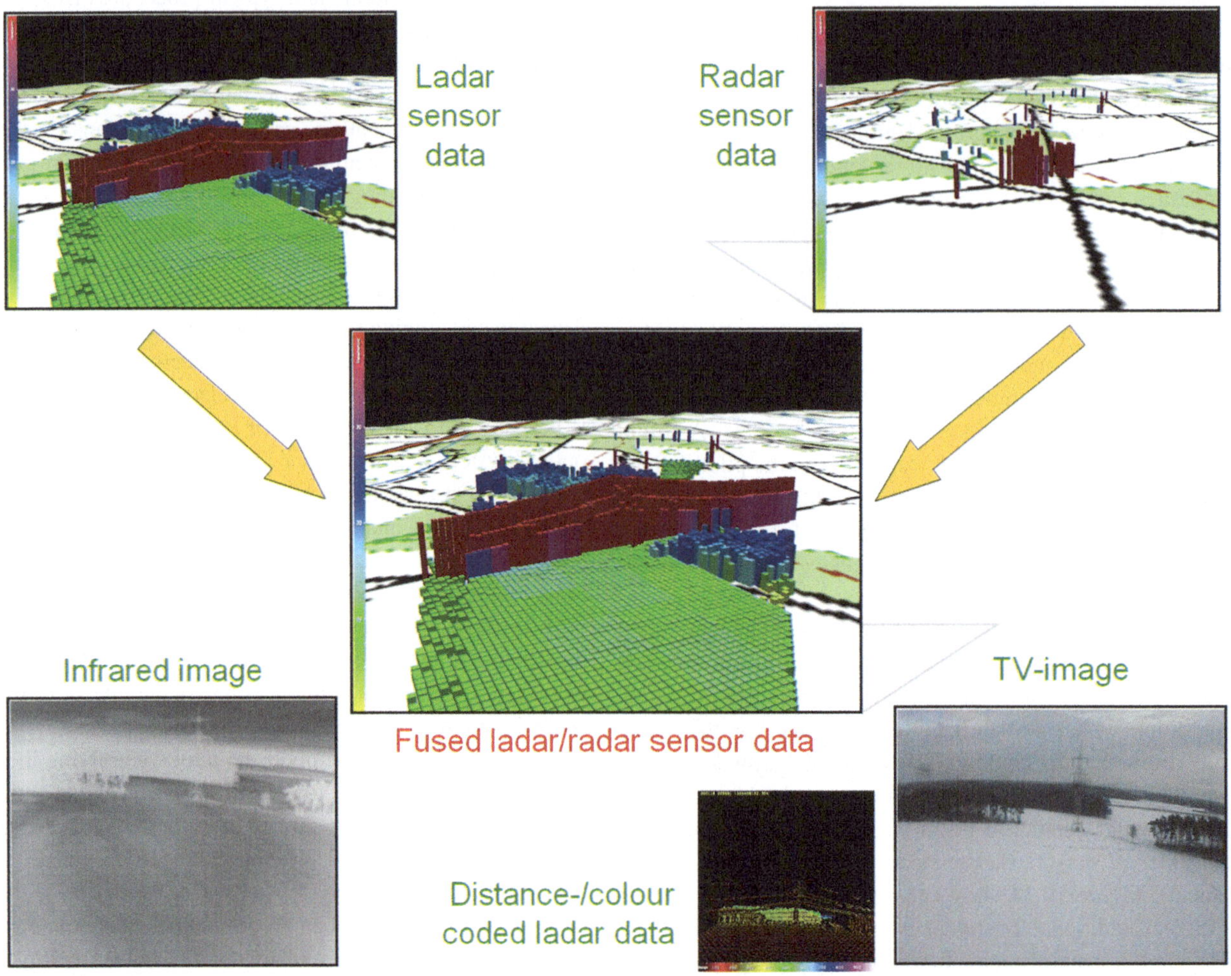

Fig. 10.36 Sensor data fusion for 3D image generation

Fig. 10.37 FHS Evaluation with JedEye™ helmet

tions by displaying mission dependent symbologies. The main focus concentrated on the extraction of relevant information (e.g. obstacles) out of the adequate visual conformal symbology. The presentation on the helmet allowed following predefined 3D trajectories, such as noise abatement flight procedures. This symbology was validated in FHS flight tests [32].

10.4.5 Automatic Stabilization and Positioning of Sling Loads

For demonstrating sling load assistant systems and for reducing the tremendous workload of pilots during sling load transport, the FHS was equipped with a rescue hoist (see Fig. 10.38). Challenges of the rescue hoist derive from the variable cable length and the disturbing rolling moment that is generated by the side installation. Main project objectives were an additional automatic stabilizing and positioning mode connected to a modern automatic flight control system (AFCS). Sling load motions are detected by an infrared camera (see Fig. 10.39). Control algorithms were derived to dampen load oscillations and to support a precise load delivery. This algorithm could become part of an AFCS for advanced utility and transport rotorcraft [33–35].

10.5 Pilot and Test Engineer Training

Following the tragic Bo 105 ATTHeS accident, the pilot training with this helicopter had to be discontinued in 1995 (see Chap. 8). However, due to the highly positive experience with the Bo 105 ATTHeS, the English Empire Test Pilots' School (ETPS) was further interested in the utilization of a helicopter in-flight simulator for pilot and flight test engineer training. Accordingly, ETPS visited DLR Flight Test Facility in Braunschweig during spring 2005 to explore the resumption of these opportunities with the FHS. Once again the ETPS was convinced of the overall set up consisting of the experimental system, data recording, monitoring, and ground-based simulator, as FHS offered flexibility and efficient hands-on training.

The first training campaign for the ETPS with FHS started in autumn 2005. As a part of their thesis work, a pilot and a flight test engineer were allowed to test and evaluate FHS. Because of the complexity of the overall system, this was a challenge for the trainee students, which they successfully absolved. In the following year, FHS was deployed on a regular basis in the EPTS test pilot courses for flying qualities training. 6–8 trainees from ETPS visited Braunschweig for training purposes (see Fig. 10.40). The task comprised of optimizing the flight control laws for a given mission and then to assess the helicopter suitability. A typical mission could be, for example, rescue operation at night under poor visibility conditions. After installation of an active sidestick in the year 2010, it opened up new areas of pilot training for ETPS. As already elaborated, the damping or force-displacement characteristics of the sidestick could be changed easily via the onboard computer. Optimization of control characteristics together with flight control laws was thus part of the training program.

The test campaigns also provided valuable insights into test vehicle and flight control law design. As such it was interesting to note that some teams preferred attitude control, whereas others the rate control for the same task. It turned out that the prior exposure to flying transport or combat helicopters had a significant impact on the pilot ratings. Another insight was that the interaction between the different flight control laws and the force-displacement characteristics of an active control device was quite important. For example, a combination of two components, individually assessed to be good, however, resulted in poor ratings for the overall

Fig. 10.38 Flight test for sling load assistance system with load motion sensor

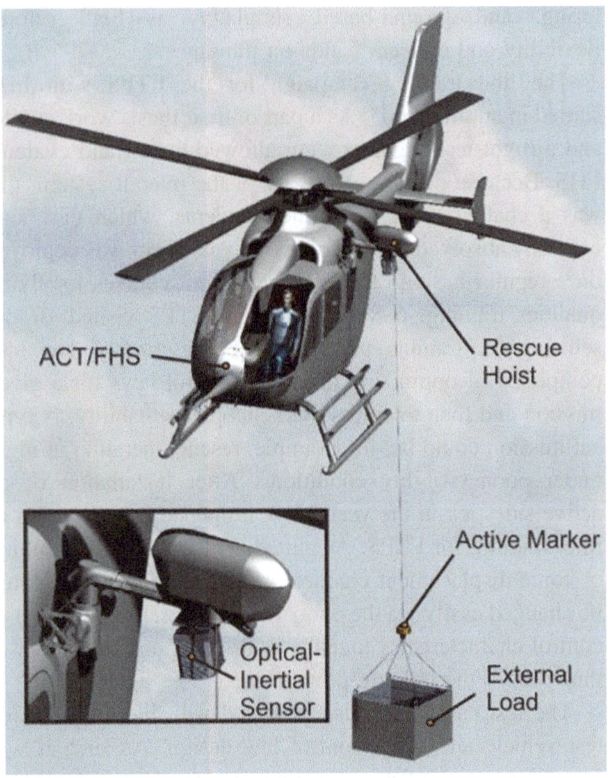

Fig. 10.39 Sensor installation for sling load assistance system

system. Even more interesting was the case when the combined overall system was rated to be good when the individual components were rated poor.

The importance of in-flight simulators for pilot and flight test engineer training was repeatedly confirmed [36, 37]. The French Test Pilot School EPNER also showed interest in its

Fig. 10.40 ETPS trainees in the year 2013 (from *left* test and safety pilot *U. Göhmann* together with *M. Mühlhäuser*, *A. Delaney*, *O. Higgins*, *I. West*, *D. Lee*, *J. Wolfram*, *M. Barnett*, *W. Krebs*, *S. Soffner* and *M. Bernhardt*)

utilization and joint work. Accordingly, an EPNER team visited DLR Braunschweig in December 2013 to explore the possibilities of future utilization of FHS for the training of flight personnel.

References

1. Butter, U., Schimke, D.: Pausder, H.-J., Gollnick, V.: ACT-Demonstrator/Fliegender Hubschrauber Simulator—Projektdefinitionsstudie—Abschlußbericht, DLR IB 111-95/26 and ECD-0045-95, July 1995 (in German)
2. Pausder, H.-J., Butter, U., Gollnick, V.: ACT-Demonstrator/flying Helicopter Demonstrator—A New Demonstrator Tested for Flight Control and Cockpit Technologies. American Helicopter Society 52nd Annual Forum, Washington D.C., USA, June 1996

3. Pausder, H.-J., Butter, U., Gollnick, V.: ACT-Demonstrator/flying helicopter demonstrator—an airborne testbed development project. In: 22nd European Rotorcraft Forum, Brighton, UK, Sept 1996
4. Kaletka, J., Kurscheid, H., Butter, U.: FHS, the new research helicopter ready for service. In: 29th European Rotorcraft Forum, Friedrichshafen, Germany, Sept 2003
5. Pausder, H.-J., Butter, U., Steinmaier, F.: ACT/FHS for the next generation technologies evaluation and demonstration. In: 25th European Rotorcraft Forum, Rome, Italy, Sept 1999
6. Ockier, C., Butter, U.: ACT/FHS—an airborne rotorcraft simulator for technology development and research. In: AIAA Modeling and Simulation Technologies, Denver, USA, AIAA-2000-4575, Aug 2000
7. Butter, U., Pausder, H.-J., Steinmaier, F.: Safety concept of technology demonstrator ACT/FHS. In: American Helicopter Society 56th Annual Forum, Virginia Beach, USA, May 2000
8. Gelhaar, B., Oertel, H., Alvermann, K., Bodenstein, M., Gandert, R., Graeber, S., Schwaneck, H.-P.: FHS—experimental system for flying helicopter simulator put into operation. In: American Helicopter Society 59th Annual Forum, Phoenix, USA, May 2003
9. Steinmaier, F., Bickel, N., Pausder, H.-J.: Survivable fly-by-light control system. In: ACT/FHS, RTO MP-047, Paper 10, April 2000
10. Bickel, N., Steinmaier, F.: Design of a primary fly-by-light control with a dissimilar redundant architecture. In: American Helicopter Society 54th Annual Forum, Washington D.C., USA, May 1998
11. Bickel, N., Butter, U., Hammerlindl, M., Eichberger, W.: Getting a primary fly-by-light system into flight. In: American Helicopter Society 59th Annual Forum, Phoenix, USA, May 2003
12. Hamers, M., Bouwer, G., von Grünhagen, W.: System Simulator for the DLR's in-flight simulator ACT/FHS—software and hardware realization. In: AHS Vertical Lift Aircraft Design Conference, San Francisco, USA, January 2000
13. Hamers, M., von Grünhagen, W.: Advanced modelling approach for ACT/FHS controller development. In: 27th European Rotorcraft Forum, Moscow, Russia, Sept 2001
14. Seher-Weiß, S., von Grünhagen, W.: EC 135 system identification for model following control and turbulence modeling. In: CEAS Conference, Berlin, Sept 2007
15. Hamers, M., Bouwer, G.: Flight director for helicopter with slung load. In: 30th European Rotorcraft Forum, Marseille, France, Sept 2004
16. Hamers, M., Bouwer, G.: Helicopter slung load stabilization using a flight director. In: American Helicopter Society 61st Annual Forum, Gaylord Resort Grapevine, USA, June 2005
17. Hamers, M., Lantzsch, R., Wolfram, J.: First control system evaluation of the research helicopter FHS. In: 33rd European Rotorcraft Forum, Kazan, Russia, Sept 2007
18. Lantzsch, R., Wolfram, J., Hamers, M.: Increasing handling qualities and flight control performance using an air resonance controller. In: American Helicopter Society 64th Annual Forum, Montreal, CA, April 2008
19. Hamers, M., et al.: Extension of the ACT/FHS experimental flight envelope. In: American Helicopter Society 65th Annual Forum, Gaylord Resort Grapevine, TX, USA, May 2009
20. Abildgaard, M., von Grünhagen, W.: Demonstration of an active sidestick in the DLR flying helicopter simulator (FHS). In: 34th European Rotorcraft Forum, Liverpool, UK, Sept 2008
21. von Grünhagen, W., Müllhäuser, M., Abildgaard, M.: Active inceptors in FHS for pilot assistance systems. In: 36th European Rotorcraft Forum, Paris, France, Sept 2010
22. von Grünhagen, W., Schönenberg, T., Lantzsch, R., Lusardi, J., Fischer, H., Lee, D.: Handling qualities studies into the interaction between active sidestick parameters and helicopter response types. In: 38th European Rotorcraft Forum, Amsterdam, The Netherland, Sept 2012
23. von Grünhagen, W., Müllhäuser, M., Höfinger, M., Lusardi, J.: In-flight evaluation of active sidestick parameters with respect to the handling qualities for rate command and attitude command types. In: AHS Rotorcraft Handling Qualities Specialists' Meeting, Huntsville, USA, Feb 2014
24. Lüken, T., Korn, B.: PAVE: assistance system to support pilots for IFR rotorcraft airport operations. In: 25th ICAS Congress, Hamburg, Sept 2006
25. Döhler, H.-U., Lüken, T., Lantzsch, R.: ALLFlight—a full scale enhanced and synthetic vision sensor suite for helicopter applications. In: SPIE Defense, Security + Sensing 2009, Orlando, USA, April 2009
26. Stromaier, T., Lantzsch, R., Greiser, S.: Assisted landing for helicopters in confined areas. In: 36th European Rotorcraft Forum, Paris, France, Sept 2010
27. Lüken, T., Döhler, H.-U., Lantzsch, R.: ALLFlight—fusion sensor information to increase helicopter pilot's situation awareness. In: 36th European Rotorcraft Forum, Paris, France, Sept 2010
28. Lüken, T., Döhler, H.-U., Lantzsch, R.: ALLFlight—fusion sensor suite with "see through" capability for helicopter applications. In: American Helicopter Society 66th Annual Forum, Phoenix, USA, May 2010
29. Lüken, T. et al: ALLFlight—helicopter flight trials under DVE conditions with an AI-130 mmW radar system. In: 37th European Rotorcraft Forum, Gallarate, Italy, Sept 2011
30. Lantzsch, R., et al.: AllFlight—helicopter pilot assistance in all phases of flight. In: 38th European Rotorcraft Forum, Amsterdam, NL, Sept 2012
31. Lantzsch, R., et al.: ALLFlight—a full scale pilot assistance test environment. In: American Helicopter Society 68th Annual Forum, Fort Worth, USA, May 2012
32. Lüken, T., Döhler, H.-U., Schmerwitz, S.: Helmet mounted display supporting helicopter missions during en route flight and landing. In: Degraded visual environments: enhanced, synthetic, and external vision solutions 2016, 9839 (11). SPIE Press. SPIE Defense + Commercial Sensing 2016, 2016, Baltimore, USA, April 2016
33. Kim, H.-M., Nonnenmacher, D., Götz, J., Weber, P., von Hinüber, E., Knedlik, S.: Initial flight tests of an automatic slung load control system for the ACT/FHS. In: 40th European Rotorcraft Forum, Southampton, UK, Sept 2014
34. Nonnenmacher, D. Jones, M.: Handling qualities evaluation of an automatic slung load stabilization system for the ACT/FHS. In: 41th European Rotorcraft Forum, Munich, Germany, Sept 2015
35. Kim, H.-M.: Design of an automatic load positioning system for hoist operations. In: American Helicopter Society, 72st Annual Forum, West Palm Beach, FL, USA, May 2016
36. Lewis, P., et al.: Learning to test. In: Flight International, pp. 39–43, Nov 28–Dec 4 2000
37. Osborne, T.: Training times. In: Aviation week and space technology, pp. 47–48, 2 Dec 2013

Author Biography

Jürgen Kaletka was a research scientist at the Institute of Flight Systems at DLR in Braunschweig (1973–2009). He received his Dipl.-Ing. degree in Aeronautical Engineering from the Technical University of Braunschweig. His main research interest are flight dynamics, in-flight simulation, modeling and system identification (helicopter). From 1987 to 1991 he was the manager of the AGARD FMP Working Group WG 18 Rotorcraft System Identification, from 1979 to 1994 the task manager for system identification in the US-German Memorandum of Understanding (MoU) Helicopter Aero Mechanics. From 1999 to 2003 he was the technical project manager for the development of ACT/FHS (helicopter in-flight simulator based on EC 135).

11 "DLR Project Cancelled"

Peter G. Hamel

11.1 Introduction

The term "Project Cancelled" acquired a special meaning in British aviation history during the postwar period. In a critical documentation (see Fig. 11.1), the British doyen of investigative aviation journalism, *Derek Wood* vividly

P.G. Hamel (✉)
Braunschweig, Germany
e-mail: p.hamel@t-online.de

P.G. Hamel (ed.), *In-Flight Simulators and Fly-by-Wire/Light Demonstrators*,
DOI 10.1007/978-3-319-53997-3_11

Fig. 11.1 Book title "Project Cancelled" in various versions

portrayed how the wrong political decisions ushered the downfall of the once leading British aeronautical industry [1]. This included the termination of the Miles M.52 supersonic project on January 31, 1946, the discontinuation of the Saunders Roe SR.177 fighter aircraft development based on the successful SR.53 in 1957, and the abandonment of the then world's most advanced Fly-by-Wire supersonic interceptor BAC TSR-2 in 1965.

The British were particularly incensed about the fact that the complete know-how of the Miles M.52 project equipped with a thin, straight wing and provided with a sharp leading edge ("Gilette") was made available to the United States for their supersonic project Bell X-1 (see Fig. 11.2). A variety of technical innovations were to be incorporated in the M.52. A key element was an all-moving tail plane ("flying tail"), which became necessary for an effective flight control in the supersonic range due to a large shift of the center of pressure. It differed from the traditional tail design with horizontal stabilizer and hinged elevator. Without this British know-how, the world's first successful supersonic flight of XS-1 on October 16, 1947 may not have become possible so quickly. As justification for the M.52 project discontinuation, *Sir Ben Lockspeiser* quoted the German knowledge about the advantages of swept wing for high-speed flight. A year earlier he had visited the Aeronautical Research Institute (LFA) at Braunschweig-Völkenrode after the collapse of the Third Reich. In the year 1977, when enquired about the root causes of M.52 project termination, *Sir Ben* replied: "*old men forget*" [2].

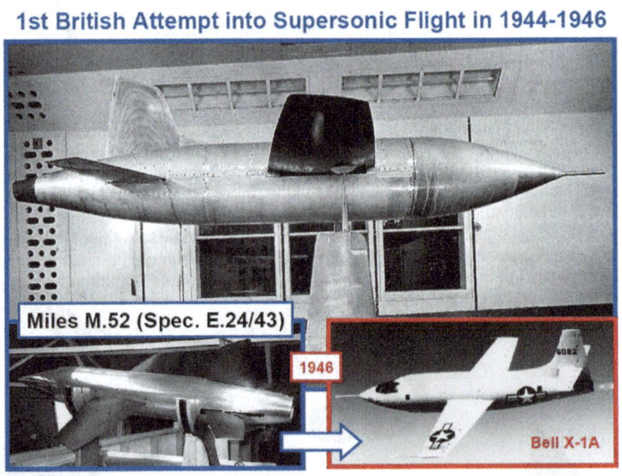

Fig. 11.2 British know-how transfer to USA

The termination of the SR.177 project, shortly before its first flight in April 1958, was also attributed to a blatant misjudgment of future air defense requirements. In a White Paper, defense minister *Duncan Sandys* issued the statement that the English Electric Lightning (see Fig. 11.3) would be the last manned interceptor ("*No more manned aircraft*"). As a result, the SR.177 variants, which were planned for the Canadian and German Air Forces, were also terminated (see Fig. 11.4).

Also the BAC TSR-2 Fly-by-Wire project, at that time technologically most advanced in the Western world, was a victim of political conflicts (see Fig. 11.5). Despite the

Fig. 11.3 English electric lightning

Fig. 11.4 SR.177—planned for German Air Force

Fig. 11.5 BAC TSR.2 first flight takeoff on September 27, 1964

ongoing successful flight testing (first supersonic flight on February 22, 1965), the project was discontinued by the newly elected Labor government in April 1965. The defense minister of the Labor government *Dennis Healy* was quoted as follow: "*The only way to make money in the aircraft industry is never to produce an aircraft* [3]".[1]

It resulted in mass layoffs of highly qualified engineers who migrated to other industries or emigrated to North America. The cost-saving alternative planned by the Labor government, namely the purchase of the US-American variable geometry swing wing aircraft General Dynamics F-111K, resulted in another disaster leading to contract termination due to serious technical deficiencies.

As a final result, a McDonnell F-4 Phantom-version from the United States was selected with British engines (RR Spey), which was characterized by particularly high maintenance efforts as "*hodgepodge*" aircraft (patchwork aircraft).

Like the British industry, the German aeronautical industry too dealt with projects which were either not realized or did not reach the flight test stage in the 60s and 70s, because the military-political scenarios had changed. Accordingly, all of the vertical takeoff demonstrator programs VJ-101, Do 31 and VAK 191 (see Sect. 6.1.3.1) were abandoned, after a thorough flight test phase in cooperation with the United States.

These events on the British side were dramatic and the consequent dependence of England on the United States in aviation policy matters was tragic. Comparably interesting are the unrealized project-initiatives of DLR in the field of Fly-by-Wire technologies and in-flight simulation, which are elaborated in the following Sects. 11.2 through 11.5.

11.2 DLR/Dornier AlphaJet CASTOR (1984)

In March 1984, together with its partners Dornier (*H. Max, H. Wünnenberg*), BWB AFB LG IV (*R. Rosenberg*) of the Federal Office of Defense Technology and Procurement (BWB), WTD-61, the German Air Force Flight Test Center, and the Institute of Flight Mechanics DFVLR compiled a proposal for the development of an in-flight simulator for combat aircraft (see Fig. 11.6). Based on the Alpha Jet prototype P 03, the Fly-by-Wire test aircraft—abbreviated CASTOR (*Combat Aircraft Simulator for Training, Operations, and Research*)—was to serve the purposes of pilot training, development and integration of new flight control and display technologies, and the assessment of flying qualities.

The partners were convinced at that time that the development of a digital-electrical flight control system for a variable stability aircraft and the conversion of an appropriate test vehicle to an in-flight simulator would be an important cornerstone for the future collaborative work. It would have been at the disposal of German Air Force, aeronautical industry and the German Aerospace Research Establishment (DFVLR). As there was no comparable airborne combat aircraft simulator in Europe, the interest of NATO partners was foreseen.

[1]See [2], pp. 219–229.

Fig. 11.6 Alpha Jet CASTOR scope proposal

The various anticipated tasks were grouped into two elements as follows: (1) systems engineering investigations to reduce the developmental risks in new flight control and guidance concepts and (2) pilot familiarization and training. The experience at Dornier and BWB LG IV, gained during the development and testing phase of direct force controller (DFC), provided a sound knowledge base (see Sect. 6.3.5). Furthermore, all necessary knowledge pertaining to systems engineering control system design, software development as well as experimental and evaluation procedures was available at DFLVR based on decades of experience in the field of in-flight simulation. Also, important know-how related to electrical flight controls and safety concepts (redundancy requirements), acquired by MBB as a part of the F-104 CCV program, could have been directly utilized in the CASTOR program (see Sect. 6.3.4).

Accordingly, the DFC system integration and evaluation were to be carried out in the first phase of the Alpha Jet P03 development project. Integration of special equipment for the in-flight simulation was planned in the second phase. The total cost for this last stage was estimated to be about 25 million DM [4].

While the individual DFC components could be implemented and tested in the first phase, the overall project had to be abandoned due to inadequate financial resources (see Sect. 6.3.5).

An interesting aspect in this context was another option of Alpha Jet utilization for civilian purposes. As a part of the European Hermes Spaceplane project, besides the Hermes Training Aircraft (HTA) based on a Dassault Falcon 900 or Grumman Gulfstream IV, an Alpha Jet with minor modifications was also contemplated as a Trajectory Training Aircraft (TTA) for "fitness training" of the astronaut-pilots (see Sect. 11.5).

11.3 DLR/MBB BK 117 HESTOR (1984–1986)

Envisioning future military and civilian rotorcraft to be fitted with Fly-by-Wire flight control systems on a regular basis, there was an increasing demand for the design and testing of rotorcraft control augmentation systems. For this purpose, supported by MBB UD (today: Airbus Helicopters, Germany), the DFVLR (today: DLR) Institute of Flight Mechanics (today: Flight Systems) conceptualized an in-flight simulator HESTOR (*Helicopter Simulator for Technology, Research and Operations*) based on a BK 117 helicopter (today: Eurocopter/Airbus Helicopters EC145/H145 in different variants). Accordingly, a proposal was put forward jointly with MBB (see Fig. 11.7) [5].

The research objectives of the HESTOR project were to obtain, under real operational conditions, reliable and generally valid evidence about the future helicopter flying qualities and system characteristics for (1) new and extended flight missions and (2) integration of new key technologies such as intelligent sensors, computer and actuation systems, and advanced displays and control devices (sidestick). At DFVLR, an in-flight simulator Bo 105 ATTHeS was already in operation for basic research purposes (see Chap. 8). The experience and knowledge gained with this testbed, particularly in the field of Fly-by-Wire/Light flight control technologies were to be utilized in the HESTOR project. ATTHeS was hitherto the only European helicopter in-flight simulator and this situation was to be extended through the acquisition of HESTOR.

BK 117 was one of the most modern helicopters with (1) an advanced hingeless rotor system with exceptionally

Fig. 11.7 BK 117 HESTOR framework proposal

good control response, (2) high power reserves for testing in an extended flight regime, and (3) large installation space for test equipment and for implementing two independent experimental cockpits. High component reliability and low maintenance cost were further important features of the helicopter. Also, the BK 117 was produced in sufficient quantities and served particularly successfully the air rescue market [6].

The HESTOR project proposal was based on the objectives of the German Working Group on Helicopter Technology AKH (Arbeitskreis Hubschrauber Technologien) being sponsored by the Federal Ministry for Research and Technology (BMFT). In a meeting with BMFT on April 22, 1986, it was agreed that DFVLR would lead the project in cooperation with MBB-UD and the Federal Ministry of Defense (BMVg). In the meantime, financing models were discussed, which also took financial participation of industry and DFVLR into account.

Meanwhile, the US Army had several times clearly expressed their interest to procure a virtually identical helicopter in-flight simulator under the existing MoU (Memorandum of Understanding) between Germany and the United States in the area of Helicopter Flight. The starting point of this US interest was the impressive comparative flight testing of the BK 117 with various helicopters of the US industry. Thereby the BK 117 excelled particularly due to its high maneuverability. Even good flying qualities were attested for aerial combat. Furthermore, a solid cooperative basis for such a project was established by the years of successful joint research between the Aeroflightdynamics Directorate of the US Army, the NASA Ames Research Center and the DFVLR Institute of Flight Mechanics in the field of in-flight simulation of rotorcraft. (see also Sect. 12.3.3). Of course, the concerted procurement of two HESTOR helicopters would have been also extremely attractive for cost reasons.

In spite of that, the well-prepared and promising project proposal failed because ultimately the common willingness of the management in research, industry and government departments lacked the commitment to undertake such a project. For nearly 10 years the project name HESTOR haunted still the DFVLR offices. Finally, the Institute of Flight Mechanics successfully managed to realize a helicopter in-flight simulator, now based on an EC 135. This time, the course was set right by the clear terms on the part of the BMVg and by the stipulated MoA (Memorandum of Agreement) on November 2, 1993 with the former French development director *Yves Richard* (see Fig. 8.36) of Eurocopter S.A. (today: Airbus Helicopters). This time an optimal constellation of personalities and decision makers was found to realize such a project. This included besides *Ives Richard* from Eurocopter, *Rolf Schreiber*, the former Deputy Section Head in the Federal Ministry of Defense (MoD), *Wieland König*, Head of Helicopter Department at the Federal Office of Defense Technology and Procurement (BWB) and later director of the German Flight Test Center (WTD-61), and *Heinz Max*, former Dornier development director and Program Director for Aeronautics at DLR (see Chap. 10).

11.4 DLR BK 117 Tele-Hestor (1986)

Considering the aforementioned HESTOR project, once again based on a DFVLR initiative, the BMFT Working Group on Helicopter Technology (AKH) had recommended

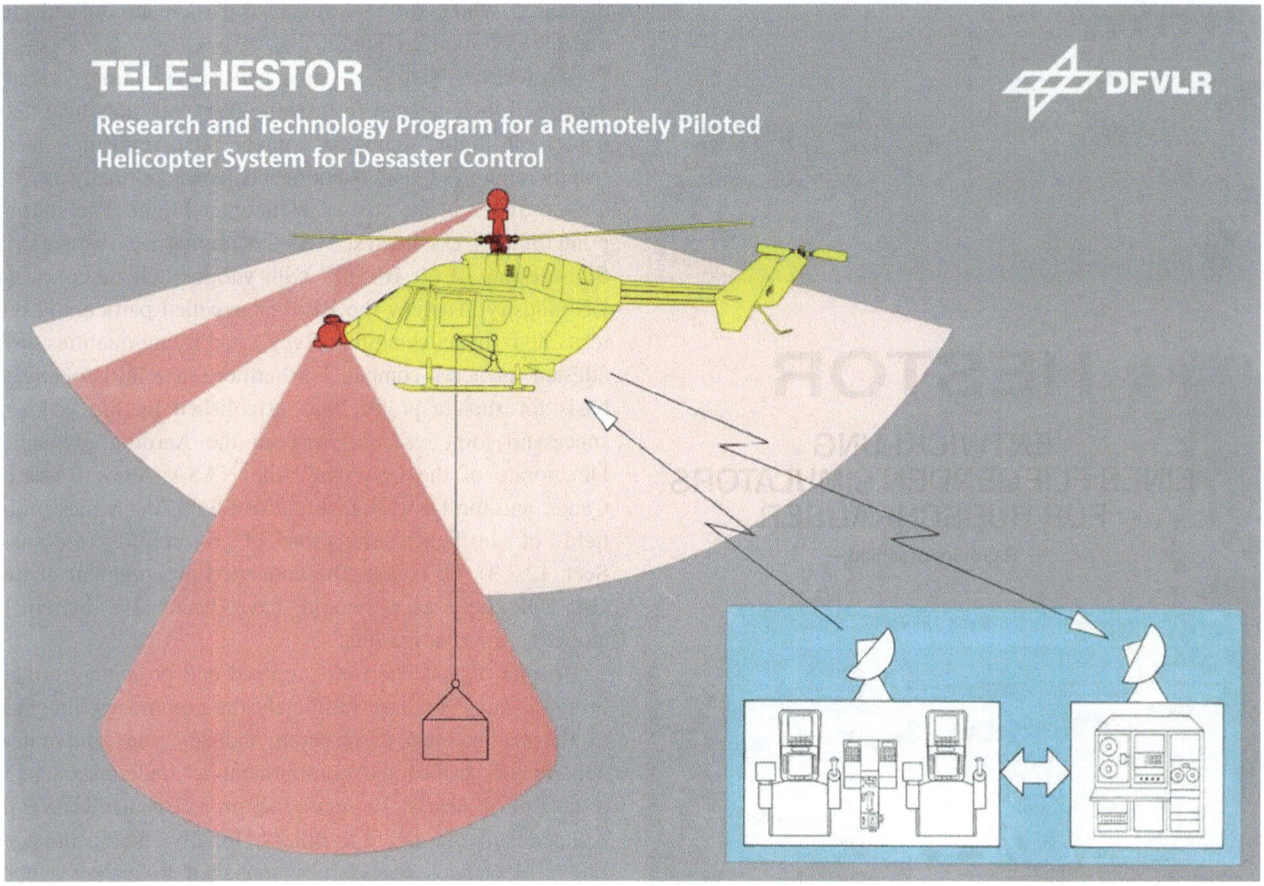

Fig. 11.8 Project proposal TELE-HESTOR

in a Meeting on July 10, 1986, to rehash the utilization potential of a HESTOR technology demonstrator for the development and operationalization of new technologies for disaster management. The Institute of Flight Mechanics of DFVLR then submitted a memorandum that was essential for the realization of a powerful, unmanned helicopter system (Telecopter) for disaster prevention and protection measures (see Fig. 11.8 [7]). The required integration issues of various high-technology areas were elaborated therein.

Telecopter missions were aimed at extending the operation-flexibility of tele-operator systems for disaster management and control (see Fig. 11.9). They included missions in hazardous, emergency and disaster areas at high risk for human beings. They included tasks for reconnaissance and monitoring, damage control as well as rescue and recovery operations. The Telecopter should be remotely flown by a "mission pilot", who manipulates in a mobile ground control station at a sufficiently safe distance from the actual place of operation. All of the visual and flight status information, required onboard the Telecopter by the mission pilots, should be gathered by exclusive sensor systems such as electro-optical sensors for all-weather conditions and transmitted via image processing and telemetry data links to the ground control station. The determination of exact positions should be provided by satellite navigation, and the control of the Telecopter via command links (see Fig. 11.10).

The Telecopter should be operated by a "mission operator" in the ground control station, who remotely operates the sensor and manipulator systems as required, for example by aligning video cameras, activating measurement systems, and dropping and lifting of loads. To meet the high standards of flight safety in European airspace, the remotely operated Telecopter will be monitored during training missions by an onboard safety pilot. Many years of experience at DFVLR in the operation of in-flight simulators with safety pilots would thus be of particular importance.

The technology demonstration program TELE-HESTOR was planned for testing the essential technologies for a future Telecopter system employing large payloads over sufficient

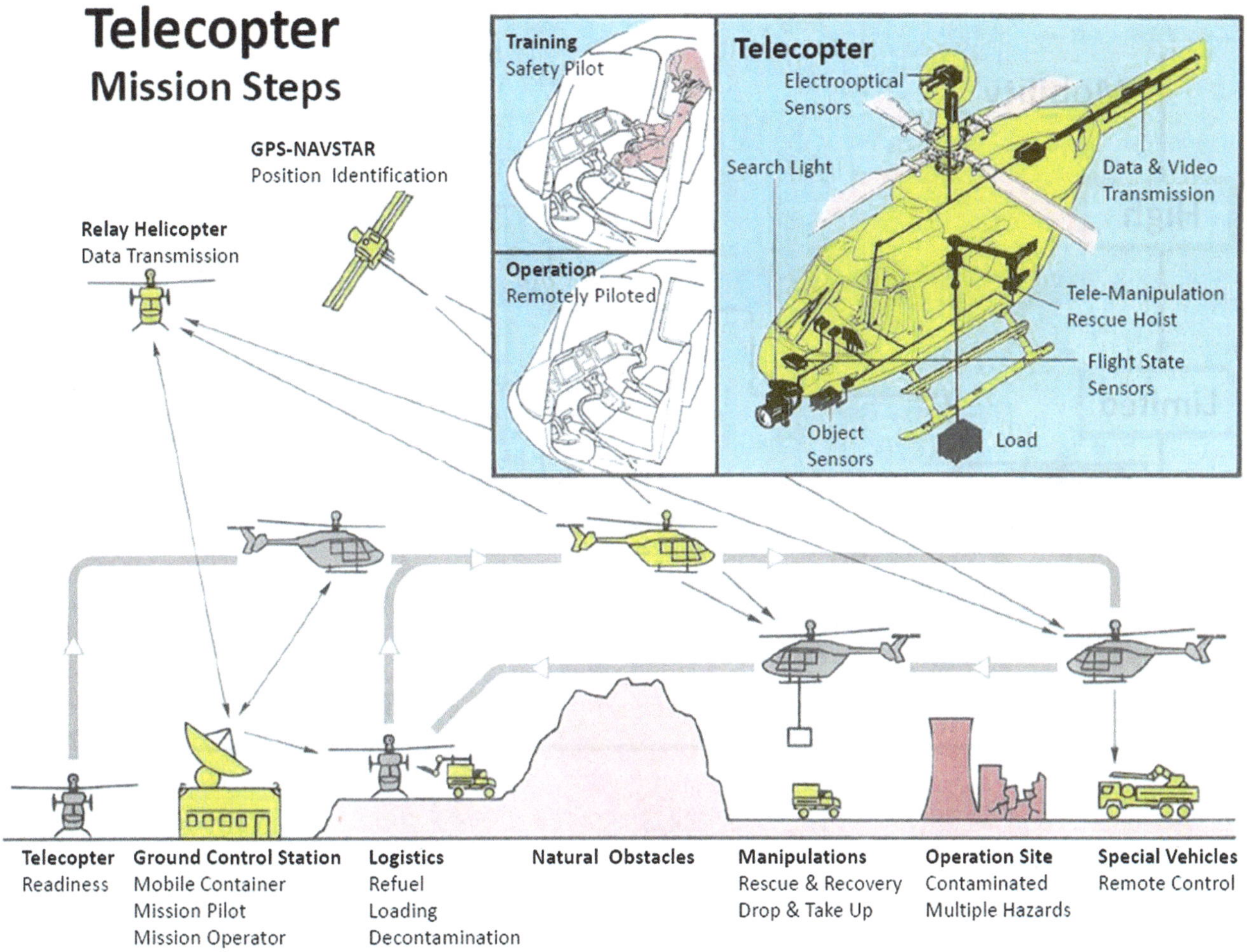

Fig. 11.9 Tele-operator systems for disaster management

ranges (minimum requirements: 1 ton, 400 km) and with an ability to carry out highly accurate remote operations such as reconnaissance and measurement missions. Besides technologies such as high-precision sensor systems for determining flight conditions and environmental variables, ruggedized robust computer systems and robust electro-optical flight control systems, an emphasis was placed on planning and optimizing the human-machine interface A further focus was on the information technologies and robotics as important components of the overall experimental TELE-HESTOR system.

The TELE-HESTOR testing concept included the in-flight simulator BK 117 HESTOR as a key element (see Sect. 11.3). Research objectives would have been to evaluate and optimize the pilot-helicopter interface through smart control devices, displays and computer support under operational conditions. A mobile control station was envisaged with workstations for the mission pilot and mission operator (see Fig. 11.11). High technological demands were placed on the visual information, which necessitated integrated electro-optical sensor systems, capable of alignment, for day and night utilization and all-round visibility. The displays were to be either with high-resolution color multi-functional and panorama screens or directly in the field-of-view by head-mounted displays. Wide vision fields for peripheral motion cues and sufficient visual depth (3D detection and object or obstacle recognition) are essential human perceptional parameters for carrying out remotely piloted helicopter missions. Another important issue was the disturbance free

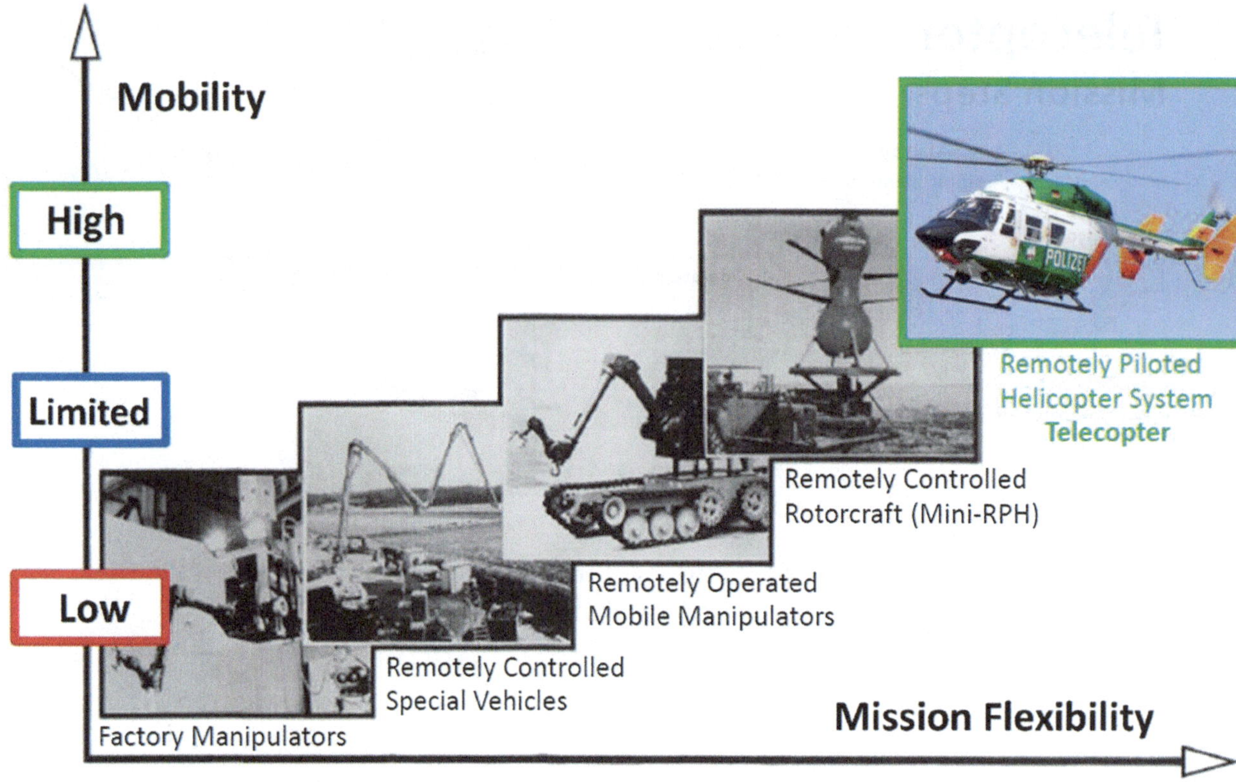

Fig. 11.10 Telecopter—operational profile

and reliable transmission of images and flight data in real time as well as of the command signals for remotely directing and manipulating the TELE-HESTOR platform.

Three implementation steps were planned for the TELE-HESTOR program, that was aimed at successive transfer of the onboard workstations to the ground control station (see Fig. 11.12). With the configuration 3/0, missions were planned with the fully manned in-flight simulator HESTOR, that is, with a safety pilot, mission pilot, and operator. The focus of the investigation was to be on the selection, integration, and evaluation of onboard sensor systems and on the optimization of crew interaction and coordination issues. Data gathering and analysis of flight and environmental data, as well as that of the mission equipment, were to be carried out on the ground.

With the configuration 2/1, the mission operator workplace should be shifted from the helicopter to the mobile ground control station. The scientific investigations should be focused during this step on the adequate visual cues for remote manipulations and expert systems to relieve the mission operator.

In the configuration 1/2, only the safety pilot should be onboard to enable a safe flight operation. Realistic flight tasks for actual disaster prevention were to be carried out remotely-manned from the mission pilot workplace. Thereby special attention is focused on the flight mechanical issues, such as controller-based adaptation of handling qualities of the remotely-manned helicopter to the skills of the mission pilot working on the ground under limited visibility and motion cues. Finally, complete remote-operator-missions were to be tested with this configuration.

Because the HESTOR-Project had not received additional public funding, even this highly regarded TELE-HESTOR project initiative had to be abandoned in 1986. More than 20 years later, US rotorcraft companies like Boeing and Sikorsky picked up an equivalent concept for a variable manned helicopter system and even patented the whole thing termed “variably manned aircraft” [8].

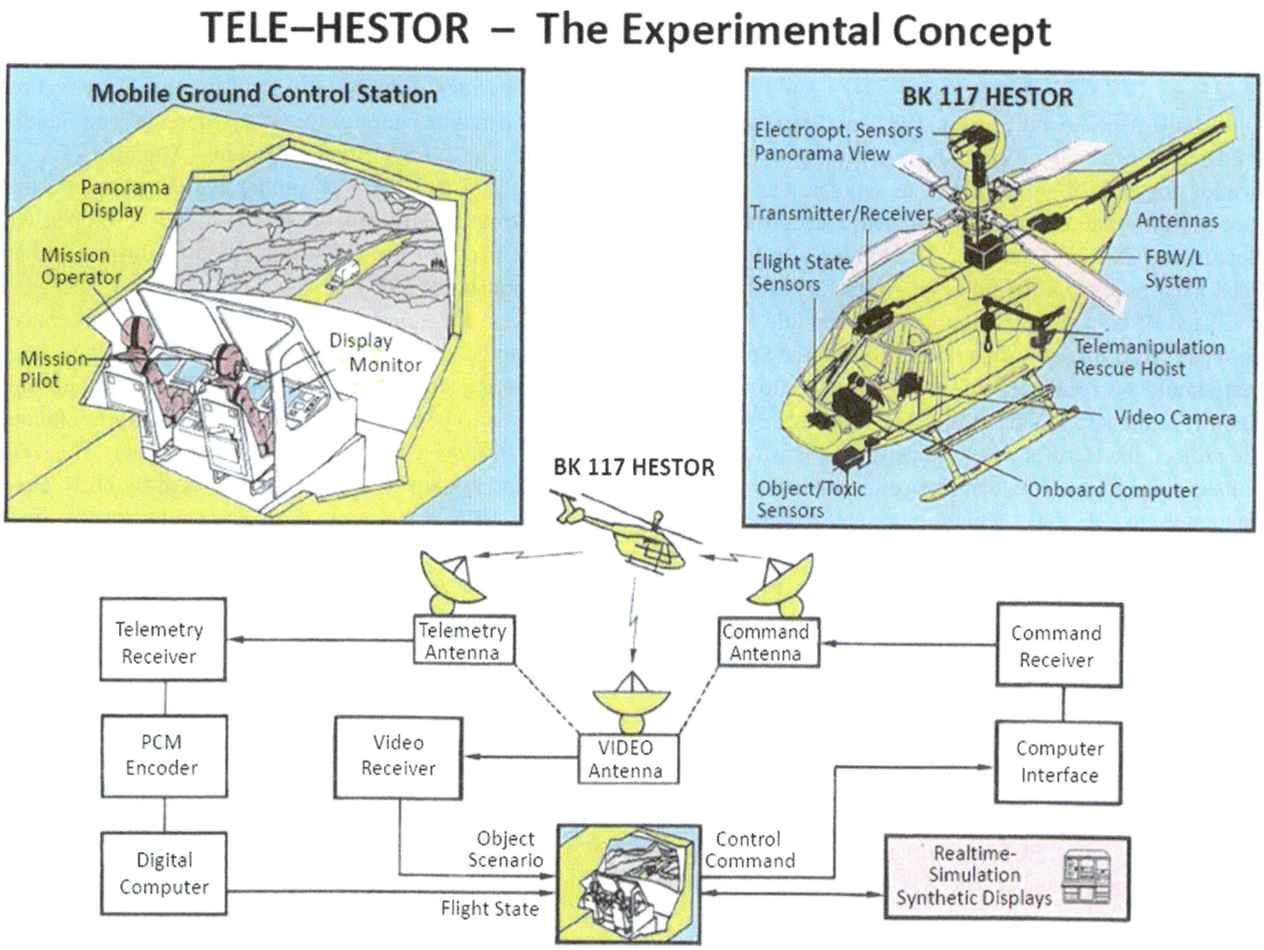

Fig. 11.11 TELE-HESTOR—the test concept

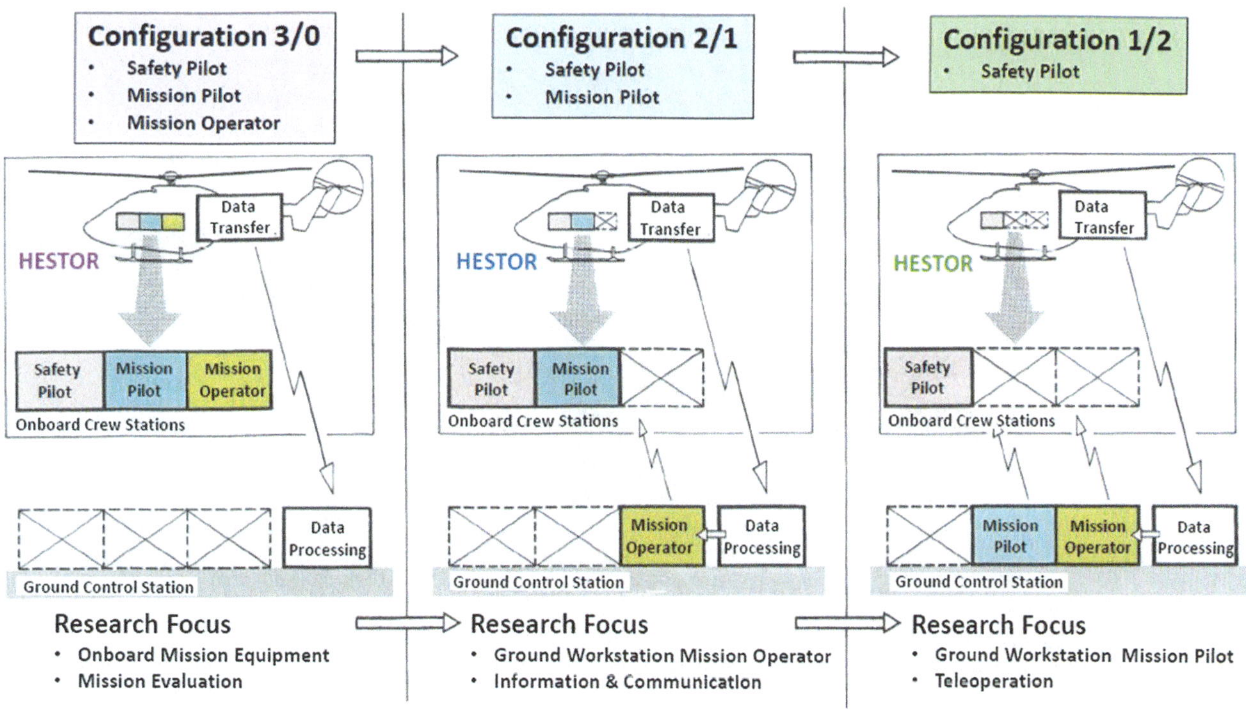

Fig. 11.12 TELE-HESTOR implementation steps

11.5 DLR/Dornier Hermes Training Aircraft (HTA) (1987–1992)

On behalf of ESA/CNES, from 1987 to 1989, the DFVLR prepared a technical concept and a complete system specification for an in-flight simulator to simulate the visual and motion information of the planned European Hermes Spaceplane (see also Sect. 9.2.2). Hermes was to be launched into space from the tip of an Ariane 5-Plus rocket and consisted of two modules: the resource module that would be separated before atmospheric reentry and the Shuttle itself that should be recovered and landed similar to the Space Shuttle. In the last version of the plan, prior to termination of the project, the Hermes was to transport three astronauts and a three-ton payload. The total mass at the takeoff would have amounted to 21 tons, which represented the maximum payload of the Ariane 5-Plus rocket.

The aim of the in-flight simulator was to provide a training aircraft for astronaut pilots, who should be able to perform a safe landing at high airspeeds after a steep descent at about 19° flight path angle. The planned flight regime of the so-called Hermes Training Aircraft (HTA) included the approach from about 12 km altitude to touchdown. Ten approaches should be possible on a training flight with a planned utilization of about 4000 sorties a year.

The glide ratios (lift over drag—L/D) of the Hermes Spaceplane and that of the HTA-host aircraft differed significantly by a factor of about 3. Hence, thrust reversal and landing gear extension on the host aircraft were indispensable besides airbrakes to simulate the Hermes flight dynamic behavior during the steep descent and landing approach. The NASA Shuttle Training Aircraft (STA, see Sect. 5.2.2.14) had to meet similar requirements. Furthermore, a stringent DLR quality criterion was developed with which the proof of the HTA simulation quality could have been demonstrated [9].

Particular attention was paid to the HTA cockpit concept to take into account different training aspects such as *Single Pilot Training* or *Crew Coordination Training*. The crew training was to be implemented by an additionally mounted cockpit (*Hermes Crew Training Flight Deck*). The HTA concept and system specifications generated by DLR served the ESA/CNES as the basis to float a tender for the development of HTA [9–19]. Proposals with detailed recommendations for the implementation were submitted by two vendors based on the Grumman Gulfstream II and the Dassault Falcon 900 (see Fig. 11.13), which were evaluated by the DLR. The functional ability of the proposed simulation concept was demonstrated in an ATTAS in-flight simulation. The achieved simulation quality is given in Fig. 11.14. It can be seen that the deviations in the roll rate between the Hermes model and the actual flown ATTAS response lie within the "permissible" mismatch boundaries of the DLR quality criterion.

Fig. 11.13 Dassault Falcon 900 chosen as HTA-host aircraft

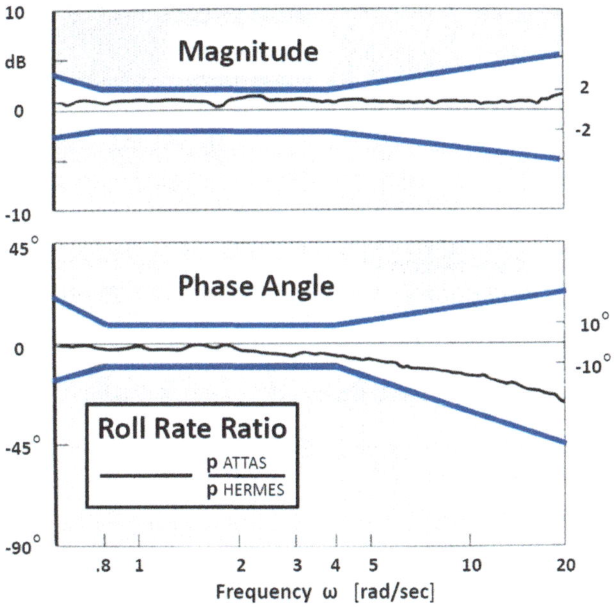

Fig. 11.14 Simulation quality of Hermes in-flight simulation with ATTAS

After the technical, financial and political scenarios had changed in Europe, the European Hermes Spaceplane project was discontinued in November 1992, giving preference to space capsules with parachute recovery.

In conclusion, from the European Hermes Spaceplane project, only a rich treasure of multinational project experience and a beautiful plastic demonstration model were left over (see Chapter title picture).

References

1. Wood, D.: Project Cancelled. Searching Criticism of the Abandonment of Britain's Advanced Aircraft Projects. Macdonald & Jane's, London (1975)
2. Gunston, B.: Plane Speaking—A Personal View of Aviation History, pp. 166–170. Patrick Stephens Limited (1991)
3. Gunston, B.: Plane Speaking—A Personal View of Aviation History, pp. 219–229. Patrick Stephens Limited (1991)
4. Anon.: CASTOR—Entwicklung eines Fliegenden Simulators für Kampfflugzeuge, Rahmenvorschlag, BWB/ErpSt61/Dornier/DFVLR, March 1984
5. Anon.: HESTOR – Entwicklung eines Fliegenden Simulators für Hubschrauber, Rahmenvorschlag, MBB/DFVLR, June 1984
6. Hamel, P., Gmelin, B., Hummes, D., Pausder, H.-J.: Fliegender Simulator HESTOR—Dokumentation zur Projektrealisierung, DFVLR IB 111-86/25, (1986)
7. Hamel, P. (ed.): TELE-HESTOR—Forschungs- und Technologieprogramm für ein Fernbemanntes Hubschraubersystem für den Katastrophenschutz, Memorandum, DFVLR IB 111-86/41, October 1986
8. Jones, R.D., Whelan, D.A., Wenberg, L.L.: Variable Manned Aircraft. US Patent No: US20090105891 A1, The Boeing Company, 23 April 2009
9. Hanke, D., Rosenau, G.: Hermes Training Aircraft. Technical Specifications (Edition 1, Revision 1), DLR IB 111-92/18 (1992)
10. Gargir, G.: Technical specification. Study on Hermes Training Aircraft HTA, CNES H-CT-5114-01-CNE, August 1987
11. Hamel, P.: Hermes Simulation and Training Aircraft—Concept Study, vol. 1, Executive Summary, DFLVR IB-111-88/02-1, July 1988
12. Hanke, D.: Hermes Simulation and Training Aircraft—Concept Study, vol. 2, General specification, DFVLR IB-111-88/02-2 (1988)
13. Köpp, J., et al.: Hermes Simulation and Training Aircraft—Concept Study, vol. 2, Chapter 4: Evaluation of Host Aircraft, Dornier Document No. H-PV-5114-001 DOR, May 1988
14. Hamel, P., Rosenau, G.: Hermes Training Aircraft—Executive Summary of Complementary Study Phase, DLR IB-111-89/17-0, October 1989
15. Hanke, D.: Hermes Training Aircraft, vol. 1, Concept Analysis, DLR IB 111-89/17-1, October 1989
16. Hanke, D., Rosenau, G.: Hermes Training Aircraft, vol. 2, Technical Specifications, DLR IB 111-89/17-2, October 1989
17. Wilhelm, K., Schafranek, D., Altenkirch, D., Rosenau, G.: Hermes Training Aircraft, vol. 3, Host Aircraft Evaluation, DLR IB 111-89/17-3, October 1989
18. Schafranek, D.: Hermes Training Aircraft, vol. 1A, Necessity of Side Force Generators, DLR IB 111-89/17-1A (1989)
19. Döler, N., Bouckaert, F.: The In-flight Trainer of the European Hermes Programme, In: [1.19], Paper 25 (1991)

Author Biography

Peter G. Hamel was the Director of the Institute of Flight Mechanics/Flight Systems of the German Aerospace Center (DLR/DFVLR) (1971–2001). He received his Dipl.-Ing. and Dr.-Ing. degrees in Aerospace Engineering from the Technical University of Braunschweig in 1963 and 1968, and his SM degree from M.I.T. in 1965. From 1970 to 1971 he was Section Head of Aeronautical Systems at Messerschmitt-Bölkow-Blohm in Hamburg. Since 1995 he is an Honorary Professor at the Technical University of Braunschweig and a founding member of three collaborative research centers at the University. He was Chairman of the National Working Group on Helicopter Technology (AKH) (1986–1994) and the appraiser for the National Aviation Research Program (LuFo) until today. He was also the Manager of DLR's Rotorcraft Technology Research Program and the German Coordinator for the former AGARD Flight Mechanics/Vehicle Integration (FMP/FVP) Panel. He is a member of the German Society for Aeronautics and Astronautics (DGLR) and of the American Helicopter Society (AHS), and a Fellow of AIAA. He is the recipient of the AGARD 1993 Scientific Achievement Award, of AGARD/RTO von Kármán Medal 1998, of AHS Dr. A. von Klemin Award 2001, and of the prestigious DGLR Ludwig-Prandtl-Ring 2007.

12 International Cooperation

Peter G. Hamel

12.1 Overview

Scientific and technical evaluations of the DLR Institute of Flight Systems are carried out regularly at intervals of about five years by national and internationally renowned experts. These evaluations confirmed the institute's outstanding and unique system competence in Europe in the field of in-flight simulation and related disciplines such as system identification, flight dynamics, and flight control systems. It was, therefore, but natural that the institute evolved into an internationally recognized cooperation partner.

The activities can be divided into four broad areas: (1) participation in international technical committees, workshops and symposia such as AGARD/RTO/STO FMP, FVP, AVT and SCI Panels, SAE Aerospace Control & Guidance Systems Committee, AIAA Technical Committees on Flight Mechanics as well as Modeling and Simulation, (2) participation in development projects such as X-31, Eurofighter, NH-90, and Hermes Spaceplane, (3) long-term

P.G. Hamel (✉)
Braunschweig, Germany
e-mail: p.hamel@t-online.de

P.G. Hamel (ed.), *In-Flight Simulators and Fly-by-Wire/Light Demonstrators*,
DOI 10.1007/978-3-319-53997-3_12

cooperation agreements such as USAF MoU, US Army MoU, and MoA Eurocopter, and (4) short-term programs with research and industrial organizations and with test pilot schools. A major part of these activities has already been elaborated in Chaps. 6 through 10 pertaining to DLR in-flight simulators and their utilization. This chapter focuses on a few selected special activities and events in which the in-flight simulation played a significant role.

12.2 International Workshops and Symposia

12.2.1 International Symposium on In-Flight Simulation (1991)

As already pointed out in the Introduction, Chap. 1, the international significance of in-flight simulation and its technological utilization and resulting benefits were highlighted in an International Symposium in Braunschweig from July 1–3, 1991 (see Fig. 12.1, [see Ref. [19] in Chap. 1, 1]). The event was discussed in the national and international

Fig. 12.1 First international symposium on in-flight simulation (1991)

Fig. 12.2 Aerospace journalist *Michael Mecham* (AW & ST) covering the symposium in Braunschweig (from *left Michael Preß*, *Wolfgang Beduhn*, *Dietrich Hanke*, *Michael Mecham*, Testpilot *Michael Parrag* (Calspan) and "*HaLu*" *Meyer*

aerospace press, such as the world's leading aviation magazine Aviation Week & Space Technology (AW&ST) [2, 3].

In the AW&ST article ("*Gathering of the In-Flight Simulation Fraternity*"), the US aerospace journalist *Michael Mecham* elucidates the special role of the in-flight simulation as a "driving force in the exploration of future flight control systems" (see Figs. 12.2 and 12.3). During this first-ever international symposium on in-flight simulation, the introductory lecture was delivered by the variable-stability aircraft pioneer, test pilot, and flying qualities specialist *Robert P. Harper* from Calspan (see Fig. 12.4). The paper was essentially based on the historical treatise by *Waldemar Breuhaus* [see Ref. [5] in Chap. 1].

In his presentation, he underlined the special role of the in-flight simulation to reduce the technological risks and costs of development, testing and modernization of flight systems ("*Get it right before the first flight*") [4]. The 25 presentations included those from research institutions, universities and industry as follows: USA (9), Germany (6), England (3), France (3), Japan (2) and Canada (1). A firmly planned lecture by a delegate from the Russian Flight Research Institute (FRI) could not be delivered at the last minute. All the more gratifying were the scientific contacts with leading scientists of this research and flight test institute two years later (Sect. 12.2.2).

12.2.2 German-Russian Workshop on In-Flight Simulation (1993)

Assisted by Hans-Heinz Lange and Bernd Krag

Initiated by the Russian Flight Research Institute FRI ("Gromov" Flight Research Institute—LII), the first contact with the DLR Institute of Flight Mechanics was established during 1990. This was followed by the first visit of DLR

AW&ST 75

AVIATION WEEK & SPACE TECHNOLOGY

A McGRAW-HILL PUBLICATION $5.00 OCTOBER 7, 1991

COMPETITION SPURS SPACE TELECOMMUNICATIONS PAGE 50

AIRBORNE SIMULATION EXPANDS PAGE 42

AIRBORNE SIMULATION EXPANDS

Airborne Simulation Is Driving Force In Advanced Control System Research

MICHAEL MECHAM/BRAUNSCHWEIG, GERMANY

The increasing sophistication of computer modeling, wind-tunnel tests and ground-based flight simulators has sometimes overshadowed the contribution that airborne flight research makes to the development of new aircraft.

In-flight simulation has been a basic tool in development of a number of high-performance or advanced technology aircraft. New in-flight simulators will be used in the 1990s.

For fixed-wing aircraft the emphasis will continue to be on basic research in such areas as high angle-of-attack. In-flight simulation will become increasingly important for helicopters as fly-by-wire technology is introduced and the military explores autonomous systems to relieve pilot workloads, particularly in nap-of-the-earth flight.

Recently, leading members of the in-flight simulation fraternity gathered at Braunschweig, Germany, to discuss programs for the 1990s in a conference sponsored by the German Aerospace Research Agency, the Royal Aeronautical Society, American Helicopter Society, Aeronautical & Astronautical Assn. of France, Aerospace Industries Assn. of America and German Society for Aeronautics and Astronautics. These articles are based on that conference and additional reporting by Bonn Bureau Chief Michael Mecham.

Fig. 12.3 "Airborne Simulation Expands" (*Credit* AW&ST 7.10.1991)

scientists (*Dietrich Hanke*, *Hans-Heinz Lange*) to Moscow. The in-flight simulator TU-154M FACT was presented, that was utilized for the airborne simulation of the steep descent and landing approach of the projected manned Russian Space Shuttle Buran (see also Sect. 5.6.1).

Based on these contacts, a bilateral workshop of DLR with the Russian Flight Research Institute was organized and held at the Hannover airport on October 29, 1993. The focus of the event was on in-flight simulation for pilot training and flying qualities evaluation of very large aircraft, which were discussed under the name *Ultra-High Capacity Aircraft* (UHCA) both in the United States and in Europe at that time. The Airbus project A3XX, from which the A380 emerged later, was still the topic of applied research at DLR during the beginning of the nineteen nineties. Of particular interest was the determination of the extent of coupling between the elastic modes of the wing and the flight dynamics of an aircraft assumed to be rigid.

FRI had already gained experience with the flight testing of the Ukrainian large aircraft Antonov An 124 and An 225. As is generally known, these large airplanes are nowadays also deployed for civilian missions and for the United Nations. The visit of the FRI under the leadership of the Director of Flight Test Techniques, *Wilgem Vid*, and the

Fig. 12.4 Calspan flying qualities expert and test pilot *Robert Harper* (*center*) with *Eckhard Wohlfeil*, *Hans-Heinz Lange*, "*HaLu*" *Meyer* and *Michael Preß* (from *left*) prior to ATTAS demonstration flight

Fig. 12.5 Su-27 with programmable flying qualities with *Hans-Heinz Lange* (*on the ladder*) and *Peter Hamel* in cockpit (*Credit* Michael Bauschat)

aeronautical expert *Sergej Boris* took place from October 25 to 30, 1993. It was covered extensively in the news media. Thereby the two in-flight simulators, FRI Tu-154M FACT and DLR VFW 614 ATTAS, were captured spectacularly in a photo during a flight on October 27, 1993 over the clouds in Hannover. Pictures were shot from an AlphaJet of the German Air Force Flight Test Center (WTD-61, (see title picture of this Chapter). The Tu-154M could not be flown to Braunschweig because the surface loading (load per unit area) of the six-wheeled main landing gear was too high for the asphalted runway at the Braunschweig airport.

During the Moscow Air Show 2003 (August 19–24), DLR scientists had the opportunity to visit the Fly-by-Wire test demonstrator Su-27 with variable flying qualities (see also Sect. 5.6.2). The Su-27 was connected with a ground-based simulator over a telemetry data link in order to provide "virtual flying" on the ground. The DLR staff were invited to "fly" with this hardware-in-the-loop device the spectacular extreme angle of attack "Cobra Maneuver" on the ground, and it was better that way (see Fig. 12.5).

12.2.3 Workshop "20 Years of ATTAS" (2001)

Finally, a special event needs to be mentioned, namely the workshop "20 Years of ATTAS", held on October 16–17, 2001, with international participation to celebrate the 20th anniversary of the in-flight simulator ATTAS. US test pilots such as *Rogers Smith* (former NASA test pilot) and *Michael Parrag* (Calspan) were among the active participants (see Figs. 12.6 and 12.7), and *Stefan Levedag* as well who took over the management of the DLR Institute of Flight Systems since 2001 (see Fig. 12.8). Ten years back, *Rogers Smith* had already participated in ATTAS flight experiments together with the DLR test pilots from September 3–7, 1990, in Braunschweig (see Fig. 12.9).

Fig. 12.6 ATTAS workshop participant ex-NASA test pilot *Rogers Smith* (*right*) with *Dietrich Hanke*

Fig. 12.7 Calspan test pilot *Michael Parrag* (*left*) during ATTAS workshop with *Peter Hamel*

Fig. 12.8 *Stefan Levedag* (in the *front*) welcoming the ATTAS workshop participants

Fig. 12.9 NASA test pilot *Rogers Smith* (3rd from right) during 1990 with DLR ATTAS flight test experts *Knut Wilhelm*, *Hans-Peter Joenck*, *Dieter Schafraneck*, *Michael Preß*, *Eckhard Wohlfeil*, *Dietrich Altenkirch*, and "*Halu*" *Meyer*

12.3 Transatlantic Cooperation

12.3.1 Introduction

Once NASA became aware of the research activities in the field of digital flight control at the Institute of Flight Guidance of DFVLR, at the invitation of DFVLR the NASA test pilot and astronaut *Neil Armstrong* undertook a fact-finding visit to Braunschweig in 1971 (see Fig. 12.10). *Armstrong* was highly impressed by the creativity and efficiency of a "handful" of scientists and engineers who developed the digital flight control and inceptor systems (see Figs. 12.11 and Sect. 6.3.1.5).

Since that time, an increasing transatlantic exchange of information evolved, which was promoted through joint programs and symposia within the Flight Mechanics Panel (FMP), the subsequent Flight Vehicle Integration Panel (FVP), and the Guidance and Control Panel (GCP) of the Advisory Group of Aerospace Research & Development (AGARD). This also conformed with the ideas of *Theodore of Kármán*, the founder of AGARD and later recipient of the Gauß medal award of the Braunschweig Scientific Society (see Fig. 12.12), namely collaboration by (1) bringing together and exchanging leading scientist and engineers, (2) jointly utilizing complementary facilities, and (3) providing mutual scientific and technical support to achieve mutually beneficial progress in aeronautical research and development at cost savings to all participants [5].

Towards the end of nineteen seventies, two intergovernmental agreements (Memorandum of Understanding—MoU) were finalized between the Federal Republic Germany and the United States of America on cooperation between the two countries in the fields of flight control and flying qualities of aircraft (*Aircraft Flight Control Concepts*) and of helicopters (*Helicopter Flight Control*) (see Fig. 12.13). The overall objective of these MoU was to enable faster and cost effective research and development in both countries through joint and coordinated research projects/tasks with the complementary utilization of scientific opportunities and experimental facilities of both countries [6].

12.3.2 US/FRG MoU Aircraft Flight Control Concepts (1979–1992)

The research program in the field of flying qualities of highly control-augmented aircraft was carried out jointly by the Flight Dynamics Laboratory of the Air Force Wright Aeronautical Laboratories (AFWAL) and the DFVLR Institute of Flight Mechanics on behalf of the United States Department of Air Force and the German Federal Ministry Defense (FMoD)

Fig. 12.10 Visit of *Neil Armstrong* during interview with *Hermann Blenk*, *Karl-Heinrich Doetsch* and *Peter Hamel* (from *left*) in 1971 to Braunschweig

Fig. 12.11 *Neil Armstrong* operating a DFVLR side arm controller on the Do 27

respectively. The overall joint research project comprised of developing the task-oriented flying qualities criteria for control-augmented aircraft. The results were immediately incorporated in the *Standard Handbook* MIL-STD-1797 "*Flying Qualities of Piloted Aircraft*" [see Ref. [13] in Chap. 2]. The utilization of the in-flight simulators TIFS, NT-33A on the US side and FLISI and ATTAS on the German side played an equally important role as the exchange of test pilots. The relevant utilization programs and also the provision of a mobile microwave landing system TALAR by the USAF for FLISI steep approach flight tests with direct lift control have already been discussed in Sect. 7.3.4.

A particularly innovative and successful activity addressed a new experimental technique for detection and evaluation of flight critical weak points in a highly control-augmented aircraft performing precision maneuvers. Appropriate testing and evaluation methods were hitherto unavailable for evaluating the overall pilot-aircraft system performance in aggressive maneuvering, such as air-to-air refueling or air-to-ground tracking. DLR scientist *Ruthard Koehler* succeeded in 1983 to develop for the first time a method for a special precision flight task, namely air-to-ground tracking, with the so-called GRATE experimental technique (*Ground Attack Test Equipment*) [7]. The experimental procedure was based on the experience of the system identification (see Fig. 3.1) and had proved successful during the AlphaJet DFC flight tests in cooperation with Dornier and the German Air Force Flight

Fig. 12.12 *Theodore von Kármán* (*center*), C.F. Gauß medal award holder, with *Hermann Blenk* (*right*) and *Otto Lutz* during 1960 in Braunschweig

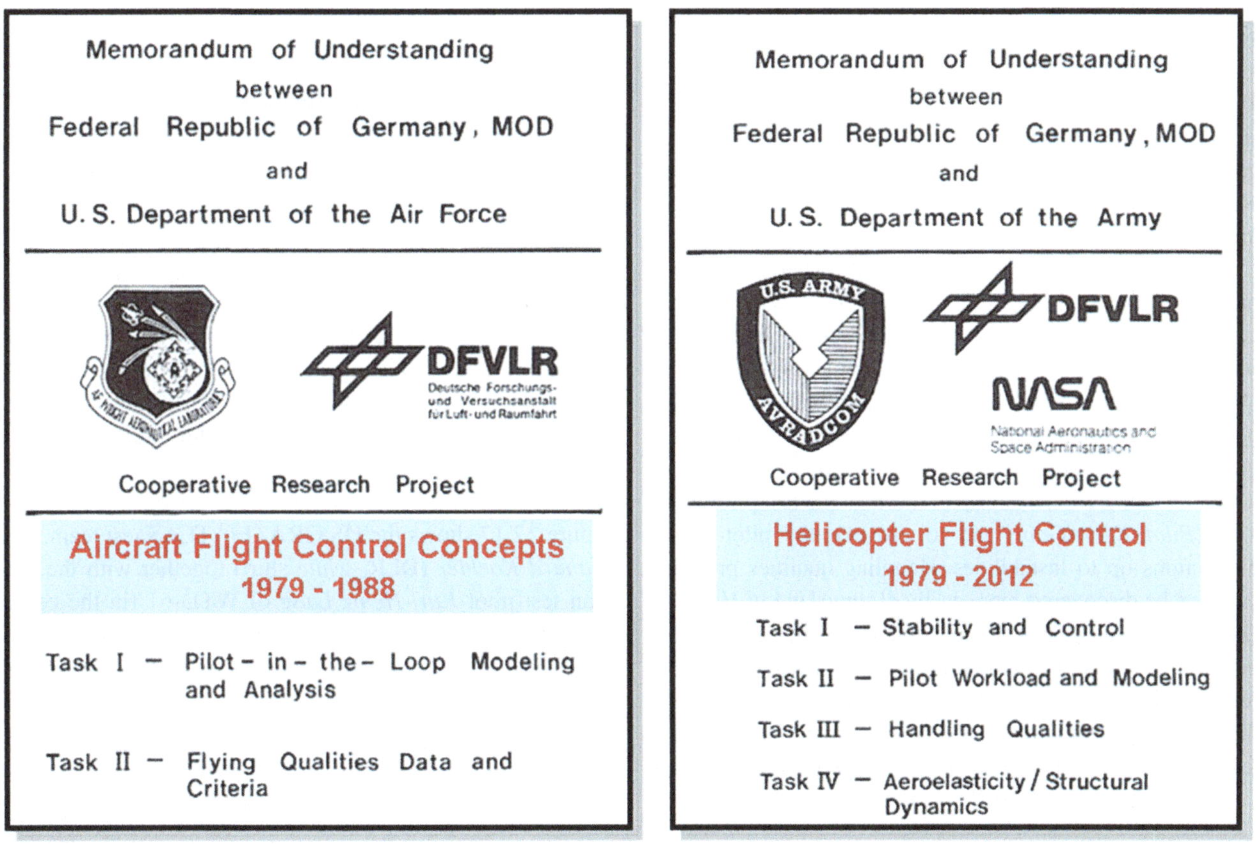

Fig. 12.13 Transatlantic cooperation

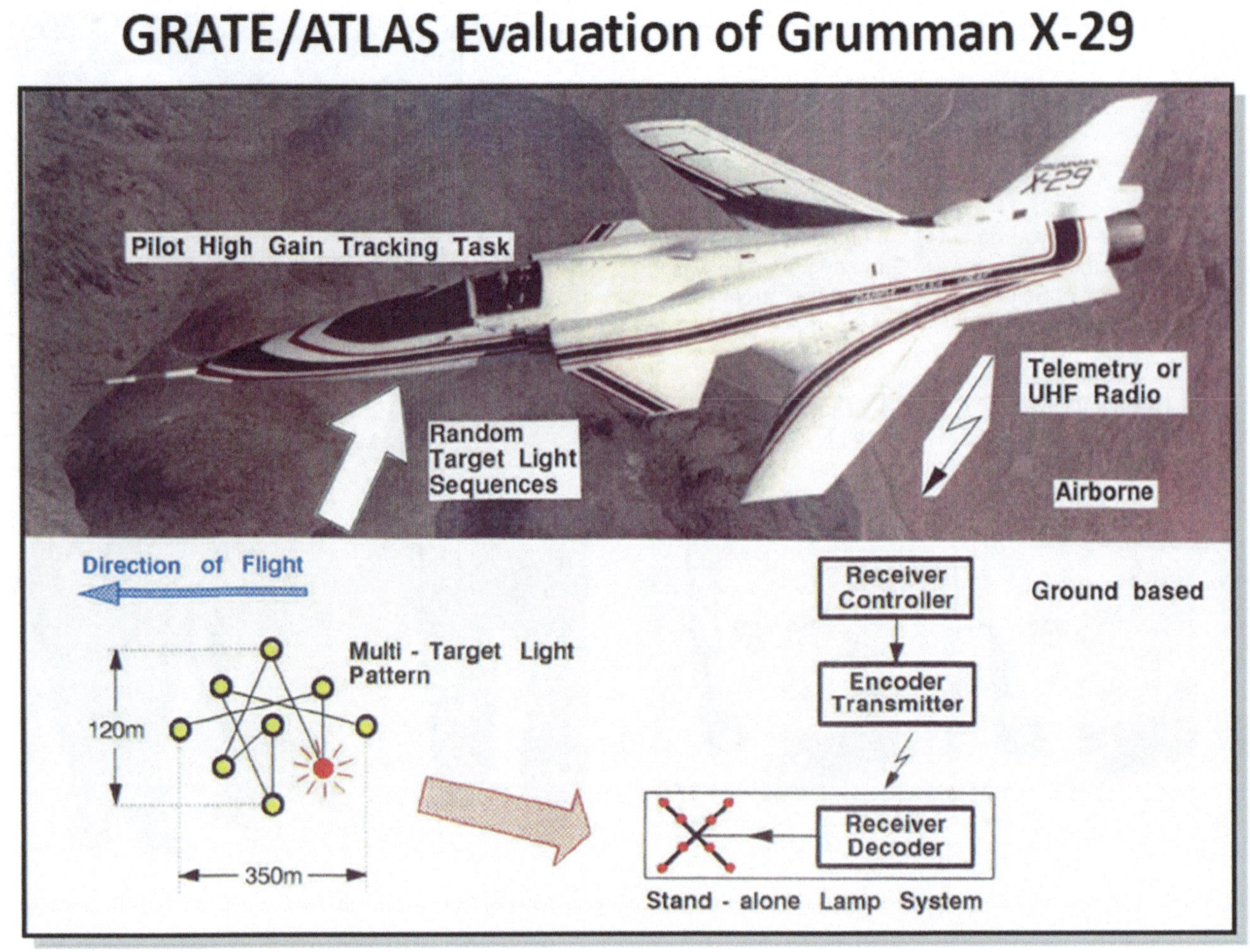

Fig. 12.14 The GRATE/ATLAS experimental technique

Test Center (WTD-61). This technique was validated also in simulator tests in the USAF AFWAL LAMARS the ground-based simulator. A similar system was reproduced by NASA under the acronym ATLAS (*Adaptable Target Lighting Array System*) [8].

The principle of the GRATE/ATLAS experimental technique is depicted Fig. 12.14. It consists of capturing and aligning the aircraft on to a randomly chosen bright light lamp (target) from an array of nine lamps arranged on the ground (*Multiple-Target Light Pattern*). The pilot can vary the frequency of light targets (*Random Target Light Sequences*) and influence the required reaction speed from the cockpit via telemetry signals. This allows to increase the test pilot workload (*Pilot Effort/Gain*) and to provoke the pilot-aircraft interactions up to instabilities. Handling qualities problems can thus be discovered more easily (*Unmasking of Handling Qualities Deficits*), that remain undetected in normal flight operations. The advantage of this technique is in the safety and repeatability of experiments and the reproducibility of the flight test results. Learning effects are virtually eliminated in terms of a pre-cognitive behavior. The experimental technique is suitable for all aircraft types and it is recommended as a standard equipment for all test pilot schools. Certain limitations can be expected in the high-speed flight regime.

Extensive flight tests with the GRATE/ATLAS equipment were performed at USAF Edwards Flight Test Center with the airborne simulator NT 33A of Calspan, besides those with the Grumman X-29. For example, it is obvious from Fig. 12.15 that a close correlation exists between the amount of time delay in the flight control system and the pilot rating. It could be found that even turbulence effects hardly affected the evaluation of the pilot-aircraft system.

The key milestones of the GRATE/ATLAS system development and utilization are summarized in Fig. 12.16. Figure 12.17 shows the US GRATE/ATLAS test team, with *Ruthard Koehler* (DLR, *white* shirt) together with the German test pilot *Karl-Heinz Lang* of WTD-61 (in the center, next to *Koehler* on right). Later, *Karl-Heinz Lang* survived the X-31A crash which was caused by icing of the air data sensor (see also Sect. 6.3.6).

During a presentation of GRATE/ATLAS experimental technique to a high-level US Air Force representative at the

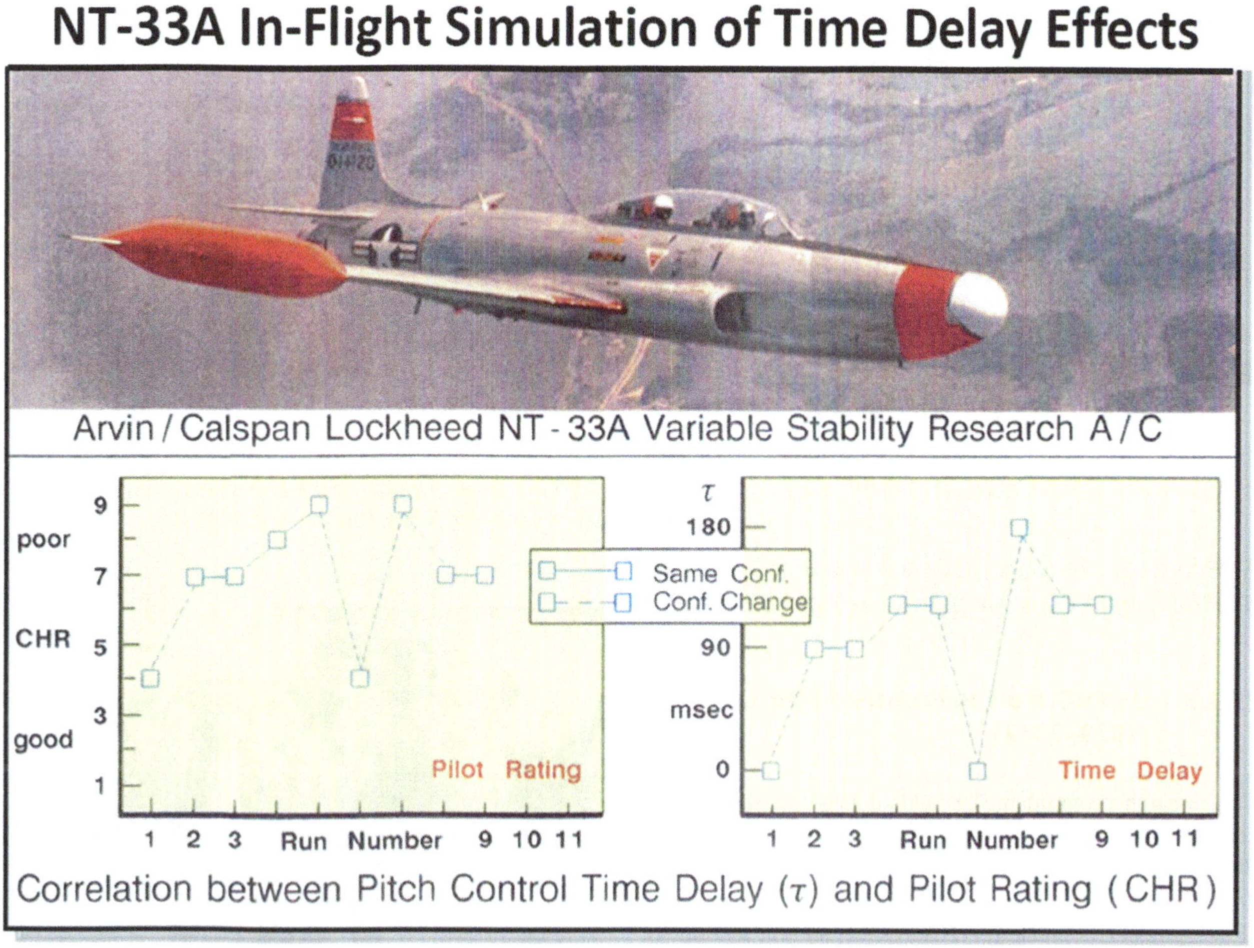

Fig. 12.15 Good correlation between time delay and pilot assessment (CHR)

GRATE / ATLAS - Road Map

1983 DLR Design of GRATE (Ground Attack Test Equipment)
at DLR - Braunschweig RC

1984 Alpha - Jet DSFC Flying Qualities Evaluation with GRATE
at GAF - Manching FTC (WTD - 61)

1985 G91 and F - 104 German Pilot Training with GRATE
at GAF - Manching FTC (WTD - 61)

1987 GRATE Implementation on LAMARS Simulator
at USAF Wright - Patterson AFB

NASA Design of ATLAS (Adaptable Target Lighting Array System - Functional Equivalent of GRATE)
at NASA Ames / Dryden RC

1988 T - 38A and TF - 104G Pilot Training with ATLAS
at USAF Edwards AFB

NT - 33A Variable Stability A / C and X - 29A Flying Qualities Assessment with ATLAS
at USAF Edwards AFB

Fig. 12.16 Milestones of GRATE/ATLAS development

Fig. 12.17 *Ruthhard Koehler* (*center*, white shirt) and *Karl-Heinz Lang* of WTD-61 (next to Koehler on *right*) with GRATE-ATLAS Team (*Credit* NASA Armstrong Flight Test Center)

DLR Executive Board, the US guest wished an additional option for the pilot, namely to knock out the individual light targets so as to obtain a "personal achievement" [9]. Many more successful results of this cooperation can be found in a final report of the MoU project leaders *Bob Woodcock* (USAF

Fig. 12.18 Calspan NT-33A with MoU project officers *Knut Wilhelm* (DFVLR, *center*) and *Bob Woodcock* (AFFDL, *front*)

Fig. 12.19 Test pilots *Ron Gerdes* (NASA, *center*), *Hannemann* (WTD-61), *Manfred Rössing* (2nd from *left*) and *Klaus Sanders* (*right*) and project leader *Jürgen Pausder* (both DFVLR)

AFWAL-Flight Dynamics Laboratory) and *Knut Wilhelm* (DLR Institute of Flight Mechanics) (see Fig. 12.18, [10]).

12.3.3 US/FRG MoU Helicopter Flight Control (1979–2012)

The research program in the field of handling qualities of control-augmented helicopters was carried out for United States Department of the Army by the Aeroflightdynamics Directorate (AFDD) of the Army Aviation Research and Development Command (AVRADCOM) and for Institute of Flight Mechanics of the German Aerospace research Center DFVLR. One of the common research objectives comprised of the development of task-related handling and flight performance criteria for control-augmented helicopter systems. The research results contributed directly to the new guidelines for rotorcraft handling qualities ADS-33E [see Ref. [28] in Chap. 8].

The exchange of scientists and test pilots (see Fig. 12.19) played as important a role as the utilization of the in-flight simulators CH-47B and UH-60 RASCAL (see Fig. 12.20) on the US side and of the Bo 105 ATTHeS and EC 135 FHS on the German side. Details about joint utilization programs have already been provided in the Sects. 8.3.3 and 8.4.1 and 10.4.3. It is, of course, unusual, if not even unique, that this transatlantic cooperation initiated and managed by *Irving Statler* (AFDD) and *Peter Hamel* (DLR) and later pursued by *David Key* and *Chris Blanken* (both AFDD) on the American side and by *Bernd Gmelin, Jürgen Pausder, Berend van der Wall* and *Marc Höfinger* (all DLR) on the German side, could prove so successful over a period of 33 years [11–13].

The partial change of generation of the involved scientists had certainly contributed to the fact, that the MoU was repeatedly fertilized with new scientific ideas and their implementations. This is also evidenced by a large number of joint publications. The MoU success story is considered as a role model of transatlantic cooperation as *Dr. John Berry*, Chief of Aviation and Missile Technologies, US Army International Technology Center-Atlantic, stated during the 30th anniversary of this MoU: "*The US-German MoU for Helicopter Aeromechanics is a textbook example of a mature, formal exchange agreement. The activities... were always of interest and have proven to be of highest value*".

Fig. 12.20 US-RASCAL Team with *Wolfgang von Grünhagen* (DLR) in the *middle*

On the same occasion, an appraisal was carried out of the history of flight test campaigns accomplished over such a long period. While Fig. 12.21 demonstrates the special role of the aforementioned four in-flight simulators in developing of handling qualities criteria for control-augmented rotorcraft, Fig. 12.22 highlights flight test campaigns at Army/NASA Ames Research Center and at the Research

Fig. 12.21 30 years of complementary research with in-flight simulators (*Credit* Chris L. Blanken)

Airport Braunschweig over a period of more than three decades [14]. The extensive exchange and the participation of scientists and test pilots were as vital as the impressive amount of joint publications. As concisely summarized by two participants [11]: "*Mostly it was hard work for the research engineers and the test pilots, but sometimes the Californian sun at the beach of Santa Cruz or a glass of German beer in a Gasthaus were very helpful and supporting!*"

During the ceremony marking the 30th anniversary on September 11, 2008, in the historic Braunschweig location "Kemenate", the German co-initiator of the MoU, *Peter Hamel*, talked about a special historical reminiscence [14]:

> Let me recount from our common MoU history one challenging episode: when DLR contemplated on a Fly-by-Wire successor helicopter for its aging in-flight simulator Bo-105 ATTHeS, AFDD was in a similar situation concerning its CH-47 operated by NASA Ames. Hence, in both countries new in-flight simulators were needed. Consultations between AFDD, NASA and DLR yielded excellent prerequisites to initiating a common approach. The host helicopter to be selected and retrofitted with a modern Fly-by-Wire flight control system should have high available power control in order to achieve high bandwidth and agility to simulate a wide spectrum of future rotorcraft systems. In 1985 the US Army, NASA, and DLR came unanimously to the conclusion that a very promising host helicopter candidate would be the BK 117 of MBB.
>
> To make it short, a Letter of Interest, signed by Dick Carlson, was sent to DLR and later to the German FMoD indicating the Army's full intentions to procure an identical helicopter from Germany if DLR pursues the effort to convert a BK 117 helicopter into a variable stability research vehicle, which we dubbed HESTOR (Helicopter Simulator for Technology, Operations, and Research). This cooperative effort would have saved funding for both partners (see Sect. 11.3).
>
> Unfortunately, we failed in this transatlantic procurement attempt—and we on the German side had to be blamed. The reasons cannot be simply explained. Not funding limits, but two other aspects played a decisive role: sophisticated bureaucracy or better known as red tape and industrial lordliness.
>
> So the US Army/NASA did it their way to develop the JUH-60A RASCAL, the Rotorcraft-Aircrew Systems Concept Airborne Laboratory, based on a UH-60 (see Sect. 5.2.3.17). It took DLR another 10 years to develop, this time together with Eurocopter and a strong support of the German FMoD, the Flying Helicopter Simulator FHS, based on an EC 135

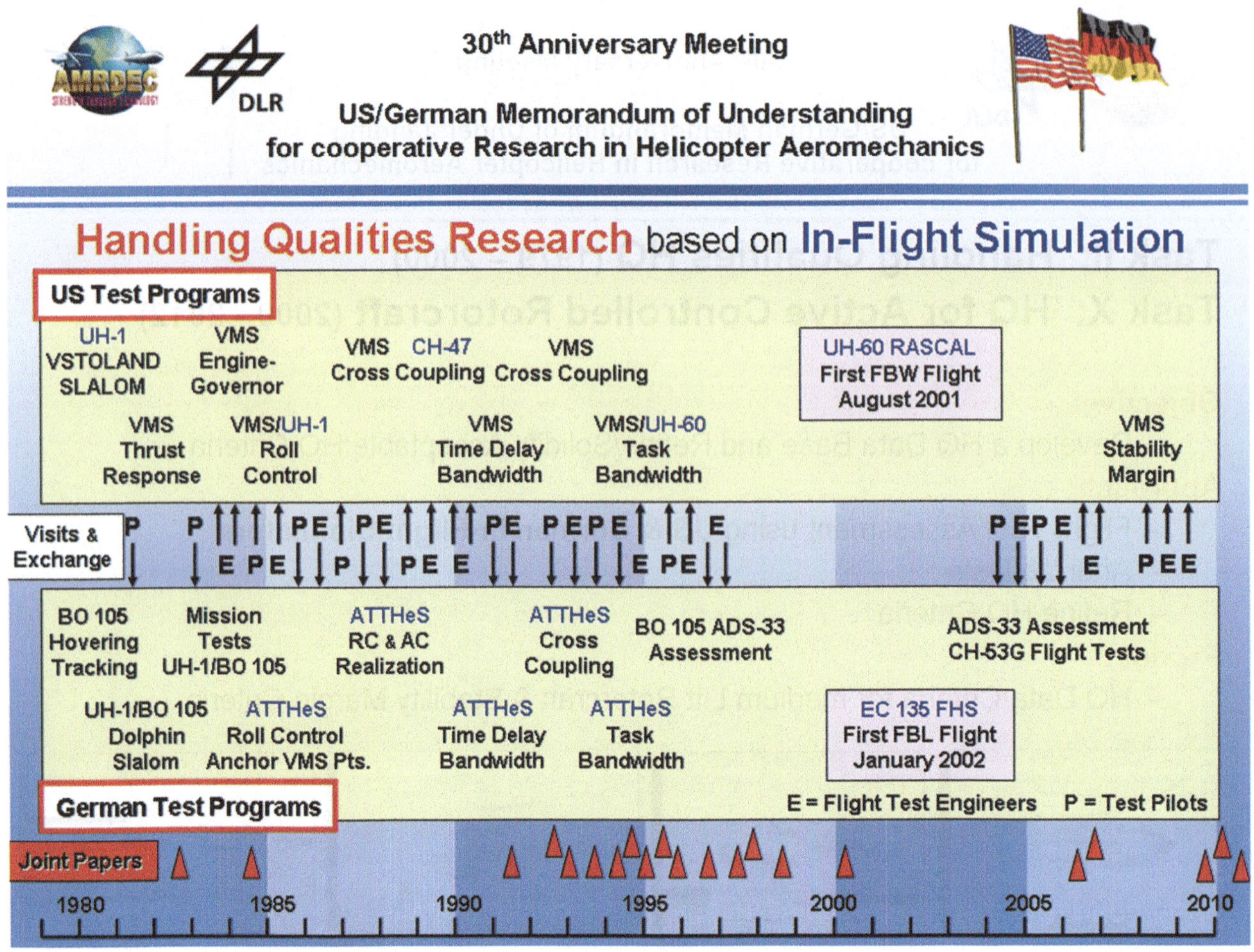

Fig. 12.22 Statistics in complementary in-flight simulator research (*Credit* Chris L. Blanken)

helicopter as host helicopter with high controllability and low life cycle costs (see Chap. 10).

A small reminiscence: today, that is twenty years since flat lining the collaborative initiative BK 117 HESTOR, the US Army is procuring with its UH-72 Lakota a militarized version of the EC 145, which in turn is a modified BK 117. Twenty years ago the only US concern about the BK 117 was if it will be still in production in the 1992–1993 timeframe for the planned HESTOR development. Today and in the coming years, a heavily modified version BK 117, for example, either the EC 145 or the UH-72 is produced in series in Germany and in the US!

What is the message? If we would have successfully proceeded with our common vision on HESTOR twenty years ago, the US and German Armies could now operate Fly-by-Wire Lakotas with all its future capabilities, potentials, and promises.

All the same, a follow-up program is currently in preparation at the U.S. Naval Test Pilot School (USNTPS) for its both NSH-60 VSC helicopters equipped with a variable stability and control system (see Fig. 5.34). The UH-72 Lakota developed by Eurocopter was envisaged as the host helicopter, which is traced back to the basic configuration of the EC 145 and BK 117. A total of five UH-72 are at the disposal of USNTPS Originally, two of them were planned to be retrofitted to in-flight simulators under the name NUH-72 VSS (variable stability system). Unfortunately, this program was cancelled in 2015 in favor of the NUH-60L VSS alternative for unknown reasons (see Sect. 5.2.1.17).

Epilogue

This unique MoU and up to the last day highly successful scientific cooperation ended in 2013 after thirty-five years, exclusively due to formal legal contract squabbles between German and American authorities.

12.4 Cooperation with Airbus (1994–1995)

As part of an Airbus contract to investigate the flight behavior of a very large aircraft (*Ultra High Capacity Aircraft—UHCA*), a mathematical model of the flight dynamics of the A3XX (precursor of A380) was programmed on the ATTAS model following controller and tested in flight. The Airbus Engineering test pilot *Claude Lelaie* and the Director of the Airbus Test Flight and Development Division *Gilles*

Fig. 12.23 Airbus chief test pilot *Claude LeLaie* and *Gilles Robert* (both in the *center*) after an A3XX inflight simulation with ATTAS

Fig. 12.24 Contemplating Airbus experts *Claude LeLaie* and *Gilles Robert* in ATTAS ground based simulator

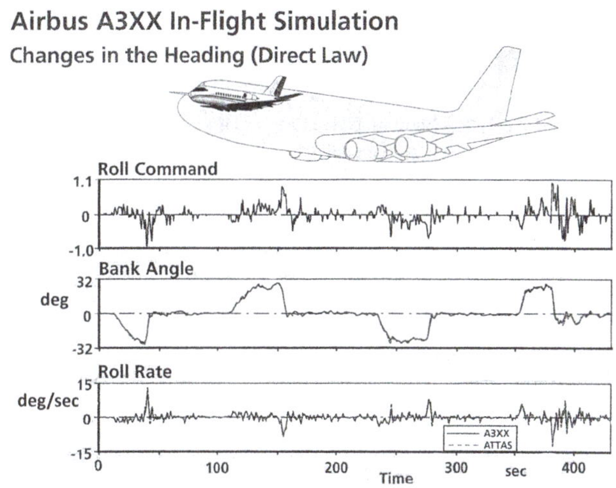

Fig. 12.25 A3XX-type simulation flight with ATTAS (direct control law)

Robert were offered an opportunity on August 23, 1995, to participate in a flight demonstration of an A3XX-type in-flight simulation with ATTAS with a subsequent debriefing (see Figs. 12.23, 12.24, 12.25, and 12.26). From the closing meeting and all the work related to this project, DLR gained important insights into the simulation of a future large commercial aircraft.

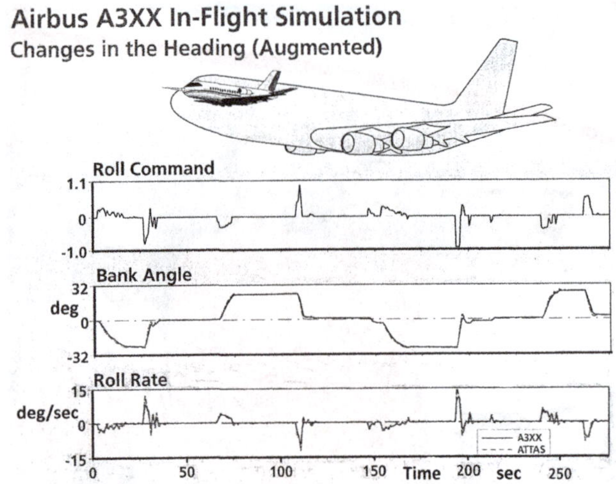

Fig. 12.26 A3XX-type simulation flight with ATTAS (with control augmentation)

References

1. Hamel, P. (ed.): In-Flight Simulation for the 90's, International Symposium, July 1–3, 1991, Braunschweig, DGLR Mitteilungen 91-05 (Proceedings, Inquiries at: DLR, Institute of Flight Systems, P.O. Box 3267, 38022 Braunschweig) (1991)
2. Hillebrand, H.L.: Fliegende Simulatoren minimieren das Risiko, vol. 35, pp. 37–38. Aerokurier (1991)
3. Mecham, M.: Airborne Simulation Expands, pp. 42–49. AW&ST (1991)
4. Harper, R.P.: The evolution of in-flight simulation at Calspan. In: Hamel, P. (ed.) In-Flight Simulation for the 90's, International Symposium, July 1–3, 1991, Braunschweig, DGLR Mitteilungen 91-05 (Proceedings, Inquiries at: DLR, Institute of Flight Systems, P.O. Box 3267, 38022 Braunschweig), Paper 91-05-1 (1991)
5. van der Bliek, J. (ed.): AGARD—The History 1952–1997. The NATO Research and Technology Organization (RTO), Spring 1999
6. Hamel, P.: Kooperative Forschungsprojekte zwischen der Bundesrepublik Deutschland und den Vereinigten Staaten von Amerika auf dem Gebiet der Flugeigenschaften, DFVLR Jahresbericht 1982, Seiten 95–98 (1983)
7. Koehler, R., et al.: GRATE—A New Flight Test Tool for Flying Qualities Evaluations, AGARD CP-452, Paper 8, Oct 1988
8. Shafer, M.F., Koehler, R., Wilson, E.M., Levy, D.R.: Initial Flight Test of a Ground-Deployed System for Flying Qualities Assessment, NASA TM-101700 and AIAA Paper 89-3359, Aug 1989
9. Hamel, P.: USAF-FMOD Memorandum of Understanding on Aircraft Flight Control Concepts—Recent Advances and Related Activities, Presentation on the Occasion of the Visit of 4-Star General Powell at DLR, Köln-Porz, 10 Nov 1989
10. Wilhelm, K., Woodcock, R.W., Legett, D.B.: Summary of Joint Program of Research into Flight Control Concepts (Period 1983–1992), USAF Flight Dynamics Directorate, Wright Laboratory, WL-TR-93-3039, May 1993
11. Key, D.L., Gmelin, B.: A Decade of Cooperation, Vertiflite, vol. 35, No. 6, pp. 48–54. Sept–Oct 1989
12. Pausder, H.-J., Blanken, C.L.: Report on Activities 1994–2003, DLR IB 111-2004/79 (2004)
13. van der Wall, B.G., Blanken, C.L. (eds.): Nine Years of Cooperation—The US/German Memorandum of Understanding (MoU) on Helicopter Aeromechanics—2003–2012, US AAMRDE TR RDMR-AF-14-01 and DLR IB 111-2013/33 (2013)
14. van der Wall, B.G., Blanken, C.L.: 30th Anniversary Celebration of the US/German MoU on Helicopter Aeromechanics, DLR IB 111-2009/1 (2009)

Author Biography

Peter G. Hamel was the Director of the Institute of Flight Mechanics/Flight Systems of the German Aerospace Center (DLR/DFVLR) (1971–2001). He received his Dipl.-Ing. and Dr.-Ing. degrees in Aerospace Engineering from the Technical University of Braunschweig in 1963 and 1968, and his SM degree from M.I.T. in 1965. From 1970 to 1971 he was Section Head of Aeronautical Systems at Messerschmitt-Bölkow-Blohm in Hamburg. Since 1995 he is an Honorary Professor at the Technical University of Braunschweig and a founding member of three collaborative research centers at the University. He was Chairman of the National Working Group on Helicopter Technology (AKH) (1986–1994) and the appraiser for the National Aviation Research Program (LuFo) until today. He was also the Manager of DLR's Rotorcraft Technology Research Program and the German Coordinator for the former AGARD Flight Mechanics/Vehicle Integration (FMP/FVP) Panel. He is a member of the German Society for Aeronautics and Astronautics (DGLR) and of the American Helicopter Society (AHS), and a Fellow of AIAA. He is the recipient of the AGARD 1993 Scientific Achievement Award, of AGARD/RTO von Kármán Medal 1998, of AHS Dr. A. von Klemin Award 2001, and of the prestigious DGLR Ludwig-Prandtl-Ring 2007.

Quo Vadis? 13

Peter G. Hamel

Aircraft controllability and flying qualities investigations have been focal points in aeronautics since the first flight by Wright brothers more than a century back. With the evolution of aerodynamically efficient aircraft configurations and modern Fly-by-Wire/Light technology, with associated advances in computer, sensor, measurement, and information technology, these issues have not only become even more important, but they have also become more complex. To reap the benefits of these modern, powerful techniques, it is required not only to take into consideration the human-vehicle interaction, but also the piloting skills and new experimental techniques. In the early phases of aircraft design, variable stability aircraft were utilized for this purpose. During the last few decades more sophisticated variable stability aircraft have evolved into in-flight simulators and being used for a far broader system oriented application spectrum. They provide a safer and economically more viable experimental approach to flying and handling qualities investigations that help to assess new aircraft designs prior to their first flight. The familiarization and training process of pilots and flight-test engineers and the testing of new control laws and subsystems hardware as part of an integrated flight control system are further important issues for airborne simulations.

This compendium elaborates on the global state of development and utilization of in-flight simulators and on their significance in optimizing the flying qualities of fixed and rotary wing aircraft. Thereby, it also provides a historical account of research, development, and testing of electronic and electro-optical flight control systems (Fly-by-Wire/Light).

The performance of modern flight vehicles can be greatly enhanced through the use of Fly-by-Wire/Light flight-control systems in conjunction with digital information systems such as multifunctional displays and advanced sensor systems. The increasing automation, however, raises the question of whether the human-automation interactions are adequately understood and taken into account during the design process. In most cases, the automation has helped the operator, but at times operator confusion as to what the automation is doing has created dangerous situations. This problem, known as mode confusion, has been difficult to analyze and thus *solutions tend to be reactive instead of proactive* (M.I.T.). By what means can this unsatisfactory situation be improved with the aim of further increasing the flight safety? Or does the ever-increasing system complexity lead rather to opposite effects? Do the hitherto automation strategies need to be reconsidered and amended?

How can the right way be ensured to deal properly with new, rapidly changing information technologies in research and development of controlled manned or unmanned flight systems with an increasing degree of automation?

Besides the widespread ground-based simulation, the in-flight simulation plays an important role in the clarification of these issues to enable the optimization of system properties and for pilot training under realistic environmental conditions, extreme flight situations and unexpected fault occurrences and disturbances.

A major issue for the future is also that of maintaining, expanding, and passing on the acquired system know-how to young scientists and engineers. An improved documentation system in the form of "*Lessons Learned*", generated after flight critical technical failures, incidents and accidents being adequately communicated, may contribute to avoid at least some equivalent future safety issues.

Will the system knowledge being generated during the design cycle be available over the aircraft lifetime and preserved for future expert generations? Are the specialists

P.G. Hamel (✉)
Braunschweig, Germany
e-mail: p.hamel@t-online.de

P.G. Hamel (ed.), *In-Flight Simulators and Fly-by-Wire/Light Demonstrators*,
DOI 10.1007/978-3-319-53997-3_13

being educated and trained appropriately, and will generalists still be sufficiently available, those who have an appreciation for the overall system, or do they fall into oblivion? The famous Space industrialist *Manfred Fuchs* (1938–2014) describes the continuous learning process very vividly: "*Learning is like rowing against the current. Once you stop, you drift back. We will keep on rowing*".

In this sense, the editor and authors hope to have made a constructive contribution, both in the supreme discipline of in-flight simulation as well as for future challenges in the design of manned and unmanned aircraft systems; a contribution in the form of a reference book for knowledge conservation, for general reading by those interested in aeronautics, and to motivate future aviation enthusiasts.

Author Biography

Peter G. Hamel was the Director of the Institute of Flight Mechanics/Flight Systems of the German Aerospace Center (DLR/DFVLR) (1971–2001). He received his Dipl.-Ing. and Dr.-Ing. degrees in Aerospace Engineering from the Technical University of Braunschweig in 1963 and 1968, and his SM degree from M.I.T. in 1965. From 1970 to 1971 he was Section Head of Aeronautical Systems at Messerschmitt-Bölkow-Blohm in Hamburg. Since 1995 he is an Honorary Professor at the Technical University of Braunschweig and a founding member of three collaborative research centers at the University. He was Chairman of the National Working Group on Helicopter Technology (AKH) (1986–1994) and the appraiser for the National Aviation Research Program (LuFo) until today. He was also the Manager of DLR's Rotorcraft Technology Research Program and the German Coordinator for the former AGARD Flight Mechanics/Vehicle Integration (FMP/FVP) Panel. He is a member of the German Society for Aeronautics and Astronautics (DGLR) and of the American Helicopter Society (AHS), and a Fellow of AIAA. He is the recipient of the AGARD 1993 Scientific Achievement Award, of AGARD/RTO von Kármán Medal 1998, of AHS Dr. A. von Klemin Award 2001, and of the prestigious DGLR Ludwig-Prandtl-Ring 2007.

Erratum to: In-Flight Simulator VFW 614 ATTAS

Dietrich Hanke and Klaus-Uwe Hahn

Erratum to:
Chapter 9 in: P.G. Hamel (ed.),
In-Flight Simulators and Fly-by-Wire/Light Demonstrators,
DOI 10.1007/978-3-319-53997-3_9

The original version of the book was inadvertently published without the following corrections:

In Chap. 9, author biography of "Bernd Gmelin" should be replaced with "Klaus-Uwe Hahn".

The erratum chapter and the book have been updated with the changes.

The updated original online version for this chapter can be found at http://dx.doi.org/10.1007/978-3-319-53997-3_9

P.G. Hamel (ed.), *In-Flight Simulators and Fly-by-Wire/Light Demonstrators*,
DOI 10.1007/978-3-319-53997-3_14

Author Index

P.G. Hamel (ed.), *In-Flight Simulators and Fly-by-Wire/Light Demonstrators*,
DOI 10.1007/978-3-319-53997-3

Aircraft Index

P.G. Hamel (ed.), *In-Flight Simulators and Fly-by-Wire/Light Demonstrators*,
DOI 10.1007/978-3-319-53997-3

Subject Index

P.G. Hamel (ed.), *In-Flight Simulators and Fly-by-Wire/Light Demonstrators*,
DOI 10.1007/978-3-319-53997-3

List of Contributors

Editor

Hamel, Peter G. was the Director of the Institute of Flight Mechanics/Flight Systems of the German Aerospace Center (DLR/DFVLR) (1971–2001). He received his Dipl.-Ing. and Dr.-Ing. degrees in Aerospace Engineering from the Technical University of Braunschweig in 1963 and 1968, and his SM degree from M.I.T. in 1965. From 1970 to 1971 he was Section Head of Aeronautical Systems at Messerschmitt-Bölkow-Blohm in Hamburg. Since 1995 he is an Honorary Professor at the Technical University of Braunschweig and a founding member of three collaborative research centers at the University. He was Chairman of the National Working Group on Helicopter Technology (AKH) (1986–1994) and the appraiser for the National Aviation Research Program (LuFo) until today. He was also the Manager of DLR's Rotorcraft Technology Research Program and the German Coordinator for the former AGARD Flight Mechanics/Vehicle Integration (FMP/FVP) Panel. He is a member of the German Society for Aeronautics and Astronautics (DGLR) and of the American Helicopter Society (AHS), and a Fellow of AIAA. He is the recipient of the AGARD 1993 Scientific Achievement Award, of AGARD/RTO von Kármán Medal 1998, of AHS Dr. A. von Klemin Award 2001, and of the prestigious DGLR Ludwig-Prandtl-Ring 2007.

Translator

Jategaonkar, Ravindra V. was a Senior Scientist at the Institute of Flight Systems of the German Aerospace Center (DLR) in Braunschweig, Germany (1986–2015). Prior to joining DLR, he was a researcher at the National Aeronautical Laboratories, Bangalore, India (1975–1986). He received his degrees in engineering from the Calicut University, India (B.Sc., Engg., 1972, and M. Sc., Engg., 1975), and the Indian Institute of Science, Bangalore, India (Ph.D., 1986). He has been the group leader at the DLR Institute and extensively involved in the development and application of nonlinear system identification methods to flight vehicle modeling. He is the recipient of the prestigious DLR award and the title "Senior Scientist" for scientific excellence (2001) and of the AIAA Sustained Service Award (2007). He is an author of the book titled "Flight Vehicle System Identification: A Time Domain Methodology", published under Progress Series by AIAA (2006, 2015). He was a member of the AIAA AFM Technical Committee for more than ten years. He is a member of the German Society for Aeronautics and Astronautics (DGLR) and an Associate Fellow of AIAA.

Authors:

Adam, Volkmar was a research scientist at the Institute of Flight Guidance of the German Aerospace Center (DLR/DFVLR) (1969–2004), where he was the Head of the Department of Flight Control Systems (1987–2000) and of the Department of Controller Assistance (2000–2004). He received his Dipl.-Ing. degree in Aeronautical Engineering from the Technical University of Berlin (1969). His research focus was on automatic flight control, 4D approach guidance, flight management, ground-based 4D arrival planning, cooperative Air Traffic Management, and UAV guidance. From 1975 to 1983, he participated in numerous flight tests with the HFB 320. He was a visiting scientist at NASA Langley Research Center (1983–84). He participated in flight tests with ATOPS B737 in 1987. He is a member of the German Society for Aeronautics and Astronautics (DGLR).

Altenkirch, Dietrich was a research scientist at the Institute of Flight Systems of the German Aerospace Center (DLR/DFVLR) (1977–2016). He received his Dipl.-Ing degree in Aeronautical Engineering from the Technical University of Braunschweig. His research focus was on dynamic simulations in the wind tunnel, flight performance, flight dynamics, flight test (fixed-wing aircraft). As a member of Akaflieg Braunschweig, he was involved in the design and construction of the gliders SB 10 and SB 11. From 1983 to 2011 he participated in a variety of flight tests with the in-flight simulators HFB 320 FLISI and the VFW 614

P.G. Hamel (ed.), *In-Flight Simulators and Fly-by-Wire/Light Demonstrators*,
DOI 10.1007/978-3-319-53997-3

ATTAS, and 2000–2008 in the DLR project ATTAS Wasla-HALE. He was several times German champion in radio-controlled model flying.

Brieger, Oliver was a research scientist at the Institute of Flight Systems of DLR in Braunschweig and was a group leader of the DLR field office Flight Testing at WTD 61 in Manching (2003–2010). Since 2010, he is the joint head of the flight operations at DLR. He received his Dipl.-Ing. degree in Aerospace Engineering from the Technical University of Munich (2000). He underwent training as a flight test engineer at the US Naval Test Pilot School and the National Test Pilot School, USA. He participated in the X-31 VECTOR program (2000–2003). He is the recipient of the SFTE Kelly Johnson Award and of the Naval Air Warfare Center Aircraft Division Area Commander's Award (2003). He is a member of SFTE, Co-Chair of the NATO STO Flight Test Technical Team and DGLR, and senior member of AIAA, and board member SAE ACGSC.

Czerlitzki, Bernhard is a research scientist at the DLR Institute of Flight Guidance, Department of Pilot Assistance (formerly Flight Control) in Braunschweig since 1989. He received his Dipl.-Math. Degree in Mathematics from the Technical University of Braunschweig (1988). His research interests include planning of 4D flight profiles for transport aircraft, flight management system integration into future ATM scenarios, CDTI/ASAS partial delegation, cooperative air traffic management, performance-based navigation, and RNP approach procedures. He was involved in numerous flight tests with VFW 614 ATTAS.

Fezans, Nicolas is a research scientist at the Institute of Flight Systems, DLR in Braunschweig (2009-to date). He received his French Engineering diploma (MSc, 2005) from the Ecole des Mines de Nantes, France and his Master's degree by research in control engineering, robotics, and signal processing from the EDSTIM Nantes, France (2005). From 2005 to 2009 he was a Ph.D. student at ONERA/DCSD Toulouse and CNES/DLA Paris and received his Ph.D. degree from ISAE SUPAERO Toulouse, France (2009). His research interests are control engineering (robust and fault tolerant flight control, automation of air-to-air refueling, fixed-wing aircraft, atmospheric re-entry, gust alleviation functions) and applied mathematics (online algorithms for filtering and optimization).

Gmelin, Bernd was a research scientist and head of the Rotorcraft Department at the Institute of Flight Mechanics of DLR at Braunschweig (1973–1999). From 1999 to 2006 he headed the DLR Helicopter Program under the German/French DLR/ONERA Common Rotorcraft Research Program. Prior to joining DLR, he was a researcher at the Institute for Rotorcraft in Stuttgart (1968–1973). He received his Dipl.-Ing. degree in Aerospace Engineering from the University of Stuttgart (1968). His primary interests are mathematical modeling, helicopter simulation in the wind tunnel, control systems, flying qualities, helicopter in-flight simulation. He is a member of the German Society for Aeronautics and Astronautics (DGLR) and of the American Helicopter Society (AHS). He is the recipient of the AHS Agusta International Fellowship Award (1996), DLR/ONERA Team Award (2006), and AHS Fellow Award (2008).

Griem, Hartmut studied Mechanical Engineering/Control and Precision Engineering at the Technical High School Aachen and the Technical University of Braunschweig. Since 1964 he worked with the Vereinigte Flugtechnische Werke in Bremen and the successor companies, initially in Flight Mechanics and Systems Engineering in the projects VAK 191B (VTOL) and VFW 614 respectively with the partial responsibility for the certification of flight control. He was the program manager for the conversion of the VFW 614 into an in-flight simulator of DFVLR. From 1985 to 1995 he was chief engineer of the FLA (Future Large Aircraft), which was the earlier version of the A400M military transporter. From 1995 to 1999 he was the program director for the Fly-by-Wire demonstrator program ATD.

Hahn, Klaus-Uwe is a research scientist at the Institute of Flight Systems (formerly Flight Mechanics) at DLR in Braunschweig since 1989. He is currently head of Flight Dynamics and Simulation. He received his Dipl.-Ing. degree in Mechanical/Production Engineering (1975) from the FH Braunschweig-Wolfenbüttel and in Mechanical/Aeronautical Engineering (1981), and Dr.-Ing. degree (1988) from the Technical University of Braunschweig. In 2008 he was appointed Professor at the Technical University of Hamburg-Harburg. His main research interests are flight performance, flight dynamics, flight control, flight test (fixed-wing aircraft). From 1989 to 2011 he was involved in a variety of flight tests with the VFW 614 ATTAS. He is a member of the DGLR, a member of the Planning and Advisory Board of the SAE Aerospace Control and Guidance Systems Committee (ACGSC), and a senior member of AIAA.

Hamel, Peter G. For details refer to Editor's profile at the beginning of this section.

Hanke, Dietrich was a research scientist at the Institute of Flight Systems at DLR in Braunschweig (1970–2005). He received his Dipl.-Ing. degree in Mechanical/Aeronautical Engineering from the Technical University of Braunschweig (1970). His main research interests are flying qualities, flight and system dynamics, flight control, active sidestick, and in-flight simulation (fixed-wing aircraft). From 1972 to 1984 he was project leader for the development of the HFB 320 FLISI as an in-flight simulator. In 1981 he became Head of the new established Flight Simulation department. In this position, he was responsible for the technical development of the VFW 614 ATTAS in-flight simulator and demonstrator aircraft, as well as for the development, operation, and maintenance of the Fly-by-Wire/Light system. Also, he headed several research projects with the HFB 320 aircraft (Direct Lift Control, Command Control), the ATTAS aircraft (Hermes Space Plane, Rate Limiting) and several international cooperation programs with the AFFDL and TsAGI, Russia. After his retirement he founded the Maui Ultra Fins company dealing with the development and production of high performance windsurfing fins.

Hofer, Hermann received his Ing.grad. degree in the field of Power Engineering from the Technical High School in Konstanz. From 1966 to 1970 he was employed as a systems engineer at Dornier to support the flight control team for the Do 31 VTOL transport aircraft. From 1971 to 2006 he worked at Liebherr-Aerospace Lindenberg in various positions for the system-embedded electronics products and electronic flight control systems. Since 2007, he is a consultant for the aeronautical research programs and quality management systems in accordance with EN 9100.

Kaletka, Jürgen was a research scientist at the Institute of Flight Systems at DLR in Braunschweig (1973–2009). He received his Dipl.-Ing. degree in Aeronautical Engineering from the Technical University of Braunschweig. His main research interest are flight dynamics, in-flight simulation, modeling and system identification (helicopter). From 1987 to 1991 he was the manager of the AGARD FMP Working Group WG 18 Rotorcraft System Identification, from 1979 to 1994 the task manager for system identification in the US-German Memorandum of Understanding (MoU) Helicopter Aero Mechanics. From 1999 to 2003 he was the technical project manager for the development of ACT/FHS (helicopter in-flight simulator based on EC 135).

Krag, Bernd was a research scientist at the Institute of Flight Systems of DLR in Braunschweig (1972–2002). From 1993 to 2002 he was head of the Fixed Wing Aircraft department. Prior to joining DLR, he was a research assistant at the Flight Mechanics Branch at the Technical University of Braunschweig (1967–1972). He received his Dipl.-Ing. degree (1967) and Dr.-Ing. degree (1976) in Aerospace Engineering from the Technical University of Braunschweig. His main research interest were control configured vehicles (CCV), active control, modeling and system identification, databases for training simulators, and the wake vortex problem. Since retiring from DLR, his interest is in the history of aeronautics research in Braunschweig and aviation history.

Kuenz, Alexander is a researcher at the Institute for Flight Guidance at DLR in Braunschweig (since 2001). He received his Dipl.-Inform. degree in Computer Sciences/Signal Processing from the Technical University of Braunschweig and a Ph.D. from Leibniz University, Hanover. His research focus is on air-ground integration, noise abatement procedures, conflict detection and avoidance. He was involved in numerous projects with flight tests on ATTAS (2001–2008).

Lange, Hans-Heinz was a research scientist at the Institute of Flight Systems (formerly Flight Mechanics) at DLR in Braunschweig (1972–2006), and leader of the group on System und Experimental Software Development in the Simulation Techniques branch. He received his Dipl.-Ing. Degree in Mechanical/Aeronautical Engineering from the Technical University of Braunschweig. His main research focus was on flight control systems for the test demonstrators HFB 320 FLISI and VFW 614 ATTAS.

Leißling, Dirk is a researcher at the Institute of Flight Systems (formerly Flight Mechanics) at DLR in Braunschweig (1999-to date) He received his Dipl.-Ing. degree in Mechanical/Aeronautical Engineering from the Technical University of Braunschweig (1999). His main interests are in the areas of in-flight simulation, flight control, flying qualities, and flight testing of fixed wing aircraft. Since 2012 he pursues the works in the field of helicopters.

Looye, Gertjan was research engineer at the Institute of Robotics and Mechatronics, Department of System Dynamics and Control, DLR in Oberpfaffenhofen (1997–2012). Since 2012, he is Head of Aircraft Systems Dynamics department of the Institute of System Dynamics and Control, which in turn is part of the Robotics and Mechatronics Centre, founded in the same year. He received his Master's and Ph.D. degrees in Aerospace Engineering from the Technical University of Delft, Netherlands (1996, 2008 respectively), as well as an honorary Diploma of the Dutch General Snijders Foundation (2010). His main interests are flight dynamics, nonlinear and robust flight control, modeling of elastic aircraft, and flight loads. He was involved in several test flights with ATTAS, such as nonlinear flight control, TECS, automatic landing, and automatic spiral approaches.

Luckner, Robert graduated from Technische Universität Braunschweig, Germany, with the degrees Dipl.-Ing. and Dr.-Ing. in Mechanical Engineering in 1976 and 1984, respectively. He is currently a full professor for Flight Mechanics, Flight Control and Aeroelasticity at TU Berlin, the Deputy Managing Director of the Institute of

Aeronautics and Astronautics and Vice Dean of the Faculty. He was a principal engineer in the domain 'Flight Control' at Airbus, Hamburg, where he worked from 1984–2004. He was responsible for flight control law design in the technology projects SAFIR and VFW-ATD. He is chair of the CEAS Technical Committee "Guidance Navigation and Control" (since 2007), chair of the DGLR Technical Committee L6 "Flight Mechanics, Flight Guidance and Control" (since 1994), member of the DGLR Senate, Associate Fellow AIAA, and was member of the "A380 Wake Vortex Group" (2003–2006). His research focus is on flight mechanics and flight control related to aeroelastic commercial transport and utility aircraft.

Mönnich, Wulf is a research scientist at the Institute of Flight Systems at DLR in Braunschweig (since 1985). He received his Dipl.-Ing. degree in Mechanical/Aeronautical Engineering from the Technical University of Braunschweig (1985). His main research interests are modeling and simulation, system identification, and flight control, particularly for nonlinear in-flight simulation. He was a member of the AIAA AFM Technical Committee for three years. He is a member of the German Society for Aeronautics and Astronautics (DGLR) and a senior member of AIAA.

Pausder, Heinz-Jürgen was a research scientist at the Institute of Flight Systems at DLR in Braunschweig (1973–2009). He was a technical project manager for the development of the ACT/FHS (helicopter in-flight simulator) (1995–1999) and head of the Helicopter Department at the DLR Institute of Flight Systems (1999–2006). From 2001 to 2006 he was German project manager for the US-German Memorandum of Understanding (MOU) on Helicopter Flight Control. He received his Dipl.-Ing. degree in Mechanical/Aeronautical Engineering from the Technical University of Braunschweig (1973). His main research interests are flight dynamics, in-flight simulation, and flying qualities (helicopter). He was a member of the Board of the Community Foundation Braunschweig (2010–2014).

Peter, Lars studied Mechanical Engineering and Aerospace Engineering at the Technical University of Munich (2009). He also pursues doctoral research at TUM on advanced flight control systems for General Aviation aircraft and contributed to the foundation of the Safety Critical Systems and Avionics group. He is the lead engineer and project manager for the DA42 MNG research aircraft and is responsible for technology transfer into further control system developments for manned and unmanned applications. As a consultant and systems engineer, he supports airframe, system and component manufacturers in developing new technologies and products.

Reichel, Reinhard studied Electrical Engineering at the Graz University of Technology. He pursued his doctoral degree at the Paderborn University (1985), with the focus on control theory. From 1980 to 1985 he was a scientific staff member at the Institute of Control Theory, University of Paderborn. From 1985–2003 he worked at Diehl Avionik Systeme as systems engineer, then as project manager and finally as head of the product division "Flight Control & Avionics". Since 2003 he is a professor and the director of the Institute of Aircraft Systems (ILS), University of Stuttgart.

Schänzer, Gunther was the Director of the Institute for Flight Mechanics and Flight Guidance of the Technical University of Braunschweig (1976–2004). Prior to joining the University, he worked at Bodensee Gerätetechnik as department head, then head of Advanced Flight Control and Navigation (1969–1976) and as a research scientist at the Institute of Flight Guidance DLR in Braunschweig (1965–1969). He received his Dipl.-Ing. degree (1965) and Dr.-Ing. degree (1969) in Aeronautical Engineering from the Technical University of Braunschweig. He was the speaker for several collaborative research projects. He was a member and deputy chairman for two years of the DLR Senate. He was also chairman of the Research Airport Braunschweig, and a member of the Academy of Science and Engineering (AKAtech), as well as chairman of the board of Aerodata.

Schwithal, Jana is a research scientist at the Institute of Flight Systems at DLR in the department Flight Dynamics and Simulation (since 2010). She received her Dipl.-Ing. degree in Aerospace Engineering from the Technical University of Braunschweig. Her research interests are flight dynamics and flight control (fixed-wing aircraft).

Tersteegen, Johannes was a research scientist at the Institute of Flight Guidance DLR in Braunschweig (1962–2001). He received his Dipl.-Ing. degree in Mechanical Engineering from the Technical High School Niederrhein in Krefeld. His research interests were development of digital electrohydraulic actuators and fast switching valves. Between 1964 and 2001 he was responsible for the design of hydraulic systems and hydraulic actuators in all research aircraft and helicopter at DLR in Braunschweig. He was a project leader for long-term testing of a digital FBW control system in the research ship Planet.

von Grünhagen, Wolfgang was a research scientist at the Institute of Flight Systems at DLR in Braunschweig (1979–2015). He received his Dipl.-Math. degree (1973) and his Dr.-Ing. degree (1979) from the Technical University of Braunschweig. His main research interests are flight dynamics, generic nonlinear modeling, system identification, and control systems for in-flight simulation (helicopter). He was the principal investigator for the task "Modeling and Simulation for Rotorcraft Systems" (2000–2008) and for the task "Handling Qualities for Active Controlled Rotorcraft—Active Inceptors" (2009–2012) in the US-German

Memorandum of Understanding (MoU) Helicopter Aero Mechanics. He was the project leader for the integration of the two active sidesticks into the EC 135 FHS (2004–2009). During the development of the ACT/FHS, he supported the project with the setup of the system-simulation. He was the leader of the SMART Rotorcraft research field in the common ONERA/DLR Rotorcraft Research (2004–2010).

Wilhelm, Knut was Professor at the Flight Mechanics and Flight Control Department of the Berlin Technical University (1993–2001). Prior to joining the University, he was a research scientist at the Institute of Flight Mechanics, DLR in Braunschweig (1966–1993), where he was head of Flight Mechanics of Fixed-Wing Aircraft department from 1972 to 1993. From 1979 to 1990 he was a project manager for the US-German Memorandum of Understanding (MoU) on Aircraft Flight Control Concepts. He received his Dipl.-Ing. degree (1966) and Dr.-Ing. degree (1968) in Mechanical/Aeronautical Engineering from the Technical University of Braunschweig. His research interests are flying qualities, flight dynamics, flight control, flight test, wind tunnel simulation, and in-flight simulation (fixed-wing aircraft).

Wünnenberg, Horst was employed at Dornier, initially in Flight Mechanics branch, later as head of the Flight Physics unit (1966–1998). Prior to joining Dornier, he was a research scientist at the Institute of Flight Guidance of the Technical University of Braunschweig. He received his Dipl.-Ing. degree in Mechanical/Aerospace Engineering from the Technical University of Braunschweig. He participated in the projects VSTOL aircraft Do 31, Alpha Jet, Do 328, and in the management of technology programs gust alleviation system OLGA and direct lateral force control on the Alpha Jet. He was a member (1980–1996) and chairman (1994–1996) of AGARD Flight Mechanics Panel. After leaving Dornier, he has been a technical consultant for the Zeppelin Airship Technology company.

The manufacturer's authorised representative in the EU is Springer Nature Customer Service Centre GmbH, Europaplatz 3, 69115 Heidelberg, Germany. If you have any concerns regarding our products, please contact ProductSafety@springernature.com

Printed and bound by CPI Group (UK) Ltd, Croydon, CR0 4YY
01/10/2026
02225885-0002